Asymptotic Methods for Wave and Quantum Problems

American Mathematical Society
TRANSLATIONS
Series 2 • Volume 208

Advances in the Mathematical Sciences — 53

(*Formerly Advances in Soviet Mathematics*)

Asymptotic Methods for Wave and Quantum Problems

M. V. Karasev
Editor

American Mathematical Society
Providence, Rhode Island

2000 *Mathematics Subject Classification*. Primary 34E20, 35Lxx, 81Qxx.

Library of Congress Card Number 91-640741
ISBN 0-8218-3336-7
ISSN 0065-9290

∞ The paper used in this book is acid-free and falls within the guidelines
established to ensure permanence and durability.
Visit the AMS home page at http://www.ams.org/

10 9 8 7 6 5 4 3 2 1 08 07 06 05 04 03

Contents

Preface vii

Quantization and intrinsic dynamics
MIKHAIL KARASEV 1

Weak asymptotics method and interaction of nonlinear waves
V. G. DANILOV, G. A. OMEL'YANOV, AND V. M. SHELKOVICH 33

Global asymptotics and quantization rules for nonlinear differential equations
M. V. KARASEV AND A. V. PERESKOKOV 165

Asymptotics of eigenfunctions in shallow potential wells and related problems
P. ZHEVANDROV AND A. MERZON 235

Preface

The present collection contains papers in rather different areas of mathematical physics, but all the papers are united by the intrinsic coherence of the asymptotic methods used. In some sense, this coherence reflects common characteristics and roots inherent in different models and problems.

The hidden relations between different models become clear after a deep analysis, but sometimes they can be felt only intuitively. They gave birth to hopes that achievements in particular problems and particular areas will result in considerable progress in our general understanding of quantization, wave processes, and geometry.

The introductory paper "Quantization and Intrinsic Dynamics" describes relations between the abstract problem of quantization of symplectic manifolds and nonlinear wave processes. It is demonstrated that in this problem both the weak asymptotics method, i.e., asymptotics in generalized functions, and the semiclassical approximation method work successfully, and moreover, an interesting dynamic geometry attends the solution.

The paper "Weak Asymptotics Method and Interaction of Nonlinear Waves" is devoted to one of the most important problems in mathematical physics, namely, the problem of propagation and interaction of nonlinear waves. In the last three decades of the twentieth century, tremendous success was achieved by using the inverse scattering transform method; however, exact analytic solutions are found only in rare special cases. For equations with variable coefficients in "nonintegrable" cases, it is possible to apply only asymptotic methods. The paper describes the weak asymptotics method, which allows one to solve the problem of wave propagation in many "nonintegrable" cases. Simultaneously with the weak asymptotics method, different generalizations of the Whitham method, i.e., the semiclassical approximation for nonlinear equations, are also exploited.

The paper "Global Asymptotics and Quantization Rules for Nonlinear Differential Equations" presents a review of the theory of semiclassical approximation (in particular, the Whitham method) for nonlinear second-order ordinary differential equations or equations reducible to them. Primary attention is paid to the construction of global asymptotics, including neighborhoods of turning points, which is an important problem for nonlinear equations.

In the last paper "Asymptotics of Eigenfunctions in Shallow Potential Wells and Related Problems," one special but very interesting question is studied. The following fact is known: if in the (linear) Schrödinger equation the potential well is sufficiently shallow, then the discrete spectrum contains exactly one point. It is possible to calculate this single eigenvalue and the single localized eigenstate by asymptotic methods if the depth of the well is taken to be a small parameter.

All the papers in this collection contain detailed references and are oriented not only to specialists in mathematical physics, but also to a wider audience including graduate students.

I wish to thank M. A. Shishkova for her help in preparing this collection and all my colleagues from the Department of Applied Mathematics of the Moscow Institute of Electronics and Mathematics for general useful discussions.

M. Karasev

Amer. Math. Soc. Transl.
(2) Vol. **208**, 2003

Quantization and Intrinsic Dynamics

Mikhail Karasev

ABSTRACT. A dynamical scheme of quantization of symplectic manifolds is described. It is based on intrinsic Schrödinger and Heisenberg type nonlinear evolutionary equations with multidimensional time running over the manifold. From these equations the semiclassical asymptotics of a unique quantum product on oscillating functions is derived explicitly via objects and operations of dynamic geometry (Ether geodesics, reflections, groupoid of symplectic translations). The weak asymptotics of the same solution to the basic equations is associated with a deformation quantization scheme and with kinematic geometry (symplectic connection, curvature). It is also discussed how the intrinsic dynamics of symplectic manifolds can be used for solving semiclassically the exterior quantum problems.

CONTENTS

1. Introduction
2. Basic wave equations
3. Zero curvature equation
4. Intrinsic dynamical objects
5. Weak asymptotics and quantum zero curvature equation
6. Semiclassical asymptotics and quantum lift
7. Intrinsic Hamilton–Jacobi and transport equations
8. Exterior quantum dynamics
References

1. Introduction

The paper deals with constructing irreducible quantum manifolds, that is, quantizing symplectic manifolds [**1, 2, 3, 4**]. The most important problem within this framework is not even to prove the existence of a noncommutative product over generic phase manifold but to find a global geometric construction for it. It was

2000 *Mathematics Subject Classification.* Primary 81S10, 81S30, 53D55.

Key words and phrases. Quantization, symplectic geometry, semiclassical approximation.

This research was supported in part by the Russian Foundation for Basic Research, Grant 02-01-00952.

observed in [**5, 6**] that the symplectic connection could be a convenient geometric instrument here, and in [**7, 8**] this idea was beautifully realized via the formal deformation approach. Later, steps in the deformation direction [**9, 10, 11**] showed that symplectic structures and symplectic connections are significant infinitesimal objects, but there is something else in the actual heart of quantization. The semiclassical approach [**4**] (which led to the discovery of symplectic groupoids) clearly demonstrated that, at least, a partial geometric summation of deformation series is possible. Also the ideas of string theory and soliton theory show their close relationship to quantization problems.

The general principal question (or point of view (see [**12**])) could be formulated as follows: can the quantization process be reduced in some way to the geometric quantization [**13**] or to its "dynamic" generalizations? One would like to answer "yes," and probably this is one of the strongest incentives in this area. The results of the present paper deal with this question. On the other hand, at least until now, one could not be overoptimistic in assuming that any quantum mechanical problem can be reduced to some geometric construction (say, via a generalization of the theory [**14**]). Probably at this point real noncommutative geometry should start working [**3**].

In the present paper, reviewing results of previous works [**15, 16**], we investigate an *intrinsic evolutionary differential equation* for the integral kernel of a quantum associative product of functions on a symplectic manifold. The "time" variable in this evolutionary equation is multidimensional and runs over the same symplectic manifold. We describe the solution of this equation in the semiclassical approximation, as well as in the sense of weak asymptotics, by purely geometric means.

It is an open question as to how to solve this equation exactly (and so to construct exact quantization) for a general symplectic manifold. However, in some nontrivial classes of examples, this turns out to be possible.

The quantization which we deal with is formal in the sense that we do not consider analytic conditions on function spaces, Hilbert norms, etc. Nevertheless, this approach is not completely formal: it is adapted to the strict quantization scheme [**17**] and directly related to operator representations. The program for the future is to construct the exact operator realization of this approach on an arbitrary symplectic manifold. An important case where the operator representation is possible is the case of Kählerian manifolds, in particular, symplectic leaves in some Poisson manifolds with partial complex structure. The equations for the $*$-product kernel in this case were used in [**18, 19**]. For examples for the non-Kählerian situation, see, for instance, [**17, 20, 21**].

In addition to the algebraic and analytic machinery, which we explain below, there is an important geometric framework around such a dynamical scheme of quantization. Here we discuss only a few objects of this *dynamic geometry*; details are given in another paper.

In order to motivate the guiding idea, we recall an important point in the prehistory of quantum mechanics. At the dawn of the twentieth century, new ideas and constructions were needed to answer the questions about what a quantum space is and what kind of underlying geometry accompanies quantum phenomena and long-range interactions.

The work [**22**] is widely known as a publication where Herman Weyl discovered generalized (affine) connections and initiated tremendous forthcoming success of gauge theory. Actually, Weyl's intention was wider and deeper than the results presented in that paper. His imagination was to a large extent inspired by the Fichtean philosophy (see [**23**]). Weyl wanted to find a geometry combining the infinitesimal properties at a point with the global properties of the whole, i.e., of the "medium of free becoming."

The physical spirit of this concept was based, of course, on the well-known phenomena of long-range interactions in gravity and electromagnetism.

At that time, physics had already switched its explanation of long-range phenomena from the Ether hypothesis to the Field hypothesis. Fields became active participants of the physical world, and the new geometry had determined them mathematically via the notion of connections. But, as a philosopher, Weyl was not satisfied: the conception of space as a medium for free activity and for translating the infinitesimal properties from point to point did not have place in this geometry.

The connection theory initiated by Weyl turned out to be, in fact, a kinematic theory. It did reflect very well all infinitesimal properties of the space by geometrical means, but did not reflect the dynamics. The dynamical characteristics were attributed not to geometry but only to objects moving inside geometrical spaces or, on the quantum mechanical level, to the Schrödinger wave functions (see Weyl's remark in [**24**] made with certain regret).

Note that the old Ether concept was *a priori* opposite in spirit. The Ether as a medium between points of space was assumed to be a dynamical substance itself even in the absence of any particles, and so, its hypothetical geometry would be dynamic rather than kinematic.

In the present paper we would like to pay attention to the fact that the Ether idea, or more precisely, the idea of intrinsic dynamics on the phase space level, is mathematically fruitful and, moreover, automatically arises from the quantization problem. Hopefully, in this way, one can realize Weyl's early objectives and his intuitive program of exploring the continuum "relationship between a part and the whole" (see [**23**]).

2. Basic wave equations

Let us look first for wave functions ψ and for energy levels λ of the stationary Schrödinger equation

$$(\widehat{p}^2 + V(\widehat{q}))\psi = \lambda\psi, \tag{2.1}$$

where $\widehat{q} = q$, $\widehat{p} = -i\hbar\partial/\partial q$. Solving this equation is difficult, and the properties of the wave functions are complicated. For example, in the semiclassical approximation we have

$$\psi \sim c_\hbar e^{\frac{i}{\hbar}S}(\rho + O(\hbar)), \qquad c_\hbar = \text{const}. \tag{2.2}$$

However, this expression is very far from the asymptotics of the wave function in the presence of focal or turning points.

On the other hand, the weak limit ψ^0 of the function ψ as $\hbar \to 0$ is very simple. Indeed, if λ^0 denotes the limit of λ, then equation (2.1) at $\hbar = 0$ becomes

$$V(q)\psi^0(q) = \lambda^0\psi^0(q).$$

The solutions are obvious:

$$\psi^0(q) = \delta_{q^0}(q), \qquad \lambda^0 = V(q^0), \tag{2.3}$$

where q^0 is arbitrary. So, *the quantum wave functions in the weak limit* $\hbar = 0$ *are just* δ*-functions* characterizing the singular distribution of the probability amplitude of a classical particle to be localized at a given point.[1]

Now let us look at the algebraic picture behind equation (2.1), namely, at the Heisenberg commutation relations

$$[\widehat{q}, \widehat{p}] = i\hbar\mathbf{I}. \tag{2.4}$$

These relations introduce a noncommutative product $*$ into the function algebra over the q, p-phase space as follows:

$$\widehat{f} \stackrel{\text{def}}{=} f(\widehat{q}, \widehat{p}), \qquad \widehat{f}\widehat{g} = \widehat{f * g}. \tag{2.5}$$

Here we use the Weyl-symmetrized functions in the noncommuting operators $\widehat{q}, \widehat{p}$ so that, if $f(q,p) = \exp\{i(\alpha q + \beta p)\}$, then $\widehat{f} = \exp\{i(\alpha\widehat{q} + \beta\widehat{p})\}$.

If $\hbar = 0$, then the algebra (2.4) becomes commutative and the product $f * g$ becomes the usual commutative product fg of functions. One can represent it via the integral kernel as

$$(fg)(z) = \iint \delta_x(z)\delta_y(z) f(x) g(y)\, dm(x) dm(y). \tag{2.6}$$

Here the arguments x, y, z run over the q, p-phase space and δ denotes the Dirac delta function with respect to a certain measure dm on the phase space.

If $\hbar \neq 0$, then the noncommutative product $f * g$ in (2.5) can be represented in integral form as well:

$$(f * g)(z) = \frac{1}{(2\pi\hbar)^{2n}} \iint K_{x,y}(z) f(x) g(y)\, dm(x) dm(y), \tag{2.7}$$

where $2n$ is the dimension of the phase space. The integral kernel here is the noncommutative product of δ-functions:

$$\frac{1}{(2\pi\hbar)^{2n}} K_{x,y} = \delta_x * \delta_y.$$

From (2.6) we see that in the weak limit as $\hbar \to 0$ the integral kernel is equal to the commutative product of delta functions:

$$\lim_{\hbar \to 0} \frac{1}{(2\pi\hbar)^{2n}} K_{x,y} = \delta_x \delta_y.$$

This is the same picture as that for the wave function ψ in (2.1) and its weak limit ψ^0 (2.3). Looking at this analogy, we conclude that *a natural interpretation of the kernel of the* $*$*-product could be a "wave function" of something.*

Of what?

---

[1]If God had stopped at the place $\hbar = 0$, then we would see that classical particles stay at the points where their potential energy is equal to the total energy and do not move even when the force (the gradient of the potential) becomes different from zero. Newtonian mechanics does not work in this picture since there is still no symplectic structure in the phase space. To get the Newtonian (Hamiltonian) equations of motion, one needs to consider not only the zero jet at $\hbar = 0$, but also the first jet when the presence of the parameter $\hbar$ in $\widehat{p}$ in (2.1) becomes significant. In this way the mechanics is being born. The commutation relations between the coordinate $\widehat{q}$ and the momentum $\widehat{p}$ turn out to be necessary even for the classical motion of particles.

In (2.4), (2.5), and (2.7), there are no particles, no *a priori* Hamiltonians. Nevertheless, one can introduce a Hamiltonian of the product $*$ itself by mimicking the Schrödinger evolutionary equation.

Let us first note that the operators $\widehat{f}$ in (2.5) depend on f linearly and so can be written as integrals

$$\widehat{f} = \frac{1}{(2\pi\hbar)^n} \int f\mathbf{S}\, dm \tag{2.8}$$

over the phase space. Here $\mathbf{S} = \{\mathbf{S}_x\}$ is a family of operators parametrized by points x of the phase space. In the case of the flat phase space $\mathbb{R}^{2n}$ one could choose the Liouville measure, $dm(x) = dx = dqdp$; then the explicit formula for the integral kernel of the operators $\mathbf{S}_x$ in the Hilbert space $L^2(\mathbb{R}^n)$ is

$$\mathbf{S}_x \sim \delta\left(q - \frac{q' + q''}{2}\right) \exp\left\{\frac{i}{\hbar} p(q' - q'')\right\}, \qquad x = (q, p).$$

An important fact about the representation (2.8) is that it is easily inverted [**25**]:

$$f = \operatorname{tr}(\widehat{f}\mathbf{S}). \tag{2.8a}$$

The integral kernel $K_{x,y}$ in (2.7) is related to $\mathbf{S}_x$ by the formula

$$\widehat{K}_{x,y} = \mathbf{S}_x\mathbf{S}_y. \tag{2.9}$$

It is not surprising that each $\mathbf{S}_x$ is self-adjoint (if one wants to associate self-adjoint operators $\widehat{f}$ with real symbols f by (2.8)), but it is remarkable that all $\mathbf{S}_x$ are almost unitary; namely, $\mathbf{S}_x$ differs from a unitary only by a constant factor

$$\mu = 2^n.$$

Thus we have

$$\mathbf{S}^* = \mathbf{S}, \qquad \mathbf{S}^2 = \mu^2 \cdot \mathbf{I}. \tag{2.10}$$

The first consequence of this almost unitarity is the representation

$$\mathbf{S} = \mu(\mathbf{I} - 2\mathbf{P}), \tag{2.11}$$

where $\mathbf{P} = \{\mathbf{P}_x\}$ is a family of orthogonal projections.

Second, if one looks at the parameter x of the almost unitary family $\mathbf{S} = \{\mathbf{S}_x\}$ as at a "time" variable, then one can derive the generator of $\mathbf{S}$ with respect to this multidimensional "time":

$$i\hbar\partial\mathbf{S} = \widehat{\mathcal{H}}^\hbar\mathbf{S}. \tag{2.12}$$

Here the differential $\partial = \partial/\partial x$ is taken with respect to the variable x running over the phase space. The operator $\widehat{\mathcal{H}}^\hbar$ is the self-adjoint generator given by

$$\widehat{\mathcal{H}}^\hbar = 2i\hbar(\partial\mathbf{P} \cdot \mathbf{P} - \mathbf{P} \cdot \partial\mathbf{P}). \tag{2.13}$$

Since $\mathbf{S}$ and $\widehat{\mathcal{H}}^\hbar$ are self-adjoint, we conclude from (2.12) that

$$\widehat{\mathcal{H}}^\hbar\mathbf{S} = -\mathbf{S}\widehat{\mathcal{H}}^\hbar \tag{2.14}$$

and so equation (2.12) can be transformed as follows:

$$i\hbar\partial\mathbf{S} = \frac{1}{2}[\widehat{\mathcal{H}}^\hbar, \mathbf{S}]. \tag{2.12a}$$

This dynamical equation looks like the Heisenberg evolution equation, or like a Lax type equation for the "$L - A$ pair."

By means of (2.8a) and (2.9), we return to the integral kernel $K_{x,y}$ and obtain from (2.12) that

$$i\hbar\partial_x K_{x,y} = \mathcal{H}_x^\hbar * K_{x,y}. \tag{2.15}$$

The Cauchy data are

$$K_{y,y} = \mu^2, \qquad \forall\, y. \tag{2.16}$$

The constant μ^2 is determined by the condition that the unit function 1 is the identity element for the product (2.7), i.e., the function

$$\frac{1}{(2\pi\hbar)^{2n}} \int K_{x,y}(z)\, dm(y) \sim \delta_x(z) \tag{2.17}$$

in arguments x, z serves as the kernel of the unity operator in the quantum function algebra, that is, it equals $\delta_x(z)$ for the Euclidean phase space.[2]

Also from the self-adjointness (2.10), we have

$$\overline{K_{x,y}(z)} = K_{y,x}(z), \quad \text{or} \quad \overline{f * g} = \overline{g} * \overline{f}. \tag{2.18}$$

In addition to these simple conditions, there is also the fundamental *cyclicity condition*, which follows from the trace representation $K_{x,y}(z) = \mathrm{tr}(\mathbf{S}_x\mathbf{S}_y\mathbf{S}_z)$, namely,

$$K_{x,y}(z) = K_{z,x}(y), \quad \text{or} \quad \int f * g\, dm = \int fg\, dm. \tag{2.19}$$

Actually, this is the condition for the choice of the measure dm.

Note that integrating (2.15) with the function $1(y)f(z)$ and using the property (2.19), one obtains the identities

$$i\hbar df(x) = (f * \mathcal{H}_x^\hbar)(x) = -(\mathcal{H}_x^\hbar * f)(x), \qquad \forall\, f. \tag{2.20}$$

These identities determine the "germ" of the $*$-product near a given point x of the phase space. Global information about the $*$-product is contained in the wave equation (2.15), which can be solved using the operator multiplicative exponential or the $*$-exponential[3] as follows:

$$\widehat{K}_{x,y} = \mu^2 \cdot \mathrm{Exp}\Big\{ -\frac{2i}{\hbar}\int_y^x \widehat{\mathcal{H}}^\hbar \Big\},$$

or

$$f * g = \frac{\mu^2}{(2\pi\hbar)^{2n}} \iint f(x)\, \mathrm{Exp}_*\Big\{ -\frac{2i}{\hbar}\int_y^x \mathcal{H}^\hbar \Big\} g(y)\, dm(x) dm(y). \tag{2.21}$$

This formula manifests the dynamical character of quantization. We see a sum of parallel-transported amplitudes multiplied by values of functions g and f at the initial and final "time" points. The role of "time" is played by points of the phase space.

We emphasize that (2.15), as well as (2.12), are Schrödinger type dynamical equations that do not require the presence of particles. We see no particles but,

---

[2]For phase manifolds diffeomorphic to $\mathbb{R}^{2n}$ one has $\mu = 2^n$, but for compact manifolds, $\mu = 1/N_0$, where $N_0 = N_+ - N_-$ is the *quantum signature* of the manifold defined by projections in (2.11): $N_+ \equiv \dim(\mathbf{I} - \mathbf{P})$, $N_- \equiv \dim \mathbf{P}$.

[3]Such exponentials were introduced in early works on deformation quantization [**5, 26**] and were intensively exploited; for instance, see [**27**] and the references therein.

nevertheless, something is "moving" and controlled by the Hamiltonian $\mathcal{H}^\hbar$. We call $\mathcal{H}^\hbar$ the *Ether Hamiltonian.*

Of course, in the case of the standard phase space $\mathbb{R}^{2n}$, the kernel K is well known independently of the calculations performed above, namely,

$$K_{x,y}(z) = \exp\left\{\frac{i}{\hbar}\Phi_{x,y}(z)\right\}, \tag{2.22}$$

where $\Phi_{x,y}(z)$ is the symplectic area of the triangle with mid-points x, y, z (see [**28, 29**] and [**5, 30**] for general flat symplectic spaces).

The Ether Hamiltonian $\mathcal{H}^\hbar$ is easily derived in the case $\mathbb{R}^{2n}$ from equation (2.12) or from (2.15), (2.16):

$$\mathcal{H}^\hbar_x(z) = \frac{i\hbar}{\mu^2}\partial_x K_{x,y}(z)\bigg|_{y=z} = 2\omega(z-x)\cdot dx. \tag{2.23}$$

Here ω is the constant matrix of the symplectic form on $\mathbb{R}^{2n}$ in the Cartesian coordinates. In the Darboux coordinates $\omega = J = \begin{pmatrix} 0 & I \\ -I & 0 \end{pmatrix}$.

Even for the simplest integral kernel (2.22) the differential equation (2.15) seems to be new. Other attempts to find some differential equations for the $*$-product kernel were initiated in [**31, 32**], but they are certainly different from our approach.

Note that the operator $\widehat{\mathcal{H}}^\hbar$ in (2.12) is defined via (2.8), and so equations (2.12), (2.12a) are actually nonlinear:

$$i\hbar\partial\mathbf{S}_x = \int \mathcal{H}^\hbar_x(y)\mathbf{S}_y\mathbf{S}_x\,dy, \tag{2.24}$$

or

$$i\hbar\partial\mathbf{S}_x = \frac{1}{2}\int \mathcal{H}^\hbar_x(y)[\mathbf{S}_x,\mathbf{S}_y]\,dy. \tag{2.24a}$$

Similarly, the product $*$ in the evolutionary equation (2.15) for the "wave function" K is determined by the same function K via (2.7), and so this equation is actually nonlinear. Explicitly, it looks as

$$i\hbar\partial_x K_{x,y}(z) = \frac{1}{(2\pi\hbar)^{2n}}\iint \mathcal{H}^\hbar_x(z')K_{z',y'}(z)K_{x,y}(y')\,dm(z')dm(y'). \tag{2.25}$$

One can "resolve" this nonlinearity by using the representation of the $*$-product via operators L of the left regular representation (see general definitions, e.g., in [**16**]). In the case of the phase space $\mathbb{R}^{2n}$ it is well known that

$$f * g = f(L)g, \tag{2.26}$$

where the operators $L = (L_q, L_p)$ are Weyl-symmetrized and are given by the formulas

$$L = x + \frac{i\hbar}{2}J\frac{\partial}{\partial x}, \qquad x = (q,p)\in\mathbb{R}^{2n}, \quad J = \begin{pmatrix} 0 & I \\ -I & 0 \end{pmatrix}.$$

So equation (2.15) reads as a linear equation:

$$i\hbar\partial_x K_{x,y} = \mathcal{H}^\hbar_x(L)K_{x,y}, \tag{2.27}$$

where $\mathcal{H}^\hbar$ is the Ether Hamiltonian.

The derivation of the wave equations for the $*$-product kernel demonstrated above is universal and could be done in the case of an arbitrary phase space. All formulas except (2.22) and (2.23), are general. Equation (2.15) can be considered as a *fundamental nonlinear equation determined by the Ether Hamiltonian.* This equation generates the $*$-product kernel K starting from the Cauchy data (2.16). Equation (2.27) can also be established by *a priori* computations of the left regular representation (2.26) of the $*$-product algebra.[4] But first of all, one needs to know what the Ether Hamiltonian $\mathcal{H}^\hbar$ is in general.

3. Zero curvature equation

In the case of a general phase space, i.e., of a nonflat symplectic manifold, the Ether Hamiltonian $\mathcal{H}^\hbar$ is a 1-form on this space (with respect to the "time" variable) whose values are functions on the same space (with respect to the "space" variable):

$$\mathcal{H}^\hbar_x(z) = \sum_{j=1}^{2n} \mathcal{H}^\hbar_x(z)_j \, dx^j. \tag{3.1}$$

Equations (2.12), (2.15), or (2.27) are actually systems of equations, and so their generators $\widehat{\mathcal{H}}^\hbar_j$ ($j = 1, \dots, 2n$) should satisfy a compatibility condition. It can be written in the form

$$i\hbar\partial\widehat{\mathcal{H}}^\hbar = \frac{1}{2}[\widehat{\mathcal{H}}^\hbar \mathop{\wedge}_{,} \widehat{\mathcal{H}}^\hbar],$$

or

$$\partial\mathcal{H}^\hbar + \frac{i}{2\hbar}[\mathcal{H}^\hbar \mathop{\wedge}_{,} \mathcal{H}^\hbar]_* = 0. \tag{3.2}$$

Here the differential ∂ is taken only with respect to the "time" variable and the brackets $[f,g]_* = f * g - g * f$ are taken only with respect to the "space" variable.

Condition (3.2) is the *zero curvature equation for the connection over the phase space determined by the Ether Hamiltonian on the bundle whose fibers are quantum function algebras over the same phase space.*

In the classical limit $\hbar = 0$, equation (3.2) reads[5]

$$\partial\mathcal{H} + \frac{1}{2}\{\mathcal{H} \mathop{\wedge}_{,} \mathcal{H}\} = 0. \tag{3.3}$$

[4]The derivation (2.7)–(2.27) in the case of the phase space $\mathbb{R}^{2n}$ with symplectic form $\omega = dp \wedge dq + \frac{1}{2}F(q)\, dp \wedge dq$ containing an additional "magnetic" summand was given in the author's talk in Manitoba University, Winnipeg (July 2000). The observation was that for the $*$-product kernel, equation (2.27) is immediately generalized to such a "magnetic" case and the operators L are explicitly known in that case [**33**]. For the properties of the kernel $K_{x,y}$ and the family $\mathbf{S}_x$ and for their relations with symplectic transformations and connections in the "magnetic" case, see [**34**].

[5]In [**35**], where the construction from work [**7**] was analyzed in the classical limit $\hbar = 0$, a zero curvature equation was obtained, which in a sense is analogous to (3.3). An essential distinction from our equation (3.3) is that, instead of the phase space Poisson brackets $\{\cdot,\cdot\}$, the commutator of formal vector fields along tangent fibers is used in [**35**]. On the quantum level, the so-called "Weyl bundle" used in [**7, 8**] consists of algebras of polynomials along tangent fibers whose algebraic structure is just the Groenewold–Moyal product, in contrast to (3.2), where the $*$-product is taken over the whole phase manifold.

Here we omit the index $\hbar$ in the Ether Hamiltonian when considering the limit $\hbar = 0$. The Poisson brackets in (3.3) correspond to the symplectic form

$$\omega(z) = \frac{1}{2}\omega_{jk}(z)\,dz^k \wedge dz^j. \tag{3.4}$$

Note that the symplectic case under study is a particular case of the general Poisson situation. In the general Poisson picture, the "wave" equations (2.12), (2.15) and the zero curvature equations (3.2), (3.3) were used in [**15, 36**] (see also [**16**]) and the 1-form $\mathcal{H}$ was called the Cartan structure. In this situation the "space" argument belongs to the Poisson manifold itself, while the "time" argument belongs to the dual manifold, i.e., to a finite-dimensional pseudogroup. In the symplectic case this pseudogroup can be identified with the "space" manifold, and we again obtain (2.12), (2.15), (3.2), (3.3). Thus one can use those "Poisson" results in order to study the solutions of our basic equations in the symplectic case.

The following geometric lemma follows from general results in Poisson geometry [**37, 38**] (see also [**16**]).

LEMMA 3.1. *Let $\mathcal{X}$ be a symplectic manifold. In a neighborhood of the zero section, the cotangent bundle $T^*\mathcal{X}$ admits a symplectic fibration ℓ over the base $\mathcal{X}$ such that*

(i) *ℓ is the identical map on the zero section $T_0^*\mathcal{X} \approx \mathcal{X}$, that is, $\ell(x,\eta)\big|_{\eta=0} = 0$;*

(ii) *the dual fibration r given by the reflection in momenta*

$$r(x,\eta) \stackrel{\text{def}}{=} \ell(x,-\eta), \qquad \eta \in T_x^*\mathcal{X},$$

is in involution with ℓ, that is

$$\{r^j, \ell^k\}_{T^*\mathcal{X}} = 0, \qquad \forall\, j,k.$$

A symplectic fibration with properties (i), (ii) will be called *reflective.*

One can present explicit formulas for reflective fibrations in terms of ω-affine structures.

In the fixed local chart, a local diffeomorphism is ω-affine if the matrix of its differential $((G_\ell^r))$ satisfies the identities

$$\underset{k,\ell,m}{\mathfrak{S}}\ \partial_k G_\ell^r \cdot {G^{-1}}_r^s \cdot \omega_{sm} = 0, \qquad \forall\, k,\ell,m.$$

Here the notation $\mathfrak{S}$ is used for cyclic summation. An *ω-affine structure* is an atlas of local charts such that each transfer from chart to chart is ω-affine. Local coordinates of this atlas can also be called ω-affine. A diffeomorphism is ω-affine if it preserves the ω-affine structure.

Let us fix an ω-affine structure over the symplectic manifold $(\mathcal{X},\omega)$. In ω-affine local coordinates $x = (x^1,\dots,x^{2n})$, one can present components of a reflective symplectic fibration ℓ over $\mathcal{X}$ by the following explicit formulas:

$$\ell^j(x,\eta) = \varphi\Big(\eta\Psi(x)\frac{\partial}{\partial x}\Big)x^j, \qquad \varphi(\xi) \stackrel{\text{def}}{=} \frac{\xi}{e^\xi - 1}.$$

At least in a neighborhood of the zero section $\{\eta = 0\}$ these components are well defined as smooth functions.

Now, let us construct the solution of the zero curvature equation (3.3). Note that in the symplectic case under study the matrix $\partial\ell/\partial\eta$ is nonsingular at least in a sufficiently small neighborhood of the zero section. Thus the following equation is solvable:

$$\ell(x,\eta) = z \quad \Longrightarrow \quad \eta = \mathcal{H}_x(z). \tag{3.5}$$

The symplecticity of ℓ means

$$\{\ell^j, \ell^k\}_{T^*\mathcal{X}} = \Psi^{jk}(\ell), \qquad \Psi \stackrel{\text{def}}{=} \omega^{-1},$$

and so it is easy to check that the solution of (3.5) satisfies

$$\partial_j \mathcal{H}_k - \partial_k \mathcal{H}_j + \{\mathcal{H}_j, \mathcal{H}_k\}_{\mathcal{X}} = 0. \tag{3.3a}$$

Here we use the notation $\partial_j = \partial/\partial x_j$ for the derivatives, and the Poisson brackets are taken with respect to $z \in \mathcal{X}$. The system (3.3a) coincides with (3.3).

From statements (i), (ii) of Lemma 3.1, we obtain

$$\mathcal{H}_x(z)\big|_{z=x} = 0, \qquad \frac{1}{2} D\mathcal{H}_x(z)\big|_{z=x} = \omega(x). \tag{3.6}$$

Here we use the notation $D = \partial/\partial z$ for the derivatives.

For the second derivatives at $z = x$, from (3.3a) we obtain the formula

$$\frac{1}{2} D^2 \mathcal{H}_x(z)\big|_{z=x} = \omega(x)\Gamma(x), \tag{3.7}$$

where Γ is a symplectic connection on $\mathcal{X}$, that is,

$$\partial\omega_{jk} - \Gamma^{\ell}_{js}\omega_{\ell k} - \Gamma^{\ell}_{ks}\omega_{j\ell} = 0, \quad \text{or} \quad \nabla\omega = 0.$$

This is a crucial place: the symplectic connection Γ has appeared automatically from the symplectic fibration ℓ of the secondary phase space (local symplectic groupoid[6]) over the original manifold $\mathcal{X}$.

Of course, instead of (3.6), (3.7), one can write direct formulas for $\Psi = \omega^{-1}$ and for Γ in terms of ℓ:

$$\Psi(x) = 2\frac{\partial \ell}{\partial \eta}(x,\eta)\bigg|_{\eta=0}, \qquad \Gamma(x) = 4\omega(x)\cdot \frac{\partial^2 \ell}{\partial\eta\partial\eta}(x,\eta)\bigg|_{\eta=0} \cdot \omega(x).$$

If the symplectic fibration ℓ corresponds to an ω-affine structure over $\mathcal{X}$, then the connection Γ can be characterized by the condition

$$\mathop{\mathfrak{S}}_{k,\ell,m} \Gamma^s_{k\ell}\omega_{sm} = 0, \qquad \forall k, \ell, m,$$

or can be given directly,

$$\Gamma^s_{k\ell} = \frac{1}{3}\Psi^{sm}(\partial_k \omega_{m\ell} + \partial_\ell \omega_{mk}),$$

in ω-affine local coordinates.

Now let us return to the problem of the identification of solution of the zero curvature equation (3.3) or (3.3a). Note that (3.6), (3.7) are boundary conditions on the diagonal $z = x$. We need to choose a convenient way to leave the diagonal.

Let us fix a point $x \in \mathcal{X}$ and a tangent vector $v \in T_x\mathcal{X}$. Denote by $\operatorname{Exp}_x(2v\tau)$ the trajectory of the Hamiltonian $v\mathcal{H}_x$ starting at the point x. Then

$$\mathcal{H}_x\big(\operatorname{Exp}_x(-v)\big) = -\mathcal{H}_x\big(\operatorname{Exp}_x(v)\big). \tag{3.8}$$

---

[6]The fibration ℓ is the target mapping (or the left restriction mapping) in the local symplectic groupoid over $\mathcal{X}$ which is realized as a neighborhood of the zero section of $T^*\mathcal{X}$ (**[36, 39]**).

THEOREM 3.2. *Let* $(\mathcal{X}, \omega, \Gamma)$ *be a symplectic manifold* $\mathcal{X}$ *with symplectic form* ω *and symplectic connection* Γ. *In a neighborhood of the diagonal* $z \equiv x$, *there is a unique solution of the zero curvature equation* (3.3), *which satisfies conditions* (3.6), (3.7), *and* (3.8). *This solution can be obtained by* (3.5) *via a unique reflective symplectic fibration* ℓ *over* $\mathcal{X}$.

We call $\mathcal{H}$ the *classical Ether Hamiltonian.* The mapping

$$\mathrm{Exp}_x : T_x\mathcal{X} \to \mathcal{X}$$

will be called the Ether exponential mapping,[7] and the trajectory $\mathrm{Exp}_x(2v\tau)$ will be called the *Ether geodesics* through the point x with velocity v.

Note that Ether geodesics correspond to some vertical curves in the fiber $T_x^*\mathcal{X}$, namely, to the curves

$$\eta(\tau) \stackrel{\text{def}}{=} \mathcal{H}_x\big(\mathrm{Exp}_x(2v\tau)\big). \tag{3.9}$$

They are perpendicular to the velocity $v \in T_x\mathcal{X}$ and satisfy the equations

$$\frac{d}{d\tau}\eta(\tau) = v\Omega^{[x]}(\eta(\tau)), \qquad \eta(\tau)\Big|_{\tau=0} = 0. \tag{3.10}$$

Here the tensor $\Omega^{[x]}$ is determined in a neighborhood of zero in $T_x^*\mathcal{X}$ by the relations

$$\Omega_{lk}^{[x]}(\mathcal{H}_x) = \{\mathcal{H}_{xl}, \mathcal{H}_{xk}\}. \tag{3.11}$$

This tensor introduces a symplectic structure to the fiber $T_x^*\mathcal{X}$ so that the *Ether mapping* $\mathcal{H}_x$ *is symplectic.* With respect to this structure, system (3.10) is Hamiltonian and the related Hamilton function is just the linear function v (i.e., the function $\langle v, \eta\rangle$, $\eta \in T_x^*\mathcal{X}$).

In addition to other properties, let us emphasize that the infinitesimal geometry of space, including its symplectic structure, connection, and curvature, is sitting inside the Ether Hamiltonian which can be considered as a "generating function" for this kinematic geometry. Using the Ether exponential coordinates $z = \mathrm{Exp}_x(v)$, we represent the components of this "generating function" as follows:

$$\mathcal{H}_j = 2\omega_{jk}(x)v^k + 2\omega_{km}(x)R^m_{l,js}(x)v^k v^l v^s + O(v^5), \tag{3.12}$$

where R is the curvature of the symplectic connection Γ. The skew-symmetry with respect to the tangent argument v corresponds to condition (3.8). Higher terms in (3.9) contain derivatives of the curvature R.

Now we demonstrate an alternative way to obtain the Ether Hamiltonian.

Let $x \in \mathcal{X}$. A symplectic mapping $s_x : \mathcal{X} \to \mathcal{X}$ is called a *reflection*[8] in x if it is an involution $s_x^2 = \mathrm{id}$, and x is an isolated fixed point $s_x(x) = x$. A symplectic manifold $\mathcal{X}$ with a smooth family of reflections $\{s_x \mid x \in \mathcal{X}\}$ will be called a *reflective* symplectic manifold.

---

[7]In general, the mapping Exp_x is not equal to the exponential mapping $\exp_x$ related to the connection Γ. Nevertheless, in a neighborhood of each point x, there is a connection whose exponential mapping at x coincides with Exp_x. This connection is not symplectic.

[8]In [**40**] such a mapping is called a symmetry under the additional condition that the fixed point is unique. However, the term "symmetry" in the classical theory of symmetric spaces carries the strong property $s_x s_y s_x = s_{s_x(y)}$, which does not hold in our case.

THEOREM 3.3. *Let $(\mathcal{X}, \omega)$ be a simply-connected reflective symplectic manifold. Then there is a natural symplectic connection on $\mathcal{X}$:*

$$\Gamma(z) = -\frac{1}{2} D^2 s_x(z)\Big|_{x=z}, \qquad z \in \mathcal{X}. \tag{3.13}$$

The Ether Hamiltonian corresponding to $(\mathcal{X}, \omega, \Gamma)$ is given by

$$\mathcal{H}_x(z) = -\int_x^z \langle \omega(z) \partial s_x(s_x(z)), \, dz \rangle. \tag{3.14}$$

Vice versa, the Ether Hamiltonian corresponding to $(\mathcal{X}, \omega, \Gamma)$ uniquely determines the reflective structure over $\mathcal{X}$ by solving the Cauchy problem

$$\partial s_x + \{\mathcal{H}_x, s_x\} = 0, \qquad s_x(z)\Big|_{x=z} = z, \tag{3.15}$$

or just by solving the implicit equation

$$\mathcal{H}_x(s_x(z)) = -\mathcal{H}_x(z) \tag{3.16}$$

(in a domain where $D\mathcal{H}$ does not degenerate). This family of reflections is related to the connection Γ by formula (3.13). *The Ether geodesics are symmetric with respect to reflections*

$$s_x(\operatorname{Exp}_x(v)) = \operatorname{Exp}_x(-v) \tag{3.17}$$

(but, in general, reflections s_x are not affine with respect to Γ).

In particular, this theorem can be applied to symplectic symmetric manifolds [**41**], i.e., to the case where the curvature of Γ is covariantly constant ($\nabla R = 0$). In this case the connection (3.13) is just the canonical Cartan–Loos connection, the axiom $s_x s_y s_x = s_{s_x(y)}$ holds, all reflections s_x are affine, and the Ether geodesics coincide with the usual geodesics of the connection Γ.

As an example, let us consider the sphere $\mathbf{S}^2 = \{|x| = 1\}$ embedded in $\mathbb{R}^3$ and endowed with standard symplectic form. In this case

$$\mathcal{H}_x(z) = 2[x \times z] \cdot dx\Big|_{\mathbf{S}^2}, \tag{3.18}$$

where the brackets $[\cdot \times \cdot]$ stay for the vector product in $\mathbb{R}^3$.

The following fact is important: the reflective structure provides the global existence of the Ether Hamiltonian via (3.14) not only in a neighborhood of the diagonal $z = x$ as in Theorem 3.2.

CONJECTURE 3.4. *A reflective structure and so an Ether Hamiltonian exists globally on any simply-connected symplectic manifolds except possibly cases with incompleteness phenomena.*

4. Intrinsic dynamical objects

Because of the zero curvature condition, the intrinsic dynamics, that is, the translation along the phase space by means of the Ether Hamiltonian, meets no obstructions.

Let us denote by $g_{x,y}$ the symplectic transformations

$$z \to g_{x,y}(z), \qquad z \in \mathcal{X},$$

obtained by shifts along the Ether dynamical system (that is, the "time" derivative $\partial_x g_{x,y}(z)$ coincides with the Ether Hamiltonian vector field at the point $g_{x,y}(z)$, and the "initial" data are $g_{y,y}(z) = z$).

These transformations satisfy the groupoid rule

$$g_{x,y} \cdot g_{y,z} = g_{x,z} \tag{4.1}$$

and produce a certain group of intrinsic transformations of the phase space. In general, this group is infinite-dimensional, and we prefer to speak about the finite-dimensional *groupoid of Ether translations.*

In terms of symplectic reflections s_x the symplectic translations $g_{x,y}$ are expressed by the formula

$$g_{x,y} = s_x \cdot s_y, \tag{4.2}$$

and so

$$s_x(y) = g_{x,y}(y).$$

Other important transformations are related to the Ether exponential mapping. Let us fix $x \in \mathcal{X}$, and let $v \in T_x\mathcal{X}$. Recall that by $\operatorname{Exp}_x(2vt)$ we denote the trajectory of the Hamiltonian $v\mathcal{H}_x$ starting at the point x. But there are other trajectories starting at other points. We consider all of them.

Denote by $e_x^{2vt}(y)$ the trajectory of the Hamiltonian $v\mathcal{H}_x$ starting at the point y. In particular, if $y = x$, then we have

$$e_x^v(x) \equiv \operatorname{Exp}_x(v). \tag{4.3}$$

LEMMA 4.1. *For each $x \in \mathcal{X}$ the pseudogroup of symplectic transformations $\{e_x^v\}$ satisfies the identity*

$$s_x \cdot e_x^v = e_x^{-v} \cdot s_x. \tag{4.4}$$

We call the e_x^v *exponential transformations.*

What is the quantum version of all these transformations?

Heuristically, we associate the reflections s_x with the operators $\mathbf{S}_x$, the translations $g_{y,x}$ with the operators $\mathbf{S}_x\mathbf{S}_y = \widehat{K}_{x,y}$, and the exponential transformations e_x^v with the operators $\exp\{\frac{i}{2\hbar} v\widehat{\mathcal{H}}_x^\hbar\}$.

More precisely, we introduce quantum mappings $\widehat{s}_x^*$, $\widehat{g}_{y,x}^*$, and $\widehat{e}_x^{v*}$, acting in the $*$-product algebra, by evaluating the Heisenberg transforms. For example, we could define the mapping $f \to \widehat{s}_x^* f$ in the following way:

$$\widehat{\widehat{s}_x^* f} = \mathbf{S}_x \widehat{f} \mathbf{S}_x^{-1}, \tag{4.5}$$

or

$$(\widehat{s}_x^* f)(z) = \operatorname{tr}\left(\mathbf{S}_x^{-1}\mathbf{S}_z\mathbf{S}_x\widehat{f}\right). \tag{4.5a}$$

If one does not know the operator realization of the $*$-product algebra, then, instead of operators $\mathbf{S}_x$, it is possible to use the differential equations with respect to the parameters $x, y, \ldots$. These equations follow from our basic relations (2.12) and (2.15).

Denote by $F(x,z) = (\widehat{s}_x^* f)(z)$ the solution of the Cauchy problem

$$\begin{cases} \partial F + \frac{i}{\hbar}[\mathcal{H}^\hbar, F]_* = 0, \\ F\big|_{\text{diagonal}} = f, \end{cases} \tag{4.6}$$

and we call the mappings $\widehat{s}_x^*$ *quantum reflections.*

In (4.6) the unknown function $F = F(x,z)$ depends on $x, z \in \mathcal{X}$, the differential ∂ is taken with respect to x, the quantum brackets $[\cdot,\cdot]_*$ are taken with respect to z, and the Cauchy data are given on the diagonal $x = z$.

If we consider the same equation (4.6) but for the function $\mathcal{F} = \mathcal{F}(x,y,z)$ depending on the additional argument $y \in \mathcal{X}$ and change the Cauchy data by the data

$$\mathcal{F}(x,y,z)\big|_{x=y} = f(z), \tag{4.7}$$

then the solution $\mathcal{F}$ determines the *quantum translations* $\widehat{g}_{y,x}^*$ as

$$\mathcal{F}(x,y,z) = (\widehat{g}_{y,x}^* f)(z). \tag{4.8}$$

Finally, the solution of the Cauchy problem

$$\frac{\partial \mathcal{E}}{\partial t} + \frac{i}{\hbar}[\mathcal{E}, v\mathcal{H}_x^\hbar]_* = 0, \qquad \mathcal{E}\Big|_{t=0} = f, \tag{4.9}$$

taken at $t = 1/2$, determines the *quantum exponentials*

$$\mathcal{E}\big|_{t=1/2} = \widehat{e}_x^{v*} f, \qquad v \in T_x\mathcal{X}. \tag{4.10}$$

The restriction of this function to the diagonal (as in (4.3)) can be called a *quantum Ether exponential mapping*:

$$f \to f\big|_{\widehat{\operatorname{Exp}_x}(v)} \stackrel{\text{def}}{=} (\widehat{e}_x^{v*} f)(x). \tag{4.11}$$

On the quantum level we have analogs of identity (4.4):

$$\widehat{e}_x^{v*} \cdot \widehat{s}_x^* = \widehat{s}_x^* \cdot \widehat{e}_x^{-v*}, \tag{4.12}$$

and of identities (3.8), (3.16):

$$\widehat{s}_x^* \mathcal{H}_x^\hbar = -\mathcal{H}_x^\hbar, \tag{4.13}$$

$$\mathcal{H}_x^\hbar\big|_{\widehat{\operatorname{Exp}_x}(-v)} = -\mathcal{H}_x^\hbar\big|_{\widehat{\operatorname{Exp}_x}(v)}. \tag{4.14}$$

All these formulas assume that one knows the $*$-product, and so it is possible to solve (4.6) and (4.9) at least asymptotically as $\hbar \to 0$.

Under the usual assumption

$$f * g \simeq fg + \sum_{k\geq 1} \hbar^k c_k(f,g), \tag{4.15}$$

where c_k are bidifferential operators of order k,

$$c_1 = -\frac{i}{2}\{\cdot,\cdot\}, \qquad c_k(f,g) = (-1)^k c_k(g,f) = \overline{c}_k(g,f), \tag{4.16}$$

the quantum mappings defined above correspond to their classical counterparts:

$$\begin{aligned} \widehat{s}_x^* &= s_x^*\big(I + O(\hbar^2)\big), \\ \widehat{e}_x^{v*} &= e_x^{v*}\big(I + O(\hbar^2)\big), \\ f\big|_{\widehat{\mathrm{Exp}}_x(v)} &= \big(I + O(\hbar^2)\big) f\big|_{\mathrm{Exp}_x(v)}. \end{aligned} \tag{4.17}$$

Here the quantum corrections $O(\hbar^2)$ are formal $\hbar^2$-series of the following type:

$$O(\hbar^2) = \hbar^2 a_1 + \hbar^4 a_2 + \dots;$$

their coefficients a_k are differential operators of order $2k+1$ easily derived from (4.6), (4.9) via the coefficients c_k (4.15).

In the next section we consider the procedure of constructing the coefficients c_k.

5. Weak asymptotics and quantum zero curvature equation

We begin to study the basic equation (2.15) in order to construct the $*$-product over $\mathcal{X}$ at least asymptotically as $\hbar \to 0$. There are two types of asymptotic expansions of $*$-products: weak and semiclassical. Both are useful and both follow from the exact fundamental nonlinear equation (2.15) for the $*$-product kernel given by the Ether Hamiltonian.

Note that the semiclassical view on the quantization problem over general symplectic manifolds has been developed starting from the first papers [**42, 43, 4**] (summarized in the book [**16**]). In [**4**] this method was called *asymptotic quantization.* The $*$-product was constructed there by matching local semiclassical expansions, and the quantization condition

$$\frac{1}{2\pi\hbar}[\omega] - \frac{1}{2}c_1 \in H^2(\mathcal{X}, \mathbb{Z}) \tag{5.1}$$

first appeared as a necessary condition for the existence of an irreducible operator representation of the $*$-product [**43, 4, 16**].

The weak asymptotics approach to the quantization problem, pioneered in the fundamental papers [**1, 5**], was a basis of all those semiclassical developments. In [**4**] the name *deformation quantization* was proposed for the weak asymptotics method. General results in this deformation direction were obtained in [**44**]–[**47**], [**7, 8**].

Now we apply our basic evolutionary equations to construct a unique $*$-product in the standard deformation form (4.15):

$$(f * g)(z) = \Big(1 + \sum_{k\geq 1} \hbar^k c_k\Big) f(x)g(y)\Big|_{x=y=z}. \tag{5.2}$$

Here c_k are certain differential operators acting on the x, y-arguments and satisfying conditions (4.16). The corresponding weak asymptotics of the integral kernel is

$$K_{x,y} = (2\pi\hbar)^{2n}\Big(1 + \sum_{k\geq 1} \hbar^k c_k'\Big)\delta_x\delta_y, \tag{5.3}$$

where the prime $'$ means the transposition with respect to the measure dm.

The easiest way to find the coefficients c_k in (5.2) or (5.3) is to substitute expression (5.2) into (2.20) and (4.14). Then we uniquely derive all c_k in terms of the Taylor expansion (3.12) of the Ether Hamiltonian as follows:

$$f * g = fg - \frac{i\hbar}{2} f \overleftarrow{\nabla} \Psi \overrightarrow{\nabla} g - \frac{\hbar^2}{8} f (\overleftarrow{\nabla} \Psi \overrightarrow{\nabla})^2 g + \dots. \tag{5.4}$$

Here $\Psi = \omega^{-1}$ is the Poisson tensor on the phase space, ∇ is the covariant derivative corresponding to the symplectic connection Γ, the lower arrows ($\leftarrow$ or $\rightarrow$) indicate the multiplier (f or g) on which the derivative acts. Higher terms of (5.4) contain the curvature of Γ.

Note that together and simultaneously with deriving formulas (5.2), (5.4), we have to satisfy the quantum zero curvature equation (3.2) in which the same $*$-product (5.2) should be used. Thus we are looking for the quantum Ether Hamiltonian in the form

$$\mathcal{H}^\hbar = \mathcal{H} + \hbar^2 \mathcal{C} + \hbar^4 \mathcal{D} + \dots, \tag{5.5}$$

where $\mathcal{H}$ is the classical Ether Hamiltonian and the quantum corrections $\mathcal{C}, \mathcal{D}, \dots$ satisfy variations of the zero curvature equation.

The first of these variation equations is

$$\begin{gathered} \partial \mathcal{C} + \frac{1}{2}\{\mathcal{H} \underset{,}{\wedge} \mathcal{C}\} + \frac{1}{2}\{\mathcal{C} \underset{,}{\wedge} \mathcal{H}\} + i c_3(\mathcal{H} \underset{,}{\wedge} \mathcal{H}) = 0, \\ \mathcal{C}\big|_{\text{diagonal}} = 0. \end{gathered} \tag{5.6}$$

Here we used formula (4.16) for c_1, which follows from (2.20), in the first order of $\hbar$ with (3.6) taken into account.

The formula for the next coefficient c_2 in (5.2) also follows from (2.20) and from its second iteration:

$$(i\hbar)^2 D^2_{jk} f = i\hbar f * \partial_k \mathcal{H}^\hbar_j + f * \mathcal{H}^\hbar_j * \mathcal{H}^\hbar_k \qquad \text{on the diagonal.} \tag{5.7}$$

Here we do not write the arguments of the functions $f = f(z)$ and $\mathcal{H}^\hbar_j = \mathcal{H}^\hbar_{jx}(z)$, but only indicate in (5.7) that the equation is taken on the diagonal $z = x$. In terms of the order of $\hbar^2$ equations (2.20) and (5.7) imply

$$\begin{gathered} c_2(f, \mathcal{H}_k) = 0 \quad \text{on the diagonal,} \\ D^2_{jk} f + c_2(f, \mathcal{H}_j \mathcal{H}_k) + \frac{1}{4}\{f, \partial_k \mathcal{H}_j + \partial_j \mathcal{H}_k\} + f c_2(\mathcal{H}_j, \mathcal{H}_k) = 0 \quad \text{on the diagonal.} \end{gathered} \tag{5.8}$$

Here we used the fact that c_1 is already known, and the zero curvature equation (3.3) was applied to simplify the formula. Both equations (5.8) are taken at $z = x$.

We are looking for the operator c_2 in the form

$$c_2 = \underset{\leftarrow}{D^2_{sl}} c_2^{sl,s'l'} \underset{\rightarrow}{D^2_{s'l'}} + \underset{\leftarrow}{D_s} c_2^{s,s'l'} \underset{\rightarrow}{D^2_{s'l'}} + \underset{\leftarrow}{D^2_{sl}} c_2^{sl,s'} \underset{\rightarrow}{D_{s'}}. \tag{5.9}$$

In view of (3.6), (3.7), from the second equation (5.8), we find $c_2^{sl,s'l'}$ and $c_2^{s,s'l'}$ as follows:

$$c_2^{sl,s'l'} = -\frac{1}{8}\Psi^{ss'}\Psi^{ll'}, \qquad c_2^{s,s'l'} = \frac{1}{8}\Gamma^s_{jk}\Psi^{js'}\Psi^{kl'}.$$

The coefficient $c_2^{sl,s'}$ can be found from the conjugate equation in (2.20) in the same way; we have

$$c_2^{sl,s'} = \frac{1}{8}\Gamma^{s'}_{jk}\Psi^{js}\Psi^{kl}.$$

But in order to find $c_2^{s,s'}$, we have to consider the first equation (5.8), which implies

$$c_2^{s,s'} = -\frac{1}{8}\Gamma^s_{jk}\Psi^{jm}\Psi^{kr}\Gamma^{s'}_{mr}.$$

Thus we obtain the whole term of the order of $\hbar^2$ in (5.2), (5.4).

In order to find the next coefficient c_3 in (5.2), we have to consider the third iteration of (2.20):

$$(5.10)\quad (i\hbar)^3 D^3_{jkl} f = (i\hbar)^2 f * \partial^2_{lk}\mathcal{H}^\hbar_j + i\hbar\Big(f * \partial_l\mathcal{H}^\hbar_j * \mathcal{H}^\hbar_k + f * \mathcal{H}^\hbar_j * \partial_l\mathcal{H}^\hbar_k + f * \partial_k\mathcal{H}^\hbar_j * \mathcal{H}^\hbar_l\Big) + f * \mathcal{H}^\hbar_j * \mathcal{H}^\hbar_k * \mathcal{H}^\hbar_l \quad \text{on the diagonal.}$$

We are looking for the operator c_3 in the form similar to (5.9), but containing also the third derivatives. From (5.10) and from its conjugate version, in the order of $\hbar^3$, in the same way as above, we easily calculate the coefficients

$$c_3^{jkl,j'k'l'},\qquad c_3^{jk,j'k'l'},\qquad c_3^{j,j'k'l'},\qquad c_3^{jkl,j'k'},\qquad c_3^{jkl,j'}.$$

Then we return to the second iteration (5.7) and find, in the order of $\hbar^3$, the coefficients

$$c_3^{jk,j'k'},\qquad c_3^{j,j'k'},\qquad c_3^{jk,j'}$$

(the last one – from the conjugate equation).

To obtain the formula for $c_3^{j,j'}$, one has to consider the first iteration (2.20) in the order of $\hbar^3$:

$$c_3(f,\mathcal{H}_k) = i\{f,\mathcal{C}_k\} \qquad \text{on the diagonal.}$$

From this equation we find

$$(5.11)\quad c_3^{j,j'} = \frac{i}{2}\Psi^{jl}\alpha_{lr}\Psi^{rj'} + \Big(c_3^{j,j'k'}D^2_{j'k'}\mathcal{H}_r + c_3^{j,j'k'l'}D^2_{j'k'l'}\mathcal{H}_r\Big)\Psi^{rj'} \qquad \text{on the diagonal,}$$

where

$$(5.12)\qquad \alpha_{lr} = \frac{1}{2}D_r\mathcal{C}_l \qquad \text{on the diagonal.}$$

The last relation can be considered as the boundary condition for the 1-form

$$(5.13)\qquad \mathcal{C}_x(z) = \sum_{l=1}^{2n}\mathcal{C}_{lx}(z)\,dx^l,$$

analogous to the boundary condition (3.6).

So we arbitrarily fix a closed 2-form α on $\mathcal{X}$ and define $c_3^{j,j'}$ by (5.11). Then the whole operator c_3 is known.

It is important that we have uncoupled the calculation of c_3 and the problem of solving the variation zero curvature equation for $\mathcal{C}$. The variation equation (5.6) is now well defined: the term $ic_3(\mathcal{H}\underset{,}{\wedge}\mathcal{H})$ represents the known "right-hand side" of the equation.

There are also two boundary conditions: the second equality in (5.6) and equality (5.12). But in order to fix $\mathcal{C}$ uniquely, one has to draw in the skew-symmetry condition (4.14). In the order $\hbar^2$ this condition implies an additional identity for $\mathcal{C}_x(z)$ which selects a unique solution at least near the diagonal $z = x$.

This procedure is continued in all orders of the deformation parameter $\hbar$ in (5.2) and in (5.5).

Say, for $c_5^{j,j'}$, we obtain an expression similar to (5.11): $c_5^{j,j'} = \frac{i}{2}\Psi^{jl}\beta_{lr}\Psi^{rj'} +$ *known summands*, where β is a closed 2-form on $\mathcal{X}$ related to the correction $\mathcal{D}$ in (5.5): $\beta_{lr} = \frac{1}{2}D_r\mathcal{D}_l$ (on the diagonal). The variation zero curvature equation for $\mathcal{D}$

has the same form as (5.6), and it is uniquely solvable near the diagonal $z = x$ if the skew-symmetry condition (4.14) in the order of $\hbar^4$ is taken into account.

THEOREM 5.1. *Let* $(\mathcal{X}, \omega, \Gamma)$ *be a symplectic simply-connected manifold, and let* $\omega^\hbar = \omega + \hbar^2\alpha + \hbar^4\beta + \dots$ *be a closed deformation of* ω. *Then the Ether structure generates uniquely, explicitly, and geometrically (via* $\omega^\hbar$ *and* Γ*) the formal* $\hbar$*-power expansions of the* $*$*-product* (5.2), (5.4) *and of the solution* (5.5) *of the quantum zero curvature equation near the diagonal. On the diagonal the boundary condition holds:*

$$\frac{1}{2} D\mathcal{H}_x^\hbar(z)\Big|_{z=x} = \omega^\hbar(x). \tag{5.14}$$

Note that we still have not found the expansion for the integral kernel K (5.3). For this we need to know the measure dm on $\mathcal{X}$. The equation for the measure is the cyclicity condition (2.19):

$$\int_{\mathcal{X}} \left(\frac{1}{2}\{f, g\} + i\hbar c_2(f, g) + \dots\right) dm = 0, \qquad \forall\, f, g.$$

Obviously, we see that

$$dm = \text{const} \cdot (1 + \hbar^2\sigma_1 + \hbar^4\sigma_2 + \dots)\, dm^\omega, \tag{5.15}$$

where dm^ω is the Liouville measure corresponding to the symplectic form ω:

$$dm^\omega = \frac{1}{n!}|\omega \wedge \dots \wedge \omega|, \qquad 2n = \dim \mathcal{X}.$$

The odd powers of $\hbar$ are absent in (5.15) because of the self-adjointness condition (2.18). The differential equations for the corrections $\sigma_1, \sigma_2, \dots$ are not studied in the present paper. The constant factor in (5.15) is obtained from condition (2.17) (see [**48**]).

In conclusion, let us make a remark about the global solvability of the quantum zero curvature equation on reflective manifolds. Consider again the first variation equation (5.6). Denote

$$\begin{aligned} \widetilde{\mathcal{C}}_x(z) &\overset{\text{def}}{=} \mathcal{C}_{jx}(s_x(z))\, dx^j, \\ \widetilde{\mathcal{R}}_x(z) &\overset{\text{def}}{=} ic_3(\mathcal{H}_{kx}, \mathcal{H}_{jx})(s_x(z))\, dx^j \wedge dx^k, \end{aligned} \tag{5.16}$$

where the s_x are the reflection mappings. Then the variation equation (5.6) reads

$$\partial\widetilde{\mathcal{C}} = \widetilde{\mathcal{R}}. \tag{5.17}$$

The local solvability condition is $\partial\widetilde{\mathcal{R}} = 0$, which is equivalent to

$$\underset{j,k,l}{\mathfrak{S}} \Big(c_3(\mathcal{H}_j, \{\mathcal{H}_k, \mathcal{H}_l\}) + \{\mathcal{H}_j, c_3(\mathcal{H}_k, \mathcal{H}_l)\}\Big) = 0.$$

This conditions follows from the Jacobi identity for the double commutators $[\mathcal{H}_j, [\mathcal{H}_k, \mathcal{H}_l]_*]_*$ in the order of $\hbar^4$.

Note that the operation c_3 in the definition of the form $\widetilde{\mathcal{R}}$ contains the part $\underset{\leftarrow}{D_j}\, c_3^{j,j'}\, \underset{\rightarrow}{D_{j'}}$, and $c_3^{j,j'}$ given by (5.11) contains the summand $\frac{i}{2}\Psi^{jl}\alpha_{lr}\Psi^{rj'}$. By procedure (5.16), this summand generates the closed 2-form

$$\frac{1}{2}\alpha_{lr}(s_x(z))\partial s_x^r(z) \wedge \partial s_x^l(z), \tag{5.18}$$

which stays as a part of $\widetilde{\mathcal{R}}$. Here α is an arbitrarily fixed closed 2-form from the boundary condition (5.12). But actually we can include this form, as well as all further forms, as quantum corrections to the input symplectic form $\omega \to \omega^\hbar$. Then the summand (5.18) disappears from the form $\widetilde{\mathcal{R}}$, and this procedure does not affect the global solvability of equation (5.17) (and of all other equations in the next orders).

CONJECTURE 5.2. *The quantum zero curvature equation* (3.2) *over any simply-connected reflective symplectic manifold is globally solvable at least in the sense of the formal $\hbar$-power expansion* (5.5).

6. Semiclassical asymptotics and quantum lift

The weak asymptotics (5.2)–(5.4) loses some important information about the "wave function" $K_{x,y}$. A more accurate approximation is the semiclassical one, say, the WKB-approximation

$$K_{x,y} = e^{\frac{i}{\hbar}\Phi_{x,y}}\varphi_{x,y} + O(\hbar), \tag{6.1}$$

which works well outside the focal set (in the semiclassically-simple domain). The difference between (6.1) and (5.3) is the same as the difference between (2.2) and (2.3).

The phase $\Phi_{x,y}$ in (6.1) carries a dynamic geometry of the $*$-product. At the "initial moment" $x = y$ this phase is zero, but its "time" derivatives are not zero:

$$\Phi_{y,y}(z) = 0, \qquad \partial_y \Phi_{x,y}(z)\Big|_{x=y} = \mathcal{H}_y(z). \tag{6.2}$$

Thus, in semiclassically-simple domains, the Ether structure is automatically generated by the WKB-phase of the $*$-product integral kernel.

The $*$-product itself, after the substitution of (6.1) into (2.7), reads

$$(f * g)(z) = \frac{1}{(2\pi\hbar)^{2n}} \iint \Big(e^{\frac{i}{\hbar}\Phi_{x,y}(z)}\varphi_{x,y}(z) + O(\hbar)\Big) f(x)g(y)\, dm(x)dm(y). \tag{6.3}$$

This is a very useful formula in the case where the functions f, g themselves are of WKB type (or similar):

$$f(x) \sim e^{\frac{i}{\hbar}A(x)}a(x), \qquad g(y) \sim e^{\frac{i}{\hbar}B(y)}b(y).$$

Oscillations of such type on the quantum level mean the presence of a nonzero momentum, and so, of a certain dynamics on the level of classical mechanics.

We see that the weak expansion (5.2) for the $*$-product is, in some sense, a kinematic formula, namely, it does not allow oscillations of f, g. On the contrary, the integral representation (6.3) is dynamical; it allows oscillations of f or g (which appear immediately in any application of $*$-products to quantum problems).

The difference between two asymptotical approaches generates a distinction of the kinematic geometry from the dynamic one. Oscillations imply momentum, and momentum implies parallel transport, that is, dynamics. The fact that weak expansions like (5.2) do not work for $\hbar$-oscillating functions actually proves that we cannot transfer the kinematic results to dynamical ones.

In order to compute the semiclassical approximation (6.1) of the $*$-product kernel, we use the Ether wave equation and Ether translations.

First, we resolve the nonlinearity of equations (2.15), (2.25) following the scheme (2.26), (2.27). For this, one needs to know operators of the left regular representation. We compute them in the same way as in [**15, 16**].

LEMMA 6.1. *The left regular representation of the $*$-product algebra is given by the formula*

$$(f * g)(x) = f^{\#}(\overset{2}{x}, -i\hbar\overset{1}{\partial_x})g(x), \tag{6.4}$$

where the function $f^{\#}$ on the secondary phase space $T^*\mathcal{X}$ is determined by the equation

$$f^{\#}(\overset{2}{x}, \overset{1}{-i\hbar\partial_x + \mathcal{H}_x^\hbar *})1 = f, \qquad \forall x. \tag{6.5}$$

*Here the integers $1, 2, \dots$ over the operators indicate (as in [**49**]) the mutual order of action of these operators.*[9]

Indeed, from (2.15), by complex conjugation, we get

$$-i\hbar\frac{\partial}{\partial x^j}K_{y,x} = K_{y,x} * \mathcal{H}_{xj}^\hbar = K_{y,x} * \Big(-i\hbar\frac{\partial}{\partial x^j} + \mathcal{H}_{xj}^\hbar *\Big)1.$$

The second derivative is

$$\begin{aligned}(-i\hbar)^2\frac{\partial^2}{\partial x^k \partial x^j}K_{y,x} &= K_{y,x} * \Big(\mathcal{H}_{xk}^\hbar * \mathcal{H}_{xj}^\hbar - i\hbar\frac{\partial}{\partial x^k}\mathcal{H}_{xj}^\hbar\Big)\\ &= K_{y,x} * \Big(-i\hbar\frac{\partial}{\partial x^k} + \mathcal{H}_{xk}^\hbar *\Big)\Big(-i\hbar\frac{\partial}{\partial x^j} + \mathcal{H}_{xj}^\hbar *\Big)1,\end{aligned}$$

and so on for higher derivatives. Thus for "any" function $f^{\#}(x,\eta)$, we have

$$f^{\#}(\overset{2}{x}, -i\hbar\overset{1}{\partial/\partial x})K_{y,x} = K_{y,x} * f^{\#}(\overset{2}{x}, \overset{1}{-i\hbar\partial/\partial x + \mathcal{H}_x^\hbar *})1 \equiv K_{y,x} * f.$$

Hence the operator of the right $*$-multiplication by f is given by

$$*f(x) = f^{\#}(\overset{2}{x}, -i\hbar\overset{1}{\partial_x})',$$

where the prime denotes the transposition with respect to the measure dm. But from the cyclicity condition (2.19) we see that

$$*f = (f*)',$$

and so, the preceding formula implies (6.4).

COROLLARY 6.2. *The asymptotic solution of equation (6.5) as $\hbar \to 0$ is given by*

$$\begin{aligned}f^{\#}(x,\eta) = f(\ell(x,\eta)) &- \frac{i\hbar}{2}\frac{\partial^2}{\partial x^k \partial \eta_k}f(\ell(x,\eta))\\ &- \frac{i\hbar}{2}\frac{\partial}{\partial \eta_s}f(\ell(x,\eta))\frac{\partial}{\partial \eta_k}\Big(\partial_k\mathcal{H}_x(\ell(x,\eta))_s\Big) + O(\hbar^2).\end{aligned} \tag{6.6}$$

Thus the operation $\#$ in the classical limit $\hbar = 0$ is just the lift of functions from $\mathcal{X}$ to $T^*\mathcal{X}$ by means of the symplectic fibration ℓ (see Section 3 above). We denote $f^{\#} = \widehat{\ell}^* f$ and call the mapping $\widehat{\ell}^*$ a *left quantum lift.*

---

[9] Note that for any j,k the operators $-i\hbar\partial/\partial x^j + \mathcal{H}_{xj}^\hbar *$ and $-i\hbar\partial/\partial x^k + \mathcal{H}_{xk}^\hbar *$ mutually commute, and so, in (6.5) we use the common vector notation for them.

REMARK 6.3. The quantum lift generates an operator representation of the $*$-product algebra:

$$f \to (\widehat{\ell}^* f)(\overset{2}{x}, -i\hbar\overset{1}{\partial_x}).$$

But this representation is, of course, reducible. The problem of constructing irreducible representations and the operator family $\mathbf{S}_x$ (2.8) is complicated. An approach to this problem might be based on the idea of the commutation relations (3.11) or, more precisely, on their quantum version

$$[\mathcal{H}_{xl}*, \mathcal{H}_{xk}*]_* = -i\hbar\Omega_{lk}^{[x,\hbar]}(\mathcal{H}_x*).$$

If the tensor $\Omega_{ls}^{[x,\hbar]}$ is well defined, then one can ask about *irreducible* representations of the algebra generated by the relations

$$[\mathbf{A}_{xl}, \mathbf{A}_{xk}] = -i\hbar\Omega_{lk}^{[x,\hbar]}(\mathbf{A}_x)$$

for each fixed $x \in \mathcal{X}$. When such a gauge field $\mathbf{A}$ is constructed and its curvature is zero, i.e., $\partial_k \mathbf{A}_{xl} - \partial_l \mathbf{A}_{xk} = \Omega_{lk}^{[x,\hbar]}(\mathbf{A}_x)$, then the operator family $\mathbf{S}_x$ could be determined by the differential equations

$$i\hbar\partial\mathbf{S}_x = \mathbf{A}_x\mathbf{S}_x = -\mathbf{S}_x\mathbf{A}_x.$$

Now let us return to our basic wave equations (2.25), (2.15) and apply (6.4). The result is the following.

THEOREM 6.4. (i) *The integral kernel of the $*$-product satisfies the linear equation*

$$i\hbar\partial_x K_{x,y}(z) = \mathcal{L}_x^\hbar(\overset{2}{z}, -i\hbar\overset{1}{\partial_z})K_{x,y}(z). \tag{6.7}$$

Here the 1-form $\mathcal{L}^\hbar$ over $\mathcal{X}$ with values in functions over $T^*\mathcal{X}$ is defined using the left quantum lift of the quantum Ether Hamiltonian: $\mathcal{L}^\hbar = \widehat{\ell}^*\mathcal{H}^\hbar$, i.e., by procedure (6.5):

$$\mathcal{L}_x^\hbar(\overset{2}{z}, -i\hbar\partial_z \overset{1}{+} \mathcal{H}_z^\hbar *)1 = \mathcal{H}_x^\hbar, \qquad \forall x, z. \tag{6.8}$$

The asymptotics of $\mathcal{L}^\hbar$ as $\hbar \to 0$ is

$$\mathcal{L}^\hbar = \mathcal{L} - \frac{i\hbar}{2}\frac{\partial^2\mathcal{L}}{\partial z^k \partial\xi_k} - \frac{i\hbar}{2}\mathcal{A} + O(\hbar^2), \tag{6.9}$$

where

$$\mathcal{L}_x = \ell^*\mathcal{H}_x, \quad \textit{or} \quad \mathcal{L}_x(z,\xi) = \mathcal{H}_x(\ell(z,\xi)), \quad \xi \in T_z^*\mathcal{X}, \tag{6.10}$$

and

$$\mathcal{A}_x(z,\xi)_j = \frac{\partial}{\partial\xi_s}\mathcal{L}_x(z,\xi)_j \frac{\partial}{\partial\xi_k}\partial_k\mathcal{H}_z(\ell(z,\xi))_s. \tag{6.11}$$

(ii) *If the arguments x, y, z are considered altogether without separation of "space" and "time" variables, then the linear equation for the kernel $K = K_{x,y}(z)$ reads*

$$i\hbar\, dK = \Delta^\hbar K. \tag{6.12}$$

Here

$$\Delta^\hbar = \mathcal{L}_x^\hbar(\overset{2}{z}, -i\hbar\overset{1}{\partial_z}) + \mathcal{L}_y^\hbar(\overset{2}{x}, -i\hbar\overset{1}{\partial_x}) + \mathcal{L}_z^\hbar(\overset{2}{y}, -i\hbar\overset{1}{\partial_y}),$$

where the 1-form $\mathcal{L}^\hbar$ is defined in (6.8) *and* (6.9).

So, for the integral kernel of the $*$-product we derived the linear(!) pseudodifferential equation (6.7) equipped with the Cauchy data (2.16) or the linear equation (6.12) with the Cauchy data

$$K\Big|_{x=y=z} = \mu^2. \tag{6.13}$$

Now it is a routine exercise to compute the semiclassical asymptotics of the solution K in the semiclassically-simple domain without focal effects. We substitute (6.1) into (6.7) or (6.12) and derive a Hamilton–Jacobi equation for the phase $\Phi = \Phi_{x,y}(z)$ and a transport equation for the amplitude $\varphi = \varphi_{x,y}(z)$.

7. Intrinsic Hamilton–Jacobi and transport equations

To the pseudodifferential equation (6.7), one has to assign the following *intrinsic* Hamilton–Jacobi equation (with multidimensional "time"):

$$\partial_x \Phi_{x,y}(z) + \mathcal{L}_x(z, \partial_z \Phi_{x,y}(z)) = 0, \qquad \Phi\big|_{x=y} = 0, \tag{7.1}$$

where $\mathcal{L}_x$ is given by (6.10) via the classical Ether Hamiltonian.

The Hamilton system (with multidimensional "time"), which is related to (7.1), is

$$\begin{cases} \partial_x z = D\mathcal{L}_x/D\xi, & z\big|_{x=y} = b \in \mathcal{X}, \\ \partial_x \xi = -D\mathcal{L}_x/Dz, & \xi\big|_{x=y} = 0. \end{cases} \tag{7.2}$$

Recall that ξ is the momentum dual to z so that

$$\xi = \partial_z \Phi_{x,y}(z), \tag{7.3}$$

and the zero Cauchy data for ξ in (7.2) follow from the zero Cauchy data for $\Phi_{x,y}$ at $x = y$.

Let us denote

$$c = \ell(z, \xi). \tag{7.4}$$

Then for this equation we obtain from (7.2) that

$$\frac{\partial c^m}{\partial x^k} = \{\ell^n, \ell^m\} D_n \mathcal{H}_x(c)_k.$$

Since ℓ is symplectic, $\{\ell^n, \ell^m\} = \Psi^{nm}(\ell)$ and

$$\frac{\partial c^m}{\partial x^k} = D_\ell \mathcal{H}_x(c)_k \Psi^{\ell m}(c), \qquad c\Big|_{x=y} = b. \tag{7.5}$$

So the point c is the Ether translation of the point b:

$$c = g_{x,y}(b). \tag{7.6}$$

Now from the first equation (7.2) it follows that

$$\frac{\partial z^r}{\partial x^k} D_l \mathcal{H}_z(c)_r = D_l \mathcal{H}_x(c)_k,$$

and from (7.6) we obtain

$$\frac{\partial c^m}{\partial x^k} = \frac{\partial z^r}{\partial x^k} D_l \mathcal{H}_z(c)_r \Psi^{lm}(c).$$

On the other hand, the reflection equation (3.15) implies

$$\frac{\partial s_z^m}{\partial x^k} = \frac{\partial z^r}{\partial x^k} \frac{\partial s_z^m}{\partial z^r} = \frac{\partial z^r}{\partial x^k} D_l \mathcal{H}_z(s_z)_r \Psi^{lm}(s_z).$$

Comparing this relation with the previous one, we see that the c^m are equal to the s_z^m if their initial data at $x = y$ coincide. Thus we prove that

$$c = s_z(b). \tag{7.7}$$

By definition (7.4), we also have

$$\xi = \mathcal{H}_z(c) = \mathcal{H}_z(s_z(b)) = -\mathcal{H}_z(b). \tag{7.8}$$

From (7.3), (7.8), and (7.1) we get the derivatives of $\Phi_{x,y}(z)$ with respect to z and with respect to x. The derivative with respect to y can be computed, say, from the skew-symmetry condition (2.18) which implies $\Phi_{x,y}(z) = -\Phi_{y,x}(z)$. So, we have $\partial_y\Phi_{x,y}(z) = -\partial_y\Phi_{y,x}(z) = \mathcal{H}_y\big(\ell(z, \partial_z\Phi_{y,x}(z))\big) = \mathcal{H}_y\big(\ell(z, -\partial_z\Phi_{y,x}(z))\big) = \mathcal{H}_y\big(\ell(z, \mathcal{H}_z(b))\big) = \mathcal{H}_y(b)$. Here we have again used (7.3), (7.8), and (7.1).

Thus we have proved the following statements.

LEMMA 7.1. (i) *The equations* (7.7), (7.8) *determine the trajectories* $z = z(x')$, $\xi = \xi(x')$ *of the Hamilton system* (7.2) *in* $T^*\mathcal{X}$ *via the trajectories* $c = c(x')$ *of the Ether system* (7.5) *in* $\mathcal{X}$.

(ii) *In the semiclassically-simple domain*[10] *the phase of the* $*$*-product kernel is determined by the Ether Hamiltonian as follows*:

$$d(\Phi_{x,y}(z)) = \mathcal{H}_x(a) + \mathcal{H}_y(b) + \mathcal{H}_z(c). \tag{7.9}$$

Here, on the left-hand side the differential is taken with respect to all variables x, y, z *and on the right-hand side the points* a, b, c *are taken from the equations*

$$c = s_z(b), \qquad b = s_y(a), \qquad a = s_x(c). \tag{7.10}$$

These points are related to one another via the pseudogroup operation $\circledast$ *on* $\mathcal{X}$ *corresponding to the "Cartan structure" as in* [**36**]:

$$z = x \overset{a}{\circledast} y, \qquad x = z \overset{b}{\circledast} y, \qquad y = z \overset{c}{\circledast} x. \tag{7.11}$$

(iii) *The Poincaré–Cartan integral representation for the solution of the Hamilton–Jacobi equation* (7.1) *is*

$$\Phi_{x,y}(z) = \int_y^x \mathcal{H}(a) - \int_b^z \mathcal{H}(b). \tag{7.12}$$

Here the first integral is taken along an arbitrary path γ *from* y *to* x, *and the second integral is taken along the corresponding path* $\{z(x') \mid x' \in \gamma\}$ *from* b *to* z (*see item* (i) *above*).

Representation (7.12) follows directly from the Hamilton–Jacobi theory applied to equation (7.1). However, there is some gap: it is not so clear why (7.12) is cyclically symmetric with respect to permutations of x, y, z. However (7.9) is cyclically symmetric. So we just have to integrate (7.9), taking into account the dependence on the argument "y" rather than freezing "y" as in (7.12).

For any triple of points x, y, z in $\mathcal{X}$, we have a triple of points in a neighborhood $\mathcal{E}$ of the zero section in the bundle $T^*\mathcal{X}$:

$$X = \big(x, \partial_x\Phi_{x,y}(z)\big), \qquad Y = \big(y, \partial_y\Phi_{x,y}(z)\big), \qquad Z = \big(z, -\partial_z\Phi_{x,y}(z)\big).$$

In view of (7.9), they are related to one another by

$$\ell(X) = r(Y), \qquad \ell(Y) = \ell(Z), \qquad r(X) = r(Z),$$

---

[10]I.e., in a domain where a solution of (7.10) exists and is unique.

where ℓ and r are dual symplectic fibrations of $\mathcal{E}$ over $\mathcal{X}$ (see Lemma 3.1). Thus, viewed as elements of the local symplectic groupoid $\mathcal{E}$ over $\mathcal{X}$, the points X, Y, Z satisfy the equation $Y \circ X = Z$, that is, the triple (Z, Y, X) belongs to the graph Λ of the groupoid multiplication (see [**16, 36, 39**]). This graph is a Lagrangian submanifold in $\mathcal{E}^{(-)} \times \mathcal{E} \times \mathcal{E}$ with respect to the form $-\omega^{\#} \oplus \omega^{\#} \oplus \omega^{\#}$, where $\omega^{\#}$ is the canonical 2-form on $T^*\mathcal{X}$. So, the function $\Phi_{x,y}(z)$ given on the configuration space $\mathcal{X} \times \mathcal{X} \times \mathcal{X}$ corresponds to this Lagrangian submanifold. It is uniquely determined by the zero Cauchy data at a single point, say at the point

$$X = (a_0, 0), \quad Y = (a_0, 0), \quad Z = (a_0, 0), \qquad a_0 \in \mathcal{X}.$$

Such zero Cauchy data correspond to the Cauchy condition in (7.1). Hence, from [**40**] it follows that $\Phi_{x,y}(z)$ can be calculated as the symplectic area of "any" triangle in $\mathcal{X}$ with vertices a, b, c and *reflective* sides whose mid-points are x, y, z (i.e., the sides are invariant under the reflections s_x, s_y, or s_z). Now recall that, in view of (3.17), any Ether geodesics is reflective with respect to its origin.

COROLLARY 7.2. *Let the triple x, y, z belong to the semiclassically-simple domain, and let $\Delta(x, y, z)$ be the triangle membrane in $\mathcal{X}$ bounded by the Ether geodesics through the points x, y, z connecting the vertices a, b, c. The solution of the intrinsic Hamilton–Jacobi equation* (7.1) *is given by the symplectic area of this triangle*:

$$\Phi_{x,y}(z) = \int_{\Delta(x,y,z)} \omega. \tag{7.13}$$

This statement specifies the result of [**40**] by fixing the choice of sides of the triangle in (7.13) and proves that formula (7.13) *actually presents the phase of the $*$-product kernel determined by* (2.15)–(2.19).

Now let us calculate the amplitude $\varphi = \varphi_{x,y}(z)$ in the semiclassical representation (6.1). It follows from (6.7), (6.9), and (2.16) that this amplitude satisfies the *intrinsic* transport equation

$$\begin{gathered} \left(\frac{\partial}{\partial x} + \frac{\partial \mathcal{L}_x}{\partial \xi}\frac{\partial}{\partial z} + \frac{1}{2}\operatorname{tr}\left(\frac{\partial^2 \mathcal{L}_x}{\partial\xi\partial\xi}\frac{\partial^2\Phi}{\partial z\partial z} + \frac{\partial^2\mathcal{L}_x}{\partial z\partial\xi}\right) + \frac{1}{2}\mathcal{A}_x\right)\varphi = 0, \\ \varphi\Big|_{x=y} = \mu^2, \end{gathered} \tag{7.14}$$

where $\mathcal{L}_x$ is determined by (6.10) via the classical Ether Hamiltonian. In this equation, instead of the argument ξ, one has to substitute the derivative of the phase Φ from (7.3).

Now we consider equation (7.14) on the trajectory $z = z(x)$ from system (7.2). Then

$$\begin{gathered} \frac{\partial}{\partial x} + \frac{\partial \mathcal{L}_x}{\partial \xi}\frac{\partial}{\partial z} \to \frac{\partial}{\partial x}, \\ \operatorname{tr}\left(\frac{\partial^2 \mathcal{L}_x}{\partial\xi\partial\xi}\frac{\partial^2\Phi}{\partial z\partial z} + \frac{\partial^2\mathcal{L}_x}{\partial z\partial\xi}\right) \to \frac{\partial}{\partial x}\left(\ln\det\frac{\partial z}{\partial b}\right). \end{gathered} \tag{7.15}$$

In order to calculate the coefficient $\mathcal{A}_x$ in (7.14), we need some identities.

LEMMA 7.3. *The derivatives of the Ether Hamiltonian satisfy the following identities*:

$$\begin{aligned} \partial_k \mathcal{H}_z(s_z(b))_j &= -\partial_j \mathcal{H}_z(b)_k, \\ \partial_k D_m \mathcal{H}_z(s_z(b))_j \cdot D\mathcal{H}_z(s_z(b))^{-1ml} &= \partial_s D_m \mathcal{H}_z(b)_k \cdot D\mathcal{H}_z(b)^{-1ml}, \end{aligned} \tag{7.16}$$

where the s_z are reflections and b is close to z in order to ensure the invertibility of the matrix $D\mathcal{H}_z(b)$.

PROOF. Both formulas in (7.16) follow from the formula $\mathcal{H}_z(s_z(b)) = -\mathcal{H}_z(b)$ (see (3.16)). Indeed, taking the derivative with respect to b, we get

$$Ds = -D\mathcal{H}(s)^{-1} \cdot D\mathcal{H} \tag{7.17}$$

(the arguments z and b are omitted and $D = \partial/\partial b$). Also taking the derivative with respect to z, we get

$$\partial\mathcal{H}(s) + D\mathcal{H}(s) \cdot \partial s = -\partial\mathcal{H}, \tag{7.18}$$

where $\partial = \partial/\partial z$. From (3.15) we have

$$\partial s = -\Psi(s) \cdot D\mathcal{H}(s)',$$

where the prime means transposition of a matrix. Then

$$D\mathcal{H}(s) \cdot \partial s = -\{\mathcal{H}, \mathcal{H}\}(s) = -\{\mathcal{H}(s), \mathcal{H}(s)\} = -\{\mathcal{H}, \mathcal{H}\} = \partial\mathcal{H}' - \partial\mathcal{H}.$$

Here we used the symplecticity of the mapping s, the property (3.16), and the zero curvature equation (3.3a). Now (7.18) becomes $\partial\mathcal{H}(s) = -\partial\mathcal{H}'$. This is the first identity of (7.16). Applying the derivative with respect to b to this identity and using (7.17), we get the second identity of (7.16). □

Now the function $\mathcal{A}$ (6.11) on the trajectory of the Hamilton system (7.2) is given by

$$\mathcal{A}_j(z, \xi) = \frac{\partial z^s}{\partial x^j} \cdot (D_m \partial_k \mathcal{H}_z)(c)_s \cdot \frac{\partial \ell^m}{\partial \xi_k}(z, \xi) = \frac{\partial z^s}{\partial x^j} (\partial_k D_m \mathcal{H}_z)(c)_s \cdot D\mathcal{H}_z(c)^{-1mk}.$$

Here we have used the equality $\mathcal{H}_z(\ell(z, \xi)) = \xi$ and its derivative with respect to ξ. Also recall that (7.4), (7.7) hold on the trajectory. So, applying the second identity of (7.16), we get

$$\begin{aligned} \mathcal{A}_j &= \frac{\partial z^s}{\partial x^j} \cdot \partial_s D_m \mathcal{H}_k \cdot D\mathcal{H}^{-1mk} \\ &= \frac{\partial z^s}{\partial x^j} \cdot \partial_s \ln\det D\mathcal{H} = \frac{\partial}{\partial x^j} \ln\det D\mathcal{H}_z(b). \end{aligned} \tag{7.19}$$

Combining (7.15) and (7.19), we make (7.14) read

$$\partial_x \varphi + \frac{1}{2} \varphi \partial_x \ln\left(\det \frac{\partial z}{\partial b} \cdot \det D\mathcal{H}_z(b)\right) = 0.$$

Here $\varphi = \varphi_{x,y}(z(x))$, and $z = z(x)$ is the trajectory (7.2) with the initial data $z(y) = b$. Integrating the last equation and using the Cauchy data (7.14) and (3.6), we derive

$$\varphi = 2^n \mu^2 \left(\frac{\det \omega(b)}{\det \partial z/\partial b \cdot D\mathcal{H}_z(b)}\right)^{1/2}. \tag{7.20}$$

Now from (7.7) we have

$$\frac{\partial c}{\partial b} = \partial s \frac{\partial z}{\partial b} + Ds, \qquad \partial = \frac{\partial}{\partial z}, \quad D = \frac{\partial}{\partial b},$$

and from (3.15)

$$\partial s = Ds \cdot \Psi \cdot D\mathcal{H}',$$

so we get

$$\frac{\partial c}{\partial b} = Ds \cdot \Big(\Psi \cdot D\mathcal{H}' \cdot \frac{\partial z}{\partial b} + I\Big),$$

or

$$D\mathcal{H}' \cdot \frac{\partial z}{\partial b} = \omega \cdot \Big((Ds)^{-1} \cdot \frac{\partial c}{\partial b} - I\Big). \tag{7.21}$$

Since s is an involution, we have $(Ds)^{-1} = (Ds)(s)$ and in view of (7.6),

$$\frac{\partial c}{\partial b} = D\big(s_x(s_y(b))\big) = Ds_x\big(s_y(b)\big) \cdot Ds_y(b).$$

Thus

$$(Ds)^{-1}\frac{\partial c}{\partial b} = Ds_z(s_z(b)) \cdot Ds_x(s_y(b)) \cdot Ds_y(b) = Ds_z(c) \cdot Ds_x(a) \cdot Ds_y(b).$$

After substitution in (7.21), we obtain

$$D\mathcal{H}_z(b)'\frac{\partial z}{\partial b} = \omega(b)\Big(D(s_z \circ s_x \circ s_y)(b) - I\Big). \tag{7.22}$$

Now formula (7.20) reads

$$\varphi_{x,y}(z) = 2^n \mu^2 \det \Big(I - D(s_z \circ s_x \circ s_y)(b)\Big)^{-1/2}. \tag{7.23}$$

In particular, we have proved the conjecture formulated in [**40**].

COROLLARY 7.4. (i) *Let $\mathcal{X}$ be a reflective simply-connected symplectic manifold. In the semiclassically-simple domain the amplitude of the $*$-product kernel* (6.1) *over $\mathcal{X}$ is given by formula* (7.23) *in which b is the fixed point*[11] *of the mapping $s_z \circ s_x \circ s_y$. The equivalent formula for the amplitude, which does not use reflections, is the following*:

$$\varphi_{x,y}(z) = 2^n \mu^2 \left(\frac{\det \partial_y D_z \Phi_{x,y}(z) \cdot \det \omega(b)}{\det D\mathcal{H}_y(b) \cdot \det D\mathcal{H}_z(b)}\right)^{1/2}. \tag{7.24}$$

(ii) *The semiclassically-multiple domain of equation* (2.25) *is formed by those triples $x, y, z \in \mathcal{X}$ for which the mapping $s_z \circ s_x \circ s_y$ has several isolated fixed points (say, on the sphere $\mathbf{S}^2$ any nonfocal triple is of multiplicity* 2). *If the triple x, y, z belongs to a semiclassically-multiple domain, then in formula* (6.1) *one has to take the sum over all possible Ether geodesic triangles $\Delta(x, y, z)$ and to multiply each summand by a suitable exponential* $\exp\{-i\frac{\pi}{2}m\}$, *where $m \in \mathbb{Z}$ is the Maslov index on the graph of the groupoid multiplication.*

(iii) *The focal set of the basic equation* (2.25), *where the asymptotics of the solution is singular as $\hbar \to 0$, consists of those triples $x, y, z \in \mathcal{X}$ for which the symplectic mapping $s_z \circ s_x \circ s_y$ has nonisolated fixed points.*

---

[11]This is equivalent to equations (7.10).

(iv) *If the second Betti number of the manifold $\mathcal{X}$ is nontrivial, then the existence of the global semiclassical solution of the basic equation* (2.25) *is ensured by the quantization condition* (5.1).

Actually, the semiclassical approximation of the kernel $K_{x,y}(z)$, globally in the phase space, including a neighborhood of the focal set, can also be obtained by applying some version of the "canonical operator" [**14**] to (6.7).

There are some interesting papers about the semiclassical approximation of the $*$-product kernel in the special symmetric case (the curvature R is covariantly constant) [**21, 32, 50**]. For instance, in [**50**] it was first demonstrated that the focal set, where the kernel $K_{x,y}(z)$ is not of the WKB-type (6.1), exists not only on such topologically nontrivial manifolds like $\mathbb{S}^2$, but even on the Lobachevski plane with its standard symplectic form.

The formal deformation quantization in the symmetric symplectic case was first constructed in [**41**]. It would be interesting to check whether the explicit $\hbar$-power series for the $*$-product on symmetric spaces obtained in [**41**] follows as a particular case from our wave equations.

There is another special case, namely, that of Kählerian manifolds. In [**51**], the presence of the complex structure allowed deriving integral representations for $*$-products via coherent states under some additional conditions. The most critical condition is that the Liouville measure should be a reproducing measure (up to a constant factor; see [**51**]). This condition certainly holds on homogeneous manifolds. Schematically, this means that the integral $*$-product kernel has the form $K = \exp\{\frac{i}{\hbar}\Phi\} \cdot \text{const}$, i.e., the amplitude φ is constant, assuming that the integration measure dm is the Liouville one. Of course, on general Kählerian manifolds φ is not constant and the entire series for the amplitude $\varphi + \hbar\varphi^1 + \dots$ can be calculated explicitly [**52**]. Thus the asymptotic quantization is geometrically well defined and the semiclassical expansion of the integral $*$-kernel is explicitly known (without the focal set problem) over any Kählerian manifold. It is easy to see that the scheme (2.12), (2.15), (2.25) works in this polarized case, although the cyclic property (2.19) fails.

Note that the tunneling effect observed in [**52**] for the kernel K over nonsimply-connected Kählerian manifolds maintains our general view on $*$-products as on "Ether wave functions." From this viewpoint, quantum manifolds themselves pretend to be actual dynamical objects of quantum mechanics and, at least, they can be subjected to all transformations and definitions usual for wave functions in quantum mechanics: tunneling, scattering, gauge transformations, reduction by symmetries, spectral expansions, constrains, adiabatic approximations, etc. The corresponding properties of $*$-products in the integral form (2.7) are topics for future investigations.

8. Exterior quantum dynamics

There is another opportunity to apply the intrinsic geometry generated by the Ether Hamiltonian. Let us consider a Hamilton function H on the phase space $\mathcal{X}$ and the corresponding "exterior" quantum equations

$$i\hbar\frac{d}{dt}G_t = H * G_t = G_t * H. \tag{8.1}$$

If the operator representation (2.5), (2.8) of the $*$-product algebra is known, then

$$G_t(x) = \operatorname{tr}\left(\mathbf{S}_x \cdot \exp\left\{-\frac{it}{\hbar}\widehat{H}\right\}\right), \qquad G_0 = 1. \tag{8.2}$$

Here the exponential $\exp\{-\frac{it}{\hbar}\widehat{H}\}$ represents the solution of the "exterior" Schrödinger evolution equation. So, the function G_t is the symbol of the quantum evolution operator.

Solving equation (8.1) means that one has to take the exponential $\exp\{-\frac{it}{\hbar}H\}$ and then replace the classical multiplication determining this exponential by the quantum multiplication $*$. The last procedure can be made using the intrinsic Ether dynamics following (2.21).

THEOREM 8.1. *Let $\mathcal{X}$ be a reflective simply-connected symplectic manifold.*

(i) *The Schrödinger evolution equation* (8.1) *is equivalent to*

$$i\hbar\frac{d}{dt}G_t = H^{\#}(\overset{2}{x}, -i\hbar\overset{1}{\partial}_x)G_t, \qquad G_0 = 1, \tag{8.3}$$

where $H^{\#} = \widehat{\ell}^* H$ is the left quantum lift of H defined by (6.5).

(ii) *Denote by γ_H^t the Hamilton flow generated by H on $\mathcal{X}$. Then the semiclassically-simple domain for problem* (8.3) *as $\hbar \to 0$ consists of those $x \in \mathcal{X}$ for which the mapping $s_x \circ \gamma_H^t$ has a unique fixed point.*

If this mapping has several (finitely many) fixed points, then x belongs to the semiclassically-multiple domain.

The focal set of problem (8.3) *consists of those x for which the mapping $s_x \circ \gamma_H^t$ has nonisolated fixed points.*

(iii) *In the semiclassically-simple domain the asymptotics of G_t is the following*

$$G_t = \exp\left\{\frac{i}{\hbar}\int_{\Sigma_t}\omega - \frac{it}{\hbar}H\right\}\varphi_t + O(\hbar). \tag{8.4}$$

Here $\Sigma_t(x)$ is a certain dynamic segment in $\mathcal{X}$ (a sickle-shaped membrane); the exterior arc of this segment is a Hamilton trajectory of H, whose time-length is t, and the other arc of the segment connecting its ends (spikes of the sickle) is given by Ether geodesics through the mid-point x. The value of H in (8.4) *is taken on the exterior arc of the segment.*

The amplitude φ_t in formula (8.4) *is given by*

$$\varphi_t(x) = 2^n\big(\det(I - D(s_x \circ \gamma_H^t)(x_0))\big)^{-1/2}, \tag{8.5}$$

where x_0 is the fixed point of the mapping $s_x \circ \gamma_H^t$.

(iv) *In the semiclassically-multiple domain on $\mathcal{X}$, the asymptotics of G_t is a sum of expressions like* (8.4) *over all fixed points of the mapping $s_x \circ \gamma_H^t$; each of the summands is multiplied by an exponential $\exp\{-i\frac{\pi}{2}m\}$, where $m \in \mathbb{Z}$ is the Maslov index on the graph of γ_H^t.*

PROOF. Equation (8.3) follows from Lemma 6.1. The Hamilton–Jacobi equation corresponding to (8.3) is

$$\frac{\partial \Phi_t}{\partial t} + H\left(\ell\left(x, \frac{\partial \Phi_t}{\partial x}\right)\right) = 0, \qquad \Phi_0 = 0, \tag{8.6}$$

where ℓ is the left symplectic fibration over $\mathcal{X}$ (see Lemma 3.1). The Hamilton system in $T^*\mathcal{X}$ is

$$\begin{cases} \dot{x} = \dfrac{\partial}{\partial \eta} H(\ell(x,\eta)), & x\big|_{t=0} = x_0, \\ \dot{\eta} = -\dfrac{\partial}{\partial x} H(\ell(x,\eta)), & \eta\big|_{t=0} = 0. \end{cases} \tag{8.7}$$

The projected Hamilton equation in $\mathcal{X}$ is

$$\dot{z} = DH(z)\Psi(z), \qquad z\big|_{t=0} = x_0, \tag{8.8}$$

where

$$z = \ell(x,\eta) = s_x(x_0), \qquad \text{or} \qquad g_H^t(x_0) = s_x(x_0). \tag{8.9}$$

The solution of (8.6) is given by

$$\Phi_t(x) = \int_0^t \eta \dot{z}\, dt - Ht, \qquad x(t) = x, \tag{8.10}$$

where the value of H is taken on the trajectory (8.8). By the Stokes theorem, we have

$$\int_0^t \eta \dot{z}\, dt = \int_\Pi d\eta \wedge dz. \tag{8.11}$$

Here Π is a "vertical" membrane in $T^*\mathcal{X}$, whose boundary consists of three pieces: the trajectory $\{x(t') \mid 0 \le t' \le t\}$ on $\mathcal{X}$ (taken with the opposite orientation from $x(t) = x$ to x_0), the trajectory $\{(x(t'), \eta(t')) \mid 0 \le t' \le t\}$ in $T^*\mathcal{X}$, and a vertical path connecting the point $(x, \eta(t))$ with the point $(x, 0)$ along the fiber $T_x^*\mathcal{X}$. For this last vertical path, we take the solution of problem (3.10) with a suitable velocity v.

Since $d\eta \wedge dz = \ell^*\omega - r^*\omega$, where $r(x,\eta) \equiv \ell(x,-\eta)$, from (8.11) we get

$$\int_0^t \eta \dot{z}\, dt = \int_{\ell(\Pi)} \omega + \int_{r(\Pi)^{(-)}} \omega = \int_{\Sigma_t(x)} \omega. \tag{8.12}$$

Here the membrane $\Sigma_t(x) = \ell(\Pi) \cup r(\Pi)^{(-)}$ was described in part (iii) of the theorem. Thus formulas (8.10) and (8.12) yield the phase in (8.4).

From Corollary 6.2 we see that the amplitude φ_t in (8.4) in the semiclassically-simple domain satisfies the following transport equation:

$$\left[\frac{\partial}{\partial t} + \frac{\partial \mathcal{F}}{\partial \eta}\frac{\partial}{\partial x} + \frac{1}{2}\operatorname{tr}\left(\frac{\partial^2 \mathcal{F}}{\partial \eta \partial \eta} \cdot \frac{\partial^2 \Phi_t}{\partial x \partial x} + \frac{\partial^2 \mathcal{F}}{\partial \eta \partial x}\right) + \frac{1}{2}\frac{\partial \mathcal{F}}{\partial \eta} \cdot \alpha\right]\varphi_t = 0, \tag{8.13}$$

where $\mathcal{F}(x,\eta) = H(\ell(x,\eta))$ and

$$\alpha(x,\eta)_s = \frac{\partial}{\partial \eta_k}\big(\partial_x \mathcal{H}_x(\ell(x,\eta))_s\big).$$

Instead of the argument η in (8.13), one has to substitute $\eta = \partial\Phi_t/\partial x$.

In the same way as in Section 7, using Lemma 7.3 we obtain an analog of (7.19):

$$\frac{\partial \mathcal{F}}{\partial \eta} \cdot \alpha = \frac{d}{dt} \ln\det D\mathcal{H}_x(x_0) \tag{8.14}$$

on the trajectory (8.7).

Of course, there is also the standard identity

$$\operatorname{tr}\left(\frac{\partial^2\mathcal{F}}{\partial\eta\partial\eta}\cdot\frac{\partial^2\Phi_t}{\partial x\partial x}+\frac{\partial^2\mathcal{F}}{\partial\eta\partial x}\right)=\frac{d}{dt}\ln\det\frac{\partial x}{\partial x_0}$$

on the trajectory (8.7).

Thus equation (8.13) with the Cauchy data is integrated as follows;

$$\varphi_t(x)=2^n\left(\frac{\det\omega(x_0)}{\det D\mathcal{H}_x(x_0)\cdot\det\partial x/\partial x_0}\right)^{1/2}, \tag{8.15}$$

where x_0 is taken from (8.9):

$$s_x\circ\gamma_H^t(x_0)=x_0.$$

In the same way as in Section 7, from the identity $z=s_x(x_0)$ (8.9), taking the derivative with respect to x_0, we get

$$\frac{\partial z}{\partial x_0}=\partial s\cdot\frac{\partial x}{\partial x_0}+Ds=Ds\Big(\Psi\cdot D\mathcal{H}'\cdot\frac{\partial x}{\partial x_0}+I\Big).$$

Here we omit the arguments $s=s_x(x_0)$, $\Psi=\Psi(x_0)$, $\mathcal{H}=\mathcal{H}_x(x_0)$. From the last equality, we obtain the matrix $\partial x/\partial x_0$ in the form analogous to (7.21), (7.22):

$$D\mathcal{H}'\cdot\frac{\partial x}{\partial x_0}=\omega\cdot\Big((Ds)(s)\cdot\frac{\partial z}{\partial x_0}-I\Big)=\omega\cdot\big(D(s\circ\gamma_H^t)-I\big).$$

Thus

$$\det\frac{\partial x}{\partial x_0}=\frac{\det\omega(x_0)}{\det D\mathcal{H}_x(x_0)}\cdot\det\big(I-D(s_x\circ\gamma_H^t)(x_0)\big).$$

This relation and (8.15) imply (8.5). □

In the case of the Euclidean phase space $\mathcal{X}=\mathbb{R}^{2n}$, formulas like (8.4) on membranes bounded by straight-line chords were first derived in [**53, 54**]. In [**33**] see also the case of the Euclidean space endowed with the "magnetic" symplectic form ω and with an additional "electric" form along space-time directions; in this case membranes were constructed by means of special "magnetic wings."

Acknowledgements. The author is grateful to V. Drinfeld, D. Sternheimer, B. Fedosov, C.-M. Marle, T. Osborn, I. Tyutin, E. Novikova, Yu. Vorobjev, O. Grigoriev for useful discussions, as well as to Manitoba University and Université de Bourgogne for kind hospitality.

References

[1] F. Bayen, M. Flato, C. Fronsdal, A. Lichnerowicz, and D. Sternheimer, *Quantum mechanics as a deformation of classical mechanics*, Lett. Math. Phys. **1** (1975/77), 521–530.

[2] F. A. Berezin, *General concept of quantization*, Comm. Math. Phys. **40** (1975), 153–174.

[3] A. Connes, *Noncommutative geometry*, Academic Press, London, 1994.

[4] M. V. Karasev and V. P. Maslov, *Asymptotic and geometric quantization*, Uspekhi Mat. Nauk **39** (1984), no. 6, 115–173; English transl., Russian Math. Surveys **39** (1984), no. 6, 133–205.

[5] F. Bayen, M. Flato, C. Fronsdal, A. Lichnerowicz, and D. Sternheimer, *Deformation theory and quantization*, Ann. Physics **111** (1978), 61–151.

[6] A. Lichnerowicz, *Deformation of quantification*, Lecture Notes Phys., vol. 106, Springer-Velrag, Berlin–Heidelberg–New York, 1979, pp. 209–219.

[7] B. Fedosov, *Formal quantization*, Some Topics of Modern Mathematics and Their Applications to Problems of Mathematical Physics, 1985, pp. 129–136. (Russian).

[8] ———, *A simple geometrical construction of deformation quantization*, J. Diff. Geom. **40** (1994), 213–238.
[9] M. Kontsevich, *Deformation quantization of Poisson manifolds*, Preprint q-alg/9709040, 1997.
[10] A. Cattaneo and G. Felder, *Poisson sigma models and symplectic groupoids*, Preprint math.sg/0003023.
[11] ———, *A path integral approach to the Kontsevich quantization formula*, Preprint math/9902090.
[12] J. Klauder, *Quantization is geometry. After all*, Ann. Phys. **188** (1988), 120–141.
[13] B. Kostant, *Quantization and representation theory*, London Math. Soc. Lecture Note Ser. **34** (1979), 287–316.
[14] V. Maslov, *Perturbation theory and asymptotic methods*, Izd. Moskov. Univ., Moscow, 1965; French transl., Dunod, Paris, 1972.
[15] M. V. Karasev, *Quantization of nonlinear Lie-Poisson brackets in semiclassical approximation*, Inst. Theor. Phys., Kiev, Preprint N ITP-85-72P, 1985. (Russian).
[16] M. V. Karasev and V. P. Maslov, *Nonlinear Poisson brackets. Geometry and quantization*, Nauka, Moscow, 1991; English transl., Amer. Math. Soc., Providence, RI, 1993.
[17] M. A. Rieffel, *Deformation quantization of Heisenberg manifold*, Comm. Math. Phys. **122** (1989), 531–562.
[18] M. V. Karasev, *Advances in quantization: quantum tensors, explicit star-products, and restriction to irreducible leaves*, Diff. Geometry Appl. **9** (1998), 89–134.
[19] M. V. Karasev and E. M. Novikova, *Non-Lie permutation relations, coherent states, and quantum embedding*, Coherent Transform, Quantization, and Poisson Geometry (M. Karasev, ed.), Amer. Math. Soc., Providence, RI, 1998.
[20] J. M. Gracia-Bondia, *Generalized Moyal quantization on homogeneous symplectic spaces*, Deformation Theory and Quantum Groups (M. Gerstenhaber and J. Stasheff, eds.), Contemp. Math., vol. 134, Amer. Math. Soc., Providence, RI, 1992, pp. 93–114.
[21] P. Bieliavsky, *Strict quantization of solvable symmetric spaces*, Preprint math.qa/0010004.
[22] H. Weyl, *Reine Infinitesimalgeometrie*, Math. Z. **2** (1918), 384–411.
[23] E. Scholz, *Hermann Weyl's "Purely Infinitesimal Geometry,"* in: *Proc. Intern. Congress of Math., Zürich, 1994*, Birkhäuser, Basel–Boston–Berlin, 1995, pp. 1592–1603.
[24] H. Weyl, *Selecta*, Birkhäuser, Basel–Boston–Berlin, 1956, p. 192.
[25] R. Stratonovich, *On distributions in representation space*, Zh. Éxper. Teoret. Fiz. **31** (1956), 1012–1020; English transl., Soviet Phys. JETP **4** (1957), no. 6, 891–898.
[26] C. Fronsdal, *Some ideas about quantization*, Rep. Math. Phys. **15** (1979), no. 1, 111–145.
[27] D. Arnal, *The $*$-exponential*, Quantum Theories and Geometry (M. Cahen and M. Flato, eds.), Kluwer, Dordrecht, 1988, pp. 23–51.
[28] A. Grossmann and P. Huguenin, *Group-theoretical aspects of the Wigner–Weyl isomorphism*, Helvetica Physica Acta **51** (1978), 252–261.
[29] F. A. Berezin *Quantization*, Izv. Akad. Nauk SSSR Ser. Mat. **38** (1974), no. 5; English transl., Math. USSR Izv. **8** (1974), no. 5, 1109–1165.
[30] I. Batalin and I. Tyutin, *Quantum geometry*, Nuclear Phys. B **345** (1990), 645–650.
[31] A. Weinstein, *Classical theta-functions and quantum tori*, Publ. RIMS, Kyoto Univ. **30** (1994), 327–333.
[32] Zhao-Hui Qian, *Groupoids, Midpoints, and Quantization*, Thesis, Univ. of California, Berkeley, 1997.
[33] M. V. Karasev and T. Osborn, *Symplectic areas, quantization, and dynamics in electromagnetic fields*, J. Math. Phys. **43** (2002), no. 2, 756–788 (quant-ph/0002041).
[34] ———, *Magnetic curvature of quantum phase space*, in: *Proc. A. Sakharov Conference, Phys. Inst. Russian Akad. Sci., Moscow, June 2002* (to appear).
[35] C. Emmrich and A. Weinstein, *The differential geometry of Fedosov's quantization*, Lie Theory and Geometry. In Honor of B. Kostant. Birkhäuser, Boston, 1994, pp. 217–240.
[36] M. V. Karasev, *Analogues of objects of the theory of Lie groups for nonlinear Poisson brackets*, Izv. Akad. Nauk SSSR **50** (1986), no. 3, 508–538; English transl., Math. USSR Izv. **28** (1987), no. 3, 497–527.
[37] ———, *Maslov quantization conditions in higher cohomology and analogs of notions developed in Lie theory for canonical fiber bundles of symplectic manifolds.* I, II. Preprint MIEM,

1981. Deposited at VINITI (March 12, 1982, No. 1092-82, 1093-82). Abstract: in "Ref. Zhurnal Matematika"(1982) No. 7A676, 7A677; English transl., Selecta Math. Soviet **8** (1989), no. 3, 213–258.

[38] A. Weinstein, *The local structure of Poisson manifolds*, J. Differential Geom. **18** (1983), 523–557.

[39] ———, *Symplectic groupoids and Poisson manifolds*, Bull. Amer. Math. Soc. **16** (1987), 101–104.

[40] ———, *Traces and Triangles in Symmetric Symplectic Spaces*, Contemp. Math. **179** (1994), 262–270.

[41] P. Bieliavsky, M. Cahen, and S. Gutt, *Symmetric symplectic manifolds and deformation quantization*, Modern Group Theor. Methods in Phys. (J. Bertrand et al., eds.), Kluwer, Dordrecht, 1995, pp. 63–73.

[42] M. V. Karasev and V. P. Maslov, *Algebras with general permutation relations and their applications*. II, Itogi Nauki i Tekhniki: Sovremennye Problemy Mat. **13** (1979), VINITI, Moscow, 145–267; English transl., J. Soviet Math. **15** (1981), no. 3, 273–368.

[43] ———, *Global asymptotic operators of regular representation*, Dokl. Akad. Nauk SSSR **257** (1981), no. 1, 33–37; English transl., Soviet Math. Dokl. **23** (1981), 228–232.

[44] M. De Wilde and P. Lecomte, *Existence of star products and of formal deformations of the Poisson Lie algebra of arbitrary symplectic manifolds*, Lett. Math. Phys. **7** (1983), 487–496.

[45] J. Huebschmann, *On the quantization of Poisson algebras*, Symplectic Geometry and Mathematical Physics, Actes du colloque en l'honneur de J.-M.Souriau (P. Donato et al., eds.), Birkhäuser, Basel–Boston, 1991, pp. 204–233.

[46] H. Omori, Y. Maeda, and A. Yoshioka, *Weyl manifolds and deformation quantization*, Adv. Math. **85** (1991), no. 2, 224–255.

[47] ———, *Deformation quantization of Poisson algebras*, Contemp. Math. **179** (1994), 213–240.

[48] B. Fedosov, *Deformation quantization and index theory*, Akademie Verlag, Berlin, 1996.

[49] V. Maslov, *Operator methods*, Nauka, Moscow, 1973; English transl., Mir, Moscow, 1976.

[50] G. Tuynman and P. Rios, *Weyl quantization from geometric quantization*, preprint.

[51] M. Cahen, S. Gutt, and J. Rawnsley, *Quantization of Kähler manifolds*. I, J. Geom. Phys. **7** (1990), 45–62; II, Trans. Amer. Math. Soc. **337** (1993), 73–98; III, Lett. Math. Phys. **30** (1994), 291–305; IV, Lett. Math. Phys. **30** (1995), 159–168.

[52] M. V. Karasev, *Quantum surfaces, special functions, and the tunneling effect*, Lett. Math. Phys. **59** (2001), 229–269.

[53] M. Berry, *Semi-classical mechanics in phase space: a study of Wigner's function*, Philos. Trans. Roy. Soc. London Ser. A **287** (1977), 237–271.

[54] M. Marinov, *An alternative to the Hamilton–Jacobi approach in classical mechanics*, J. Phys. A **12** (1979), no. 1, 31–47.

Applied Mathematics Department, Moscow Institute of Electronics and Mathematics, Moscow 109028, Russia

E-mail address: karasev@miem.edu.ru

Amer. Math. Soc. Transl.
(2) Vol. **208**, 2003

Weak Asymptotics Method and Interaction of Nonlinear Waves

V. G. Danilov, G. A. Omel′yanov, and V. M. Shelkovich

ABSTRACT. The paper presents an asymptotic method for studying the evolution and interaction of nonlinear waves in the "nonintegrable" case.

CONTENTS

Introduction
1. Auxiliary formulas of the weak asymptotics method
2. Asymptotic subalgebras
3. Dynamics of solitary waves
4. Multidimensional waves
5. Confluence of shock waves
6. Soliton interaction in the KdV model
7. Soliton interaction in KdV type models
8. Confluence of free boundaries in the modified Stefan problem
References

Introduction

The goal of this paper is to present an asymptotic method for studying the evolution and interaction of nonlinear waves in the "nonintegrable" case. By nonlinear waves we mean solutions of nonlinear equations with localized "fast" variation. The most typical examples of such waves are solitons of equations with small dispersion $\varepsilon \ll 1$ and smoothed shock waves in equations with small viscosity. Our asymptotic method is based on the procedure in which nonlinear waves are assigned functions of the time variable with values in the space of generalized functions $\mathcal{D}'(\mathbb{R}^n_x)$.

Asymptotic methods construct approximate solutions which obey equations up to small terms. The discrepancy is said to be small if, in some norm, it admits

2000 *Mathematics Subject Classification.* Primarily 54H25, 49J35, 41A65.

Key words and phrases. Asymptotics, weak solution, generalized function, nonlinear equation, interaction, soliton, shock wave.

The third author was supported by the DFG Project 436 RUS 113/593/3-1 and by the Russian Federation for Basic Research, Grant 02-01-00483.

the estimate $O(\varepsilon^\alpha)$, where $\alpha > 0$ and $\varepsilon \to 0$ is a small parameter. The norm in the space C^k is used most often. The small parameter ε can be contained in the solution and/or in the equation. The solution is sought in the form of an ansatz whose form depends on the method. For instance, in the well-known WKB-method the approximate solution is sought in the form

$$u_{\mathrm{as}} = \exp\Big\{\frac{i}{\varepsilon}S(x,t)\Big\}\varphi(x,t,\varepsilon),$$

where $S(x,t)$ and $\varphi(x,t,\varepsilon)$ are smooth functions. This method appeared in the nineteenth century and in the early twentieth century in physics literature. Then it attracted attention of mathematicians and became the starting point of profound mathematical theories concerning linear equations. It will suffice to mention the solvability theory for linear hyperbolic equations, the Maslov canonical operator and its versions, the integral Fourier operators, the geometric quantization, and more general objects of symplectic geometry related to geometric quantization (see, e.g., [**55**] and the references therein).

A generalization of the WKB-method to nonlinear partial differential equations is the Whitham method [**128**]. In this method the asymptotic solution is sought in the form

$$u_{\mathrm{as}} = f(S/\varepsilon, x, t, \varepsilon),$$

where the function $f(\tau, x, t, \varepsilon)$ is smooth in all variables and is periodic in the variable τ with the period $T = \mathrm{const}$. For linear equations, the Whitham method leads to the WKB-method and the function f obeys a linear equation. In the general case, the function $f_0(\tau,x,t) = f(\tau,x,t,0)$ obeys a nonlinear ordinary differential equation with respect to τ called the *standard* or *model* equation. This standard equation can have either periodic or stabilizing solutions; the latter means that the solution satisfies the condition

$$\frac{\partial^\alpha f_0(\tau,x,t)}{\partial\tau^\alpha} = O(|\tau|^{-N}), \qquad |\tau| \to \infty, \quad \forall\alpha \gg 1,$$

where N is a sufficiently large positive number and the O-estimate is locally uniform in x and t.

Such stabilizing solutions correspond to *nonlinear waves with localized fast variation.* Among them are solitons of equations with small dispersion, shock waves, kinks, concentration waves, etc. By the terminology introduced in [**99**], these objects are elementary structures in a nonlinear world which is described as the evolution (propagation) and interaction of these elementary structures, as well as of more complicated structures arising in the interaction.

In the seventies and eighties of the last century, V. P. Maslov and his collaborators (see [**18, 91**] and the references therein) modified the Whitham method, which allowed them to describe propagation of nonlinear waves in rather general situations.

At about the same time, a new powerful method for constructing exact solutions of "integrable" partial differential equations describing the interaction of nonlinear solitary waves was developed. The appearance of this *inverse scattering transform method* (sometimes called the method of the inverse scattering problem) can be considered as the greatest event in mathematical physics in the twentieth century.

However, by the efforts of many famous mathematicians in the last few decades, almost all "integrable" problems were solved. Nevertheless, it is clear that the variety of models in which it is important to study the interaction of nonlinear waves are far from being exhausted by integrable models.

In this paper we describe a new approach to constructing asymptotic solutions which describe propagation and interaction of nonlinear waves. We call this approach the *weak asymptotics method.* In this method the approximate solutions are sought in the same form as in the Whitham method modified for nonlinear waves with localized fast variation [**18, 91**], but the discrepancy is assumed to be small in the sense of the space of functionals $\mathcal{D}'_x$ over test functions depending only on the "space" variable x. As we shall see below, this rather trivial modification allows us to reduce the problem of describing interaction of nonlinear waves to solving some system of ordinary differential equations (instead of solving partial differential equations).

In particular, in the framework of the weak asymptotics method we shall present formulas describing the confluence of shock waves (Section 5), the interaction of solitons for nonintegrable KdV type equations (Section 7 and see also Section 6, where the interaction of KdV solitons is considered). We also present formulas describing the confluence of free boundaries in the phase transition problems (see [**107**] and Section 8). The weak asymptotics method also allows us to describe the processes of shock wave formation and decay, as well as the interaction of multidimensional shock waves for scalar conservation laws (see [**16**]). Another example of the use of the weak asymptotics method is the description of interaction of delta-shock waves [**27**].

As already mentioned, in the framework of the weak asymptotics method, the discrepancy is calculated and estimated in $\mathcal{D}'_x$. This means that the result of substitution of the asymptotic solution into the equation is treated in the sense of $\mathcal{D}'_x$. In fact, this means that we study generalized solutions of nonlinear equations. One of the first examples of this study was the method proposed by V. P. Maslov [**87**] for deriving the Rankine–Hugoniot conditions for shock waves. Let us recall this method for the Hopf equation

$$\frac{\partial u}{\partial t} + \frac{\partial u^2}{\partial x} = 0.$$

A *shock wave* is defined to be a solution of this equation of the form

$$u = u_0 + u_1 H(x - Vt),$$

where $H(z)$ is the Heaviside function. For simplicity, we assume that u_0 and u_1 are constants. Substituting this expression into the Hopf equation, we obtain

$$\big[- Vu + (2u_0u_1 + u_1^2)\big]\delta(x - Vt) = 0.$$

For this relation to hold, it is necessary and sufficient that the expression in the square brackets be equal to zero. We immediately obtain

$$V = \frac{2u_0u_2 + u_1^2}{u_1} = \frac{[u^2]}{[u]},$$

where $[f]$ denotes the jump of the function f on the line $x = Vt$.

Moreover, from the general viewpoint, this approach is based on the relation

$$(u_0 + u_1 H)^2 = a + bH, \tag{0.1}$$

where $a = u_0^2$ and $b = u_1^2 + 2u_0u_1$. Moreover, a similar relation holds for any continuous function $g(z)$:

$$g(u_0 + u_1 H) = A + BH, \tag{0.2}$$

where $A = g(u_0)$ and $B = g(u_0 + u_1) - g(u_0)$. Thus the set of linear combinations of the form

$$u_0 + u_1 H$$

is closed with respect to nonlinear functional operations. In other words, the pair $(1; H)$ generates a subalgebra of generalized functions over the ring of smooth functions. This idea was the foundation of Maslov's hypothesis that smooth self-similar solutions of nonlinear equations correspond to subalgebras with finitely many generators. Moreover, Maslov claimed that there are only three such algebras [**88**].

The most famous construction that allows enlarging the space of generalized functions to an algebra is the construction introduced by Colombeau [**13**]. Here we do not describe this construction; it can be found in the well-known works [**13, 100, 4**]. We only note that elements of the Colombeau algebra are equivalence classes of smooth functions $f_\varepsilon(x)$ whose derivatives $\partial^\alpha f/\partial x^\alpha$ increase at most as some power of $\varepsilon^{-N(\alpha)}$.

The difficulties in the use of the Colombeau algebra for constructing solutions of differential equations are related to the fact that the equivalence relation

$$f_\varepsilon^1(x) \sim f_\varepsilon^2(x)$$

means that the following relation holds:

$$f_\varepsilon^1 - f_\varepsilon^2 = O(\varepsilon^N) \qquad \text{for all } N.$$

In turn, if such a condition is chosen, then regularizations of generalized functions must be constructed by using the convolution

$$f_\varepsilon(x) = \frac{1}{\varepsilon}\omega_c\left(\frac{x}{\varepsilon}\right) * f(x)$$

with a special mollifier $\omega_c(z)$ having *zero moments*

$$\int \omega_c(z) z^k \, dz = 0 \qquad \text{for all } k.$$

This property implies that the regularization of a smooth function f constructed by using such a mollifier "almost" coincides with the original function

$$\frac{1}{\varepsilon}\omega_c\left(\frac{x}{\varepsilon}\right) * f(x) = f(x) + O(\varepsilon^N)$$

for all $N > 0$. The same is also valid for generalized functions with accuracy $O(\varepsilon^N)$ in the weak sense.

So, in the standard version of the theory of Colombeau generalized functions, there is no scale with respect to powers of the small parameter ε, as it is used in the theory of asymptotic methods.

Moreover, from the viewpoint of differential equations, the equation itself is "choosing" a regularization of the generalized functions contained in its solution (regularizations with respect to small viscosity, small dispersion, etc.), which is inconsistent with the Colombeau approach. The regularizations obtained in solving the differential equations do not satisfy the zero moment condition.

To make the Colombeau approach closer to asymptotic methods and to obtain constructive formulas, an algebraic extension of the linear span of the set of homogeneous and associated homogeneous generalized functions was proposed in [**17, 23**]. As an example, the harmonic regularization of the delta function was taken to be a mollifier with nonzero moments. A one-to-one correspondence was established between the algebra generated by the harmonic regularizations of the elements of this linear span and the asymptotic Laurent–Hartogs type series called *asymptotic distributions* [**17, 23**]

$$f_\varepsilon \sim \sum_i \sum_j \varepsilon^{\alpha_i} (\ln \varepsilon)^{\beta_{ij}} f_{ij}(x).$$

Here f_{ij} are homogeneous or associated homogeneous generalized functions, $\{\alpha_i\}$ is a bounded below monotone increasing sequence of real numbers, and $\{\beta_{ij}\}$ is a finite set of real numbers for each fixed i.

The product of two asymptotic distributions is defined as an asymptotic distribution associated with the product of the corresponding elements in the algebra of harmonic regularizations. This also allows introducing asymptotic subalgebras in the algebra of regularizations modulo $O(\varepsilon^\alpha)$ in a weak sense for an appropriate choice of α. These constructions can easily be generalized to the case of arbitrary smooth regularizations (rather than to the case of harmonic regularizations).

The notion "small in the weak sense" mentioned above is one of the most important in the weak asymptotics method. We say that $f(x,\varepsilon)$ is of order $O_{\mathcal{D}'}(\varepsilon^\alpha)$ if for all test functions $\psi(x)$ we have the relation

$$\langle f(\cdot,\varepsilon), \psi(x)\rangle = O(\varepsilon^\alpha).$$

For example, we consider the set $\mathcal{H}$ of regularizations $H(x,\varepsilon)$ of the Heaviside function $H(x)$. Here $H(x,\varepsilon) = \omega(x/\varepsilon)$, $\omega(\infty) = 1$, $\omega(-\infty) = 0$, and $\omega'(z) \in S(\mathbb{R}^1)$, where $S(\mathbb{R}^1)$ is the Schwartz space. It is easy to see that

$$H(x,\varepsilon) = H(x) + O_{\mathcal{D}'}(\varepsilon)$$

(for details, see Section 1). Since $[H(x,\varepsilon)]^m$ again belongs to $\mathcal{H}$ for any m, we have

$$[H(x,\varepsilon)]^m = H(x) + O_{\mathcal{D}'}(\varepsilon).$$

This readily implies that, in general, for any continuous function $g(z)$ we have

$$g\big(a_1 + b_1 H(x,\varepsilon)\big) = a_2 + b_2 H(x) + O_{\mathcal{D}'}(\varepsilon),$$

i.e., we have repeated formulas (0.1)–(0.2) using regularizations and have verified that elements of $\mathcal{H}$ generate an asymptotic subalgebra $(\operatorname{mod} O_{\mathcal{D}'}(\varepsilon))$ with generators $(1; H(x))$. This construction is the subject of Sections 1 and 2, and all the asymptotic subalgebras are listed there.

It is precisely these subalgebras generated by arbitrary smooth regularizations that are used in the weak asymptotics method. Namely, the asymptotics is constructed as a linear combination of regularizations of generators of some subalgebra.

The fact that the set of all linear combinations considered is closed up to $O_{\mathcal{D}'}(\varepsilon^\alpha)$ with respect to nonlinear mappings leads to a finite system of equations for the coefficients of the original linear combination exactly as in the above example of the shock wave.

The only tricky place in this approach is the choice of a definition of a weak solution admitting passage to the "right" limit problem.

This problem is the well-known stumbling block in the theory of nonlinear differential equations. Particular cases, the zero viscosity limit, and the zero dispersion limit were studied in the well-known works [**75, 76, 15**].

Note that our notation $O_{\mathcal{D}'}(\varepsilon^\alpha)$ in the right-hand side of the equation can be either a small dispersion or a small viscosity, but the solutions corresponding to these two regularizations are quite different. Therefore, in our construction a well-posed definition of a weak solution admitting passage to the limit problem as $\varepsilon \to 0$ is very important.

Recall that the standard method for constructing the weak solution consists in multiplying the original equation by a test function and then integrating several times by parts. However, under this approach, the definitions of weak solutions up to terms of order $O_{\mathcal{D}'}(\varepsilon)$ in the right-hand side of the Bürgers equation with small viscosity

$$\frac{\partial u}{\partial t} + \frac{\partial u^2}{\partial x} = \varepsilon \frac{\partial^2 u}{\partial x^2},$$

of the Korteweg–de Vries equation with small dispersion

$$\frac{\partial u}{\partial t} + \frac{\partial u^2}{\partial x} + \varepsilon \frac{\partial^3 u}{\partial x^3} = 0,$$

and of the Hopf equation

$$\frac{\partial u}{\partial t} + \frac{\partial u^2}{\partial x} = 0$$

turn out to be equivalent, but this, of course, does not hold. More precisely, this is correct only for the Hopf and Bürgers equations, while a more complicated construction should correspond to the KdV equation (see Section 3).

Apparently, this situation was first pointed out in [**22, 21**], where the passage to the limit (from the phase field system to the Stefan–Gibbs–Thompson problem) was studied. In this paper, a definition of the weak solution of the nonlinear phase field system was given, which is different from that obtained according to the above standard approach. As shown in [**22, 21**], this definition admits passage to the limit problem. The construction of [**22, 21**] is close to the weak asymptotics method described in the present paper.

Roughly speaking, the logic of the method is the following. We consider both the limit problem (without ε) and the regularized problem (containing ε) simultaneously. We ask: What is a weak solution of the regularized problem corresponding to the limit problem? The word "corresponding" means that the weak limit of the weak solution should obey the limit problem.[1]

So, how can one obtain a definition of the weak solution that would automatically satisfy this correspondence condition?

Let us look at nonlinear solitary waves in the regularized problem (kinks, solitons, concentration waves, etc.) constructed by the above-mentioned method [**91, 18**]. All of these waves in the leading term as $\varepsilon \to 0$ are determined by a solution of a certain standard (model) equation. This equation itself does not contain ε. Its smooth solutions can usually be interpreted as the shape of a δ-approximating sequence or as the shape of a Heaviside-approximating sequence, etc. One can say that the model equation generates a smooth framework, which contains significant

[1]In particular, the singular support of the weak solution must coincide with the position of the front of the limit wave.

ε-independent information about the ε-dependence of solutions of the generalized problem.

But the model equation is not the whole story: the first variation of this equation should also be considered in order to construct ε-asymptotics for the regularized problem in higher orders in ε.

The variation equation is ε-independent (as the model equation itself), and, as a rule, its solvability is guaranteed by the orthogonality condition with the null-kernel of the adjoint problem.

Actually, this orthogonality condition determines the position of the front of the nonlinear wave. Obviously, this condition, in some way, must be an essential part of the "right" definition of the weak solution which we are interested in.

Therefore, we give the definition of a weak solution admitting passage to the limit in such a way that, in the weak limit, this definition becomes the solvability condition for the linearized problem.

The number of orthogonality conditions is equal to the dimension of the kernel of the adjoint linearized operator. Hence the definition of a weak solution for a single scalar equation can contain just one integral identity (if the dimension of the kernel equals 1) or two integral identities (as for KdV type equations, see below).

This approach allows us to solve the problem of propagation of solitary nonlinear waves easily. Some examples are collected in Section 3. Here we only note that, although all the examples in Section 3 can be investigated by the method of [**18, 91**], the weak asymptotics method allows a significant simplification of computations.

As the main advantage, the weak asymptotics method offers an opportunity to describe the interaction of nonlinear waves analytically without using exact integration methods (such as the inverse scattering transform method, group theory methods, etc.).

This approach is possible because of the special property of asymptotic subalgebras of generalized functions whose elements are used to construct weak asymptotic solutions. This is described in Section 2 in more detail. Here we just point at a simple example, which is the technical base of the analytic description of the interaction (confluence) of shock waves.

Let us consider the set $\mathcal{H}$ of regularizations of the Heaviside function introduced above. We take two elements $H_i(x,\varepsilon) \in \mathcal{H}$, $i = 1, 2$, and consider the product

$$H_1(x,\varepsilon)H_2(x-\phi,\varepsilon), \tag{0.3}$$

where ϕ is a parameter. In our example, the elements of $\mathcal{H}$ converge to $H(x)$ not only in the weak sense but also pointwise. Hence, as $\varepsilon \to 0$, we have

$$H_1(x,\varepsilon)H_2(x-\phi,\varepsilon) \to H(x)H(x-\phi)$$

in the sense of pointwise convergence. What can we say about the product in the right-hand side? For example, it is obvious that if $\phi > 0$, then

$$H(x)H(x-\phi) = H(x-\phi)$$

and if $\phi < 0$, then

$$H(x)H(x-\phi) = H(x).$$

But there does not exist any formula for the product $H(x)H(x-\phi)$ such that the corresponding function smoothly depends on ϕ in the sense of generalized functions.

If we admit a nonsmooth dependence on ϕ, then such a formula can easily be rewritten, e.g., as

$$H(x)H(x-\phi) = H(x-\phi)H(\phi) + H(x)(1-H(\phi)). \tag{0.4}$$

Approximating the Heaviside functions $H(\phi)$ in the right-hand side by elements of the set $\mathcal{H}$, we obtain a function continuous in ϕ in the right-hand side. However, it is still required to verify whether the difference between the regularization obtained and the original product is small in the sense of $O_{\mathcal{D}'}$-estimates. In Section 1 we give a simple derivation of general formulas for the weak asymptotics of expressions like (0.3) and similar formulas constructed using generators of other asymptotic subalgebras.

For instance, as will be shown in Section 1, the weak asymptotics of the product (0.3) has the form

$$H_1(x,\varepsilon)H_2(x-\phi,\varepsilon) = H(x-\phi)B(\phi/\varepsilon) + H(x)(1-B(\phi/\varepsilon)) + O_{\mathcal{D}'}(\varepsilon), \tag{0.5}$$

where

$$B(\phi/\varepsilon) = 1 - \int_{\mathbb{R}^1} \omega_1'(\eta)\omega_2(\eta-\phi/\varepsilon)\,d\eta$$

and the functions ω_i correspond to the elements $H_i \in \mathcal{H}$, i.e., $H_i(x,\varepsilon) = \omega_i(x/\varepsilon)$. This formula implies an analog of relation (0.1) in the form

$$\begin{aligned}\big(u_0 + u_1H_1(x-\phi_1,\varepsilon) + u_2H_2(x-\phi_2,\varepsilon)\big)^2 \\ = a + bH(x-\phi_1) + cH(x-\phi_2) + O_{\mathcal{D}'}(\varepsilon),\end{aligned} \tag{0.6}$$

where (cf. (0.1))

$$\begin{aligned} a &= u_0^2, \qquad b = u_1^2 + 2u_0u_1 + B\big((\phi_1-\phi_2)/\varepsilon\big)u_1u_2, \\ c &= u_2^2 + 2u_0u_2 + \big(1 - B\big((\phi_1-\phi_2)/\varepsilon\big)\big)u_1u_2, \qquad u_i = \text{const}. \end{aligned}$$

Formula (0.6) allows us almost literally to repeat the above Maslov construction but for two-wave solutions of the Hopf equation and thus to solve the problem of shock wave interaction.

In fact, formula (0.6) determines (in the special case of quadratic nonlinearity) a nonlinear superposition law. In this case the contribution of nonlinearity is contained only in the coefficients of the weak asymptotics. That is why we can "separate" singularities and find an equation for functions determining the solution both in the case of a single singularity and in the case of many singularities.

The laws of nonlinear superposition for general nonlinearities and for other singularities allow us to solve the problem of interaction of nonlinear waves in rather general cases.

In what follows, in the description of the interaction of nonlinear waves, we assume that $\phi_i = \phi_i(t,\varepsilon)$, where the functions $\phi_i(t,\varepsilon)$ are the coordinates of the fronts of interacting nonlinear waves. Hence the above formulas of nonlinear superposition turn out to be uniform with respect to time.

Now we can also explain the importance of the above remark that time is assumed to be a parameter of generalized functions. Indeed, if we consider the asymptotics of the product

$$H_1(x,\varepsilon)H_2(x-\phi,\varepsilon)$$

in $\mathcal{D}'_{x,t}$ (i.e., calculating t in the coordinate space), then we must take the right-hand side of expression (0.4) instead of (0.5); but expression (0.4) is not continuous in ϕ, and so in t, as was already mentioned above.

In what follows, we present a short review of well-known methods used to solve problems close to those studied in this paper.

1. The problem of multiplication of distributions (generalized functions) arose when the theory of distributions was constructed. Using the counterexample [**119**]

$$0 = 0 \cdot P(x^{-1}) = \Big(\delta(x) \cdot x\Big) \cdot P(x^{-1}) \neq \delta(x) \cdot \Big(x \cdot P(x^{-1})\Big) = \delta(x),$$

L. Schwartz showed that there is no associative and commutative multiplication of distributions consistent with multiplication by smooth functions. One of the approaches to the problem of multiplication of distributions and, consequently, an approach to the substitution of distributions into nonlinear equations, is associated with the name of Colombeau. Colombeau [**13**] constructed an *associative and commutative algebra* $\mathcal{G}$ of "new generalized functions," which contains the Schwartz space of distributions $\mathcal{D}'(\mathbb{R}^n)$ and the space $C^\infty(\mathbb{R}^n)$. Moreover, the latter is embedded in $\mathcal{G}$ as a subalgebra. In the Colombeau algebra $\mathcal{G}$ the product of distributions is defined and thus a new generalized function from $\mathcal{G}$ is presented. A remarkable contribution to the development of the Colombeau theory was made by Oberguggenberger [**100**].

The theory of "new generalized functions" is widely used in different problems in mathematical physics, where products of distributions should be defined, for example, in the theory of the nonlinear partial differential equations (Oberguggenberger) [**100**], (Oberguggenberger and Rosinger) [**101**], (Biagioni and Oberguggenberger) [**6, 5**], in stochastic analysis (Oberguggenberger and Russo) [**102**], in quantum mechanics and quantum field theory (Grosse, Oberguggenberger, and Todorov) [**41**], and in elasticity (Colombeau and Le Roux) [**14**].

In the framework of the Colombeau theory, some special tools and new functional spaces, which are used to study general solutions of linear and nonlinear equations, were introduced by Pilipović with coauthors [**112, 113, 114, 98, 115, 29, 97**] and Marti [**83, 84**], and some new theorems on the existence of solutions to nonlinear equations in these spaces were proved.

Another approach[2] to this problem is associated with the names of Livchak [**81**], Li Bang-He [**79, 80**], and Ivanov [**45**]. The main idea of this approach is that the product $f_1(x)f_2(x)$ of distributions $f_1(x)$ and $f_2(x)$ from a certain class is defined to be an asymptotic approximation of the product of the regularizations $f_1(x,\varepsilon)$ and $f_2(x,\varepsilon)$ of these distributions as $\varepsilon \to +0$ considered in the weak sense, where $\varepsilon > 0$ is a regularization parameter.

In the framework of this approach [**120, 121, 122, 123**] an *associative and commutative differential algebra* of *asymptotic distributions* generated by delta function, its derivatives, the principal value of the functions x^{-m}, and the powers x^m, $m \in \mathbb{N}$, was constructed, so that this algebra included the space of generating distributions. In the papers [**17, 23**], the results from [**120, 121, 122, 123**] were essentially developed, namely, a similar algebra of *asymptotic distributions* generated by homogeneous and associated homogeneous distributions was constructed.

For further development of this construction, see [**64, 65, 66**].

---

[2]Note that there exists one more approach to the "algebraization" of distributions, which was developed by Egorov [**32**].

The use of algebras of asymptotic distributions in [**17, 23**] made it possible to obtain some substantial results about the structure of singular self-similar solutions to strictly hyperbolic systems of conservation laws.

In [**23, 24, 25, 26, 27, 16, 106**], using the above ideas, the *weak asymptotics method* was developed. It is a new asymptotic method for investigating the dynamics of propagation and interaction of singularities of solutions to hyperbolic systems of conservation laws. In the framework of the weak asymptotics method, the dynamics of singular (nonsmooth) solutions of various types was obtained. In the papers [**23, 26**], nonlinear solitary waves of a new type (infinitely narrow P-solitons) were discovered for quasilinear equations. In [**26**] the weak asymptotics method was used to study an infinitely narrow δ-soliton related to KdV type equations. In [**24, 25**] the *weak asymptotic method* was used to study the interaction of shock waves for quasilinear equations in the one-dimensional case. In [**27**] the propagation and interaction of delta-shock waves of hyperbolic systems of conservation laws were considered. In [**16**] some examples about the formation and decay of shock waves are collected, and the phase field system is considered in [**107**].

2. The bibliography on the inverse transform method is very extensive. After the famous work by Gardner, Green, Kruskal, and Miura [**36**], numerous papers by Ablowitz, Dubrovin, Faddeev, Its, Krichever, Matveev, McLaughlin, Newell, Novikov, Zakharov, and other famous mathematicians appeared in this field. A detailed bibliography can be found in the books and papers [**11, 31, 7, 73, 82, 70, 95**].

3. The discovery of the inverse transform method stimulated interest in this subject of many mathematicians and physicists. The problem of stability of soliton type solutions with respect to relatively small (in one or other sense) perturbations was one of the first problems. The perturbation theory for solitons (and kinks) was developed in two directions. One of them, closely related to the inverse transform method, originates from the papers of Ablowitz, Gurevich, Karpman, E. Maslov, McLaughlin, Pitaevskii, Scott, Segur [**1, 31, 34, 56, 57, 60, 59, 70, 61, 68, 69, 85, 86**] and was further developed in the works by Il′in, Kalyakin, and other mathematicians (see, e.g., [**44, 50, 54, 49, 51, 52, 53**]). A remarkable feature of this theory is the constructive possibility of describing the interaction of distorted solitons (kinks) including the "radiation" effect, i.e., the appearance of a small rapidly oscillating correction of the Airy function type. Nevertheless, the possibility of using this theory is rigidly restricted by the set of equations, which, in the leading term, can be reduced to problems integrable by the inverse transform method.

The so-called "direct" methods, which do not appeal to the inverse transform method, are free of this restriction. The main idea of this passage is to adapt the method of two-scaled expansions (the Kuzmak–Whitham method [**72, 128**]) to the class of sufficiently rapidly stabilizing solutions. Apparently, the first results in this direction were obtained in the works of physicists (Ostrovskii, Pelinovskii, and others, see, e.g., [**110**]). The systematic mathematical theory of perturbations for solitons was developed in the works by Maslov, Omel′yanov, and Tsupin (see [**91**] and the references therein). The papers by Molotkov and Vakulenko [**96**] on the same topic appeared independently, but somewhat later. These ideas for dissipative problems (the Kolmogorov–Petrovskii–Piskunov equation, Zeldovich equation, etc.) were developed in the works by Danilov, Maslov, and Volosov, which were summarized in the monograph [**18**].

This perturbation theory, later called the *method of geometric asymptotics for nonlinear equations* [**91**], allows calculating the dynamics of nonlinear waves for practically all nonlinear equations. Nevertheless, in the framework of this theory, rather rigid conditions on the Cauchy data were imposed: a special form of the initial data is required not only in the leading term of the asymptotic approximation but also in all lower-order corrections (more exactly, there is some freedom in the lower-order corrections, but this freedom is finite-dimensional). This restriction was removed by Il′in and Kalyakin [**44**], however only for problems integrable by the inverse transform method.

Another drawback of this method is the fact that the class of problems, in which it is possible to describe the interaction of nonlinear waves, is rather narrow. Of course, this range of problems contains problems somehow reducible (in the leading term) to problems integrable by the inverse transform method [**91**]. However, all other problems, which were solved, are restricted by special equations for which it is possible, by one or another method, to prove that there exists an exact multiphase solution of the corresponding model partial differential equation and, which is not less important, to establish the finite dimensionality of the discrete part of the spectrum of the corresponding linearized operator [**91**]. The same holds for the method [**18**] applicable to dissipative problems.

4. Shock waves are a special case of generalized solutions to systems of hyperbolic conservation laws (both scalar and vector). In the case of a single conservation law, Oleinik [**103, 104**] posed some stability conditions and proved the uniqueness of the generalized solution under these conditions.

The general existence theorem for solutions with bounded variation for systems of conservation laws was proved by Glimm [**38**]. Glimm's proof was based on a piecewise constant approximation of the solution and on the study of evolution of this approximation.

There is another method for studying hyperbolic conservation laws, namely, the method of regularization by small viscosity [**104, 71**].

The notions of measure-valued solutions to systems of conservation laws [**15, 71**] and the entropy conditions are closely related to this method.

Using the regularization by small viscosity, Il′in [**43**] solved the problem of describing the appearance of a discontinuity (a shock wave) from a smooth initial condition for a scalar conservation law. To solve this problem, he used the well-known method of sewing asymptotic approximations. Solutions of smoothed shock wave type for equations of gas dynamics with small viscosity were studied in detail by physicists (see, e.g., [**12**] and [**126**]).

It is well known that general results of existence of entropy weak solutions are obtained only for initial values with small total variation [**38, 15, 77, 62**]. But it is recognized that most of the physical systems do not fit into the standard theory of conservation laws [**77, 62**]. The Riemann problem in this situation does not possess a weak L^∞-solution except for some particular initial data even if they are assumed to be small [**77**]. This fact contrasts with the standard results of existence of weak solutions to strictly hyperbolic systems [**38, 15**].

In order to solve the Cauchy problem arising in this nonclassical situation, it is necessary to introduce new elementary singularities called delta-shocks, which are solutions of hyperbolic systems, where some components of the solution can have the form $v(x,t) = v_0 + v_1 H(-x+ct) + e(t)\delta(-x+ct)$.

In **[27]** the weak asymptotics method is used to investigate both propagation and interaction of delta-shock waves for some systems considered in the works mentioned above **[62, 77, 33]**

1. Auxiliary formulas of the weak asymptotics method

In this section we present formulas for weak asymptotic expansions of some expressions. These formulas are used to construct solutions describing propagation of nonlinear waves and, which is most important, their interaction. Moreover, the formulas presented below readily imply some statements about asymptotic algebras mentioned in the Introduction and described in Section 2 below.

1.1. Regularizations of generalized functions. We recall that a family of functions $f(x, \varepsilon)$ smooth for $\varepsilon > 0$ and satisfying the condition

$$\operatorname*{w-lim}_{\varepsilon \to 0} f(x, \varepsilon) = f(x)$$

is called a *regularization* of the generalized function $f(x)$. By definition, this relation can be rewritten as

$$\lim_{\varepsilon \to 0} \langle f(x, \varepsilon), \varphi(x) \rangle = \langle f, \varphi \rangle$$

for any test function $\varphi(x)$ (here and in the following, $\langle\, ,\, \rangle$ denotes the action of a generalized function on a test function). Sometimes it is convenient to write the family of functions $f(x, \varepsilon)$ as an asymptotic series in powers of ε.

1.2. Weak asymptotic expansions. Let $\omega(\eta)$ be a continuous function decreasing sufficiently rapidly as $|\eta| \to \infty$ (so that the integrals below are meaningful), and let β be independent of x. We consider the expression $\omega(\beta x/\varepsilon)$ as an element of $\mathcal{D}'$. For any test function $\varphi(x)$, we have

$$\begin{aligned}\left\langle \omega\left(\beta \frac{x}{\varepsilon}\right), \varphi(x) \right\rangle &= \int_{\mathbb{R}^1} \omega\left(\beta \frac{x}{\varepsilon}\right) \varphi(x)\, dx = \frac{\varepsilon}{\beta} \int_{\mathbb{R}^1} \omega(\eta) \varphi\left(\varepsilon \frac{\eta}{\beta}\right) d\eta \\ &= \sum_{k=0}^{n} \frac{\varepsilon^{k+1}}{\beta^{k+1} k!} \varphi^{(k)}(0) \int_{\mathbb{R}^1} \omega(\eta) \eta^k \, d\eta + O(\varepsilon^{n+2}).\end{aligned}$$

In the theory of algebras of the Colombeau generalized functions, this formula is well known as the momentum decomposition.

Using the notation $O_{\mathcal{D}'}(\varepsilon^\alpha)$ introduced above, we can rewrite this relation as

$$\omega\left(\beta \frac{x}{\varepsilon}\right) = \sum_{k=0}^{n} \frac{\varepsilon^{k+1}}{\beta^{k+1} k!} (-1)^k \Omega_k \delta^{(k)}(x) + O_{\mathcal{D}'}(\varepsilon^{n+2}) \tag{1.1}$$

where $\Omega_k = \int \eta^k \omega(\eta)\, d\eta$ and $\delta(x)$ is the Dirac delta function. In what follows, we shall study only the leading term in (1.1). We have

$$\omega\left(\beta \frac{x}{\varepsilon}\right) = \varepsilon \frac{\Omega_0}{\beta} \delta(x) + O_{\mathcal{D}'}(\varepsilon^2)$$

in the general case, and the last relation can be made more precise by specifying the properties of the function $\omega(\eta)$. Namely, the term $O_{\mathcal{D}'}(\varepsilon^2)$ has the form

$$O_{\mathcal{D}'}(\varepsilon^2) = -\varepsilon^2 \delta'(x) \beta^{-2} \int_{\mathbb{R}^1} \eta \omega(\eta)\, d\eta + O_{\mathcal{D}'}(\varepsilon^3).$$

Thus if $\omega(\eta)$ is an even function, then we have

$$\omega\left(\beta\frac{x}{\varepsilon}\right)=\varepsilon\frac{\Omega_0}{\beta}\delta(x)+O_{\mathcal{D}'}(\varepsilon^3).$$

Next, we introduce the family $f(x,\varepsilon)$ by the formula

$$f_\varepsilon(x)=f(x)*\omega\left(\frac{x}{\varepsilon}\right),$$

where $\omega(\eta)\in S(\mathbb{R}^1)$ and $\int_{\mathbb{R}^1}\omega(\eta)\,d\eta=1$. Then

$$f_\varepsilon(x)=f(x)+\sum_{k\geq 1}(-1)^k\frac{\varepsilon^k\Omega_k}{k!}f^{(k)}(x).$$

For example, the regularization of the Heaviside function $H(x)$ constructed in this way is

$$H(x,\varepsilon)=\frac{1}{\varepsilon}\int_{-\infty}^{x}\omega\left(\frac{\eta}{\varepsilon}\right)d\eta=\int_{-\infty}^{x/\varepsilon}\omega(\eta)\,d\eta.$$

Let us prove that the family $H(x,\varepsilon)$ approximates the Heaviside H-function. To this end, we calculate $\langle H(x,\varepsilon),\varphi(x)\rangle$ for $\varepsilon>0$. We have

$$\begin{aligned}\langle H(x,\varepsilon),\varphi(x)\rangle&=\int H(x,\varepsilon)\varphi(x)\,dx=-\int\left[H(x,\varepsilon)\frac{d}{dx}\int_x^\infty\varphi(y)\,dy\right]dx\\&=-H(x,\varepsilon)\int_x^\infty\varphi(y)\,dy\Big|_{-\infty}^{\infty}+\int\left[\frac{d}{dx}H(x,\varepsilon)\int_x^\infty\varphi(y)\,dy\right]dx\\&=\varepsilon^{-1}\int\omega(x/\varepsilon)\mathcal{F}(x)\,dx,\end{aligned}$$

where $\mathcal{F}(x)=\int_x^\infty\varphi(y)\,dy$. Although $\mathcal{F}(x)$ is not a test function, the last integral can be calculated by using the momentum decomposition precisely as above. We thus obtain

$$\langle H(x,\varepsilon),\varphi(x)\rangle=\sum_{k\geq 0}\frac{\varepsilon^k}{k!}\Omega_k\mathcal{F}^{(k)}(0).$$

The leading term of this asymptotic expansion is

$$\Omega_0\mathcal{F}(0)=\int_0^\infty\varphi(y)\,dy=\langle H(x),\varphi(x)\rangle.$$

By $\widehat{\omega}(x/\varepsilon)$ we denote the function

$$\widehat{\omega}(z)=\int_{-\infty}^{z}\omega(\eta)\,d\eta.$$

Obviously, this function has the following properties:

a) $\lim_{z\to\infty}\widehat{\omega}(z)=1$;

b) the derivatives $d\widehat{\omega}/dz$ tend to zero as $|z|\to\infty$ sufficiently rapidly (like the original function).

It is clear that any function with properties a) and b) can be chosen to construct the regularization $\widehat{\omega}(x/\varepsilon)$ of $H(x)$.

1.3. The formulas obtained in this subsection will be used in Section 3 to construct weak asymptotics of noninteracting waves.

We consider the expression

$$G\big(A+\varepsilon B\delta(x,\varepsilon)\big),$$

where $G=G(z)$ is a smooth function and $\varepsilon\delta(x,\varepsilon)=\omega(x/\varepsilon)$ as above, and calculate its weak asymptotics.

First, we note that, from the viewpoint of our calculations, the expression

$$G\big(A+\varepsilon B\delta(x,\varepsilon)\big)-G(A)\stackrel{\text{def}}{=}\breve{\omega}(x/\varepsilon)$$

has the same properties as the original function $\omega(x/\varepsilon)$. Acting as in Subsection 1.2, we obtain

$$G\big(A+\varepsilon B\delta(x,\varepsilon)\big)=G(A)+\sum_{k\ge 0}\frac{\varepsilon^{k+1}}{k!}(-1)^k\Omega_k\delta^{(k)}(x).$$

The first term in this asymptotic series is $\varepsilon\Omega_0\delta(x)$, and Ω_0 can be calculated by the formula

$$\Omega_0=\int\breve{\omega}(z)\,dz=\int\big[G(A+B\omega(z))-G(A)\big]\,dz.$$

Similarly, assuming that $G(A+B)\ne G(A)$, for the expression

$$G\big(A+BH(x,\varepsilon)\big)$$

we have

$$G\big(A+BH(x,\varepsilon)\big)-G(A)\stackrel{\text{def}}{=}\widehat{\omega}(x/\varepsilon),$$

where the function $\widehat{\omega}(z)/\big(G(A+B)-G(A)\big)$ has the same properties as the function $\widehat{\omega}(z)$ in Subsection 1.2:

a) $\lim_{z\to\infty}\widehat{\omega}(z)/\big(G(A+B)-G(A)\big)\to 1$;

b) the derivatives $d^\alpha/dz^\alpha\{\widehat{\omega}(z)/\big(G(A+B)-G(A)\big)\}$ tend to zero as $|z|\to\infty$ sufficiently rapidly. Hence we obtain

$$G\big(A+BH(x,\varepsilon)\big)=G(A)+\widehat{\omega}(x/\varepsilon)\big(G(A+B)-G(A)\big).$$

In particular, we have

$$G\big(A+BH(x,\varepsilon)\big)=G(A)+H(x)\big(G(A+B)-G(A)\big)+O_{\mathcal{D}'}(\varepsilon).$$

1.4. Now let us consider a more complicated expression, which we shall use in Sections 6 and 7 to describe the soliton interaction,

$$f\Big(\omega_1\Big(\beta_1\frac{x-\phi_1}{\varepsilon}\Big)+\omega_2\Big(\beta_2\frac{x-\phi_2}{\varepsilon}\Big)\Big),$$

where $f(\eta)$ is a smooth function and $\omega_i(\eta)$, $i=1,2$, possess the same properties as $\omega(\eta)$ in Subsection 1.2, $\beta_1,\beta_2=\text{const}>0$, and ϕ_1,ϕ_2 are independent of x. We have

$$\begin{aligned}&\Big\langle f\Big(\omega_1\Big(\beta_1\frac{x-\phi_1}{\varepsilon}\Big)+\omega_2\Big(\beta_2\frac{x-\phi_2}{\varepsilon}\Big)\Big),\varphi\Big\rangle\\&=\Big\langle\frac{\varepsilon}{\beta_1}\delta(x-\phi_1)\int f_1\Big(\omega_1(\eta)+\omega_2\Big(\frac{\beta_2}{\beta_1}\eta+\frac{\beta_2}{\varepsilon}\Delta\phi\Big)\Big)\,d\eta+f(0),\varphi\Big\rangle+O(\varepsilon^2),\end{aligned}$$

where $f_1(\eta)=f(\eta)-f(0)$ and $\Delta\phi=\phi_1-\phi_2$.

On the other hand, introducing the variable η by the formula $\beta_2(x-\phi_2)=\varepsilon\eta$, we obtain

$$
\begin{aligned}
&f\Big(\omega_1\Big(\beta_1\frac{x-\phi_1}{\varepsilon}\Big)+\omega_2\Big(\beta_2\frac{x-\phi_2}{\varepsilon}\Big)\Big)\\
&\qquad=f(0)+\frac{\varepsilon}{\beta_2}\delta(x-\phi_2)\int_{\mathbb{R}^1}f_1\Big(\omega_1\Big(\frac{\beta_1}{\beta_2}\eta-\frac{\beta_1}{\varepsilon}\Delta\phi\Big)+\omega_2(\eta)\Big)\,d\eta+O_{\mathcal{D}'}(\varepsilon^2).
\end{aligned}
$$

It is easy to see that the coefficients of $\delta(x-\phi_1)$ and $\delta(x-\phi_2)$ in the last formulas are equal to each other:

$$
\begin{aligned}
&\beta_1^{-1}\int_{\mathbb{R}^1}f_1\Big(\omega_1(\eta)+\omega_2\Big(\frac{\beta_2}{\beta_1}\eta+\frac{\beta_2}{\varepsilon}\Delta\phi\Big)\Big)\,d\eta\\
&\qquad=\beta_2^{-1}\int_{\mathbb{R}^1}f_1\Big(\omega_1\Big(\frac{\beta_1}{\beta_2}\eta-\frac{\beta_1}{\varepsilon}\Delta\phi\Big)+\omega_2(\eta)\Big)\,d\eta\overset{\text{def}}{=}\Omega_{\Delta\phi}.
\end{aligned}
$$

This implies the relation

$$
\begin{aligned}
&f\Big(\omega_1\Big(\beta_1\frac{x-\phi_1}{\varepsilon}\Big)+\omega_2\Big(\beta_2\frac{x-\phi_2}{\varepsilon}\Big)\Big)\\
&\qquad=f(0)+\varepsilon\Omega_{\Delta\phi}(\lambda\delta(x-\phi_1)+\nu\delta(x-\phi_2))+O_{\mathcal{D}'}(\varepsilon^2),
\end{aligned}
$$

where λ and ν are arbitrary constants, $\lambda+\nu=1$.

1.5. To describe the shock wave interaction, we need the weak asymptotics of the expression

$$
f\big(A+BH_1(x-\phi_1,\varepsilon)+CH_2(x-\phi_2,\varepsilon)\big),\qquad A,B,C=\text{const},
$$

which is calculated by the same method as the weak asymptotics of the regularization $H(x,\varepsilon)$ to the Heaviside function in Subsection 1.2. We have

$$
\begin{aligned}
&\langle f\big(A+BH_1(x-\phi_1,\varepsilon)+CH_2(x-\phi_2,\varepsilon)\big),\varphi\rangle\\
&\quad=-\Big\langle f\big(A+BH_1(x-\phi_1,\varepsilon)+CH_2(x-\phi_2,\varepsilon)\big),\frac{d}{dx}\int_x^\infty\varphi(y)\,dy\Big\rangle\\
&\quad=-\Big\langle f\big(A+BH_1(x-\phi_1,\varepsilon)+CH_2(x-\phi_2,\varepsilon)\big),\int_x^\infty\varphi(y)\,dy\Big\rangle\Big|_{-\infty}^{\infty}\\
&\qquad+\Big\langle f'\big(A+BH_1(x-\phi_1,\varepsilon)+CH_2(x-\phi_2,\varepsilon)\big)\\
&\qquad\times\Big(B\frac{dH_1(x-\phi_1,\varepsilon)}{dx}+C\frac{dH_2(x-\phi_2,\varepsilon)}{dx}\Big),\int_x^\infty\varphi(y)\,dy\Big\rangle.
\end{aligned}
$$

Obviously, the last term is the sum of two terms. Now we consider the first term

$$
\begin{aligned}
I_1&=\Big\langle f'(A+BH_1(x-\phi_1,\varepsilon)+CH_2(x-\phi_2,\varepsilon))B\frac{dH_1(x-\phi_1,\varepsilon)}{dx},\int_x^\infty\varphi(y)\,dy\Big\rangle\\
&=\varepsilon^{-1}A\int\Big[f'\Big(A+B\widehat{\omega}_1\big((x-\phi_1)/\varepsilon\big)+C\widehat{\omega}_2\big((x-\phi_2)/\varepsilon\big)\Big)\\
&\qquad\times\widehat{\omega}_1'\big((x-\phi_1)/\varepsilon\big),\int_x^\infty\varphi(y)\,dy\Big]\,dx.
\end{aligned}
$$

Next, after the change of variables $(x-\phi_1)/\varepsilon=z$, we obtain

$$
I_1=B\int\Big[f'\Big(A+B\widehat{\omega}_1(z)+C\widehat{\omega}_2\Big(z+\frac{\Delta\phi}{\varepsilon}\Big)\Big)\widehat{\omega}_1'(z)\int_{\phi_1+\varepsilon z}^\infty\varphi(y)\,dy\Big]\,dz
$$

$$= B_1\Big(\frac{\Delta\phi}{\varepsilon}\Big) B\langle H(x-\phi_1), \varphi(x)\rangle + O(\varepsilon),$$

where

$$B_1\Big(\frac{\Delta\phi}{\varepsilon}\Big) = \int f'\Big(A + B\widehat{\omega}_1(z) + C\widehat{\omega}_2\Big(z + \frac{\Delta\phi}{\varepsilon}\Big)\Big)\widehat{\omega}_1'(z)\,dz.$$

Similarly, we obtain

$$\Big\langle f'(A + BH_1(x-\phi_1,\varepsilon) + CH_2(x-\phi_2,\varepsilon))C\frac{dH_2(x-\phi_2,\varepsilon)}{dx}, \varphi\Big\rangle$$

$$= B_2\Big(\frac{\Delta\phi}{\varepsilon}\Big) C\langle H(x-\phi_2), \varphi(x)\rangle + O(\varepsilon),$$

where

$$B_2\Big(\frac{\Delta\phi}{\varepsilon}\Big) = \int f'\Big(A + B\widehat{\omega}_1\Big(z - \frac{\Delta\phi}{\varepsilon}\Big) + C\widehat{\omega}_2(z)\Big)\widehat{\omega}_2'(z)\,dz.$$

Obviously, we have

$$B_1\Big(\frac{\Delta\phi}{\varepsilon}\Big) + B_2\Big(\frac{\Delta\phi}{\varepsilon}\Big) \equiv 1.$$

Finally, we obtain

$$f\big(A + BH_1(x-\phi_1,\varepsilon) + CH_2(x-\phi_2,\varepsilon)\big)$$

$$= f(A) + B_1\Big(\frac{\Delta\phi}{\varepsilon}\Big) BH(x-\phi_1) + B_2\Big(\frac{\Delta\phi}{\varepsilon}\Big) CH(x-\phi_2) + O_{\mathcal{D}'}(\varepsilon).$$

1.6. Asymptotic linear independence. If we want to consider linear combinations of generalized functions with accuracy $O_{\mathcal{D}'}(\varepsilon^\alpha)$, then we need to modify the notion of linear independence. This modification plays the key role in considerations related to the soliton interaction problem; see Sections 6 and 7.

Indeed, let $\phi_1 \neq \phi_2$ be independent of x. We consider the relation

$$g_1\delta(x-\phi_1) + g_2\delta(x-\phi_2) = O_{\mathcal{D}'}(\varepsilon^\alpha), \qquad \alpha > 0,$$

where g_i are independent of ε. Obviously, we obtain the relations

$$g_i = O_{\mathcal{D}'}(\varepsilon^\alpha), \qquad i = 1, 2,$$

which, by our assumption, imply

$$g_i = 0, \qquad i = 1, 2.$$

Everything is different if we assume that the coefficients g_i can depend on ε. Here we consider only a special case of such dependence, which we shall use later. Namely, let

$$g_i = A_i + S_i(\Delta\phi/\varepsilon), \qquad i = 1, 2,$$

where the A_i are independent of ε and the $S_i(\rho)$ decrease sufficiently rapidly as $|\rho| \to \infty$.

Let us find out what properties of the coefficients g_i follow from the relation

$$g_1\delta(x-\phi_1) + g_2\delta(x-\phi_2) = O_{\mathcal{D}'}(\varepsilon).$$

Applying both sides of the equality to a test function φ, we obtain

$$g_1\varphi(\phi_1) + g_2\varphi(\phi_2) = O(\varepsilon)$$

or, which is the same,

$$[A_1\varphi(\phi_1) + A_2\varphi(\phi_2)] + [S_1\varphi(\phi_1) + S_2\varphi(\phi_2)] = O(\varepsilon). \tag{1.2}$$

Let us consider the expression in the second brackets. Using Taylor's formula, we obtain

$$[S_1\varphi(\phi_1)+S_2\varphi(\phi_2)] = S_1\varphi(\phi_1)+S_2\varphi(\phi_1)+S_2(\phi_2-\phi_1)\varphi'(\phi_1+\theta\phi_2), \quad 0<\theta<1.$$

Now we see that

$$S_2(\Delta\phi/\varepsilon)(\phi_2-\phi_1) = \{-\rho S_2(\rho)\}\big|_{\rho=\Delta\phi/\varepsilon}\cdot\varepsilon = O(\varepsilon),$$

since the function $\rho S_2(\rho)$ is bounded uniformly in $\rho\in\mathbb{R}^1$.

So we can rewrite relation (1.2) as

$$A_1\varphi(\phi_1)+A_2\varphi(\phi_2)+(S_1+S_2)\varphi(\phi_1) = O(\varepsilon).$$

Hence, as the coefficients A_i are independent of ε, we, as usual, obtain

$$A_1=0, \qquad A_2=0, \qquad S_1+S_2=0. \tag{1.3}$$

Another method for analyzing relation (1.2) is the following. We assume that $\phi_i(t)$ are smooth functions, the relation $\phi_1(t^*)=\phi_2(t^*)$ holds for some $t=t^*$, and, moreover, $\phi_1'(t^*)\neq\phi_2'(t^*)$. Then

$$\begin{aligned}&\langle S_1\delta(x-\phi_1),\varphi\rangle+\langle S_2\delta(x-\phi_2),\varphi\rangle\\ &\quad= S_1\varphi(x^*)+S_2\varphi(x^*)+S_1O(t-t^*)+S_2O(t-t^*), \qquad x^*=\phi_1(t^*)=\phi_2(t^*).\end{aligned}$$

But $O(t-t^*)\sim O(\Delta\phi)$. Therefore, we have

$$\begin{aligned}&(A_1+S_1)\delta(x-\phi_1)+(A_2+S_2)\delta(x-\phi_2)\\ &\quad= A_1\delta(x-\phi_1)+A_2\delta(x-\phi_1)+(S_1+S_2)\delta(x-x^*)+O_{\mathcal{D}'}(\varepsilon).\end{aligned}$$

We again obtain relations (1.3).

1.7. Asymptotic expansions related to the Heaviside functions in the general case. To study the interaction of shock waves in the general case (which is not the piecewise constant case), we shall use an analog of the final formula in Subsection 1.5.

LEMMA 1.1. *Suppose that*

$$H_k(-x,\varepsilon)=\omega_{0k}(-x/\varepsilon)=\int_{-\infty}^{-x/\varepsilon}\omega_k(\eta)\,d\eta,$$

$k=1,2$, *are regularizations of the Heaviside function* $H(-x)$, $\omega_k(z)\in C^\infty(\mathbb{R})$ *are nonnegative and either have compact supports or decrease sufficiently rapidly as* $|z|\to\infty$, *and* $\int\omega_k(z)\,dz=1$, $k=1,2$.

Let $f(u)$ *be a smooth function, and let* $u_0(x,t)$, $e_1(x,t)$, *and* $e_2(x,t)$ *be bounded functions. Then the following formula holds*:

$$\begin{aligned}&f\big(u_0(x,t)+e_1(x,t)H_1(-x,\varepsilon)+e_2(x,t)H_2(-x+a,\varepsilon)\big)\\ &\quad= f(u_0(x,t))+\Big[f(u_0(x,t)+e_1(x,t))-f(u_0(x,t))\Big]H(-x)\\ &\quad+\Big[f(u_0(x,t)+e_2(x,t))-f(u_0(x,t))\Big]H(-x+a)\\ &\quad+B_1(x,t,a/\varepsilon)H(-x)+B_2(x,t,-a/\varepsilon)H(-x+a)+O_{\mathcal{D}'}(\varepsilon), \quad \varepsilon\to+0,\end{aligned}\tag{1.4}$$

where the estimate $O_{\mathcal{D}'}(\varepsilon)$ *is uniform in* a.

The functions $B_k(x,t,\rho)$, $k=1,2$, called "interaction switch functions" have the form

$$(1.5)\quad \begin{aligned} B_1(x,t,\rho) &= \int \Big[f'\big(u_0(x,t)+e_1(x,t)\omega_{01}(-\eta)+e_2(x,t)\omega_{02}(-\eta+\rho)\big) \\ &\qquad - f'\big(u_0(x,t)+e_1(x,t)\omega_{01}(-\eta)\big)\Big]e_1(x,t)\omega_1(-\eta)\,d\eta, \\ B_2(x,t,-\rho) &= \int \Big[f'\big(u_0(x,t)+e_1(x,t)\omega_{01}(-\eta-\rho)+e_2(x,t)\omega_{02}(-\eta)\big) \\ &\qquad - f'\big(u_0(x,t)+e_2(x,t)\omega_{02}(-\eta)\big)\Big]e_2(x,t)\omega_2(-\eta)\,d\eta. \end{aligned}$$

In addition, the interaction switch functions satisfy the relations

$$(1.6)\quad \begin{aligned} B_1(x,t,\rho)+B_2(x,t,-\rho) &= f\big(u_0(x,t)+e_1(x,t)+e_2(x,t)\big) \\ &\quad - f\big(u_0(x,t)+e_1(x,t)\big) - f\big(u_0(x,t)+e_2(x,t)\big) + f\big(u_0(x,t)\big), \\ \lim_{\rho\to+\infty} B_k(x,t,\rho) &= f\big(u_0(x,t)+e_1(x,t)+e_2(x,t)\big) \\ &\quad - f\big(u_0(x,t)+e_1(x,t)\big) - f\big(u_0(x,t)+e_2(x,t)\big) + f\big(u_0(x,t)\big), \\ \lim_{\rho\to-\infty} B_k(x,t,\rho) &= 0, \quad k=1,2. \end{aligned}$$

REMARK 1.1. For simplicity, we have selected the approximating functions $\omega_k(z)$ so that they either have compact supports or decrease sufficiently rapidly as $|z|\to\infty$; for example, $|\omega_k(z)|\le C_k(1+|z|)^{-N}$ for all positive integer N, $k=1,2$. Therefore, we have

$$\begin{aligned} \omega_{0k}(z) &= \int_{-\infty}^{z} \omega_k(\eta)\,d\eta = 1+O(z^{-N}), \qquad z\to+\infty, \\ \omega_{0k}(z) &= O(|z|^{-N}), \qquad z\to-\infty. \end{aligned}$$

Consequently, applying the Lagrange theorem, we obtain the estimate

$$\begin{aligned} &\Big[f'\big(u_0+e_1\omega_{01}(-\eta)+e_2\omega_{02}(-\eta+\rho)\big) - f'\big(u_0+e_1\omega_{01}(-\eta)\big)\Big]e_1\omega_1(-\eta) \\ &\quad = f''\big([u_0+e_1\omega_{01}(-\eta)]+\Theta e_2\omega_{02}(-\eta+\rho)\big)e_1e_2\omega_1(-\eta)\omega_{02}(-\eta+\rho), \end{aligned}$$

where $0<\Theta<1$.

It follows from this estimate and (1.5) that

$$\begin{aligned} B_1(x,t,\rho) &= f(u_0+e_1+e_2) - f(u_0+e_1) - f(u_0+e_2) \\ &\qquad + f(u_0) + O(\rho^{-N}), \qquad \rho\to+\infty, \quad N=1,2,\dots, \\ B_1(x,t,\rho) &= O(|\rho|^{-N}), \qquad \rho\to-\infty, \quad N=1,2,\dots. \end{aligned}$$

PROOF. First, we consider the case where $f(u)$ is an analytic function. The first step is to find the weak asymptotics of the product

$$(1.7)\qquad g(x,a,\varepsilon) = \omega_{01}^m(-x/\varepsilon)\omega_{02}^n(-(x+a)/\varepsilon),$$

which, according to Subsection 1.5, approximates the product of distributions

$$H^m(-x)H^n(-x+a), \qquad \text{where } m,n=1,2,\dots$$

It follows from the formula obtained in Subsection 1.5 that

$$(1.8)\quad \begin{aligned} J(a,\varepsilon) &= \big\langle g(x,a,\varepsilon),\varphi(x)\big\rangle \\ &= B_1^{m,n}(a/\varepsilon)\big\langle H(-x),\varphi(x)\big\rangle + B_2^{m,n}(-a/\varepsilon)\big\langle H(-x+a),\varphi(x)\big\rangle \end{aligned}$$

$$+O(\varepsilon), \quad \varepsilon \to +0,$$

where

$$(1.9)\qquad \begin{aligned} B_1^{m,n}(\rho) &= m \int \omega_{01}^{m-1}(-\eta)\omega_{02}^{n}(-\eta+\rho)\omega_1(-\eta)\,d\eta, \\ B_2^{m,n}(-\rho) &= n \int \omega_{01}^{m}(-\eta-\rho)\omega_{02}^{n-1}(-\eta)\omega_2(-\eta)\,d\eta. \end{aligned}$$

Formula (1.8) can be rewritten in the sense of the weak asymptotics (1.7) as follows:

(1.10)

$$g(x,a,\varepsilon) = B_1^{m,n}(a/\varepsilon)H(-x) + B_2^{m,n}(-a/\varepsilon)H(-x+a) + O_{\mathcal{D}'}(\varepsilon), \quad \varepsilon \to +0.$$

Using (1.7), (1.10), the obvious relation

$$[H(x,\varepsilon)]^r = H(x) + O_{\mathcal{D}'}(\varepsilon), \quad \varepsilon \to +0, \qquad r = 2,3,\dots,$$

and the binomial formula, we find the weak asymptotics

$$(1.11)\qquad \begin{aligned} &\big[e_1(x,t)\omega_{01}(-x/\varepsilon) + e_2(x,t)\omega_{02}\big((-x+a)/\varepsilon\big)\big]^n \\ &\quad = e_1^n(x,t)H(-x) + e_2^n(x,t)H(-x+a) \\ &\qquad + \tilde{B}_1(x,t,a/\varepsilon)H(-x) + \tilde{B}_2(x,t,-a/\varepsilon)H(-x+a) \\ &\qquad + O_{\mathcal{D}'}(\varepsilon), \quad \varepsilon \to +0. \end{aligned}$$

Here, as follows from (1.9),

$$\begin{aligned} \tilde{B}_1(x,t,\rho) &= \sum_{k=1}^{n-1} C_n^k e_1^k(x,t)e_2^{n-k}(x,t)B_1^{k,n-k}(\rho) \\ &= \sum_{k=1}^{n-1} nC_{n-1}^{k-1} e_1^k(x,t)e_2^{n-k}(x,t) \int \omega_{01}^{k-1}(-\eta)\omega_{02}^{n-k}(-\eta+\rho)\omega_1(-\eta)\,d\eta, \end{aligned}$$

where C_n^k are binomial coefficients.

The last expression can be rewritten as

$$\begin{aligned} \tilde{B}_1(x,t,\rho) = n \int \Big[&\big(e_1(x,t)\omega_{01}(-\eta) + e_2(x,t)\omega_{02}(-\eta+\rho)\big)^{n-1} \\ &- \big(e_1(x,t)\omega_{01}(\eta)\big)^{n-1}\Big] e_1(x,t)\omega_1(-\eta)\,d\eta. \end{aligned}$$

Similarly, we obtain

$$\begin{aligned} \tilde{B}_2(x,t,-\rho) = n \int \Big[&\big(e_1(x,t)\omega_{01}(-\eta-\rho) + e_2(x,t)\omega_{02}(-\eta)\big)^{n-1} \\ &- \big(e_2(x,t)\omega_{02}(-\eta)\big)^{n-1}\Big] e_2(x,t)\omega_2(-\eta)\,d\eta. \end{aligned}$$

Expanding the function $f(u)$ in the Taylor series and using the weak asymptotics (1.11) and the formulas for $\tilde{B}_1(x,t,\rho)$ and $\tilde{B}_2(x,t,-\rho)$, we find the weak asymptotics of $f\big(u_0 + e_1H_1(-x,\varepsilon) + e_2H_2(-x+a,\varepsilon)\big)$ as $\varepsilon \to +0$, that is, we obtain expansion (1.4) and formula (1.5) for the interaction switch functions.

To prove the first formula in (1.6), we rewrite $B_1(x,t,\rho)$ given by (1.5) as

$$B_1(x,t,\rho) = \int f'\big(u_0 + e_1\omega_{01}(-\eta) + e_2\omega_{02}(-\eta+\rho)\big)\big[e_1\omega_1(-\eta) + e_2\omega_2(-\eta+\rho)\big]\,d\eta$$

$$
\begin{aligned}
&- \int \Big[f'\big(u_0 + e_1\omega_{01}(-\eta) + e_2\omega_{02}(-\eta+\rho)\big) \\
&\qquad - f'\big(u_0 + e_2\omega_{02}(-\eta+\rho)\big)\Big] e_2\omega_2(-\eta+\rho)\,d\eta \\
&- \int f'\big(u_0 + e_1\omega_{01}(-\eta)\big) e_1\omega_1(-\eta)\,d\eta \\
&\qquad - \int f'\big(u_0 + e_2\omega_{02}(-\eta+\rho)\big) e_2\omega_2(-\eta+\rho)\,d\eta.
\end{aligned}
$$

Note that, after the substitution $-\eta + \rho \to -\eta'$, the second integral converges to the function $B_2(x,t,-\rho)$ from (1.5). Because $\lim_{\eta\to+\infty}\omega_{0k}(\eta) = 1$ and $\lim_{\eta\to-\infty}\omega_{0k}(\eta) = 0$, we see that the first integral is equal to $f(u_0+e_1+e_2) - f(u_0)$, while the third and the forth integrals are equal to $f(u_0) - f(u_0+e_1)$ and $f(u_0) - f(u_0+e_2)$, respectively.

The first formula in (1.6) is thus proved.

Since $\lim_{\rho\to+\infty}\omega_{02}(-\eta+\rho) = 1$, relation (1.5) implies

$$
\begin{aligned}
B_1(x,t,\infty) &= \lim_{\rho\to+\infty} B_1(x,t,\rho) \\
&= \int \Big[f'\big(u_0 + e_1\omega_{01}(-\eta) + e_2\big) - f'\big(u_0 + e_1\omega_{01}(-\eta)\big)\Big] e_1\omega_1(-\eta)\,d\eta.
\end{aligned}
$$

Integrating the last expression and taking into account that $\lim_{\eta\to+\infty}\omega_{01}(\eta) = 1$ and $\lim_{\eta\to-\infty}\omega_{01}(\eta) = 0$, we obtain

$$
B_1(x,t,\infty) = \lim_{\rho\to+\infty} B_1(x,t,\rho) = f(u_0+e_1+e_2) - f(u_0+e_1) - f(u_0+e_2) + f(u_0),
$$

which is the second relation in (1.6). The other relations in (1.6) can be proved similarly.

To complete the proof of Lemma 1.1, we use the theorem on approximating a smooth function on a finite interval by analytic functions. □

LEMMA 1.2. *Let $\delta_k(x,\varepsilon) = \varepsilon^{-1}\omega_k(x/\varepsilon)$, $k = 1,2$, be a regularization of the delta function $\delta(x)$, and let $H_k(x,\varepsilon) = \omega_{0k}(x/\varepsilon)$ be a regularization of the Heaviside function $H(x)$. Suppose that the function $f(u)$ is smooth and the functions $u_0(x,t)$, $g_k(t)$, and $e_k(x,t)$ are bounded. Then we have the asymptotics*

$$
\begin{aligned}
&f\big(u_0(x,t) + g_1(t)\varepsilon\delta_1(x,\varepsilon) + g_2(t)\varepsilon\delta_2(x-a,\varepsilon) \\
&\qquad + e_1(x,t)\varepsilon H_1(x,\varepsilon) + e_2(x,t)\varepsilon H_2(x-a,\varepsilon)\big) \\
&\qquad\quad = f(u_0(x,t)) + f'(u_0(x,t))e_1(x,t)\varepsilon H(x) + f'(u_0)e_2(x,t)\varepsilon H(x-a) \\
&\qquad\qquad + A_1(x,t)\varepsilon\delta(x) + A_2(x,t)\varepsilon\delta(x-a) \\
&\qquad\qquad + B(a/\varepsilon,x,t)\big(\varepsilon\delta(x) + \varepsilon\delta(x-a)\big) + O_{\mathcal{D}'}(\varepsilon^2), \quad \varepsilon \to +0,
\end{aligned}
$$

where the estimate $O_{\mathcal{D}'}(\varepsilon^2)$ is uniform with respect to a,

$$
A_k(x,t) = \int \Big[f\big(u_0(x,t) + g_k(t)\omega_k(\eta)\big) - f\big(u_0(x,t)\big)\Big]\,d\eta, \quad k = 1,2,
$$

and the functions $B_k(\rho,x,t)$, $k = 1,2$, called interaction switch functions have the form

$$
\begin{aligned}
B(\rho,x,t) = \frac{1}{2}\int \Big[&f\big(u_0(x,t) + g_1(t)\omega_1(\eta) + g_2(t)\omega_2(\eta-\rho)\big) \\
&- f\big(u_0(x,t) + g_1(t)\omega_1(\eta)\big)
\end{aligned}
$$

$$- f\big(u_0(x,t) + g_2(t)\omega_2(\eta - \rho)\big) + f\big(u_0(x,t)\big)\Big]\,d\eta.$$

In addition, for all $N = 1, 2, \dots,$ *we have* $B(\rho, x, t) = O(|\rho|^{-N})$ *as* $|\rho| \to \infty$.

The proof of this lemma is based on the same calculations as those carried out above, and we omit them here.

2. Asymptotic subalgebras

In this section we describe the asymptotic subalgebras mentioned in the Introduction and discuss their relation to algebras of generalized functions. The construction of weak asymptotic solutions is based on these algebras. For shock waves and their interaction, we use subalgebras from Lemmas 2.1 and 2.6. For solitons and their interaction, we use subalgebras from Lemmas 2.3 and 2.8.

2.1. Now we introduce *asymptotic subalgebras* of *associated homogeneous* distributions briefly mentioned in the Introduction.

First, we recall some notions of the theory of distributions.

DEFINITION 2.1. A distribution $f(x) \in \mathcal{D}'(\mathbb{R})$ is said to be *associated homogeneous* of degree λ and order k, $k = 0, 1, \dots,$ if the relation

$$\langle f(x), \varphi(x/a)\rangle = a^{\lambda+1}\langle f(x), \varphi(x)\rangle + a^{\lambda+1}\log a\langle f_0(x), \varphi(x)\rangle$$

holds for all $a > 0$ and for all $\varphi(x) \in \mathcal{D}(\mathbb{R})$, where $f_0(x)$ is some associated homogeneous distribution of degree λ and order $k-1$ (here we set $f_0(x) = 0$ for $k = 0$) (see [**37**, Ch.I, §3.11, §4]).

Associated distributions of order zero coincide with homogeneous distributions. For each λ there exist two linearly independent associated homogeneous distributions of degree λ and order k, $k = 0, 1, \dots$ (see [**37**, Ch.I, §3.11, §4]).

We introduce a linear space $E_0 \subset \mathcal{D}'(\mathbb{R})$ generated by linear combinations of associated homogeneous distributions, i.e.,

$$E_0 = \operatorname{span}\big\{(x \pm i0)^\lambda \log^p(x \pm i0) : \lambda \in \mathbb{R},\ p = 0, 1, \dots\big\},$$

where span denotes the linear span. (For $\lambda > 0$, we set $(x \pm i0)^\lambda \equiv x^\lambda$.) For example, the space E_0 contains distributions such as $\delta^{(m-1)}(x)$, $P(x^{-m})$, $m = 1, 2, \dots,$ and $x_\pm^\lambda \log^p(x_\pm)$, $p = 0, 1, \dots$.

We consider the set $\Omega(\mathbb{R})$ of *all mollifiers* $\omega(z)$ having the following properties:

a) $\omega(z) \in C^\infty(\mathbb{R})$;

b) $\omega(z)$ has a compact support or decreases sufficiently rapidly as $|z| \to \infty$; for example, $|\omega(z)| \le C(1 + |z|)^{-N}$ for all positive integers N;

c) $\int \omega(z)\,dz = 1$.

Let

$$h_0 = \operatorname{span}\{f(x, \varepsilon)\}$$

be the space of regularizations of the distributions $f(x) \in E_0$ given by

$$f(x, \varepsilon) = f(x) * \frac{1}{\varepsilon}\omega\Big(\frac{x}{\varepsilon}\Big)$$

for all $\omega(z) \in \Omega(\mathbb{R})$.

Let G_0^* be an *associative commutative differential algebra over the ring of smooth functions* $C^\infty(\mathbb{R})$ generated by the space

$$G_0 = \operatorname{span}\big\{f_{1,i}(x, \varepsilon), \varepsilon f_{2,j}(x, \varepsilon) : f_{1,i}(x, \varepsilon), f_{2,j}(x, \varepsilon) \in h_0\big\}.$$

The elements $f^*(x,\varepsilon)$ of the algebra G_0^* are linear combinations of finite products of functions $f(x,\varepsilon) \in G_0$ with coefficients from $C^\infty(\mathbb{R})$.

2.2. Asymptotic subalgebras with a single point of singularity. It follows from the results of Section 1 that some elements $f^*(x,\varepsilon)$ from the algebra G_0^* up to $O_{\mathcal{D}'}(\varepsilon^N)$ have the representation in the form of a weak asymptotics:

$$f^*(x,\varepsilon) = A_1(x)f_1(x) + A_2(x)\varepsilon f_2(x) + O_{\mathcal{D}'}(\varepsilon^N),$$

where $f_k(x) \in E_0$ and $A_k(x) \in C^\infty(\mathbb{R})$, $k = 1,2$, $N = 1$ (in this case we can set $A_2 = 0$) or $N = 2$. Then it seems natural to introduce the following definition.

Let i)

$$\mathcal{B} = \mathcal{B}\{f_1(x,\varepsilon),\dots,f_{M1}(x,\varepsilon);\varepsilon g_1(x,\varepsilon),\dots,\varepsilon g_{M2}(x,\varepsilon)\}$$

be a subalgebra of G_0^* generated by the elements $f_1(x,\varepsilon),\dots,f_{M1}(x,\varepsilon), g_1(x,\varepsilon),\dots, g_{M2}(x,\varepsilon) \in h_0$. Here $f_j(x,\varepsilon)$ (or $g_k(x,\varepsilon)$) denotes a set of regularizations of the limit distribution $f_j(x)$ (or $g_k(x)$) with the help of all the mollifiers $\omega \in \Omega(\mathbb{R})$.

ii) For any element $f^*(x,\varepsilon) \in \mathcal{B}$, we have an asymptotic expansion up to $O_{\mathcal{D}'}(\varepsilon^N)$ of the form

$$f^*(x,\varepsilon) = \sum_j A_{1j}(x)f_{1j}(x) + \varepsilon\sum_k A_{2k}(x)f_{2k}(x) + O_{\mathcal{D}'}(\varepsilon^N),$$

where $N = 1,2$, $A_{1j}(x)$ and $A_{2k}(x)$ are smooth functions, the distributions $f_{1j}(x)$ and $f_{2k}(x)$ belong to E_0 and are independent of the choice of $f^*(x,\varepsilon)$, and $\sum_j$ and $\sum_k$ are finite.

DEFINITION 2.2. We say that a subalgebra $\mathcal{B} \in G_0^*$ is a *finitely generated asymptotic distributional differential subalgebra* modulo $O_{\mathcal{D}'}(\varepsilon^N)$ if the set $\mathcal{A}_N = \{(f_{1j}(x), f_{2k}(x))\}$ is a *finite-dimensional* vector subspace of $\mathcal{D}'(\mathbb{R})$. The basis elements $(f_{1j}(x); f_{2k}(x))$ are called *distributional generators* of the *asymptotic subalgebra* $\mathcal{B}$.

In what follows, the subalgebras $\mathcal{B}$ from Definition 2.1 are called *asymptotic subalgebras*.

The asymptotic subalgebras can be described by using calculations similar to those carried out in Section 1.

LEMMA 2.1. *Let $H(x,\varepsilon)$ be regularizations of the Heaviside function $H(x)$. Then*

$$\mathcal{B}_1 = \mathcal{B}_1\{1, H(x,\varepsilon)\}$$

is the asymptotic differential subalgebra mod $O_{\mathcal{D}'}(\varepsilon)$ *with distributional generators $(1, H(x))$. The following relations hold for all $H_k(x,\varepsilon) \in \mathcal{B}_1$, $k = 1,2,\dots$:*

$$1\cdot H_k(x,\varepsilon) = H(x) + O_{\mathcal{D}'}(\varepsilon),$$

$$\prod_{k=1}^{s}\big(H_k(x,\varepsilon)\big)^{m_k} = H(x) + O_{\mathcal{D}'}(\varepsilon), \quad m_k \in \mathbb{N}_0,$$

$$\left(a_1(x) + \sum_{k=1}^{s} b_{1k}(x)H_k(x,\varepsilon)\right)^n = a_2(x) + b_2(x)H(x) + O_{\mathcal{D}'}(\varepsilon), \quad n \in \mathbb{N},$$

where $a_1(x)$, $b_{1k}(x)$, $a_2(x) = (a_1(x))^n$, and $b_2(x) = \big(a_1(x) + \sum_{k=1}^{s} b_{1k}(x)\big)^n - (a_1(x))^n$ are smooth functions.

Lemma 1.1 can be easily proved if we take the relation $\sum_{k=1}^{s} b_{1k} H_k(x,\varepsilon) = H(x)\sum_{k=1}^{s} b_{1k} + O_{\mathcal{D}'}(\varepsilon)$ (see Section 1) into account.

LEMMA 2.2. *Let $\delta(x,\varepsilon) = \varepsilon^{-1}\omega(x/\varepsilon)$ be regularizations of the Dirac delta function $\delta(x)$, $\omega(z) \in \Omega(\mathbb{R})$. Then*

$$\mathcal{B}_2 = \mathcal{B}_2\{1;\ \varepsilon\delta(x,\varepsilon)\}$$

is the asymptotic differential subalgebra mod $O_{\mathcal{D}'}(\varepsilon^2)$ *with distributional generators* $(1;\delta(x))$. *The following relations hold for all $\varepsilon\delta_k(x,\varepsilon) \in \mathcal{B}_2$, $k = 1,2,\dots$:*

$$1\cdot\varepsilon\delta_k(x,\varepsilon) = \varepsilon\delta(x) + O_{\mathcal{D}'}(\varepsilon^2),$$

$$\prod_{k=1}^{s}\big(\varepsilon\delta_k(x,\varepsilon)\big)^{m_k} = a_{m_1,\dots,m_s}\varepsilon\delta(x) + O_{\mathcal{D}'}(\varepsilon^2), \quad m_k \in \mathbb{N}_0,$$

$$\Big(a_1(x) + \sum_{k=1}^{s} b_{1k}(x)\varepsilon\delta_k(x,\varepsilon)\Big)^n = a_2(x) + b_2(0)\varepsilon\delta(x) + O_{\mathcal{D}'}(\varepsilon^2), \quad n \in \mathbb{N},$$

where $a_{m_1,\dots,m_s} = \int \prod_{k=1}^{s}\big(\omega_k(\eta)\big)^{m_k}\,d\eta$ *and* $b_{1k}(x)$, $a_2(x) = (a_1(x))^n$, *and* $b_2(x) = \int\big[(a_1(x) + \sum_{k=1}^{s} b_{1k}(x)\omega_k(\eta))^n - (a_1(x))^n\big]\,d\eta$ *are smooth functions.*

The proof follows from the results of Section 1.

LEMMA 2.3. *Let $\delta(x,\varepsilon)$ be regularizations of the Dirac delta function $\delta(x)$, and let $H(x,\varepsilon)$ be regularizations of the Heaviside function $H(x)$. Then*

$$\mathcal{B}_3 = \mathcal{B}_3\{1;\ \varepsilon\delta(x,\varepsilon),\ \varepsilon H(x,\varepsilon)\}$$

is the asymptotic differential subalgebra mod $O_{\mathcal{D}'}(\varepsilon^2)$ *with distributional generators* $(1;\delta(x),H(x))$. *The following relations hold for all $\varepsilon\delta_j(x,\varepsilon), \varepsilon H_k(x,\varepsilon) \in \mathcal{B}_3$, $j = 1,2,\dots$, $k = 1,2,\dots$:*

$$1\cdot\varepsilon H_j(x,\varepsilon) = \varepsilon H(x) + O_{\mathcal{D}'}(\varepsilon^2),$$

$$1\cdot\big(\varepsilon H_j(x,\varepsilon)\big)^m = O_{\mathcal{D}'}(\varepsilon^2), \quad m \ge 2,$$

$$\big(\varepsilon\delta_k(x,\varepsilon)\big)^m\big(\varepsilon H_j(x,\varepsilon)\big)^n = O_{\mathcal{D}'}(\varepsilon^2), \quad n \ge 1,$$

$$\Big(a_1(x) + \sum_{j=1}^{s_1} b_{1j}(x)\varepsilon\delta_j(x,\varepsilon) + \sum_{k=1}^{s_2} c_{1k}(x)\varepsilon H_k(x,\varepsilon)\Big)^n = a_2(x) + b_2(0)\varepsilon\delta(x) + c_2(x)\varepsilon H(x) + O_{\mathcal{D}'}(\varepsilon^2),$$

where $a_1(x)$, $b_{1j}(x)$, $c_{1k}(x)$, $a_2(x) = (a_1(x))^n$,

$$b_2(x) = \int\Big[\Big(a_1(x) + \sum_{j=1}^{s_1} b_{1j}(x)\omega_j(\eta)\Big)^n - (a_1(x))^n\Big]\,d\eta,$$

and $c_2(x) = n(a_1(x))^{n-1}\sum_{k=1}^{s_2} c_{2k}(x)$ *are smooth functions and* $m,n \in \mathbb{N}$.

The proof follows from the results of Section 1.

Now we introduce a regularization of the principal value of the function x^{-1} as follow:

$$P(x^{-1},\varepsilon) = P(x^{-1}) * \frac{1}{\varepsilon}\omega\Big(\frac{x}{\varepsilon}\Big) = \frac{1}{\varepsilon}\int_0^\infty \frac{\omega\big((x-t)/\varepsilon\big) - \omega\big((x+t)/\varepsilon\big)}{t}\,dt.$$

Making the change of variables $z = t/\varepsilon$, we obtain

$$P(x^{-1}, \varepsilon) = \frac{1}{\varepsilon}\mathcal{P}\Big(\frac{x}{\varepsilon}\Big),$$

where

$$\mathcal{P}(\eta) = \int_0^\infty \frac{\omega(\eta - z) - \omega(\eta + z)}{z}\, dz, \tag{2.1}$$

the function $\omega(x) \in \Omega(\mathbb{R})$ is assumed to be even. Then the regularizations $P(x^{-1}, \varepsilon)$ are odd with respect to x.

LEMMA 2.4. *Let $P(x^{-1}, \varepsilon) = \varepsilon^{-1}\mathcal{P}(x/\varepsilon)$ be regularizations of the principal value of the function x^{-1}, where $\mathcal{P}(\eta)$ is given by* (2.1)*. Then*

$$\mathcal{B}_4 = \mathcal{B}_4\{1;\ \varepsilon P(x^{-1}, \varepsilon)\}$$

is the asymptotic differential subalgebra $\operatorname{mod} O_{\mathcal{D}'}(\varepsilon^2)$ *with distributional generators $(1; \delta(x), P(x^{-1}))$. The following relations hold for all $\varepsilon P_k(x^{-1}, \varepsilon) \in \mathcal{B}_4$, $k = 1, 2, \dots s$:*

$$1 \cdot \varepsilon P_k(x^{-1}, \varepsilon) = \varepsilon P(x^{-1}) + O_{\mathcal{D}'}(\varepsilon^2),$$
$$\big(\varepsilon P_k(x^{-1}, \varepsilon)\big)^{2n} = a_{2n}\varepsilon\delta(x) + O_{\mathcal{D}'}(\varepsilon^2),$$
$$\big(\varepsilon P_k(x^{-1}, \varepsilon)\big)^{2n-1} = O_{\mathcal{D}'}(\varepsilon^2), \quad n \ge 2,$$
$$\Big(a_1(x) + \sum_{k=1}^{s} b_{1k}(x)\varepsilon P_k(x^{-1}, \varepsilon)\Big)^n = a_2(x) + b_2(x)\varepsilon P(x^{-1})$$
$$+ c_2(x)\varepsilon\delta(x) + O_{\mathcal{D}'}(\varepsilon^2),$$

where $a_{2n} = \int_{-\infty}^{\infty} \big(\mathcal{P}(\eta)\big)^{2n} d\eta$ are constants and $a_1(x)$, $b_{1k}(x)$, $a_2(x) = (a_1(x))^n$, $b_2(x) = n(a_1(x))^{n-1}\sum_{k=1}^{s} b_{1k}(x)$, and $c_2(x) = \int \big[\big(a_1(x) + \sum_{k=1}^{s} b_{1k}(x)\mathcal{P}_k(\eta)\big)^n - (a_1(x))^n - n(a_1(x))^{n-1}\sum_{k=1}^{s} b_{1k}(x)\mathcal{P}_k(\eta)\big]\, d\eta$ are smooth functions, $n \in \mathbb{N}$.

This lemma can be proved similarly to the results of Section 1.

LEMMA 2.5. *Let $\delta(x, \varepsilon)$ be regularizations of the Dirac delta function $\delta(x)$, and let $P(x^{-1}, \varepsilon)$ be regularizations of the principal value of the function x^{-1}. Then*

$$\mathcal{B}_5 = \mathcal{B}_5\{1;\ \varepsilon\delta(x, \varepsilon),\ \varepsilon P(x^{-1}, \varepsilon)\}$$

is the asymptotic differential subalgebra mod $O_{\mathcal{D}'}(\varepsilon^2)$ *with distributional generators $(1; \delta(x), P(x^{-1}))$.*

2.3. Asymptotic subalgebras with two points of singularity. Let G_1^* be an *associative commutative differential algebra over the ring of smooth functions* $C^\infty(\mathbb{R})$ generated by the space

$$G_1 = \operatorname{span}\{f_1(x - a^j, \varepsilon), \varepsilon f_2(x - a^j, \varepsilon) : f_k(x, \varepsilon) \in h_0,\ a^j \in \mathbb{R},\ k, j = 1, 2\}.$$

The elements $f^*(x, \varepsilon)$ of the algebra G_1^* are linear combinations of finite products of functions $f(x - a^j, \varepsilon) \in G_1$ with coefficients from $C^\infty(\mathbb{R})$.

It follows from the results of Section 1 that some elements $f^*(x, \varepsilon)$ from the algebra G_1^* up to $O_{\mathcal{D}'}(\varepsilon^N)$ $(N = 1, 2)$ have the representation in the form of a weak asymptotics:

$$f^*(x, \varepsilon) = B_1\big(x, (a^2 - a^1)/\varepsilon\big) f_1(x - a^1) + B_2\big(x, (a^2 - a^1)/\varepsilon\big) f_2(x - a^2)$$

$$+ B_3\big(x, (a^2 - a^1)/\varepsilon\big)\varepsilon f_3(x - a^1) + B_4\big(x, (a^2 - a^1)/\varepsilon\big)\varepsilon f_4(x - a^2) + O_{\mathcal{D}'}(\varepsilon^N),$$

where $f_k(x) \in E_0$, the *interaction switch functions* $B_k(x, \rho)$ are smooth functions and can be found explicitly, $k = 1, 2, 3, 4$; $N = 1$ (in this case we can put $B_3(x, \rho) = B_4(x, \rho) = 0$) or $N = 2$. Then it seems natural to introduce the following definition.

Let

$$\mathcal{B} = \mathcal{B}\{f_1(x, \varepsilon), \dots, f_{M1}(x, \varepsilon); \varepsilon g_1(x, \varepsilon), \dots, \varepsilon g_{M2}(x, \varepsilon)\}$$

be a subalgebra of G_1^* generated by the elements[3]

$$f_1(x, \varepsilon), \dots, f_{M1}(x, \varepsilon),\ g_1(x, \varepsilon), \dots, g_{M2}(x, \varepsilon)$$

so that, for any element $f^*(x, \varepsilon) \in \mathcal{B}$, we have an asymptotic expansion up to $O_{\mathcal{D}'}(\varepsilon^N)$ in the form

$$f^*(x, \varepsilon) = \sum_{r=1}^{2} \Big\{ \sum_j B_{1jr}(x, \rho) f_{1jr}(x - a^r) + \varepsilon \sum_k B_{2kr}(x, \rho) f_{2kr}(x - a^r) \Big\} + O_{\mathcal{D}'}(\varepsilon^N),$$

where $\rho = (a^2 - a^1)/\varepsilon$, $B_{1jr}(x, \rho)$ and $B_{2jr}(x, \rho)$ are smooth functions, $f_{1jr}(x)$ and $f_{2kr}(x)$ belong to E_0 and are independent of the choice of $f^*(x, \varepsilon)$, and $\sum_j$, $\sum_k$ are finite; $N = 1$ or $N = 2$.

DEFINITION 2.3. We say that a subalgebra $\mathcal{B} \in G_1^*$ is a *finitely generated asymptotic distributional differential subalgebra* modulo $O_{\mathcal{D}'}(\varepsilon^N)$ if the set $\mathcal{A}_N = \{(f_{1j}(x),\ f_{2k}(x))\}$ is a *finite-dimensional* vector subspace in $\mathcal{D}'(\mathbb{R})$. The basis elements $(f_{1j}(x); f_{2k}(x))$ are called *distributional generators* of an *asymptotic subalgebra*.

One can describe *asymptotic subalgebras* with two points of singularity.

LEMMA 2.6. *Let $H_k(x - a_k, \varepsilon)$ be regularizations of the Heaviside functions $H(x - a^k)$, $k = 1, 2$. Then*

$$\mathcal{B}_6 = \mathcal{B}_6\{1,\ H_1(x - a^1, \varepsilon),\ H_2(x - a^2, \varepsilon)\}$$

is the asymptotic differential subalgebra mod $O_{\mathcal{D}'}(\varepsilon)$ *with distributional generators* $(1, H(x - a^1), H(x - a^2))$. *That is,*

$$\begin{aligned}
&\big(H_1(x - a^1, \varepsilon)\big)^{m_1} \big(H_2(x - a^2, \varepsilon)\big)^{m_2} \\
&\qquad = B_1^{m_1, m_2}(\rho) H(x - a^1) + B_2^{m_1, m_2}(-\rho) H(x - a^2) + O_{\mathcal{D}'}(\varepsilon), \\
&\big(a_1(x) + b_1(x) H_1(x - a^1, \varepsilon) + b_2(x) H_2(x - a^2, \varepsilon)\big)^n \\
&\qquad = a_2(x) + B_1(x, \rho) H(x - a^1) + B_2(x, \rho) H(x - a^2) + O_{\mathcal{D}'}(\varepsilon),
\end{aligned}$$

where $\rho = (a^2 - a^1)/\varepsilon$, $B_k^{m_1, m_2}((-1)^k \rho)$ is given by (1.9), *$B_k(x, \rho)$ are smooth functions and can be found explicitly* (*see Lemma* 1.1), *$k = 1, 2$, $a_1(x)$, $b_1(x)$, $b_2(x)$, and $a_2(x) = (a_1(x))^n$ are smooth functions, and $m_1, m_2, n \in \mathbb{N}$.*

The proof follows from the results of Lemma 1.1.

LEMMA 2.7. *Let $\delta_k(x - a^k, \varepsilon) = \varepsilon^{-1} \omega_k((x - a^k)/\varepsilon)$ be regularizations of the Dirac delta functions $\delta(x - a^k)$, $\omega_1(z), \omega_2(z) \in \Omega(\mathbb{R})$, $k = 1, 2$. Then*

$$\mathcal{B}_7 = \mathcal{B}_7\{1;\ \varepsilon\delta_1(x - a^1, \varepsilon),\ \varepsilon\delta_2(x - a_2, \varepsilon)\}$$

---

[3]Here the notation has the same meaning as in Definition 2.2.

is the asymptotic differential subalgebra mod $O_{\mathcal{D}'}(\varepsilon^2)$ *with distributional generators* $(1; \delta(x-a^1), \delta(x-a^2))$. *That is,*

$$\begin{aligned}&\big(\varepsilon\delta_1(x-a^1,\varepsilon)\big)^{m_1}\big(\varepsilon\delta_2(x-a_2,\varepsilon)\big)^{m_2}\\&\qquad = B^{m_1,m_2}(\rho)\big(\varepsilon\delta_1(x-a^1)+\varepsilon\delta_2(x-a^2)\big)+O_{\mathcal{D}'}(\varepsilon^2),\\&\big(a_1(x)+b_1(x)\varepsilon\delta_1(x-a^1,\varepsilon)+b_2(x)\varepsilon\delta_2(x-a^2,\varepsilon)\big)^n\\&\qquad = a_2(x)+B_1(x,\rho)\varepsilon\delta(x-a^1)+B_2(x,\rho)\varepsilon\delta(x-a^2)+O_{\mathcal{D}'}(\varepsilon^2),\end{aligned}$$

where $\rho = (a^2-a^1)/\varepsilon$,

$$B^{m_1,m_2}(\rho) = \frac{1}{2}\int \big(\omega_1(\eta)\big)^{m_1}\big(\omega_2(-\rho+\eta)\big)^{m_2}\,d\eta$$

is a constant, $a_1(x)$, $b_1(x)$, $b_2(x)$, *and* $a_2(x) = (a_1(x))^n$ *are smooth functions, and* $B^{m_1,m_2}(\rho)$ *are smooth functions and can be obtained explicitly* (*see Section* 1), $k=1,2$, $n \in \mathbb{N}$.

The proof follows from the results of Section 1.

LEMMA 2.8. *Let* $\delta_k(x-a^k,\varepsilon)$ *be regularizations of the Dirac delta functions* $\delta(x-a^k)$, *and let* $H_k(x-a^k,\varepsilon)$ *be regularizations of the Heaviside function* $H(x)$, $k=1,2$. *Then*

$$\mathcal{B}_8 = \mathcal{B}_8\{1;\ \varepsilon\delta_k(x-a^k,\varepsilon),\ \varepsilon H_k(x-a^k,\varepsilon),\ \ k=1,2\}$$

is the asymptotic differential subalgebra mod $O_{\mathcal{D}'}(\varepsilon^2)$ *with distributional generators* $(1; \delta(x-a^k), H(x-a^k),\ k=1,2)$. *That is,*

$$\begin{aligned}&\Big(a_1(x)+\sum_{k=1}^{2}\big(b_k(x)\varepsilon\delta(x-a^k,\varepsilon)+c_k(x)\varepsilon H_k(x-a^2,\varepsilon)\big)\Big)^n\\&\qquad = a_2(x)+\sum_{k=1}^{2}\big(B_k(x,\rho)\varepsilon\delta(x-a^k)+d_k(x)\varepsilon H(x-a^k)\big)+O_{\mathcal{D}'}(\varepsilon^2),\end{aligned}$$

where $\rho = (a^2-a^1)/\varepsilon$, $a_1(x)$, $b_k(x)$, $c_k(x)$, $a_2(x) = (a_1(x))^n$, *and* $d_k(x) = n(a_1(x))^{n-1}c_k(x)$ *are smooth functions, and* $B_k(x,\rho)$ *are smooth functions and can be calculated explicitly* (*see Section* 1), $k=1,2$, $n\in\mathbb{N}$.

The proof follows from the results of Section 1.

Obviously, to prove Lemmas 2.1–2.8, it suffices to consider linear combinations of regularizations with constant coefficients.

Now we discuss in which sense our asymptotic subalgebras are algebras and to what algebra they belong.

In the sense of the algebra of asymptotic distributions and its natural generalizations (see the Introduction and Subsection 2.4), the product of the generators of subalgebras is, in general, not defined, since these generators, generally speaking, are not asymptotic distributions (they are sums of infinitely many asymptotic series, which are asymptotic distributions).

In the framework of the Colombeau construction, the products of distributional generators are defined as the equivalence class of the product of the smooth representatives of cosets of the factors. Hence we can study the algebra generated by the distributional generators. However, if the result of multiplication is considered with accuracy $O(\varepsilon^\alpha)$ for any $\alpha > 0$, then the algebra obtained is not associative (see the example below). But if we consider the subalgebra generated by the generators in the sense of the Colombeau theory, then this subalgebra will not be finitely generated.

EXAMPLE 2.1.

$$\begin{aligned} \varepsilon\delta(x,\varepsilon) &= \varepsilon\delta(x) + O_{\mathcal{D}'}(\varepsilon^2), \\ \big(\varepsilon\delta(x,\varepsilon)\big)^2 &= A\varepsilon\delta(x) + O_{\mathcal{D}'}(\varepsilon^2), \\ \big(\varepsilon\delta(x,\varepsilon)\big)^3 &= B\varepsilon\delta(x) + O_{\mathcal{D}'}(\varepsilon^2), \end{aligned}$$

where $A = \int \omega^2(\eta)\,d\eta$, $B = \int \omega^3(\eta)\,d\eta$, and we have

$$\varepsilon\delta(x)\big(\varepsilon\delta(x)\big)^2 \neq \big(\varepsilon\delta(x)\big)^3 + O_{\mathcal{D}'}(\varepsilon^2).$$

2.4. Algebra of asymptotic distributions. In [**17**] we constructed an associative algebra E_0^* of *asymptotic distributions* with identity and without zero divisors generated by the space E_0 (with one singular point). To construct the algebra E_0^*, we calculate the weak asymptotics up to $O_{\mathcal{D}'}(\varepsilon^\infty)$ of the product of regularizations of elements from h_0^*, that is, an *associative commutative differential algebra with identity and without zero divisors* generated by the space h_0. The elements $f^*(x,\varepsilon)$ of this algebra are finite sums of finite products of functions $f(x,\varepsilon) \in h_0$, $\varepsilon > 0$.

Here we outline the construction of the algebra E_0^*.

In [**17**] the space h_0^* of harmonic regularizations to elements of h_0 was introduced. The elements $f(x,\varepsilon)$ of h_0^* were defined as solutions of the Laplace equation

$$\left(\frac{\partial^2}{\partial x^2} + \frac{\partial^2}{\partial \varepsilon^2}\right) f(x,\varepsilon) = 0, \qquad f(x_0,0) = f(x) \in h_0.$$

Recall that

$$h_0 = \operatorname{span}\big\{(x - c_k \pm i\varepsilon)^\lambda \log^p(x - c_k \pm i\varepsilon), \lambda \in \mathbb{R}, p = 0, 1, \dots, \varepsilon > 0\big\},$$

where span denotes the linear span, $z = x + i\varepsilon$, $\overline{z} = x - i\varepsilon$, $\log z = \log|z| + i\arg z$, $\operatorname{Im} z \neq 0$, and $\arg z \in [0,\pi)$. The elements of this space belong to the Vladimirov algebras [**127**].

The product of distributions from the space E_0 of distributions is defined as the weak asymptotic expansion of the product of regularizations to the distributions as $\varepsilon \to +0$.

THEOREM 2.1. *Each element $f^*(x,\varepsilon)$ of the algebra h_0^* has a unique canonical representation in the form of a weak asymptotic series*

$$f^*(x,\varepsilon) = \sum_{m=0}^{\infty} \sum_{n=0}^{n_0(m)} f_{\alpha_m,n}(x)\varepsilon^{\alpha_m}\log^{n_0(m)-n}\varepsilon, \tag{2.2}$$

where $\alpha_m \in \mathbb{R}$ is an increasing sequence, $n_0(m) \in \mathbb{N}_0$, and the coefficients $f_{\alpha_m,n}(x)$ are distributions from the space E_0.

Relation (2.2) is understood in the weak sense, and we have

$$\langle f_{\alpha_m,n}(x), \varphi(x)\rangle = \lim_{\varepsilon\to+0} \Big[\langle f^*(x,\varepsilon), \varphi(x)\rangle$$

$$- \sum_{k=0}^{m-1}\sum_{j=0}^{n-1}\langle f_{\alpha_k,j}(x),\varphi(x)\rangle\varepsilon^{\alpha_k}\log^{n_0-j}\varepsilon\Big]\varepsilon^{-\alpha_m}\log^{-(n_0-n)}\varepsilon$$

for all $\varphi(x)\in\mathcal{D}(\mathbb{R})$ and all $m\geq 0$, $n=0,1,\dots,n_0$.

In particular, if an element $f^*(x,\varepsilon)\in h_0^*$ is such that $f^*(x,\varepsilon)\in h_0$, then its canonical representation given by (2.2) is its Taylor expansion.

This theorem was proved in [**17**].

DEFINITION 2.4. 1) Either the weak limit

$$f_\varepsilon^*(x)=A_0^*[f^*(x,\varepsilon)]=\lim_{\varepsilon\to+0}f^*(x,\varepsilon)\in E_0,\qquad f^*(x,\varepsilon)\in h_0,$$

2) or, for $f^*(x,\varepsilon)\in h_0^*\setminus h_0$, the asymptotic expansion

$$f_\varepsilon^*(x)=A_0^*[f^*(x,\varepsilon)]=\sum_{m=0}^{\infty}\sum_{n=0}^{n_0}f_{\alpha_m,n}(x)\varepsilon^{\alpha_m}\log^{n_0-n}\varepsilon,\quad \varepsilon\to+0,$$

given by Theorem 2.1, is called an *asymptotic distribution* [**17, 23**].

Here the subscript ε in $f_\varepsilon^*(x)$ means that the expansion depends on an asymptotic sequence, which is a function of the regularization parameter ε.

By $E_0^*=A_0^*[h_0^*]$ we denote the *linear space of all asymptotic distributions* $f_\varepsilon^*(x)=A_0^*[f^*(x,\varepsilon)]$, $f^*(x,\varepsilon)\in h_0^*$.

It was proved in [**17, 23**] that the mapping $A_0^*:h_0^*\to E_0^*$ is an *isomorphism* and $A_0^*[h_0]=E_0$, i.e., the linear span of the set E_0 of associated homogeneous distributions is a subspace of the space E_0^*.

Now we introduce the operation of multiplication on the space E_0^* of asymptotic distributions, i.e., for distributions from E_0.

DEFINITION 2.5. The product of asymptotic distributions $f_{\varepsilon,k}^*$, $k=1,2$, defined as the asymptotic distribution

$$f_\varepsilon^*(x)=f_{\varepsilon,1}^*(x)\circ f_{\varepsilon,2}^*(x)=A_0^*\big[f_1^*(x,\varepsilon)f_2^*(x,\varepsilon)\big]\in E_0^*, \tag{2.3}$$

where $f_k^*(x,\varepsilon)\in h_0^*$ are regularizations of the asymptotic distributions $f_{\varepsilon,k}^*$, $k=1,2$, is called the *product of asymptotic distributions* $f_{\varepsilon,1}^*(x)$ and $f_{\varepsilon,2}^*(x)$ from E_0^*.

Since h_0^* is an associative and commutative algebra of functions and $A_0^*:h_0^*\to E_0^*$ is an isomorphism, the multiplication (2.3) is associative and commutative. The multiplication (2.3) turns the space of asymptotic distributions E_0^* into an algebra.

EXAMPLE 2.2 (see [**120, 123**]). 1)

$$\begin{aligned}(x+i0)^{-m}\circ(x-i0)^{-n}&=A_0^*[(x+i\varepsilon)^{-m}(x-i\varepsilon)^{-n}]\\
&=(-1)^{m-1}\Big(\frac{i}{2}\Big)^{m+n-2}\pi\sum_{s=0}^{m+n-2}B_s^+\frac{i^s}{s!}\delta^{(s)}(x)\varepsilon^{-m-n+s+1}\\
&\quad+(-1)^{n-1}\frac{i\pi}{(m+n-1)!}\big[(-1)^m+2^{-m-n+2}B_{m+n-1}^+\big]\delta^{(m+n-1)}(x)\\
&\quad+P(x^{-m-n})+(-1)^{m-1}\Big(\frac{i}{2}\Big)^{m+n-1}\sum_{s=m+n}^{\infty}i^s\big[(-1)^s(x+i0)^{-s-1}B_s^+\\
&\quad-(x-i0)^{-s-1}B_s^-\big]\varepsilon^{-m-n+s+1},\quad \varepsilon\to+0,\end{aligned}$$

where $m, n = 1, 2, \dots$; $B_s^+ = B_s(m,n)$, $B_s^- = B_s(n,m)$, and the coefficients $B_s(m,n)$ have the form

$$B_s(m,n) = \begin{cases} \sum_{k=0}^{s} C_{m+n-k-2}^{n-1} C_s^k (-2)^k, & 0 \le s \le m-1, \\ \sum_{k=0}^{m-1} C_{m+n-k-2}^{n-1} C_s^k (-2)^k, & s \ge m. \end{cases}$$

2)

$$\begin{aligned} x \circ \delta(x) &= A_0^*\left[\frac{1}{2}\big((x+i\varepsilon)+(x-i\varepsilon)\big)\cdot\frac{-1}{2\pi i}\left(\frac{1}{x+i\varepsilon}-\frac{1}{x-i\varepsilon}\right)\right] \\ &= A_0^*\left[\frac{\varepsilon}{2\pi}\left(\frac{1}{x+i\varepsilon}+\frac{1}{x-i\varepsilon}\right)\right] \\ &= \sum_{k=0}^{\infty}(-1)^k\left(\frac{1}{\pi}P\left(\frac{1}{x^{2k+1}}\right)\varepsilon^{2k+1}+\frac{1}{(2k+1)!}\delta^{(2k+1)}(x)\varepsilon^{2k+2}\right), \quad \varepsilon \to +0. \end{aligned}$$

3)

$$\begin{aligned} [\delta(x)]^2 &= A_0^*\left[\left(\frac{-1}{2\pi i}\left(\frac{1}{x+i\varepsilon}-\frac{1}{x-i\varepsilon}\right)\right)^2\right] \\ &= -\frac{1}{2\pi^2}A_0^*\left[\frac{1}{2}\left(\frac{1}{x+i\varepsilon}+\frac{1}{x-i\varepsilon}\right)-\frac{1}{(x+i\varepsilon)(x-i\varepsilon)}\right] \\ &= -\frac{1}{2\pi^2}A_0^*\left[\frac{1}{2}\left(\frac{1}{x+i\varepsilon}+\frac{1}{x-i\varepsilon}\right)\right]+\frac{1}{2\pi^2}A_0^*\left[\frac{-1}{2i\varepsilon}\left(\frac{1}{x+i\varepsilon}-\frac{1}{x-i\varepsilon}\right)\right] \\ &= \frac{1}{2\pi\varepsilon}\delta(x)+\frac{1}{2\pi^2}\sum_{k=1}^{\infty}(-1)^k\left(\frac{\pi}{(2k)!}\delta^{(2k)}(x)\varepsilon^{2k-1}+P\left(\frac{1}{x^{2k+2}}\right)\varepsilon^{2k}\right), \quad \varepsilon \to +0. \end{aligned}$$

Similarly to the approach used in [**120, 121, 122, 17, 23**], one can consider asymptotic distributions $f_\varepsilon^*(x)$ from E_0^* as continuous linear vector functionals on the space $\mathcal{D}_*(\mathbb{R})$ of test function-asymptotics,

$$\varphi_{*,\varepsilon}(x) = \sum_{m=0}^{\infty}\sum_{n=0}^{\infty}\varphi_{\alpha_m,n}(x)\varepsilon^{\alpha_m}\log^{-n}\varepsilon, \quad \varepsilon \to +0, \tag{2.4}$$

where $\varphi_{\alpha_m,n}(x) \in \mathcal{D}(\mathbb{R})$ and the sums in (2.4) are upper and lower finite, i.e., $\varphi_{\alpha_\infty,\infty}(x) = 0$.

In this case, for all $\varphi_{*,\varepsilon}(x) \in \mathcal{D}_*(\mathbb{R})$, we assume

$$\langle f_\varepsilon^*(x), \varphi_{*,\varepsilon}(x)\rangle = P.f. \lim_{\varepsilon\to+0}\langle f^*(x,\varepsilon), \varphi_*(x,\varepsilon)\rangle,$$

where $\varphi_*(x,\varepsilon)$ is understood as an expression of the form (2.4) ($\varepsilon > 0$ is a parameter) and $P.f.$ is the finite part of the weak limit in the sense of Hadamard.

Obviously, the last sum is equal to

$$\langle f_\varepsilon^*(x), \varphi_{*,\varepsilon}(x)\rangle = \sum_{m=-\infty}^{\infty}\sum_{n=0}^{\infty}\langle f_{\alpha_m,n}(x), \varphi_{-\alpha_m,-n}(x)\rangle,$$

i.e., it is finite, since it contains finitely many terms.

Let us consider the space E_0^* of asymptotic distributions $f_\varepsilon^*(x)$ as an infinite-dimensional linear space of *finite-from-the-left* vector-valued distributions $f^*(x) =$

$(\dots, f_{\alpha_m,n}(x), \dots)$, where $f_{\alpha_m,n}(x) \in E_0$, $m = 0, 1, \dots$, $n = 0, 1, \dots n_0(m)$, $\alpha_m \in \mathbb{R}$ is an increasing sequence, and $n_0(m) \in \mathbb{N}_0$ (see Theorem 2.1).

Each distribution $f(x)$ from the subspace E_0 can be identified with a vector-valued distribution $f^*(x)$ from E_0^* all of whose components are equal to zero except for $f_{0,0}(x) = f(x)$.

Elements $f^*(x) \in E_0^*$ are continuous linear vector functionals over the space $\mathcal{D}_*(\mathbb{R})$ of test vector functions of the form $\varphi_*(x) = (\dots, \varphi_{\alpha_m,n}(x), \dots)$, which have only finitely many nonzero components $\varphi_{\alpha_m,n}(x) \in \mathcal{D}(\mathbb{R})$. Further generalizations can be found in [**64, 65, 66**].

3. Dynamics of solitary waves

In this section we present some examples of asymptotic solutions describing the propagation of solitary nonlinear waves (namely, shock waves, solitons, and kinks). The last is used to describe the phase transition problem in the framework of its phase field system regularization. These problems were solved earlier by using the method of [**18, 91**]. Here we only want to demonstrate our new technique.

3.1. Dynamics of shock waves. We start a detailed consideration of the weak asymptotics method from the simplest problem of describing the evolution of a shock wave. The first model example is the Hopf equation

$$\frac{\partial u}{\partial t} + \frac{\partial u^2}{\partial x} = 0 \tag{3.1}$$

with piecewise constant initial data

$$u\big|_{t=0} = u^0 + e^0 H(-x + x^0). \tag{3.2}$$

The exact solution of this problem is well known:

$$u = u^0 + e^0 H(-x + x^0 + Vt), \tag{3.3}$$

where $V = 2u^0 + e^0$ and the inequality $e^0 > 0$ is a stability condition.

Let us obtain the same result by the weak asymptotics method. Following the general approach (see the Introduction), we regularize the initial value and write the solution of the regularized problem in the form

$$u(x, t, \varepsilon) = u_0 + eH(-x + \varphi, \varepsilon), \tag{3.4}$$

where $u_0 = u_0(x, t)$, $e = e(x, t)$, and $\varphi = \varphi(t)$ are unknown smooth functions. By formulas from Subsection 1.3, we readily obtain

$$\begin{aligned} u(x, t, \varepsilon) &= u_0 + eH(-x + \varphi) + O_{\mathcal{D}'}(\varepsilon), \\ u^2(x, t, \varepsilon) &= u_0^2 + (2u_0 e + e^2)H(-x + \varphi) + O_{\mathcal{D}'}(\varepsilon). \end{aligned} \tag{3.5}$$

By substituting the weak asymptotics of u and u^2 into (3.1), we obtain

$$\begin{aligned} &\left(e\frac{d\varphi}{dt} - (2u_0 e + e^2) \right) \delta(-x + \varphi) \\ &\quad + \left(\frac{\partial e}{\partial t} + \frac{\partial}{\partial x}(2u_0 e + e^2) \right) H(-x + \varphi) + \frac{\partial u_0}{\partial t} + \frac{\partial u_0^2}{\partial x} = O_{\mathcal{D}'}(\varepsilon). \end{aligned} \tag{3.6}$$

Let us equate the coefficients of δ, H, and 1 in (3.6) with zero:

$$e\frac{d\varphi}{dt} = 2u_0 e + e^2 \qquad \text{for } x = \varphi(t), \tag{3.7}$$

$$u_{0t} + (u_0^2)_x = 0 \qquad \text{for } x > \varphi(t), \tag{3.8}$$

$$(3.9) \qquad (u_0+e)_t + \big((u_0+e)^2\big)_x = 0 \qquad \text{for } x < \varphi(t).$$

Obviously, the solutions of Eqs. (3.8), (3.9) with constant initial data are constant, $u_0 = u^0$, $e = e^0$, while the assumption $e^0 > 0$ arises as a consequence of the condition that Eq. (3.7) is well defined; namely, both of the functions u_0 and e can be defined on the line $x = \varphi(t)$. It is also clear that (3.7) is precisely the Rankine–Hugoniot condition

$$(3.10) \qquad \frac{d\varphi}{dt} = \frac{[u^2]}{[u]}.$$

It is very important that the limiting equations (3.7)–(3.9) do not depend on the way in which the regularization $H(x,\varepsilon)$ of the Heaviside function $H(x)$ is used.

Let us consider the natural generalization of problem (3.1), (3.2):

$$(3.11) \qquad \frac{\partial u}{\partial t} + \frac{\partial}{\partial x} f(u) = 0.$$
$$u\big|_{t=0} = u^0(x) + e^0(x)H(-x+x^0),$$

where we assume that the functions f, u^0, and e^0 are smooth, the function $f(u)$ is convex ($f''(u) \geq 0$), and $e^0(x) > 0$. The solution of this problem is also well known:

$$(3.3') \qquad u = u_0(x,t) + e(x,t)H\big(-x+\varphi(t)\big),$$

where u_0 and e are determined by the following equations similar to (3.8) and (3.9):

$$(3.8') \qquad \frac{\partial u_0}{\partial t} + \frac{\partial}{\partial x} f(u_0) = 0 \qquad \text{for } x > \varphi(t),$$
$$(3.9') \qquad \frac{\partial}{\partial t}(u_0+e_0) + \frac{\partial}{\partial x} f(u_0+e_0) = 0 \qquad \text{for } x < \varphi(t),$$

and the motion of the shock wave front is governed by the Hugoniot condition

$$(3.10') \qquad \frac{d\varphi}{dt} = \frac{[f(u)]}{[u]}.$$

The above elementary calculations show that for problem (3.11) the weak asymptotics method gives the same result as classical methods. Technically, the only new point here is that we use the asymptotic expansion (see Subsection 1.3)

$$f\big(u(x,t,\varepsilon)\big) = f\big(u_0(x,t)\big) + \big\{f\big(u_0(x,t)+e(x,t)\big) - f\big(u_0(x,t)\big)\big\}H\big(-x+\varphi(t)\big) + O_{\mathcal{D}'}(\varepsilon),$$

instead of the second formula in (3.5).

To be precise, we note that Eqs. (3.8′), (3.9′) supplemented with the initial conditions

$$u_0\big|_{t=0} = u^0(x) \quad \text{for } x > \varphi(0), \qquad (u_0+e)\big|_{t=0} = u^0(x) + e^0(x) \quad \text{for } x < \varphi(0)$$

have the classical solution for $t \in (0,T)$ under the assumption that

$$-\inf_{x>\varphi(0)} \big({u^0}'_x(x) f''_u(u^0(x))\big) < T^{-1},$$
$$-\inf_{x<\varphi(0)} \big\{\big(u^0(x)+e^0(x)\big)'_x f''_u\big(u^0(x)+e^0(x)\big)\big\} < T^{-1}.$$

Precisely as in problem (3.1), (3.2), the entropy condition $e^0 > 0$ is necessary for the well-posedness of Eq. (3.10′), which, in contrast to (3.7), is now nonlinear.

3.2. Dynamics of ε–δ-solitons. We continue our detailed consideration of the weak asymptotics method by studying the problem of motion of a soliton type solution of the following KdV type equation:

$$\frac{\partial u}{\partial t}+\frac{\partial u^m}{\partial x}+\varepsilon^2\frac{\partial^3 u}{\partial x^3}=0, \qquad m\geq 2. \tag{3.12}$$

For $m=2$, Eq. (3.12) is the KdV equation whose exact solution with localized fast variation is well known. This is the one-parametric family of solitary waves

$$u=\frac{6\beta^2}{\cosh^2(\beta(x-Vt-x^0)/\varepsilon)}, \qquad V=4\beta^2, \quad \beta>0, \tag{3.13}$$

which is invariant under translations. As $\varepsilon\to 0$ the weak limit of u/ε is equal to $12\beta\delta(x-(Vt+x^0))$. Hence it is natural to treat the function u as a special regularization of the ε–δ-singularity. Obviously, the exact solution (3.13) describes the dynamics of the support of this singularity

$$x=4\beta^2 t+x^0.$$

Our nearest aim is to obtain this result by using our method. As mentioned in the Introduction, here the key point is the definition of a weak asymptotic solution, which admits passing to the limit in the weak sense as $\varepsilon\to 0$; see [**26**].

DEFINITION 3.1. A *weak asymptotic solution* mod $O_{\mathcal{D}'}(\varepsilon^2)$ of Eq. (3.12) is defined to be a sequence $u(x,t,\varepsilon)\in C^\infty\big((0,T);C^\infty(\mathbb{R}^1_x)\big)$ with $\varepsilon>0$ such that for any function $\psi(x)\in\mathcal{D}(\mathbb{R}^1)$ the following relations hold:

$$\frac{d}{dt}\int u\psi\,dx-\int u^m\frac{\partial\psi}{\partial x}\,dx=O(\varepsilon^2), \tag{3.14}$$

$$\frac{d}{dt}\int u^2\psi\,dx-\frac{2m}{m+1}\int u^{m+1}\frac{\partial\psi}{\partial x}\,dx+3\int\Big(\varepsilon\frac{\partial u}{\partial x}\Big)^2\frac{\partial\psi}{\partial x}\,dx=O(\varepsilon^2). \tag{3.15}$$

We start from the simplest possible situation: $m=2$ and $u\big|_{t=0}=O(\varepsilon^\infty)$ outside the $\varepsilon^{1-\mu}$-neighborhood of the point x^0, $\mu>0$. We write the asymptotic solution of (3.12), which is smooth for $\varepsilon\neq 0$, in the form

$$u(x,t,\varepsilon)=A\omega\Big(\beta\frac{x-\varphi(t)}{\varepsilon}\Big), \tag{3.16}$$

where the constants $A,\beta>0$ and the function φ, $\varphi\big|_{t=0}=x^0$, are to be found, $\omega(\eta)\in C^\infty(\mathbb{R}^1)$, and

$$\int\omega(\eta)\,d\eta=1, \qquad \frac{1}{\varepsilon}\omega\Big(\frac{x}{\varepsilon}\Big)\to\delta(x)\quad\text{in } \mathcal{D}' \text{ as } \varepsilon\to 0. \tag{3.17}$$

As stated in the Introduction, to describe the ε–δ-solitons, it is necessary to choose a special sequence from all possible δ type sequences. For $m=2$ such a special sequence is

$$\omega(\eta)=1/2\cosh^2(\eta). \tag{3.18}$$

Elementary calculations (see Subsection 1.3) show that

(3.19)
$$u^n(x,t,\varepsilon) = \varepsilon a_n \frac{A^n}{\beta}\delta(x-\varphi) + O_{\mathcal{D}'}(\varepsilon^2), \qquad a_n = \int \omega^n(\eta)\,d\eta,$$
$$\Bigl(\varepsilon\frac{\partial}{\partial x}u(x,t,\varepsilon)\Bigr)^2 = \varepsilon a_2' A^2\beta\delta(x-\varphi) + O_{\mathcal{D}'}(\varepsilon^2), \qquad a_2' = \int \bigl(\omega_\eta'(\eta)\bigr)^2\,d\eta.$$

These relations and (3.14), (3.15) with $m=2$ imply

(3.20)
$$\varepsilon\Bigl\{-\varphi_t'\frac{A}{\beta} + a_2\frac{A^2}{\beta}\Bigr\}\delta'(x-\varphi) = O_{\mathcal{D}'}(\varepsilon^2),$$
$$\varepsilon\Bigl\{-\varphi_t' a_2\frac{A^2}{\beta} + \frac{4}{3}a_3\frac{A^3}{\beta} - 3a_2'A^2\beta\Bigr\}\delta'(x-\varphi) = O_{\mathcal{D}'}(\varepsilon^2).$$

Thus we arrive at the relations

$$\varphi_t' = a_2 A, \qquad \varphi_t' = \Bigl(\frac{4}{3}a_3 A - 3a_2'\beta^2\Bigr)a_2^{-1} \tag{3.21}$$

compatible with each other for

$$\beta^2 = \gamma A, \qquad \gamma = \Bigl(\frac{4}{3}a_3 - a_2^2\Bigr)\frac{1}{3a_2'}. \tag{3.22}$$

By calculating the numerical values of the coefficients a_2, a_3, a_2' in the case of (3.18), we obtain $a_2 = 1/3$, $a_3 = 2/15$, $a_2' = 4/15$, and $\gamma = 1/12$. Hence we have the same relations between the amplitude, the velocity of motion of the solitary wave, and the coefficient β as for the exact solution (3.13). It is also easy to verify that, for our choice of ω, (3.16) is the exact solution of the KdV equation.

Practically, the case $m > 2$ does not differ from that studied above. We only need to introduce a special regularization of the ε–δ-singularity. To this end, we define the function $z = z(\eta)$ as a positive solution of the problem

$$\frac{d^2 z}{d\eta^2} - z + z^m = 0, \qquad z \to 0 \quad \text{as} \quad \eta \to \pm\infty. \tag{3.23}$$

It is easy to see that the function z is even and infinitely smooth for integer $m \geq 2$ and its maximal value is $z(0) = ((m+1)/2)^{1/(m-1)}$. We set

$$\omega(\eta) = \frac{z(\eta)}{\int z(\eta')\,d\eta'} \tag{3.24}$$

and represent the asymptotic solution of (3.12) in the form (3.16), where A is assumed to be positive for even m and different from zero for odd m. By using formula (3.19), we arrive at the relations similar to (3.20),

$$\varphi_t' = a_m A^{m-1}, \qquad \varphi_t' = \frac{2m}{m+1}\frac{a_{m+1}}{a_2}A^{m-1} - 3\frac{a_2'}{a_2}\beta^2, \tag{3.25}$$

and the compatibility conditions for (3.25),

$$\beta^2 = \gamma^{m-1}A^{m-1}, \qquad \gamma^{m-1} = \Bigl(\frac{2m}{m+1}a_{m+1} - a_2 a_m\Bigr)\frac{1}{3a_2'}. \tag{3.26}$$

To prove that conditions (3.26) hold, it is necessary to prove that

$$\frac{2m}{m+1}a_{m+1} > a_2 a_m. \tag{3.27}$$

Note that multiplying (3.23) by z and z'_η and integrating with respect to η, we obtain the relations

$$\int (z'_\eta)^2\, d\eta + \int z^2\, d\eta = \int z^{m+1}\, d\eta, \qquad \int z^2\, d\eta = \frac{2}{m+1}\int z^{m+1}\, d\eta.$$

Now (3.24) and the relation $\int z\, d\eta = \int z^m\, d\eta$, which obviously follows from (3.23), imply

$$\begin{aligned}\frac{2m}{m+1}a_{m+1} - a_2 a_m &= \Big(\int z\, d\eta\Big)^{-(m+1)} 2\frac{m-1}{m+1}\int z^{m+1}\, d\eta \\ &= 2\frac{m-1}{m+1}a_{m+1}.\end{aligned} \tag{3.28}$$

As for the KdV equation, it is easy to show that the solution (3.16), (3.24)–(3.26) obtained above is exact. Moreover, simple calculations show that relations (3.25), just as (3.21) for $m = 2$, are exactly the energy relations

$$\begin{aligned}&\frac{d}{dt}\int xu\, dx = \int u^m\, dx, \\ &\frac{d}{dt}\int xu^2\, dx = \frac{2m}{m+1}\int u^{m+1}\, dx - 3\int \Big(\varepsilon\frac{\partial u}{\partial x}\Big)^2 dx,\end{aligned} \tag{3.29}$$

also calculated for (3.16).

A more interesting situation arises when we describe a solitary wave moving against a variable background, which distorts the soliton so that its amplitude, velocity, and the "width" β become functions of t. Moreover, there arises a "tail," i.e., a smoothed shock wave of amplitude $\sim \varepsilon$ moving after the distorted soliton. The causes for the appearance of such effects are discussed in detail in [**91, 89, 90, 93, 94**]. Therefore, we simply write the leading term of the asymptotics:

$$u(x,t,\varepsilon) = u_0(x,t) + A\omega\Big(\beta\frac{x-\varphi(t)}{\varepsilon}\Big) + \varepsilon u^-(x,t)H(\varphi(t)-x,\varepsilon) + O_{\mathcal{D}'}(\varepsilon^2). \tag{3.30}$$

Here u_0, u^- are assumed to be smooth, A, $\beta > 0$, φ are some functions of t, $\omega(\eta)$ is defined in the form (3.24), and $H(\tau,\varepsilon)$ is a regularization of the Heaviside function.

By carrying out the simple calculations described in Subsection 1.3, we find

$$\begin{aligned}u^n(x,t,\varepsilon) &= u_0^n(x,t) + \varepsilon L_n(t)\delta(x-\varphi) \\ &\qquad + \varepsilon u_0^{n-1}(x,t)u^-(x,t)H(\varphi - x) + O_{\mathcal{D}'}(\varepsilon^2), \\ \Big(\varepsilon\frac{\partial}{\partial x}u(x,t,\varepsilon)\Big)^2 &= \varepsilon a_2'\beta A^2\delta(x-\varphi) + O_{\mathcal{D}'}(\varepsilon^2),\end{aligned} \tag{3.31}$$

where

$$L_n(t) = \frac{A}{\beta}\{a_n A^{n-1} + nu_0^{n-1} + Au_0\ell_n\}\Big|_{x=\varphi(t)}, \qquad \ell_n = \sum_{i=2}^{n-1} c_n^i a_i A^{i-2}u_0^{n-i-1},$$

and c_n^i are binomial coefficients.

This and (3.14), (3.15) imply relations similar to (3.20):

$$\begin{aligned}&\varepsilon\Big\{-\varphi_t\frac{A}{\beta} + L_m\Big\}\delta'(x-\varphi) + \varepsilon\Big\{\frac{d}{dt}\Big(\frac{A}{\beta}\Big) + (\varphi_t - mu_0^{m-1})u^-\Big\}\delta(x-\varphi) \\ &\quad + \varepsilon\Big\{\frac{\partial u^-}{\partial t} + m\frac{\partial}{\partial x}(u_0^{m-1}u^-)\Big\}H(\varphi - x) + \Big\{\frac{\partial u_0}{\partial t} + \frac{\partial u_0^m}{\partial x}\Big\} = O_{\mathcal{D}'}(\varepsilon^2),\end{aligned}$$

$$\varepsilon\Big\{-\varphi_t L_2+\frac{2m}{m+1}L_{m+1}-3a_2'\beta A^2\Big\}\delta'(x-\varphi)$$
$$+\varepsilon\Big\{\frac{dL_2}{dt}+2(\varphi_t-mu_0^{m-1})u_0u^-\Big\}\delta(x-\varphi)$$
$$+\varepsilon\Big\{\frac{\partial}{\partial t}(u_0u^-)+2m\frac{\partial}{\partial x}(u_0^m u^-)\Big\}H(\varphi-x)$$
$$+\Big\{\frac{\partial}{\partial t}(u_0)^2+\frac{2m}{m+1}\frac{\partial}{\partial x}(u_0^{m+1})\Big\}=O_{\mathcal{D}'}(\varepsilon^2).$$

Thus the definition of a weak asymptotics mod $O_{\mathcal{D}'}(\varepsilon^2)$ of the solution implies the equation for the background

$$\frac{\partial u_0}{\partial t}+\frac{\partial u_0^m}{\partial x}=0,\qquad x\in\mathbb{R}^1, \tag{3.32}$$

and the equation for the "tail"

$$\frac{\partial u^-}{\partial t}+m\frac{\partial}{\partial x}(u_0^{m-1}u^-)=0,\qquad x<\varphi(t), \tag{3.33}$$

as well as the relations on the front

$$\frac{d\varphi}{dt}=a_mA^{m-1}+m\check{u}_0^{m-1}+A\check{u}_0\ell_{m-1}, \tag{3.34}$$

$$-L_2\frac{d\varphi}{dt}+\frac{2m}{m+1}L_{m+1}=3a_2'\beta A^2, \tag{3.35}$$

$$\frac{d}{dt}\Big(\frac{A}{\beta}\Big)+(\varphi_t-\check{u}_0^{m-1})\check{u}^-=0, \tag{3.36}$$

$$\frac{dL_2}{dt}+2(\varphi_t-m\check{u}_0^{m-1})\check{u}_0\check{u}^-=0. \tag{3.37}$$

Here and in what follows, the accent ˇ stands for the substitution $x=\varphi(t)$. We assume that Eq. (3.32) (supplemented with the Cauchy condition for $t=0$) has a smooth solution for $t\le T$.

Let us study relation (3.35). By excluding $d\varphi/dt$ from (3.35) with the help of (3.34) and carrying out algebraic transformations, we obtain

$$\beta^2=\gamma^{m-1}A^{m-1}+\check{u}_0P_m(\check{u}_0,A), \tag{3.38}$$

where γ is calculated by the same formula as was used in (3.26) and

$$P_m=\frac{1}{a_2'}\sum_{i=3}^{m}c_m^{i-1}(2a_i-a_{i-1})A^{i-2}\check{u}_0^{m-i}-2\frac{a_m}{a_2'}A^{m-2}.$$

In the special case $m=2$, we have $P_2=0$ and, to make (3.38) meaningful, we need to assume that $A>0$. For $m>2$ we are forced to assume that

$$\gamma^{m-1}A^{m-1}+\check{u}_0P_m(\check{u}_0,A)>0. \tag{3.39}$$

This condition is necessarily satisfied for sufficiently small u_0. Moreover, it holds for a special relation between the signs of A and $\check{u}_0$. For example, for $m=3$ relation (3.39) is reduced to

$$A^2/8\pi-6A\check{u}_0>0$$

and is necessarily satisfied for $A\check{u}_0\le 0$.

Let us consider Eq. (3.37). Taking into account relation (3.36) and the explicit form of L_2, we can readily transform this equation as follows:

$$\frac{d}{dt}\frac{A^2}{\beta} + \frac{2}{a_2}\frac{A}{\beta}\frac{d\check{u}_0}{dt} = 0. \tag{3.40}$$

In the special case $m = 2$, Eq. (3.40) can be integrated: $A + 4\check{u}_0 = \text{const}$. In the general case, by using (3.32), (3.34), and (3.38), we can readily transform Eq. (3.40) to the standard form

$$\frac{dA}{dt} = F(A, \varphi, t).$$

However, the function F is very cumbersome here.

It remains to study relation (3.36). We assume that

$$A\check{u}_0\ell_{m-1} > 0. \tag{3.41}$$

Then, using (3.34) once more, we transform (3.36) to the form of a boundary condition for u^-:

$$u^-\Big|_{x=\varphi} = -(a_m A^{m-1} + A\check{u}_0\ell_{m-1})^{-1}\frac{d}{dt}\frac{A}{\beta}, \qquad t > 0. \tag{3.42}$$

It remains to note that if (3.41) holds, then the characteristics of Eq. (3.33) issued from the points $x^0 < \varphi(0)$ at $t = 0$ do not intersect the curve $x = \varphi(t)$. Thus the Cauchy problem for (3.33) in which the initial conditions are posed on the curve $\Gamma = \{t = 0 \text{ for } x \le \varphi(0),\ x = \varphi(t) \text{ for } t > 0\}$ is solvable for all t for which Eq. (3.32) has a smooth solution. The function u^- will be sufficiently smooth if the functions $u^-\big|_{t=0}$ for $x \le \varphi(0)$ and $u^-\big|_{x=\varphi(t)}$ for $t \ge 0$ are matched at the point $(\varphi(0), 0)$.

Summarizing the above arguments, we obtain the following theorem.

THEOREM 3.1. *Suppose that for $t \le T$ there exists a smooth solution of* (3.32) *and conditions* (3.39), (3.41) *are satisfied. Suppose also that there exists a solution of system* (3.34), (3.40) *and the compatibility conditions are satisfied. Then for $t \le T$ the curve $x = \varphi(t)$ is the trajectory of motion of the ε–δ-singularity against the background $u_0(x, t)$.*

REMARK 3.1. The formulas for $m = 2$ obtained above coincide exactly with the soliton type asymptotics of the KdV equation derived by the usual method [**91, 89, 90, 93, 94**].

REMARK 3.2. Equations (3.32) and (3.33) show the natural fact that outside the support of the ε–δ-singularity the function (3.30) has the form

$$u = u_0(x, t) \quad \text{for } x > \varphi(t), \qquad u = u_0(x, t) + \varepsilon u^-(x, t) \quad \text{for } x < \varphi(t)$$

and satisfies the formal zero-dispersion limit of the KdV type equation (3.12) (i.e., a Hopf type equation). The relations on the front, (3.34) and (3.35), are the energy relations (3.29) calculated up to $O(\varepsilon^2)$ for the function (3.30) under the assumption that $u_0 \to 0$ as $x \to \pm\infty$ and $u^- \to 0$ as $x \to -\infty$. Finally, it is easy to verify that (3.36) and (3.37) are, up to $O(\varepsilon^2)$, the conservation laws

$$\frac{d}{dt}\int u\,dx = 0, \qquad \frac{d}{dt}\int u^2\,dx = 0$$

calculated for (3.30) under the same assumptions on u_0 and u^-.

3.3. Other examples. In this section we continue the detailed presentation of the weak asymptotics method by the examples of sufficiently simple and well-studied problems such as the dynamics of a weak discontinuity and the passage to the limit in the phase field system in the case of a single spatial variable. The complete statement of the second problem is very informative in both the physical sense (the free interface in two-phase media are considered) and the mathematical sense (in the limit, the solution of this problem contains both a weak discontinuity and a Heaviside function type singularity). However, in the case studied here ($x \in \mathbb{R}$, and there is no interaction), this problem is rather trivial. Hence different aspects of this problem are discussed in Subsections. 4.3 and 5.1 (there we also present the results of numerical computations). Here we restrict ourselves to rather elementary calculations.

1. To illustrate how our method can be used for describing the dynamics of a weak singularity, we consider the simplest example of the Hopf equation

$$\frac{\partial u}{\partial t} + \frac{\partial}{\partial x}u^2 = 0, \qquad u\Big|_{t=0} = u_0(x), \tag{3.43}$$

with a continuous piecewise smooth function $u_0(x)$. Let x_0 be the point of weak discontinuity of u_0 such that

$$\frac{\partial u_0}{\partial x} = \frac{\partial u_0^-}{\partial x}H(x_0 - x) + \frac{\partial u_0^+}{\partial x}H(x - x_0),$$

and let $(u_0^-)_x(x_0) \neq (u_0^+)_x(x_0)$. We choose a regularization $H(x,\varepsilon)$ of the Heaviside function $H(x)$ and pass from the initial value u_0 to its regularization

$$u_0(x,\varepsilon) = u_{0c}^-(x) + \big(u_{0c}^+(x) - u_{0c}^-(x)\big)H(x - x_0, \varepsilon). \tag{3.44}$$

Here $u_{0c}^{\pm}(x) \in C^\infty(\mathbb{R}^1)$ are some functions coinciding with $u_0(x)$ for $x \geq x_0$ and $x \leq x_0$, respectively.

We write the asymptotic solution of problem (3.43), (3.44) as

$$u(x,t,\varepsilon) = u_c^-(x,t) + \big(u_c^+(x,t) - u_c^-(x,t)\big)H(x - \varphi(t), \varepsilon) + O_{\mathcal{D}'}(\varepsilon). \tag{3.45}$$

Here $\varphi(t)$ and $u_c^{\pm}(x,t)$ are the desired smooth functions such that

$$u_c^+\big|_{x=\varphi(t)} = u_c^-\big|_{x=\varphi(t)}, \qquad t \geq 0. \tag{3.46}$$

Simple calculations discussed in Subsection 3.1 imply

$$u(x,t,\varepsilon) = u_c^-(x,t) + \big(u_c^+(x,t) - u_c^-(x,t)\big)H(x - \varphi(t)) + O_{\mathcal{D}'}(\varepsilon),$$
$$u^2(x,t,\varepsilon) = (u_c^-(x,t))^2 + \big((u_c^+(x,t))^2 - (u_c^-(x,t))^2\big)H(x - \varphi(t)) + O_{\mathcal{D}'}(\varepsilon).$$

The weak asymptotic solution of Eq. (3.43) is understood in the sense that the relation

$$\frac{d}{dt}\int u\psi\,dx - \int u^2\frac{\partial \psi}{\partial x}\,dx = O(\varepsilon)$$

must hold for any test function $\psi = \psi(x) \in \mathcal{D}$. Then it follows from the relations obtained above and condition (3.46) that the function (3.44) is a weak asymptotics of the Hopf equation (3.43) if and only if

$$\frac{\partial u_c^-}{\partial t} + \frac{\partial (u_c^-)^2}{\partial x} + \left\{\frac{\partial}{\partial t}(u_c^+ - u_c^-) + \frac{\partial}{\partial x}\big((u_c^+)^2 - (u_c^-)^2\big)\right\}H(x - \varphi) = O_{\mathcal{D}'}(\varepsilon).$$

Obviously, we thus obtain the relations

$$\frac{\partial u_c^-}{\partial t} + \frac{\partial (u_c^-)^2}{\partial t} = 0, \qquad x \in \mathbb{R}^1, \quad t > 0, \tag{3.47}$$

$$\frac{\partial u_c^+}{\partial t} + \frac{\partial (u_c^+)^2}{\partial t} = 0, \qquad x > \varphi(t), \quad t > 0. \tag{3.48}$$

We continuously extend Eq. (3.45) to the domain $x \ge \varphi(t)$ and consider $u_c^\pm$ on the characteristics $x = X^\pm(t, x_0)$ issued from $x_0 = \varphi(0)$. Since

$$\frac{d}{dt}X^\pm(t, x_0) = 2u_c^\pm(X^\pm(t, x_0), t), \qquad \frac{d}{dt}u_c^\pm(X^\pm(t, x_0), t) = 0,$$

and $u_c^+(x_0, 0) = u_c^-(x_0, 0)$, the characteristics $X^+(t, x_0)$ and $X^-(t, x_0)$ coincide and condition (3.46) holds on them. Conversely, it is obvious that if $\varphi(t) \ne X^\pm(t, x_0)$ for $t > 0$, then, for the class of initial data under study, relation (3.46) cannot hold.

So the results of our elementary calculations are well known: a weak discontinuity propagates along the characteristics. It is also clear that if the solution is described locally, relation (3.47) is considered for $x \le \varphi(t)$ and (3.48) for $x \ge \varphi(t)$, then the functions $u^\pm$ need not be continued. For further investigations in this direction, see [**16**].

2. The simplest version of the phase field equations has the form

$$\begin{aligned} &\frac{\partial \theta}{\partial t} + \frac{\partial u}{\partial t} = \frac{\partial^2 \theta}{\partial x^2}, \qquad x \in \Omega \subset \mathbb{R}^1, \quad t > 0, \\ &\varepsilon\frac{\partial u}{\partial t} = \varepsilon\frac{\partial^2 u}{\partial x^2} + \frac{1}{\varepsilon}(u - u^3) + \varkappa\theta(1 - u^2). \end{aligned} \tag{3.49}$$

The system of phase field equations was proposed by Caginalp [**8, 9, 10**] for describing two-phase media microscopically within the framework of the Ginzburg–Landau theory [**74**]. It is well known (e.g., see [**8, 9, 10**]) that in the limit as $\varepsilon \to 0$ Eqs. (3.49) become the modified Stefan problem

$$\frac{\partial \theta^\pm}{\partial t} = \frac{\partial^2 \theta^\pm}{\partial x^2}, \qquad x \in \Omega \subset \mathbb{R}^1, \quad t > 0, \tag{3.50}$$

$$\left[\frac{\partial \theta^\pm}{\partial x}\right]\Big|_{\Gamma_t} = -2V, \qquad [\theta^\pm]|_{\Gamma_t} = 0, \qquad \varkappa_1\theta^\pm|_{\Gamma_t} = V, \tag{3.51}$$

which describes the distribution of the temperature $\theta^\pm$ in the domains $\Omega_t^\pm$ filled with the phases "+" and "−" (for instance, water and ice) and separated by the free boundary $\Gamma_t \overset{\text{def}}{=} \{x = \varphi(t)\}$ moving with velocity $V = \dot\varphi(t)$. System (3.49) can be treated as a regularization of problem (3.50), (3.51), where $\theta^-(x,t)$ is a regularization of the function $\overline{\theta}(x,t)$ ($\overline{\theta} = \theta^+$ for $x \in \Omega_t^+$ and $\overline{\theta} = \theta^-$ for $x \in \Omega_t^-$) having a weak singularity on Γ_t and u is a regularization of a Heaviside type function $\overline{u}(x,t)$ ($\overline{u} = 1$ for $x \in \Omega_t^+$ and $\overline{u} = -1$ for $x \in \Omega_t^-$). In (3.49)–(3.51) $\varkappa$ and $\varkappa_1$ are some constants. It is assumed that these problems are supplemented with natural initial and boundary (on the external boundary of $\partial\Omega$) conditions. We also assume that

$$\overline{\Omega}_t^+ \cup \overline{\Omega}_t^- = \overline{\Omega}, \qquad \overline{\Omega}_t^+ \cap \overline{\Omega}_t^- = \Gamma_t, \qquad \operatorname{dist}(\Gamma_t, \partial\Omega) > 0, \tag{3.52}$$

uniformly with respect to $t \in [0, T]$.

We show that the derivation of the limit problem (3.50), (3.51) from (3.49) can be reduced to simple algebraic calculations by using the weak asymptotics method. The key role in our constructions is played by the definition of a weak asymptotics of the solution of Eqs. (3.49), which admits passing to the limit as $\varepsilon \to 0$; see **[21, 22]**

DEFINITION 3.2. A pair of functions $\theta = \theta(x,t,\varepsilon)$, $u = u(x,t,\varepsilon)$, smooth for $\varepsilon \neq 0$, is called an asymptotic solution mod $O_{\mathcal{D}'}(\varepsilon)$ of the phase field system (3.49) if the following relations hold for any test function $\psi_i = \psi_i(x) \in \mathcal{D}(\Omega)$, $i = 1, 2$:

$$\frac{d}{dt}\int \theta\psi_1\,dx + \frac{d}{dt}\int u\psi_1\,dx + \int \frac{\partial\theta}{\partial x}\frac{\partial\psi_1}{\partial x}\,dx = O(\varepsilon), \tag{3.53}$$

$$\begin{aligned}\frac{\varepsilon}{2}\int\Big(\frac{\partial u}{\partial x}\Big)^2\frac{\partial\psi_2}{\partial x}\,dx - \frac{1}{4\varepsilon}\int(u^2-1)^2\frac{\partial\psi_2}{\partial x}\,dx + \varepsilon\int\frac{\partial u}{\partial t}\frac{\partial u}{\partial x}\psi_2\,dx \\ + \varkappa\int\Big(u - \frac{1}{3}u^3\Big)\frac{\partial}{\partial x}(\theta\psi_2)\,dx = O(\varepsilon).\end{aligned} \tag{3.54}$$

It is easy to see that relation (3.53) corresponds to the standard $\mathcal{D}'$-definition of the solution of the first equation in (3.49). The left-hand side of (3.54) for smooth functions u and θ is obtained by multiplying the Allen–Cahn equation (the second equation in (3.49)) by $u_x\psi_2$ and integrating by parts. One can also see that the dissipative term in (3.54) (the term corresponding to εu_{xx} in the Allen–Cahn equation) plays a significant role for the function u varying as $1/\varepsilon$.

Now let us construct a weak asymptotics. We slightly modernize the above-described ansatz and represent the leading term of the solution of our problem as

$$\begin{aligned}\theta(x,t,\varepsilon) &= \gamma^+(x,t) + \gamma^-(x,t)\omega_1\Big(\beta\frac{x-\varphi}{\varepsilon}\Big) + O_{\mathcal{D}'}(\varepsilon),\\ u(x,t,\varepsilon) &= \omega_0\Big(\beta\frac{x-\varphi}{\varepsilon}\Big) + O_{\mathcal{D}'}(\varepsilon),\end{aligned} \tag{3.55}$$

where $\beta = \beta(t) > 0$, $\varphi = \varphi(t)$, γ^+ and γ^- are the desired smooth functions, and, moreover,

$$\gamma^-\big|_{x=\varphi(t)} = 0, \tag{3.56}$$

$\omega_i(x/\varepsilon)$, $i = 0, 1$, are regularizations of the function $\operatorname{sign}(x)$, $\operatorname{sign}(x) = 1$ for $x > 0$ and $\operatorname{sign}(x) = -1$ for $x < 0$. Let $\omega_1(x/\varepsilon)$ be an arbitrary odd-in-x regularization of $\operatorname{sign} x$, and let ω_0 be a special regularization, which, just as above, is chosen from the condition that the classical solution (if it exists) and the weak solution coincide. In our case, this implies that $\omega_0(\eta)$ must satisfy the equation (e.g., see **[117, 20, 116]**)

$$\Big(\frac{d\omega_0}{d\eta}\Big)^2 = \frac{1}{2}(\omega_0^2 - 1)^2, \tag{3.57}$$

whence we obtain $\omega_0(\eta) = \tanh(\eta/\sqrt{2})$. Since ω_0 is odd, it is easy to calculate the weak asymptotics

$$\theta(x,t,\varepsilon) = \gamma^+(x,t) + \gamma^-(x,t)\operatorname{sign}(x-\varphi) + O_{\mathcal{D}'}(\varepsilon^2), \tag{3.58}$$

$$u^{2l}(x,t,\varepsilon) = 1 + O_{\mathcal{D}'}(\varepsilon), \qquad u^{2l+1}(x,t,\varepsilon) = \operatorname{sign}(x-\varphi) + O_{\mathcal{D}'}(\varepsilon^2),$$

$$\varepsilon\Big(\frac{\partial u}{\partial x}\Big)^2 = 4\beta a_2\delta(x-\varphi) + O_{\mathcal{D}'}(\varepsilon^2), \qquad \varepsilon\frac{\partial u}{\partial t}\frac{\partial u}{\partial x} = -4\beta\varphi_t a_2\delta(x-\varphi) + O_{\mathcal{D}'}(\varepsilon^2),$$

$$\theta\frac{\partial}{\partial x}(u^\ell) = 2\gamma^+\delta(x-\varphi) + O_{\mathcal{D}'}(\varepsilon^2).$$

Moreover, for any even function $F(u)$, $F(\pm 1) = 0$, we have

$$F(u) = \varepsilon a_F \delta\big(\beta(x-\varphi)\big) + O_{\mathcal{D}'}(\varepsilon^3). \tag{3.59}$$

Here

$$a_2 = \int \omega^2(\eta)\, d\eta = \frac{1}{3\sqrt{2}}, \qquad a_F = \int F(\omega(\eta))\, d\eta,$$

where $2\omega(\eta) \overset{\text{def}}{=} \omega_0'(\eta) = (1/\sqrt{2})\cosh^{-2}(\eta/\sqrt{2})$.

Applying these asymptotic expansions, we see that (3.54) implies the relation

$$\begin{aligned}\Big\{ -2\beta a_2 + \frac{a_{(u^2-1)^2}}{4\beta} \Big\} &\delta'(x-\varphi) \\ &+ \Big\{ -4\beta\varphi_t a_2 - \frac{4}{3}\varkappa\gamma^+(\varphi,t) \Big\}\delta(x-\varphi) = O_{\mathcal{D}'}(\varepsilon).\end{aligned} \tag{3.60}$$

Hence, calculating $a_{(u^2-1)^2} = 4\sqrt{2}/3$, we determine $\beta = 1$ and obtain the condition

$$\varphi_t = -\sqrt{2}\varkappa\gamma^+(\varphi,t). \tag{3.61}$$

Furthermore, relation (3.53) implies

$$\frac{\partial}{\partial t}\{\gamma^+ + \gamma^- \operatorname{sign}(x-\varphi) + \operatorname{sign}(x-\varphi)\} = \frac{\partial^2}{\partial x^2}\{\gamma^+ + \gamma^- \operatorname{sign}(x-\varphi)\}, \tag{3.62}$$

and, taking (3.56) into account, we see that the relations

$$\frac{\partial\gamma^+}{\partial t} - \frac{\partial^2\gamma^+}{\partial x^2} + \Big\{ \frac{\partial\gamma^-}{\partial t} - \frac{\partial^2\gamma^-}{\partial x^2} \Big\} \operatorname{sign}(x-\varphi) = 0, \tag{3.63}$$

$$\frac{d\varphi}{dt} + \frac{\partial\gamma^-}{\partial x}\Big|_{x=\varphi} = 0 \tag{3.64}$$

need to be satisfied. We set

$$\gamma^\pm(x,t) = \frac{1}{2}\big(\theta_c^+(x,y) \pm \theta_c^-(x,y)\big), \tag{3.65}$$

where $\theta_c^\pm(x,y)$ are some smooth functions such that

$$\theta_c^+\big|_{x=\varphi} = \theta_c^-\big|_{x=\varphi}. \tag{3.66}$$

Then it follows from (3.63) that $\theta_c^\pm$ must be smooth continuations of the functions $\theta^\pm$, which, in turn, satisfy the heat equation (3.50) in the domains $\Omega_t^\pm$. Moreover, since

$$\frac{\partial\gamma^-}{\partial x}\Big|_{x=\varphi} = \frac{1}{2}\Big(\frac{\partial\theta_c^+}{\partial x} - \frac{\partial\theta_c^-}{\partial x}\Big) \overset{\text{def}}{=} \frac{1}{2}\Big[\frac{\partial\theta^\pm}{\partial x}\Big]\Big|_{x=\varphi},$$

we see that (3.64) is exactly the Stefan condition in (3.51). Finally, formulas (3.65) and (3.66) show that relation (3.66) is just the one-dimensional version of the Gibbs–Thompson condition in (3.51) with the constant $\varkappa_1 = -\sqrt{2}\varkappa$.

So we have obtained the desired result: the modified Stefan problem (3.50), (3.51) is the limit problem as $\varepsilon \to 0$ for the phase field system (3.49) in the one-dimensional case.

In conclusion, we note that the sign "minus" in the relation $\varkappa_1 = -\sqrt{2}\varkappa$ obtained above indicates only that the domain Ω_t^- lying to the left of Ω_t^+ decreases for positive temperature. If the domain Ω_t^- lies to the right of Ω_t^+, then we have the sign "plus" in this relation. To avoid the description of the relative position of the domains Ω_t^- and Ω_t^+, the Stefan and Gibbs–Thompson conditions are usually written in terms of the normal velocity Γ_t and the normal derivative on Γ_t, that is, in the form (3.51).

4. Multidimensional waves

In this section we continue to study the propagation of solitary nonlinear waves but already in the multidimensional case.

In fact, only some small technical modifications are required to pass from the one-dimensional case to the multidimensional case. Here the only new and actually interesting point is that the geometric nature of regularized singularities reveals itself; namely, the singularity evolution depends on the geometry of its support. This effect is most conspicuous in free boundary problems in two-phase media with surface tension; namely, in the multidimensional case the Gibbs–Thompson law directly relates the velocity of motion of the boundary not only to the temperature as in the one-dimensional case but also to the mean curvature of the interface between the phases. This will be shown in Subsection 4.3 in the study of the passage to the limit for the phase field system. Other examples considered in this section are the shock wave for three-dimensional gas dynamics equations (Subsection 4.1) and the two-dimensional distorted soliton in Kadomtsev–Petviashvili type models (Subsection 4.2). The latter is a narrow (in the direction orthogonal to its front Γ_t) crest whose height and velocity slowly vary with time along Γ_t. Hence the weak limit of a two-dimensional soliton is an ε–δ-singularity whose support is a smooth curve in $\mathbb{R}^2_x$.

For multidimensional shock waves as well as for two-dimensional KP-solitons, the normal velocity of motion of the front Γ_t depends on the curvature of Γ_t [**126, 91**]. For an ideal gas and the stable version of a KP type equation, this dependence means that the initial corrugation of the front (sinusoid type perturbations of Γ_0) becomes smoother with increasing t and simultaneously generates waves propagating along Γ_t.

For a nonideal gas (the case of ionization or the case of shock waves in solids) and the nonstable version of a KP type equation, the corrugation amplitude increases with t, which, in the end, results in the bifurcation of the solution. For KP type equations (Subsection 4.2), we study these processes analytically in sufficient detail and present the results of numerical experiments. As the shock wave is concerned, these results are well known (e.g., see [**126, 35**]), and hence we only demonstrate how the weak asymptotics method can be used for deriving the Hugoniot conditions in the case of ideal gases.

4.1. Dynamics of the shock wave in a nonisothermal gas. Let us consider equations of ideal gas dynamics in divergence form

$$
\begin{aligned}
&\frac{\partial \rho}{\partial t} + \langle \nabla, \rho u\rangle = 0,\\
&\frac{\partial}{\partial t}(\rho u_i) + \langle \nabla, \rho u u_{(i)}\rangle + \frac{\partial p}{\partial x_i} = 0, \quad i = 1,2,3, \qquad (4.1)\\
&\frac{\partial}{\partial t}\Big(\frac{1}{2}\rho u^2 + \rho\mathcal{E}\Big) + \Big\langle \nabla, \rho\Big(\frac{1}{2}u^2 + w\Big)u\Big\rangle = 0,
\end{aligned}
$$

where ρ is the gas density, $u = (u_{(1)}, u_{(2)}, u_{(3)})$ is the gas velocity, and p is the gas pressure. Under the assumption that the gas is polytropic with adiabatic coefficient

$\gamma > 1$, we shall use the standard formulas for the gas intrinsic energy $\mathcal{E}$ and the gas enthalpy w

$$\mathcal{E} = \frac{p}{(\gamma-1)\rho}, \qquad w = \frac{\gamma p}{(\gamma-1)\rho}, \tag{4.2}$$

which correspond to the state equation

$$p = \rho T, \tag{4.3}$$

relating p, ρ and the temperature T.

The problem of motion of the shock wave for Eqs. (4.1), (4.2) was studied in detail [**42, 118, 128**]. In particular, it is well known that the shock wave front behaves according to the Hugoniot conditions

$$\begin{aligned} [\rho(v_\nu - u_\nu)] &= 0, \\ [\rho(v_\nu - u_\nu)^2 + p] &= 0, \\ [\tfrac{1}{2}\rho(v_\nu - u_\nu)^2 + w] &= 0, \end{aligned} \tag{4.4}$$

where v_ν is the velocity of motion of the front Γ_t of the shock wave along the normal ν to Γ_t, $u_\nu = \langle \nu, u \rangle$ is the normal component of the vector velocity u, and $[\cdot]$ denotes, as usual, a jump across Γ_t.

Traditionally, the physicists derive conditions (4.4) by analyzing the integral conservation laws for (4.4). V. P. Maslov [**87**] proposed an approach that permits deriving the Hugoniot conditions directly from the differential equations considered in the weak sense. This approach, which was called the method of asymptotic expansion with respect to smoothness, admits very wide generalizations (e.g., see [**30, 106**]) and, for shock waves without interaction, gives the same results as the weak asymptotics method developed here. We also note that the Hugoniot conditions can be derived for the equations of nonideal gas dynamics (for large Reynolds numbers) by constructing a one-phase asymptotic solution ([**92**]). In this case, it is possible to analyze the profile of the shock wave in detail, including the fine effects arising for different ratios between the Reynolds and Prandtl numbers [**91**].

The main goal of this subsection is to show that the use of the weak asymptotics method developed here allows obtaining (4.4) from Eqs. (4.1), (4.2) by simple algebraic calculations.

We write the regularized solution in the form

$$\begin{aligned} u_{(i)}(x,t,\varepsilon) &= u_{0(i)}(x,t) + U_{(i)}(x,t)H_i(\varphi(x) - t, \varepsilon), \quad i = 1,2,3, \\ \rho(x,t,\varepsilon) &= \rho_0(x,t) + R(x,t)H_4(\varphi(x) - t, \varepsilon), \\ p(x,t,\varepsilon) &= p_0(x,t) + P(x,t)H_5(\varphi(x) - t, \varepsilon), \end{aligned} \tag{4.5}$$

where $u_{0(i)}, U_{(i)}, \rho_0, R, p_0, P, \varphi$ are the desired smooth functions, $H_j(x,\varepsilon)$ are regularized Heaviside functions,

$$H_j(x,\varepsilon) = H(x) + O_{\mathcal{D}'}(\varepsilon), \quad j = 1, \dots, 5,$$

and ε is the regularization parameter.

Similarly to Subsection 4.1, it is easy to calculate

$$\begin{aligned} \rho(x,t,\varepsilon)u(x,t,\varepsilon) &= \rho_0 u_0 + \{\rho u\} H(\varphi - t) + O_{\mathcal{D}'}(\varepsilon), \\ \rho(x,t,\varepsilon)u(x,t,\varepsilon)u_i(x,t,\varepsilon) &= \rho_0 u_0 u_{0(i)} + \{\rho u u_i\} H(\varphi - t) + O_{\mathcal{D}'}(\varepsilon), \\ \rho(x,t,\varepsilon)u^2(x,t,\varepsilon) &= \rho_0 u_0^2 + \{\rho u^2\} H(\varphi - t) + O_{\mathcal{D}'}(\varepsilon), \end{aligned} \tag{4.6}$$

$$\rho(x,t,\varepsilon)u^2(x,t,\varepsilon)u_{(i)}(x,t,\varepsilon) = \rho_0 u_0^2 u_{0(i)} + \{\rho u^2 u_{(i)}\}H(\varphi - t) + O_{\mathcal{D}'}(\varepsilon),$$
$$p(x,t,\varepsilon)u(x,t,\varepsilon) = p_0 u_0 + \{pu\}H(\varphi - t) + O_{\mathcal{D}'}(\varepsilon),$$

where we use the notation

$$\begin{aligned}
\{\rho u\} &= (\rho_0 + R)(u_0 + U) - \rho_0 u_0,\\
\{\rho u^2\} &= (\rho_0 + R)(u_0 + U)^2 - \rho_0 u_0^2,\\
\{\rho u u_i\} &= (\rho_0 + R)(u_0 + U)(u_{0(i)} + U_{(i)}) - \rho_0 u_0 u_{0(i)},\\
\{\rho u^2 u_{(i)}\} &= (\rho_0 + R)(u_0 + U)^2(u_{0(i)} + U_{(i)}) - \rho_0 u_0^2 u_{0(i)},\\
\{pu\} &= (p_0 + P)(u_0 + U) - p_0 u_0.
\end{aligned} \tag{4.7}$$

We substitute (4.6) into Eqs. (4.1) whose solutions we understand as smooth functions for $\varepsilon \neq 0$ and smooth functions of t ranging in $\mathcal{D}'(\mathbb{R}_x^3)$ as $\varepsilon \to 0$ (see Subsection 4.3). Let us use (4.2). Then, after differentiating, we obtain

$$\begin{aligned}
&\frac{\partial \rho_0}{\partial t} + \langle \nabla, \rho_0 u_0 \rangle + H(\varphi - t)\Big(\frac{\partial R}{\partial t} + \langle \nabla, \{\rho u\}\rangle\Big)\\
&\qquad + (-R + \langle [\rho u], \nabla\varphi\rangle)\delta(\varphi - t) = O_{\mathcal{D}'}(\varepsilon),\\
&\frac{\partial}{\partial t}\rho_0 u_{0(i)} + \langle \nabla, \rho_0 u_0 u_{0(i)}\rangle + \frac{\partial p_0}{\partial x_i}\\
&\qquad + H(\varphi - t)\Big(\frac{\partial}{\partial t}\{\rho u_{(i)}\} + \langle \nabla, \{\rho u u_{(i)}\}\rangle + \frac{\partial P}{\partial x_i}\Big)\\
&\qquad + \Big(- [\rho u_{(i)}] + \langle [\rho u u_{(i)}], \nabla\varphi\rangle + P\frac{\partial \varphi}{\partial x_i}\Big)\delta(\varphi - t) = O_{\mathcal{D}'}(\varepsilon),\\
&\frac{\partial}{\partial t}\Big(\frac{1}{2}\rho_0 u_0^2 + \frac{1}{\gamma - 1}P_0\Big) + \Big\langle \nabla, \Big(\frac{1}{2}\rho_0 u_0^2 + \frac{\gamma}{\gamma - 1}P_0\Big)\Big\rangle\\
&\qquad + H(\varphi - t)\Big(\frac{\partial}{\partial t}\Big(\frac{1}{2}\{\rho u^2\} + \frac{1}{\gamma - 1}P\Big) + \Big\langle \nabla, \frac{1}{2}\{\rho u^2 u\} + \frac{\gamma}{\gamma - 1}\{pu\}\Big\rangle\Big)\\
&\qquad + \Big(- \frac{1}{2}\{\rho u^2\} - \frac{1}{\gamma - 1}P + \Big\langle \frac{1}{2}\{\rho u^2\} + \frac{\gamma}{\gamma - 1}\{pu\}, \nabla\varphi\Big\rangle\Big)\delta(\varphi - t)\\
&\qquad = O_{\mathcal{D}'}(\varepsilon),
\end{aligned} \tag{4.8}$$

Here $[\rho u], \dots, [pu]$ are the same expressions as in (4.7) but calculated on the surface $\Gamma_t = \{x, \varphi(x) = t\}$, $\delta(s)$ is the delta function on the surface $S = 0$ [**37**], i.e., a functional acting by the rule

$$(\delta(s), \psi(x)) = \int_{S=0} \psi(x)\sigma, \tag{4.9}$$

and the 1-form σ (the Leray measure) is a solution of the equation $dS \wedge \sigma = dx_1 \wedge dx_2 \wedge dx_3$.

Let $\Omega_t^- = \{x, \varphi(x) < t\}$ and $\Omega_t^+ = \{x, \varphi(x) > t\}$ denote the domains ahead of the shock wave front and behind it, and let $u^\pm, \rho^\pm, p^\pm$ be the states of gas in the domains $\Omega_t^\pm$. Obviously, $u^- = u_0$, $\rho^- = \rho_0$, $p^- = p_0$ for $x \in \Omega_t^-$ and $u^+ = u_0 + U$, $\rho^+ = \rho_0 + R$, $p^+ = p_0 + P$ for $x \in \Omega_t^+$. Therefore, relations (4.8) are satisfied if we define $u^\pm, \rho^\pm, p^\pm$ as solutions of the equations of gas dynamics in the domains $\Omega_t^\pm$,

$$\begin{aligned}
&\frac{\partial \rho^\pm}{\partial t} + \langle \nabla, \rho^\pm u^\pm \rangle = 0, \qquad x \in \Omega_t^\pm, \quad t \in (0, t_0), \\
&\frac{\partial}{\partial t}(\rho^\pm u_{(i)}^\pm) + \langle \nabla, \rho^\pm u^\pm u_{(i)}^\pm \rangle + \frac{\partial p^\pm}{\partial x_i} = 0, \qquad i = 1, 2, 3, \\
&\frac{\partial}{\partial t}\Big(\frac{1}{2}\rho^\pm (u^\pm)^2 + \frac{1}{\gamma - 1} p^\pm\Big) + \Big\langle \nabla, \Big(\frac{1}{2}\rho^\pm (u^\pm)^2 + \frac{\gamma}{\gamma - 1} p^\pm\Big) u^\pm \Big\rangle = 0,
\end{aligned} \tag{4.10}$$

and if the following relations are satisfied on the front Γ_t:

$$\begin{aligned}
[\rho] &= [\rho\langle u, \nabla\varphi\rangle], \\
[\rho u] &= [\rho u \langle u, \nabla\varphi\rangle + p\nabla\varphi], \\
\frac{1}{2}[\rho u^2] + \frac{1}{\gamma - 1}[p] &= \Big[\Big(\frac{1}{2}\rho u^2 + \frac{\gamma}{\gamma - 1}p\Big)\langle u, \nabla\varphi\rangle\Big].
\end{aligned} \tag{4.11}$$

Assuming that the functions $\rho^\pm, u^\pm, p^\pm$ remain smooth on the time interval $(0, t_0)$, we set

$$\rho_0 = \rho_0^-, \qquad u_0 = u^-, \qquad p_0 = p^- \qquad \text{for } x \in \Omega_t^-$$

and extend these functions to the entire $\mathbb{R}^3$ so that they are infinitely differentiable. In a similar way, we introduce

$$R = \rho^+ - \rho_0, \qquad U = u^+ - u_0, \qquad P = p^+ - p_0 \qquad \text{for } x \in \Omega_t^+.$$

Extending R, U, P smoothly to Ω_t^-, we obtain the coefficients of expansions (4.5).

Let us consider conditions (4.11). Clearly, the vector $\nu = \nabla\varphi/|\nabla\varphi|$ is normal to Γ_t and the normal velocity of motion of Γ_t is $v_\nu = 1/|\nabla\varphi|$. Therefore, dividing both parts of the first relation in (4.11) by $|\nabla\varphi|$, we obtain the first Hugoniot condition in (4.4). Next, multiplying the second relation in (4.11) by $\nu/|\nabla\varphi|$ and by the vectors $\xi_i/|\nabla\varphi|$ orthogonal to ν, $i = 1, 2$, we obtain

$$[\rho(u_\nu^2 - v_\nu u_\nu) + p] = 0, \qquad [\rho(v_\nu - u_\nu)u_{\xi_i}] = 0, \quad i = 1, 2, \tag{4.12}$$

where $u_{\xi_i} = \langle u, \xi_i \rangle$. It follows from these relations and the condition of continuity of the mass flow (the first Hugoniot condition in (4.4)) that the tangential components of the vector $u\big|_{\Gamma_t}$ are continuous,

$$[u_{\xi_i}] = 0, \qquad i = 1, 2. \tag{4.13}$$

In turn, the first relation in (4.12) can be easily rewritten as the condition of continuity of the momentum flow (the second Hugoniot condition in (4.4)). Now the third condition in (4.11) can be easily transformed as

$$\big(\rho(u_\nu - v_\nu)\big)\Big|_{\Gamma_t}\Big[\frac{1}{2}u_\nu^2 + w - v_\nu\Big(u_\nu - \frac{1}{2}v_\nu\Big)\Big] = 0$$

if we take the continuity of the mass flow $\rho(u_\nu)$ across the boundary Γ_t into account. So we obtain the last Hugoniot condition in (4.4), i.e., the condition of continuity of the energy flow.

The reader interested in this subject can find a detailed discussion of Hugoniot conditions, including the geometric properties of the dynamics of Γ_t in [**126, 35, 128**].

4.2. Dynamics of the two-dimensional soliton in a Kadomtsev–Petviashvili type model. In general, the dynamics of the two-dimensional soliton for the Kadomtsev–Petviashvili type equation [**47**]

$$\frac{\partial}{\partial x}\left\{\frac{\partial u}{\partial t}+\frac{\partial u^m}{\partial x}+ve^2\frac{\partial^3 u}{\partial x^3}\right\}+\sigma\frac{\partial^2 u}{\partial y^2}=0, \tag{4.14}$$

where $m \geq 2$ and the parameter σ is either $\sigma = 1$ or $\sigma = -1$, can be described in the same way as that for the KdV type equation (see Subsection 3.2). We only need to define a weak solution of (4.14) in an appropriate way.

DEFINITION 4.1. A pair (u, v) will be called a *weak asymptotic solution* (mod $O_{\mathcal{D}'(\mathbb{R}^2_{xy})}(\varepsilon^2)$) of Eq. (4.14) if the following relations hold for any $\psi, \psi_1, \psi_2 \in \mathcal{D}(\mathbb{R}^2_{xy})$:

$$\int \{-u_t\psi_x + u^m\psi_{xx} + \sigma u\psi_{yy}\}\,dxdy = O(\varepsilon^2), \tag{4.15}$$

$$\begin{aligned}\int \Big\{(u^2)_t\psi_1 - 2\frac{m}{m+1}u^{m+1}\psi_{1x} + 3(\varepsilon u_x)^2\psi_{1x}&\\ + \sigma(\varepsilon v_y)^2\psi_{1x} - 2\sigma u\varepsilon v_y\psi_{1y}\Big\}\,dxdy &= O(\varepsilon^2),\end{aligned} \tag{4.16}$$

$$\int \{\varepsilon v\psi_{2x} + u\psi_2\}\,dxdy = O(\varepsilon^2), \tag{4.17}$$

where the integrals are taken over $\mathbb{R}^2_{xy}$.

Relations (4.15) and (4.17) follow from the standard definition of the $\mathcal{D}'$-solution of Eq. (4.14) and the problem

$$\varepsilon\frac{\partial v}{\partial x} = u, \qquad v \to 0 \quad \text{as} \quad x \to \infty. \tag{4.18}$$

We obtain relation (4.16) on smooth functions by rewriting (4.14) as the system consisting of the KdV type equation

$$\frac{\partial u}{\partial t}+\frac{\partial u^m}{\partial x}+\varepsilon^2\frac{\partial^3 u}{\partial x^3}+\varepsilon\sigma\frac{\partial^2 v}{\partial y^2}=0 \tag{4.19}$$

and Eq. (4.18), multiplying (4.19) by $u\psi_1$, and integrating by parts.

To simplify the formulas, we restrict ourselves to the case in which a solitary wave moves against the zero background.

First, we introduce some auxiliary functions. Precisely as in Section 3, we define $z = z(\eta)$ as the solution of the model equation

$$\frac{d^2z}{d\eta^2} - z + z^m = 0, \qquad z \to 0 \quad \text{as} \quad \eta \to \pm\infty,$$

and set

$$\omega(\eta) = \frac{z(\eta)}{\int z(\eta')\,d\eta'}. \tag{4.20}$$

Recall that ω is an even function exponentially decreasing as $\eta \to \pm\infty$ and its integrals

$$a_n = \int \omega^n(\eta)\,d\eta$$

satisfy the relation

$$\frac{2m}{m+1}a_{m+1} - a_2 a_m = 2\frac{m-1}{m+1}a_{m+1}.$$

We define the number γ by the formula

$$\gamma = \Big(2\frac{m-1}{m+1}\frac{a_{m+1}}{a_2'}\Big)^{1/(m-1)}, \qquad a_2' = \int \big(\omega'(\eta)\big)^2\,d\eta. \tag{4.21}$$

Next, for $m \neq 5$, we assume that for $t \in [0,T]$ there exists a smooth solution $\rho = \rho(t,y) > 0$, $w = w(t,y)$ of the equations

$$\frac{\partial \rho}{\partial t} + \frac{\partial}{\partial y}(\rho w) = 0, \qquad \rho\Big(\frac{\partial w}{\partial t} + w\frac{\partial w}{\partial y}\Big) + \frac{\partial p}{\partial y} = 0, \tag{4.22}$$

where the "pressure" p is related to the "density" ρ as follows:

$$p = c\rho^{\varkappa}, \qquad \varkappa = \frac{m+3}{5-m}, \qquad c = 4\sigma a_m \frac{m-1}{m+3}\gamma^{(m-1)^2/(5-m)}. \tag{4.23}$$

For $m \neq 5$ we define the functions $A = A(t,y)$ and $\varphi = \varphi(t,y)$ as follows:

$$A = s_m \gamma^{(m-1)/(5-m)}\rho^{2/(5-m)}, \qquad -2\sigma\frac{\partial \varphi}{\partial y} = w, \tag{4.24}$$

where $s_m = 1$ for even m and $s_m = \pm 1$ for odd m. Finally, we introduce the functions $u^- = u^-(x,y,t)$ and $v^- = v^-(x,y,t)$ such that

$$\frac{\partial^2}{\partial x \partial t}u^- + \sigma\frac{\partial}{\partial y^2}u^- = 0, \qquad \frac{\partial v^-}{\partial x} = -u^-, \qquad (x,y) \in \Omega_t^-, \quad t > 0, \tag{4.25}$$

$$u^-\Big|_{\Gamma_t} = \mu, \qquad \frac{\partial u^-}{\partial \nu}\Big|_{\Gamma_t} = \zeta, \qquad v^-\Big|_{\Gamma_t} = r, \tag{4.26}$$

where $\Omega_t^- = \{(x,y), x < \varphi(y,t)\}$, $\Gamma_t = \{(x,y), x = \varphi(y,t)\}$,

$$r = a_1\frac{A}{\beta}, \qquad \mu = -\frac{A^{1-m}}{a_m}\Big\{\frac{\partial}{\partial t} + w\frac{\partial}{\partial y} + \frac{1}{2}\frac{\partial w}{\partial y}\Big\}r, \tag{4.27}$$

$$\zeta = \sigma\Big(\mu\frac{\partial^2 \varphi}{\partial y^2} + \frac{\partial^2 r}{\partial y^2}\Big), \qquad \beta = \gamma^{(m-1)/2}A^{(m-1)/2}, \tag{4.28}$$

and $\partial u^-/\partial \nu$ is the derivative of u^- along the vector $\nu = (1, -\varphi_t, -2\sigma\varphi_y) \in \mathbb{R}^3_{txy}$ transversal to $\bigcup_{t>0}\Gamma_t$.

THEOREM 4.1. *Let*

$$(5-m)\sigma > 0. \tag{4.29}$$

Suppose for $t \in (0,T]$ *there exists a smooth solution of the gas dynamics equations* (4.22). *Then the weak asymptotic solution* $\operatorname{mod} O_{\mathcal{D}'(\mathbb{R}^2)}(\varepsilon^2)$ *of the Kadomtsev–Petviashvili type equation* (4.19) *has the form*

$$\begin{aligned} u &= A\omega\Big(\beta\frac{x-\varphi}{\varepsilon}\Big) + \varepsilon u^-\omega_1\Big(\beta\frac{x-\varphi}{\varepsilon}\Big) \\ &= \varepsilon\frac{A}{\beta}\delta(x-\varphi) + \varepsilon u^- H(\varphi - x) + O_{\mathcal{D}'}(\varepsilon^2), \end{aligned} \tag{4.30}$$

where $\omega_1(\eta) = \int_{-\infty}^{\eta} \omega(\eta')\,d\eta'$. *In this case the solution* $(\operatorname{mod} O_{\mathcal{D}'(\mathbb{R}^2)}(\varepsilon^2))$ *of the auxiliary problem* (4.18) *has the form*

$$v = -v^- \omega_1\Big(\beta \frac{x-\varphi}{\varepsilon}\Big) = -v^- H(\varphi - x) + O_{\mathcal{D}'}(\varepsilon^2). \tag{4.31}$$

Prior to proving this theorem, let us discuss it. First, we note that, just as for KdV type equations (see Subsection 3.2), the limit equations (4.22) with $m \neq 5$ are closely related to the energy identities for the solution of a KP type equation. Namely, in what follows, we shall prove the following assertion.

THEOREM 4.2. *Suppose that the assumptions of Theorem* 4.1 *are satisfied and the function* $u^-(x,y,t)$ *decreases sufficiently rapidly as* $x \to -\infty$. *Then it follows from relations* (4.22)–(4.28) *that the energy relations*

$$\frac{\partial}{\partial t}\int u\,dx + \varepsilon\sigma \frac{\partial^2}{\partial y^2}\int v\,dx = 0, \tag{4.32}$$

$$\frac{\partial}{\partial t}\int xu\,dx + \varepsilon\sigma \frac{\partial^2}{\partial y^2}\int xv\,dx - \int u^m\,dx = 0, \tag{4.33}$$

$$\frac{\partial}{\partial t}\int u^2\,dx + 2\varepsilon^2\sigma \frac{\partial}{\partial y}\int \frac{\partial v}{\partial x}\frac{\partial v}{\partial y}\,dx = 0, \tag{4.34}$$

$$\begin{aligned}\frac{\partial}{\partial t}\int xu^2\,dx &+ 2\varepsilon^2\sigma \frac{\partial}{\partial y}\int x\frac{\partial v}{\partial x}\frac{\partial v}{\partial y}\,dx + 3\int \Big(\varepsilon\frac{\partial u}{\partial x}\Big)^2\,dx \\ &+ \sigma\int\Big(\varepsilon\frac{\partial v}{\partial y}\Big)^2\,dx - 2\frac{m}{m+1}\int u^{m+1}\,dx = 0\end{aligned} \tag{4.35}$$

hold with accuracy $O(\varepsilon^2)$ *for the solution of the KP type equation* (4.18), (4.19), *where the integrals are calculated for the functions* u *and* v *of the form* (4.30), (4.31). *If* $A\big|_{t=0} = \text{const}$ *and* $\varphi\big|_{t=0}$ *is a linear function of* y, *then these relations are exact.*

Let us discuss the Hugoniot type conditions (4.22), which describe the evolution of the basic parameters of the two-dimensional soliton (the amplitude and the velocity of motion—see formulas (4.24)) and, simultaneously, the parameters of the limit ε–δ-singularity as $\varepsilon \to 0$. First, it should be noted that there are strong foundations for the fact that the nonlinearity is of critical character for $m = 5$, which readily follows from the formula for $\varkappa$ in system (4.23). It is easy to see that system (4.22) is hyperbolic because of inequality (4.29). For $\sigma = 1$ and $m < 5$, just as for $\sigma = -1$ and $m > 5$, the Cauchy problem for the gas dynamics equations is well posed, which naturally follows from the initial data for the KP type equation (4.14). For $\sigma(5-m) < 0$, system (4.22) is elliptic, and hence any problem with the Cauchy conditions for $t = 0$ becomes ill posed. In this case, the solution of the original KP type equation is also unstable. For the KP-equation ($m = 2$), this result is well known [**130, 131**]. We illustrate these facts by the results of numerical experiments, which will be discussed later. Finally, a degeneration occurs for $m = 5$ and, instead of (4.22), we obtain the limit problem

$$\frac{\partial \varphi}{\partial t} - \sigma\Big(\frac{\partial \varphi}{\partial y}\Big)^2 = a_5 A^4, \qquad \frac{\partial^2 \varphi}{\partial y^2} = 0. \tag{4.36}$$

The condition that the amplitude A remains bounded as $y \to \pm\infty$ implies that the front is the function $\varphi = \varphi_0(t) + yW$ linear in y, where the constant W is determined by the initial condition for (4.14). However, Eqs. (4.36) imply only one relation

$$\frac{d\varphi_0}{dt} = a_5 A^4 + \sigma W^2$$

for $\varphi_0(t)$ and $A(t)$, and hence the solution is not unique.

On the other hand, the restriction $m < 5$ is obtained by estimating the integral $\int u^{m+1}\,dx$ (for instance, in the energy identity (4.35)). Namely, only for $m < 5$, this integral can be estimated by the product $\|u; L^2(\mathbb{R}^1_x)\|^{m+1-\theta}\|u_x; L^2(\mathbb{R}^1_x)\|^{\theta}$, where $\theta < 2$.

For $m \geq 5$, we obtain $\theta \geq 2$ and thus the KP type equation cannot be analyzed by functional methods any more (e.g., see [**111**]). So the preceding analysis shows that the causes of the difficulties in the functional analysis of (4.14) for $m \geq 5$ are by no means technical.

Nevertheless, for any m and σ, Eqs. (4.22), (4.36) have the special solutions $A = \text{const}$, $\varphi = \varphi_0 + V(A)t + Wy$, where $\varphi_0, V, W = \text{const}$, corresponding to the exact soliton solution of the KP type equation (4.14). We also note that in this case $p = \text{const}$ and $\mu = \zeta = 0$ by (4.27), (4.28), and hence it follows from (4.25), (4.26) that $u^- \equiv 0$, $v^- \equiv A/\beta$, i.e., the "tail" does not appear in this case.

Let condition (4.29) be satisfied. Then for $\sigma = 1$, the gas dynamics equations (4.22) correspond to the media with "normal" thermodynamical properties, since $\partial^2 p/\partial\rho^2 > 0$. Moreover, for $m = 2$ we have $\varkappa = 5/3$, i.e., this is the case of a monatomic gas. For $\sigma = -1$, as is easy to calculate, $\partial^2 p/\partial\rho^2 < 0$, and the gas thermodynamic properties are anomalous [**133**].

Furthermore, to analyze the model equations, we can use all the facts known for the equations of gas dynamics (e.g., see [**118**]). In particular, together with the energy relation, we have the conservation laws for the mass and momentum,

$$\frac{d}{dt}\int_{-\infty}^{\infty} \rho\,dy = 0, \qquad \frac{d}{dt}\int_{-\infty}^{\infty} \rho v\,dy = 0.$$

It is not difficult to calculate the Riemann invariants r and s and to find that

$$r = v + 3/\sqrt{2}\rho^{1/3} = \text{const} \quad \text{along the characteristics } Y_r : \frac{dY_r}{dt} = \frac{1}{3}(2r+s),$$

$$s = v - 3/\sqrt{2}\rho^{1/3} = \text{const} \quad \text{along the characteristics } Y_s : \frac{dY_s}{dt} = \frac{1}{3}(r+2s).$$

As is well known, the solution of the gas dynamics equations can be simplified significantly by choosing special initial data, which correspond to the so-called Riemann waves, i.e., for either $r\big|_{t=0} = \text{const}$ or $s\big|_{t=0} = \text{const}$. For example, let $r\big|_{t=0} = r_0$ be independent of y. Then, obviously, we have

$$v = \frac{1}{2}(r_0 + s_0(y^0))\Big|_{y^0=y^0(y,t)}, \qquad \rho = \left(\frac{r_0 - s_0(y^0)}{3\sqrt{2}}\right)^3\Bigg|_{y^0=y^0(y,t)}.$$

Here $y^0(y,t)$ is the solution of the equation $y = Y_s(t, y^0)$, where

$$Y_s = \frac{1}{3}(r_0 + 2s_0(y^0))t + y^0,$$

and $s_0(y) = (v - 3/(\sqrt{2}\rho^{1/3}))\big|_{t=0}$. Clearly, this solution can exist only if the Jacobian does not vanish,

$$J = 1 + \frac{2}{3}\frac{\partial s_0}{\partial y^0}t \neq 0.$$

Let

$$\rho^{1/3}\Big|_{t=0} = \rho_*^{1/3} + (\rho^0(y))^{1/3}, \qquad v\Big|_{t=0} = -\frac{3}{2}k + v^0(y),$$

where $\rho_* = A^{3/2} = \text{const}$ and $\rho^0, v^0 = 0$ for $|y| \geq Y$. Then the assumption that $r = \text{const}$ implies that $v^0 = c - 3/\sqrt{2}(\rho^0)^{1/3}$, where $c = \text{const}$. Hence

$$\frac{\partial s_0}{\partial y} = -\frac{\sqrt{2}}{(\rho^0)^{2/3}}\frac{\partial \rho^0}{\partial y},$$

and originating from the points y^0 such that $\partial\rho^0/\partial y|_{y=y^0} \leq 0$, a rarefaction wave is formed on the characteristics Y_s, whereas for

$$\frac{\partial \rho^0}{\partial y}\Big|_{y=y^0} > 0,$$

a compression wave is formed. So in the first case the characteristics diverge and, in the rarefaction wave, the gradients of the functions ρ and v decrease in absolute value with time t. On the contrary, in the compression wave, the characteristics converge and the gradients of the functions ρ and v increase in absolute value with time t. One can easily calculate the first time moment T at which the characteristics intersect

$$T = \frac{3}{2\sqrt{2}}\left\{\max_y (\rho^0)^{-2/3}\frac{\partial \rho^0}{\partial y}\right\}^{-1}.$$

Let y_*^0 denote the value of y at which $\rho_y^0(\rho^0)^{-2/3}$ attains its maximum. For $t < T$ the amplitude A and the function φ specifying the position of the front are given by the formulas

$$A = \rho^{2/3}(y,t), \quad \varphi = ky + \frac{1}{2}\Big(A_* + \frac{3}{2}k^2\Big)t - \frac{2}{3}\int_{-\infty}^{y}\Big(v(y',t) + \frac{3}{2}k\Big)\,dy'.$$

As $t \to T$, the so-called gradient catastrophe occurs, because the gradients of the functions ρ and v increase unboundedly as $y \to y_* = Y_s(T, y_*^0)$. Such a behavior of the solution of the model equations means that at the point (T, y_*) the functions A and φ cease to be smooth: the front breaks and the amplitude of the soliton has a jump discontinuity at the breakpoint.

Prior to constructing asymptotics, we discuss the results of solving numerically the KP-equation (the computations were carried out together with A. Titov).

In numerical modeling we replace the Cauchy problem by the initial boundary value problem in the square $[0,10]\times[0,10]$ with the Neumann conditions posed on its boundary. We seek the solution by using the finite difference scheme constructed with the help of the explicit Lax–Wendroff scheme.

For the initial value we take the distorted soliton

$$u\Big|_{t=0} = A_* \cosh^{-2}\left(\sqrt{\frac{A_*}{2}}\,\frac{x - \varphi^0(y)}{\varepsilon}\right)$$

with unit amplitude ($A_* = 1$) concentrated near the front $\Gamma_0 = \{x = \varphi^0(y)\}$ (curve 0 in Figure 3), where

$$\varphi^0 = 4.5 - \frac{1}{2}\cosh^{-2}\big((y-5)/0.25\sqrt{2}\big).$$

Figure 1 shows the three-dimensional plot $u\big|_{t=0}$ with $\varepsilon = 0.4$, curve a in Figure 2 shows the projection of the three-dimensional plot of $u\big|_{t=0}$ on the (u,x)-plane

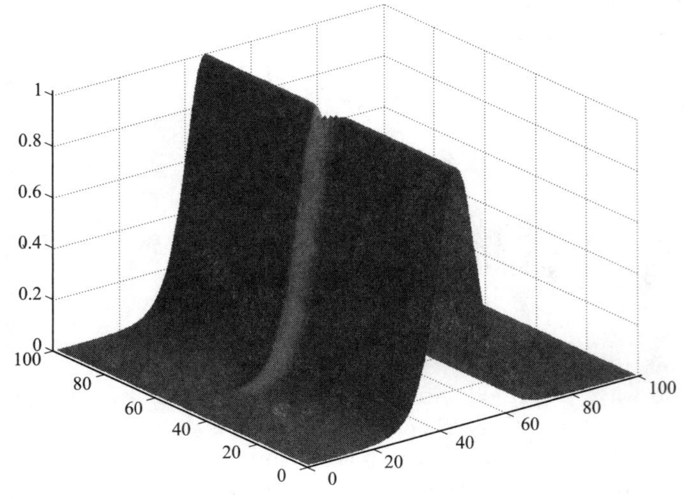

FIGURE 1. Spatial profile of the initial data.

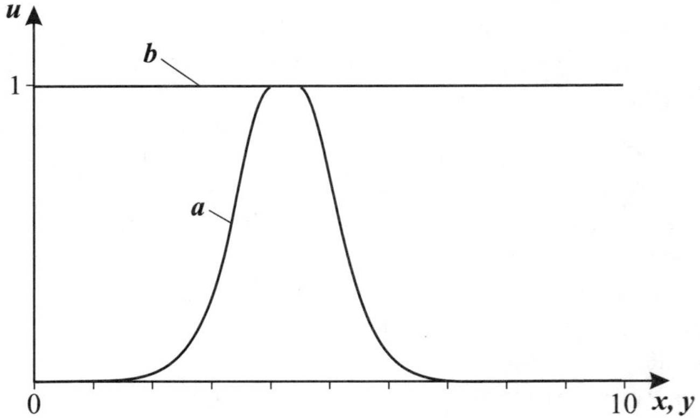

FIGURE 2. Profiles of the initial data. Curve a is the projection of u on the (u, x)-plane, and curve b is the amplitude of u at $t = 0$ on the line $x = 4.5$.

(along the y-axis), and curve b in Figure 2 is the amplitude of $u\big|_{t=0}$ on the front of the nonperturbed soliton, i.e., along the line $\Gamma'_0 = \{x = 4.5, y \in [0, 10]\}$.

The plots in Figures 3–7 correspond to the solution of the KP-equation in the stable case ($\sigma = -1$). Figure 3 illustrates the dynamics of the front of a distorted soliton. One can readily see that the perturbation domain of the front Γ_t is expanding with increasing t. However, in this case the perturbation amplitude decreases and the front has a clear tendency to become straight in a smooth way.

The tree-dimensional plots of the solution at time moments $t = 0.1$ and $t = 0.2$ are shown in Figures 4 and 6. For better understanding, Figures 5 and 7 show the projection of the three-dimensional plot of u on the (u, x)-plane (along the y-axis)

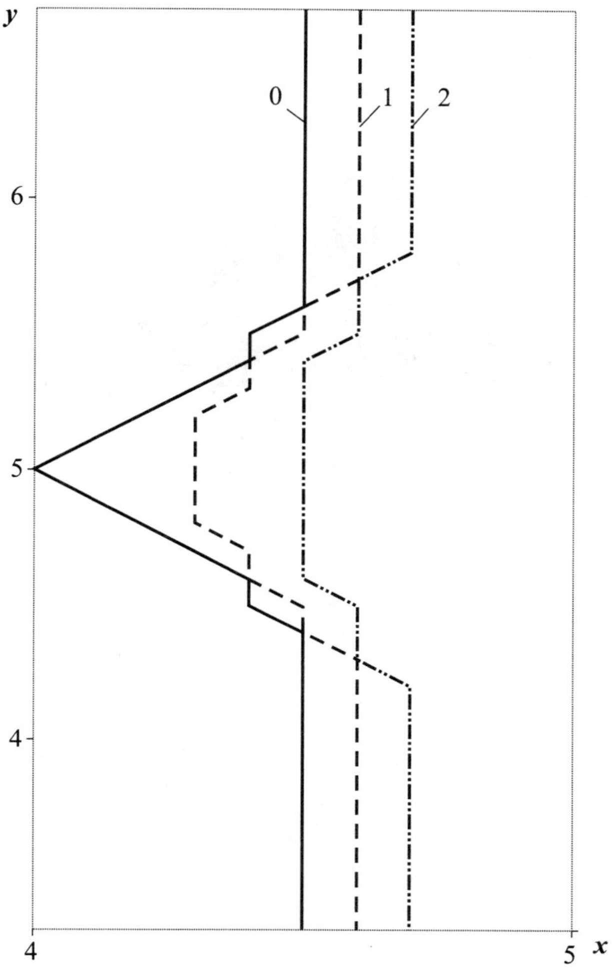

FIGURE 3. Dynamics of the front of the distorted soliton in the stable case ($\sigma = -1$). Curve 0 is the initial position; curves 1 and 2 are the front positions at $t = 0$ and $t = 0.2$.

(curve a) and the amplitude of u along the front of the nonperturbed soliton (i.e., along the line $x = 4.5 + t/2$) at $t = 0.1$ and $t = 0.2$. We see that a break in the front Γ_0 results in a soliton type perturbation of the amplitude (curve b in Figure 5). This perturbation decreases with time and simultaneously generates waves on the crest of the soliton, which propagate in opposite directions (curve b in Figure 7). One can clearly see that a tail appears behind the front of the distorted soliton (see Figures 4 and 6 and curves a in Figures 5 and 7). With increasing t the tail region expands along both the y- and x-axes, which includes the direction opposite to that of the crest motion. For small x there are oscillations in the tail, which can be explained by reflection from the boundary $x = 0$ of the computation region. The amplitude of these oscillations decreases if the boundary moves to the left. A perturbation in the solution ahead of the front of the distorted soliton is related to the error in

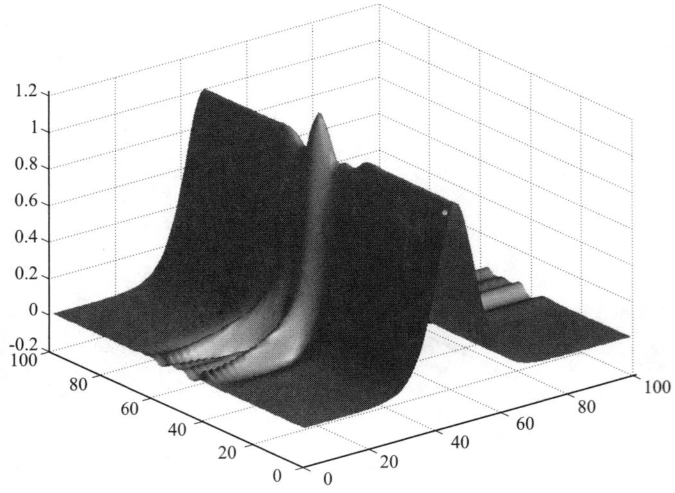

FIGURE 4. Stable case ($\sigma = -1$). The solution at $t = 0.1$.

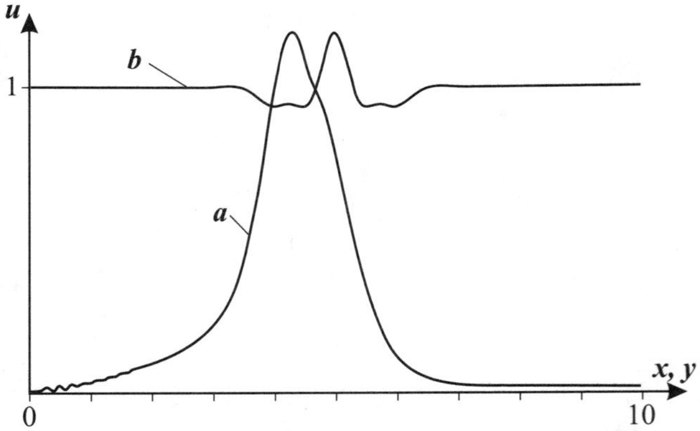

FIGURE 5. Profiles of the solution at $t = 0$ in the stable case. Curve a is the projection of the solution on the (u, x)-plane, and curve b is the amplitude of the solution on the front of the nonperturbed soliton.

numerical modeling. This perturbation is of the same order of magnitude as the local error of the difference scheme and decreases with discretization steps.

The results of numerical computations for the KP-equation with the same initial data (Figure 1) in the unstable case ($\sigma = 1$) are shown in Figures 8–12.

Because of the change in the sign of σ, two regions on the wave front become steeper during the time $\Delta t = 0.1$, while in the stable case there was no such effect (compare curves 1 in Figures 3 and 8). Nevertheless, at $t = 0.1$ the solutions obtained for $\sigma = 1$ and $\sigma = -1$ remain close to each other qualitatively and quantitatively (compare the plots in Figures 4 and 5 and in Figures 9 and 10). However,

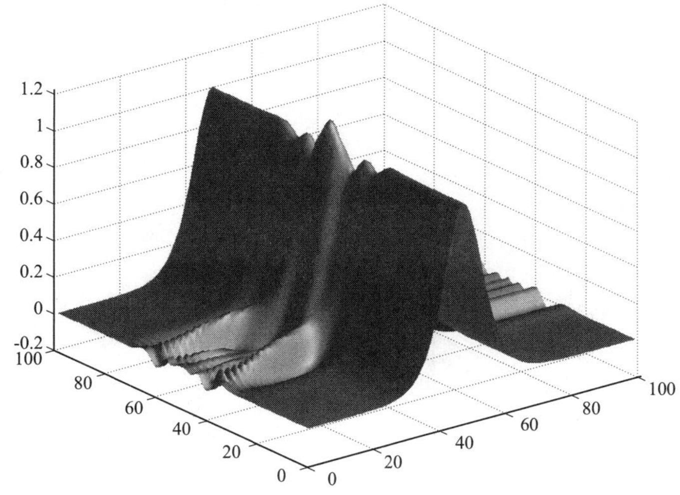

FIGURE 6. Stable case ($\sigma = -1$). The solution at $t = 0.2$.

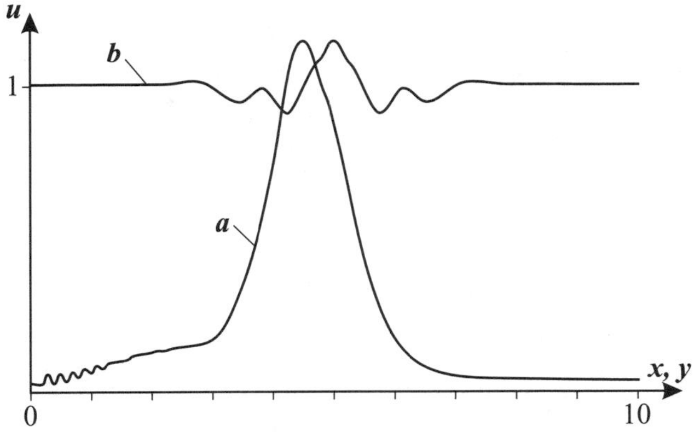

FIGURE 7. Profiles of the solution at $t = 0.2$ in the stable case.

qualitative distinctions between these solutions become clearer with increasing t. Namely, on the wave front there appear soliton type perturbations, which propagate rather fast in opposite directions (curve 2 in Figure 8).

As a whole, we see that the plots of the "exact" solution of the KP-equation agree surprisingly well with the description of the solution obtained by using the asymptotic formulas.

In conclusion, we note that the existence problem for traveling-wave type solutions of KP type equations was studied in numerous papers (e.g., see [**111, 2**] and the reference therein). Nevertheless, the functional approach to this problem does not permit describing the traveling wave evolution (the dynamics of its amplitude and front) in detail. As we see, the asymptotic approach to the KP type equation with a small dispersion (or, which is the same, to the equation with a

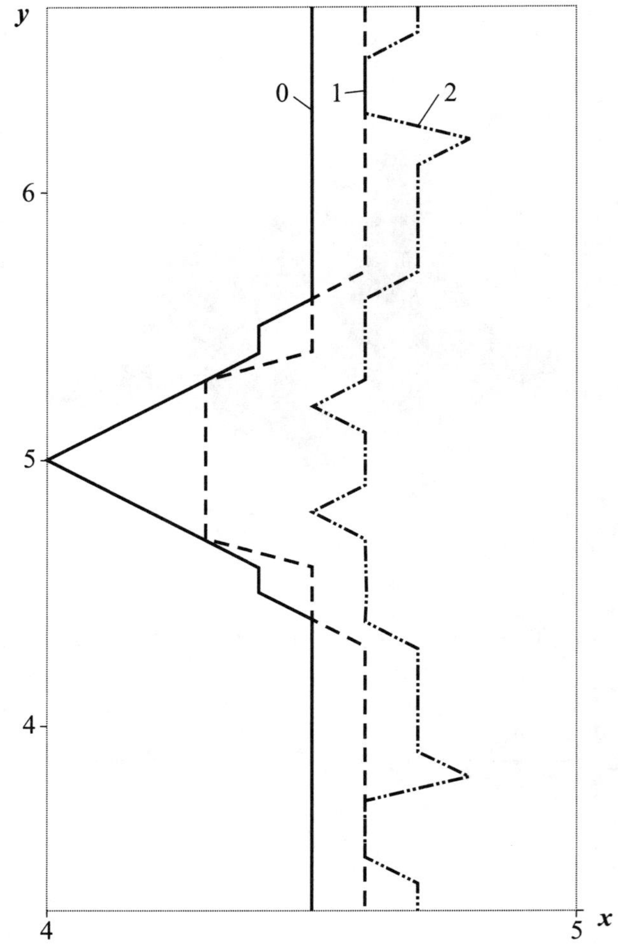

FIGURE 8. Dynamics of the front of the distorted soliton in the unstable case ($\sigma = 1$). Curve 0 is the initial position; curves 1 and 2 are the front positions at $t = 0.1$ and $t = 0.2$.

finite dispersion, but for large space and time variables and with sufficiently slowly varying amplitude and front) allows us to obtain such a description. In this case the classical asymptotic theory requires rather complicated constructions [**91, 89**], while the weak asymptotics method, as we show later, permits obtaining the same result by simple algebraic calculations.

PROOF OF THEOREM 4.1. We begin constructing a weak asymptotics of the solution of the KP type equation (4.14). We write the asymptotic solution as follows:

$$u = A\omega\Big(\beta\frac{S}{\varepsilon}\Big) + \varepsilon u^{-}(x,y,t)\omega_1\Big(-\beta\frac{S}{\varepsilon}\Big), \tag{4.37}$$

$$v = -v^{-}(x,y,t)\omega_1\Big(-\beta\frac{S}{\varepsilon}\Big), \tag{4.38}$$

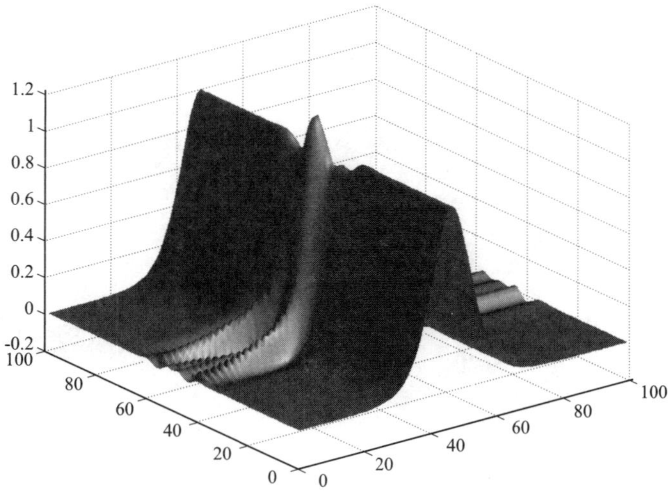

FIGURE 9. Unstable case ($\sigma = 1$). The solution at $t = 0.1$.

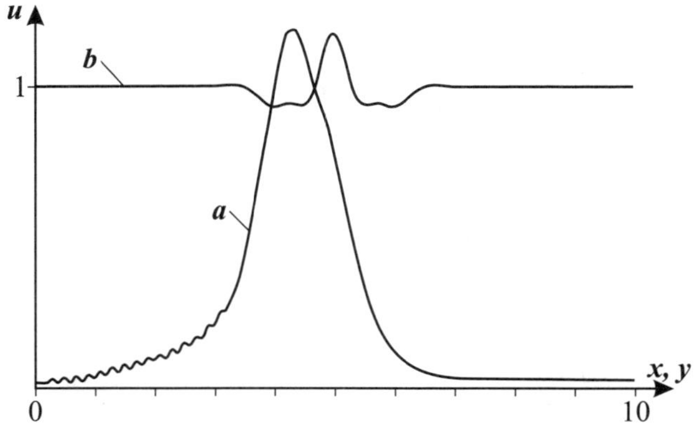

FIGURE 10. Profiles of the solution at $t = 0.1$ in the unstable case.

where $A = A(y,t) > 0$ for even m and $A \neq 0$ for odd m, $\beta = \beta(y,t) > 0$, u^-, v^-, and $S = S(x,y,t)$ are some arbitrary smooth functions, and ω, ω_1 are defined in (4.20), (4.21). By $\Gamma_0 = \{(x,y), S(x,y,0) = 0\}$ we denote the initial front of the distorted soliton $A\omega$. We assume that Γ_0 is a smooth one-dimensional curve without self-crossing. Since the solution of Eq. (4.14) does not depend on the variables symmetrically, as was assumed while deriving the Kadomtsev–Petviashvili equation, it is natural to assume that the curve Γ_0 has a single-valued projection on the y-axis. Thus it is natural to specify the initial front of the solution in the form $\Gamma_0 = \{(x,y), x = \varphi^0(y)\}$. It is also natural to assume that the property of the front to be single-valued when projected on the y-axis holds for some time $t \in [0,T]$. Consequently, for $t \leq T$, the wave front Γ_t can be described by a smooth function

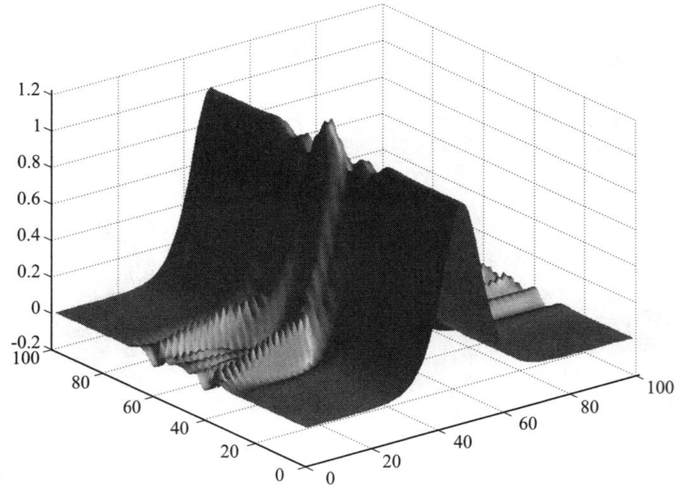

FIGURE 11. Unstable case ($\sigma = 1$). The solution at $t = 0.2$.

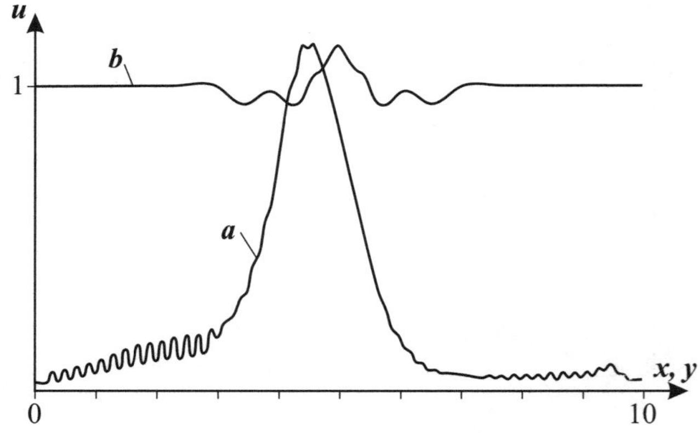

FIGURE 12. Profiles of the solution at $t = 0.2$ in the unstable case.

$\varphi(y, t)$ so that $\Gamma_t = \{(x, y), x = \varphi(x, t)\}$. This means that we can set

$$S = x - \varphi(y, t)$$

in (4.37), (4.38).

Carrying out simple calculations just as in Section 1 and Subsection 4.2 and taking into account the fact that $\omega(\eta)$ is even, we can readily see that

$$u = \varepsilon \frac{A}{\beta}\delta(S) + \varepsilon u^- H(-S) + O_{\mathcal{D}'}(\varepsilon^3), \qquad u^n = \varepsilon a_n \frac{A^n}{\beta}\delta(S) + O_{\mathcal{D}'}(\varepsilon^2),$$
$$v = -v^- H(-S) + O_{\mathcal{D}'}(\varepsilon^2), \tag{4.39}$$

where a_n, as usual, stands for the integral of ω^n, $n \geq 1$, and $\delta(S)$ is the delta function on the surface $S = 0$. Since we have $\partial S/\partial x = 1$, we choose dy to be the

1-form σ (see formula (4.9)) and obtain

$$(\delta(x-\varphi), \psi(x,y)) = \int \psi(\varphi(y,t), y)\, dy.$$

Now it readily follows from (4.39) that relation (4.15) holds for any test function if

$$\left(\frac{\partial^2}{\partial x \partial t} + \sigma \frac{\partial^2}{\partial y^2}\right)\left(\frac{A}{\beta}\delta(S) + u^- H(-S)\right) + \frac{\partial^2}{\partial x^2}\left(a_m \frac{A^m}{\beta}\delta(S)\right) = 0. \tag{4.40}$$

Differentiating and equating the coefficients of δ'', δ', and δ with zero, we obtain

$$\varphi_t - \sigma\varphi_y^2 = a_m A^{m-1}, \tag{4.41}$$

$$(\varphi_t - \sigma\varphi_y^2)u^-\Big|_{\Gamma_t} + \left\{\frac{\partial}{\partial t} - 2\sigma\varphi_y \frac{\partial}{\partial y} - \sigma\varphi_{yy}\right\}\frac{A}{\beta} = 0, \tag{4.42}$$

$$\left\{-\frac{\partial}{\partial t} + \varphi_t \frac{\partial}{\partial x} + 2\sigma\varphi_y \frac{\partial}{\partial y}\right\}u^-\Big|_{\Gamma_t} + \sigma\left\{u^-\Big|_{\Gamma_t}\varphi_{yy} + \left(\frac{A}{\beta}\right)_{yy}\right\} = 0. \tag{4.43}$$

Moreover, by equating the coefficient of H in (4.40) with zero, we obtain the first equation in (4.25) in the domain Ω_t^-, i.e., behind the front Γ_t. Similar simple calculations and relation (4.17) imply the second equation in (4.25) in the domain Ω_t^- and the relation $v^-|_{\Gamma_t} = r$ on the front, where r is defined in (4.27).

Now let us consider relations (4.16). Simple calculations show that

$$\begin{aligned}
(\varepsilon u_x)^2 &= \varepsilon A^2 \beta a_2' \delta(S) + O_{\mathcal{D}'}(\varepsilon^2), \\
(\varepsilon v_y)^2 &= \varepsilon\left(v^-|_{\Gamma_t}\varphi_y\right)^2 \beta a_2 \delta(S) + O_{\mathcal{D}'}(\varepsilon^2), \\
\varepsilon u v_y &= -\varepsilon v^-|_{\Gamma_t} A \varphi_y a_2 \delta(S) + O_{\mathcal{D}'}(\varepsilon^2).
\end{aligned}$$

Hence, taking into account the relation $v^-|_{\Gamma} = r$, we obtain

$$\begin{aligned}
\frac{\partial}{\partial t}\left(a_2 \frac{A^2}{\beta}\delta(S)\right) + \frac{\partial}{\partial x}\left\{\left[2\frac{m}{m+1}a_{m+1}\frac{A^{m+1}}{\beta} - 3A^2\beta a_2' - \sigma\frac{A^2}{\beta}\varphi_y^2 a_2\right]\delta(S)\right\}& \\
- 2\sigma\frac{\partial}{\partial y}\left(\frac{A^2}{\beta}\varphi_y^2 a_2 \delta(S)\right) &= 0.
\end{aligned}$$

Thus, equating the coefficients of δ' and δ with zero, we obtain

$$\varphi_t - \sigma\varphi_y^2 = \frac{1}{a_2}\left\{2\frac{m}{m+1}a_{m+1}A^{m-1} - 3a_2'\beta^2\right\}, \tag{4.44}$$

$$\frac{\partial}{\partial t}\left(\frac{A^2}{\beta}\right) - 2\sigma\frac{\partial}{\partial y}\left(\varphi_y \frac{A^2}{\beta}\right) = 0. \tag{4.45}$$

Now it should be noted that Eqs. (4.41) and (4.44) are mutually compatible only if

$$a_2 a_m A^{m-1} = \left\{2\frac{m}{m+1}a_{m+1}A^{m-1} - 3a_2'\beta^2\right\}.$$

This condition also implies the properties of the coefficients a_n and allows us to find β as a function of A in the form (4.28), where the constant γ is determined in (4.21).

Now we see that (4.44) and (4.45) form a closed system of equations for the functions A and φ. We rewrite this system in a more descriptive form. To this end, for $m \neq 5$, we introduce auxiliary functions ρ and w via the formulas $\rho = A^2/\beta$ and $w = -2\sigma\varphi_y$. We see that (4.45) is exactly the continuity equation in (4.22).

Then, we differentiate (4.44) with respect to y and multiply it by $2\sigma\rho$. Determining the "pressure" p by formula (4.23), we obtain the equation of motion in (4.22).

Returning from ρ, w to the initial functions A, φ, we must take into account the fact that the amplitude A can be negative for odd m. The factor s_m in formula (4.24) appears just because of this phenomenon.

In the case $m = 5$, we obtain $A^2/\beta = \text{const}$. Hence Eq. (4.35) is degenerate. It only shows that φ is a linear function of y. So in this case we consider Eqs. (4.36).

To complete the proof of the theorem, it remains to note that conditions (4.42) and (4.43) determine the values of u^- and its derivative along ν on the front Γ_t, i.e., they lead to the first two boundary conditions in (4.26). □

PROOF OF THEOREM 4.2. We begin with relation (4.32). Determining u and v by formulas (4.37) and (4.38), we calculate

$$(4.46)\qquad \int u\,dx = \varepsilon\frac{A}{\beta} + \varepsilon\int_{-\infty}^{\varphi} u^-\,dx + O(\varepsilon^2), \qquad \int v\,dx = -\int_{-\infty}^{\varphi} v^-\,dx + O(\varepsilon^2).$$

After differentiation, we transform (4.32) to the form

$$(4.47)\qquad \begin{aligned} &\frac{\partial}{\partial t}\frac{A}{\beta} - \sigma\varphi_y\frac{\partial}{\partial y}\big(v^-\big|_{x=\varphi}\big) + \{\varphi_t u^- - \sigma(\varphi_{yy}v^- + \varphi_y v_y^-)\}\big|_{x=\varphi} \\ &\quad + \int_{-\infty}^{\varphi}(u_t^- - \sigma v_{yy}^-)\,dx = O(\varepsilon^2). \end{aligned}$$

The integrand is equal to zero by virtue of (4.35). Next, the boundary condition (4.26) implies $v^-\big|_{x=\varphi} = A/\beta$. Hence, taking the second equation in (4.25) into account, we obtain

$$\frac{\partial}{\partial y}\big(v^-\big|_{x=\varphi}\big) = \frac{\partial}{\partial y}\frac{A}{\beta} + \varphi_y u^-\big|_{x=\varphi}.$$

Now it is easy to see that, up to $O(\varepsilon^2)$, relation (4.47) coincides with (4.42) and thus with the boundary condition (4.26), (4.27).

Let us consider relation (4.33). Easy calculations show that

$$(4.48)\qquad \begin{aligned} &\int xu\,dx = \varepsilon\varphi\frac{A}{\beta} + \varepsilon\int_{-\infty}^{\varphi} xu^-\,dx + O(\varepsilon^2), \\ &\int xv\,dx = -\int_{-\infty}^{\varphi} xv^-\,dx + O(\varepsilon). \end{aligned}$$

It follows from these formulas and the same argument as in the study of (4.32) that relations (4.32) take the form

$$\begin{aligned} &\varepsilon\frac{A}{\beta}\{\varphi_t - \sigma\varphi_y^2 - a_m A^{m-1}\} + \varepsilon\int_{-\infty}^{\varphi} x(u_t^- - \sigma v_{yy}^-)\,dx \\ &\quad + \varepsilon\varphi\bigg\{(\varphi_t - \sigma\varphi_y^2)u^-\Big|_{x=\varphi} + \Big(\frac{\partial}{\partial t} - 2\sigma\varphi_y\frac{\partial}{\partial y} - \sigma\varphi_{yy}\Big)\frac{A}{\beta}\bigg\} = O(\varepsilon^2). \end{aligned}$$

Relations (4.25), (4.42), and (4.43) thus imply (4.33) with accuracy $O(\varepsilon^2)$.

To consider (4.34) and (4.35), we obtain the following relations for the functions (4.37) and (4.38):

$$\varepsilon^2\int v_x v_y\,dx = -\varepsilon a_2\beta(v^-)^2\varphi_y + O(\varepsilon^2), \qquad \int u^n\,dx = \varepsilon\frac{A^n}{\beta}a_n + O(\varepsilon^2), \quad n \geq 1,$$

$$\int xu^2\,dx = \varepsilon a_2\varphi\frac{A^2}{\beta} + O(\varepsilon^2), \qquad \varepsilon^2\int xv_xv_y\,dx = -\varepsilon a_2\frac{A^2}{\beta}\varphi\varphi_y + O(\varepsilon^2),$$
$$\int(\varepsilon u_x)^2\,dx = \varepsilon a_2'\beta A^2 + O(\varepsilon^2), \qquad \int(\varepsilon v_y)^2\,dx = \varepsilon a_2\frac{A^2}{\beta}\varphi_y^2 + O(\varepsilon^2). \tag{4.49}$$

Now simple calculations show that relation (4.45) implies (4.37) up to $O(\varepsilon^2)$ and relations (4.44) and (4.45) imply (4.38) with the same accuracy.

Obviously, the last assertion of the theorem follows from the fact that, in the case of a constant amplitude and a linear front, all the remainders in formulas (4.46)–(4.49) are zero. □

4.3. Passage to the limit in the phase field system for media with surface tension. Historically, a description of the interface between phases in two-phase media (e.g., water/ice) was proposed by Stefan as the following problem with a free boundary $\Gamma_t = \overline{\Omega}_t^+ \cap \overline{\Omega}_t^-$:

$$\frac{\partial\theta^\pm}{\partial t} = k^\pm\Delta\theta^\pm, \qquad x\in\Omega_t^\pm, \quad t>0,$$
$$\left[k^\pm\frac{\partial\theta^\pm}{\partial\nu}\right]\Big|_{\Gamma_t} = -2V_\nu, \qquad \theta^\pm\Big|_{\Gamma_t} = 0, \tag{4.50}$$

where the temperature $\theta^\pm$ of both phases on Γ_t is assumed to be known and equal to the melting temperature. Here and in what follows, by $\Omega_t^\pm$ we denote the subdomains of $\Omega\subset\mathbb{R}^n$ filled with the phases "+" and "−," respectively, $\overline{\Omega}_t^+\cup\overline{\Omega}_t^- = \overline{\Omega}$, $k^\pm$ are the thermal conductivities of these phases, ν is the normal to the interface Γ_t, and V_ν is the normal velocity of Γ_t.

However, later on, fine physical experiments showed that, for some media, the temperature on the free interface can be somewhat different than the melting temperature. Moreover, it turned out that this temperature depends on the geometry of the boundary. To take this fact into account, the physicists proposed to abandon the assumption that the temperature is constant on Γ_t and to assume that the temperature obeys the so-called Gibbs–Thompson law. As a result, the following modified version of the Stefan problem (4.50) called the *Stefan problem with surface tension* was obtained:

$$\frac{\partial\theta^\pm}{\partial t} = k^\pm\Delta\theta^\pm, \qquad x\in\Omega_t^\pm, \quad t>0,$$
$$\left[k^\pm\frac{\partial\theta^\pm}{\partial\nu}\right]\Big|_{\Gamma_t} = -2V_\nu, \qquad [\theta^\pm]\Big|_{\Gamma_t} = 0, \qquad \varkappa_1\theta^\pm\Big|_{\Gamma_t} = V_\nu + \mathcal{K}, \tag{4.51}$$

where $\mathcal{K}$ is the mean curvature of Γ_t and $\varkappa_1$ is the constant characterizing the physical properties of the medium. In the modified Stefan problem, it is assumed that the temperature is continuous on Γ_t. For methodological reasons, we write this condition explicitly.

On the other hand, the Gibbs–Thompson condition was postulated by physicists only from physical considerations. Therefore, there is a problem of deriving this condition, as well as of the entire problem (4.51), accurately from some fundamental model. This problem can be solved in the framework of the Ginzburg–Landau theory of phase transitions. Taking all real parameters into account, one obtains very complicated models (e.g., see [**125, 3**]), which, nevertheless, can be simplified under sufficiently adequate restrictions on the number of parameters.

Recently, it has been universally recognized that such a simplified, but physically adequate, model is the so-called phase field system

$$\begin{aligned} \frac{\partial \theta}{\partial t} + \frac{\partial u}{\partial t} &= \langle \nabla, k(u)\nabla\theta \rangle, \qquad x \in \Omega, \quad t > 0, \\ \varepsilon \frac{\partial u}{\partial t} &= \varepsilon \Delta u + \frac{1}{\varepsilon} f(u) + \varkappa(1-u^2)\theta, \end{aligned} \tag{4.52}$$

proposed by Caginalp [**8, 9, 10**]. Here u is the so-called *order function* equal to ± 1 in the domains filled with pure phases, $k(u) > 0$ is a smooth function such that $k(\pm 1) = k^{\pm}$, $\varkappa = \text{const}$, and $f(u)$ is a smooth function satisfying the condition $f(\pm 1) = 0$. As a rule, f and k are taken to be polynomial,

$$f(u) = u - u^{2m+1}, \qquad m \geq 1, \quad k(u) = \frac{k^+ + k^-}{2} + \frac{k^+ - k^-}{2} u. \tag{4.53}$$

From the physical viewpoint, the phase field system (4.52) describes the substance microscopically, and the parameter ε, which is assumed to be small, characterizes the width of the transient zone between the domains filled with pure phases.

From the mathematics viewpoint, we can treat problem (4.52) as a regularization of the modified Stefan problem (4.51) and $\varepsilon \ll 1$ as the parameter of regularization. Correspondingly, under this approach, $\theta \in C^{\infty}$ is a regularization of the continuous function $\theta^{\pm}$, $x \in \Omega_{+}^{\pm}$, and u is a regularization of the discontinuous function $\overline{u}$ equal to 1 in Ω_t^+ and equal to -1 in Ω_t^-.

The phase field system (supplemented with natural boundary conditions on the external boundary $\partial\Omega$ and with initial data for $t = 0$) has been studied by many authors who used both the asymptotic theory (see [**8, 9, 10, 117, 20, 91, 108, 109**] and the references therein) and the functional methods (see [**116, 124**] and the references therein). It was established that if there is no contact between the external boundary and the free interface, i.e., if

$$\Gamma_t \cap \partial\Omega = \varnothing, \qquad t \geq 0, \tag{4.54}$$

then the passage to the limit in (4.52) as $\varepsilon \to 0$ gives exactly the modified Stefan problem (4.51) and the parameter $\varkappa_1$ in the Gibbs–Thompson law is uniquely determined by the substance microcharacteristics such as the latent heat of melting l, the relaxation time t_0, the surface tension ξ, and the width of the transient zone a (in (4.52) these parameters are "hidden" in dimensionless variables). We also note that for other relations between l, t_0, ξ, and a and for a different arrangement of the parameter ε in the phase field system, the passage to the limit as $\varepsilon \to 0$ yields different limit problems, in particular, the Stefan problem (4.50) [**10**].

Our aim is to study the situations in which the domain Ω contains several domains $\Omega_{t,i}^{\pm}$ separated by the boundaries $\Gamma_{t,i}$ and to describe the confluence of free boundaries. We shall use the weak asymptotics method. As has already been pointed out several times, the central point here is to postulate a well-conditioned definition of a weak solution that admits passing to the limit as $\varepsilon \to 0$. In Section 3 we tested our definition in the one-dimensional case. A natural generalization of our definition to the multidimensional case is the following definition, which we give for $\Omega \subset \mathbb{R}^2$ in order to simplify the formulas.

DEFINITION 4.2. A sequence of pairs $\big(u = u(x,t,\varepsilon), \theta = \theta(x,t,\varepsilon)\big)$ with $\varepsilon > 0$ belonging to $C^{\infty}(0,T; C^{\infty}(\Omega))$ is called a *weak asymptotic solution of the phase field system* (4.52) with accuracy $O(\varepsilon)$ if the following relations hold for any test functions ψ and $\vec{\psi} = (\psi_1, \psi_2) \in \mathcal{D}(\Omega)$:

$$\int \Big(\frac{\partial u}{\partial t} + \frac{\partial \theta}{\partial t}\Big)\psi\, dx + \int k(u)\langle \nabla\theta, \nabla\psi\rangle\, dx = O(\varepsilon), \tag{4.55}$$

$$\begin{aligned} 2\varepsilon \int \frac{\partial u}{\partial t}\langle \vec{\psi}, \nabla u\rangle\, dx &+ \varepsilon \int \Big(\frac{\partial \psi_1}{\partial x_1} - \frac{\partial \psi_2}{\partial x_2}\Big)\Big(\Big(\frac{\partial u}{\partial x_1}\Big)^2 - \Big(\frac{\partial u}{\partial x_2}\Big)^2\Big)\, dx \\ &+ 2\varepsilon \int \Big(\frac{\partial \varphi_1}{\partial x_2} + \frac{\partial \varphi_2}{\partial x_1}\Big)\frac{\partial u}{\partial x_1}\frac{\partial u}{\partial x_2}\, dx - \frac{1}{\varepsilon(m+1)}\int F(u)\langle \nabla, \vec{\psi}\rangle\, dx \\ &+ 2\varkappa \int \Big(u - \frac{1}{3}u^3\Big)\langle \nabla, \theta\vec{\psi}\rangle = O(\varepsilon), \end{aligned} \tag{4.56}$$

where $F(u) = u^{2(m+1)} - (m+1)u^2 + m$ and the integrals are taken over $\mathbb{R}^2_x$.

For smooth functions u, the left-hand side in (4.56) is obtained by multiplying the second equation in (4.52) by $\langle \vec{\psi}, \nabla u\rangle$ and integrating by parts. Relation (4.55) follows from the standard definition of a weak solution if it is understood as a sufficiently smooth function of t ranging in $\mathcal{D}'(\Omega)$.

REMARK 4.1. Selecting special test functions from $\mathcal{D}(\mathbb{R}^2_x)$, it is easy to modify Definition 4.2 so that the boundary conditions on the external boundary of the domain Ω are taken into account. We do not do this on purpose in order to concentrate our attention on studying the evolution and the interaction of free boundaries.

We shall discuss the confluence of boundaries in Subsection 8.3. Now we test Definition 4.2 in the case of a single free boundary and show that, in this case, the weak asymptotics method gives the same results as other approaches. Namely, we prove the following assertion.

THEOREM 4.3. *Suppose that for $t \in (0, T]$ there exists a smooth solution of the initial boundary value problem for the modified Stefan problem* (4.51) *and condition* (4.54) *is satisfied. Then the weak solution of the phase field system* (4.52), (4.53) *with accuracy $O_{\mathcal{D}'} * (\varepsilon)$ has the form*

$$\begin{aligned} u &= \omega_0\Big(\beta\frac{S}{\varepsilon}\Big) = \operatorname{sign}(S) + O_{\mathcal{D}'}(\varepsilon^2), \\ \theta &= \frac{1}{2}(\theta_c^+ + \theta_c^-) + \frac{1}{2}(\theta_c^+ - \theta_c^-)\omega_1\Big(\beta\frac{S}{\varepsilon}\Big) \\ &= +\frac{1}{2}(\theta_c^+ + \theta_c^-) + \frac{1}{2}(\theta_c^+ - \theta_c^-)\operatorname{sign}(S) + O_{\mathcal{D}'}(\varepsilon^2). \end{aligned} \tag{4.57}$$

Here $S = t + \Phi(x)$, $\beta = 1/|\nabla\Phi|$, $\theta_c^\pm$ are smooth continuations of $\theta^\pm$ to the entire domain Ω, $\omega_0(\eta)$ is a solution of the equation

$$\Big(\frac{d\omega_0}{d\eta}\Big)^2 - \frac{1}{m+1}F(\omega_0) = 0 \tag{4.58}$$

such that $\omega_0 \to \pm 1$ as $\eta \to \pm\infty$, and $\omega_1(z/\varepsilon)$ is an even regularization of the function $\operatorname{sign}(z)$.

Prior to proving this theorem, which is a rather simple task, we discuss our results obtained by the numerical simulation of the solution of the phase field system.

First, we consider the results illustrating the stability of the solution of system (4.52) under the choice of a regularization of the functions $\overline{u}|_{t=0}$ and $\overline{\theta}^{\pm}|_{t=0}$. The point is that, for each $m \geqslant 1$, there exists a special regularization $u^0(x, \varepsilon, m)$ determined by the solution of the model equation (4.58) and corresponding to the self-similar solution, i.e., to the solution whose profile does not change with increasing t. For $m = 1$ the leading term $u^0(x, \varepsilon, m)$ has the form

$$\begin{aligned} \omega_0 &= \tanh\big(\beta(x - x^0)/\varepsilon\big) \quad \text{for } x \in \mathbb{R}^1, \\ \omega_0 &= \tanh\big(\beta\varphi^0(x)/\varepsilon\big) \quad \text{for } x \in \mathbb{R}^n, \end{aligned} \tag{4.59}$$

where $\beta = 1/\sqrt{2}$ and $\varphi^0 \in C^\infty$ satisfies the condition $\varphi^0\big|_{\Gamma_0} = 0$. Constructing the asymptotics solution, it is easy to see that the choice of the regularization $\theta^0(x, \varepsilon, m)$ is not essential, which is related to the fact that the functions $\overline{\theta}^{\pm}$ on Γ_t have weak discontinuities of type x_+ (e.g., see [**21, 117**]). On the other hand, $\overline{u}$ has a Heaviside function type singularity on Γ_t, and the choice of another regularization means that the characteristics of the profile vary by a value of $O(1)$ in the pointwise sense (however, this means "by a value of $O(\varepsilon)$ in the $\mathcal{D}'(\Omega)$-sense"). It is clear *a priori* that the choice of $u^0(x, \varepsilon)$ in some neighborhood of $u^0(x, \varepsilon, m)$ does not lead to a significant change in the solution. Nevertheless, as far as we know, the process of establishing the "self-similar" behavior of the solution has never been studied in detail.

Figure 13, where the wave profiles shifted to the conventional point $x = 0$ are depicted, presents some results of numerical simulation of the solution of system (4.52) in the one-dimensional case with $u\big|_{t=0} = u^0(x, \varepsilon)$ different from the self-similar initial data (4.59). Namely, for $m = 1, 2$ we set

$$\theta\Big|_{t=0} \stackrel{\text{def}}{=} 5, \qquad u^0 = \frac{2}{\pi}\arctan\big(\beta(x_0 - x)/\varepsilon\big),$$

and for $m = 2, 3$ we choose $\theta^0 = 5$, determine u^0 by formula (4.59), and perform calculations varying the parameter β. In all computations, we set $\varepsilon = 0.1$.

It is easily seen that a temporary boundary layer arises: in the time interval $\Delta t \sim O(\varepsilon^2)$ the wave profile becomes self-similar. A similar result is also obtained for $n = 2$. So the initial regularization u^0 can be chosen arbitrarily.

The same figures clearly show that the qualitative picture of the kink motion remains the same with varying m. There are changes only in the steepness of the kink profile and in the velocity of its motion (see also Subsection 8.4, Figure 19).

However, it is more natural to avoid the boundary layer and to choose the regularization u^0 in a self-similar form, i.e., as a solution of Eq. (4.58).

The next series of numerical experiments deals with the dynamics of the free boundary prior to the moment of interaction. In the one-dimensional case the evolution of the temperature, as well as of the kink and Γ_t, is sufficiently uniform. The character of this evolution is illustrated by the plots given in Figure 13, and in Subsection 8.4, Figures 16 and 17.

In the multidimensional case, as it follows from the Gibbs–Thompson law, the dynamics of the free boundary is determined not only by the temperature but also by the geometry of Γ_t.

To find the geometric properties of the dynamics of Γ_t, we calculated the motion of the free boundary in the following cases: a) the curvature of Γ_0 is everywhere positive, b) Γ_0 has segments of negative mean curvature.

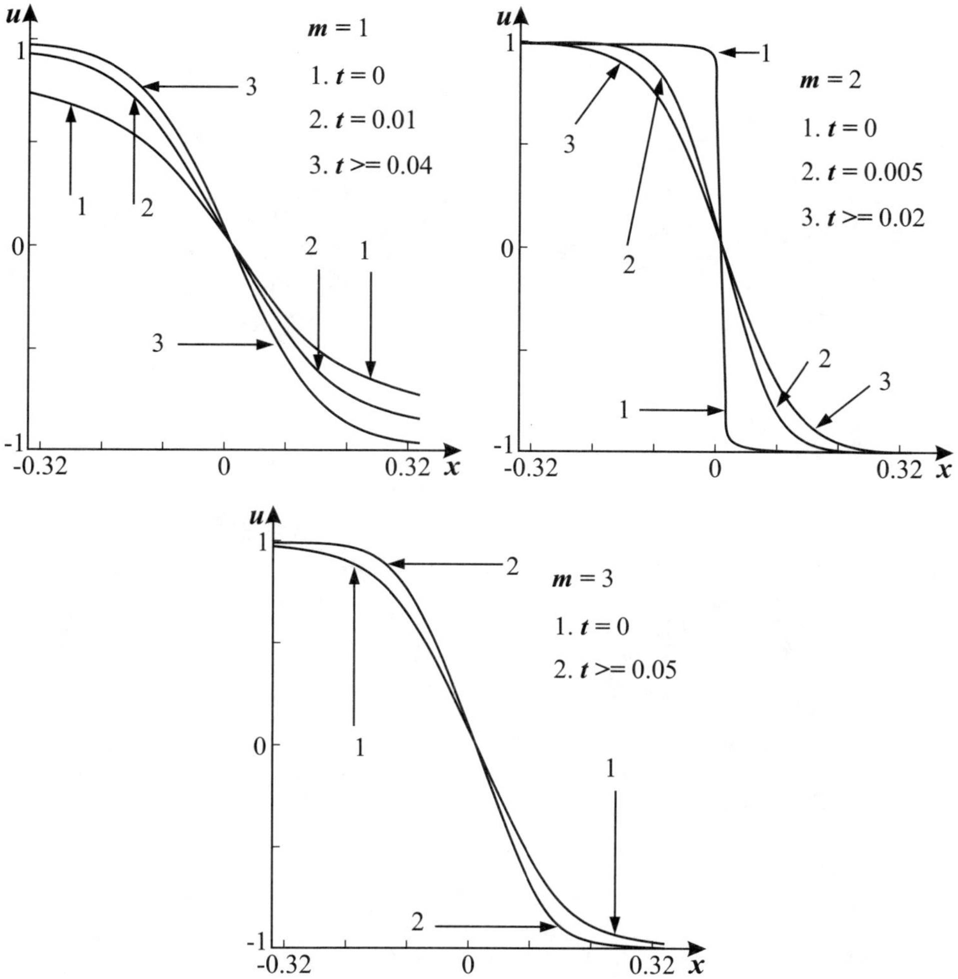

FIGURE 13. Outcome of the self-similar regime.

Namely, in case a) problem (4.52), (4.53) with $m = 1$ was solved with the initial data

$$\theta^0 = 5, \qquad u^0 = \tanh\left(\frac{\psi(x) - |x - x^0|}{\sqrt{2}\varepsilon}\right), \tag{4.60}$$

where $\psi(x) = r_1 + r_2\big(a(x_1 - x_{10}) + b(x_2 - x_{20})\big)$, $r_1 = 0.32$, $r_2 = 0.3$, $a = 2$, and $b = 0.1$. Figure 14 shows that the evolution of the free boundary is sufficiently smooth and the curvature of Γ_t decreases with increasing t.

In case b) the initial data are given by formula (4.60) with the function ψ of the form

$$\psi(x) = r_1 + r_2 \cos\left(2\arctan\left(\frac{x_1 - x_{10}}{x_2 - x_{20}}\right)\right), \qquad r_1 = 0.32, \quad r_2 = 0.2.$$

Figure 15 clearly shows that, at the initial stage of evolution, the velocity of the segments of Γ_t of negative curvature is much larger than that of the segments

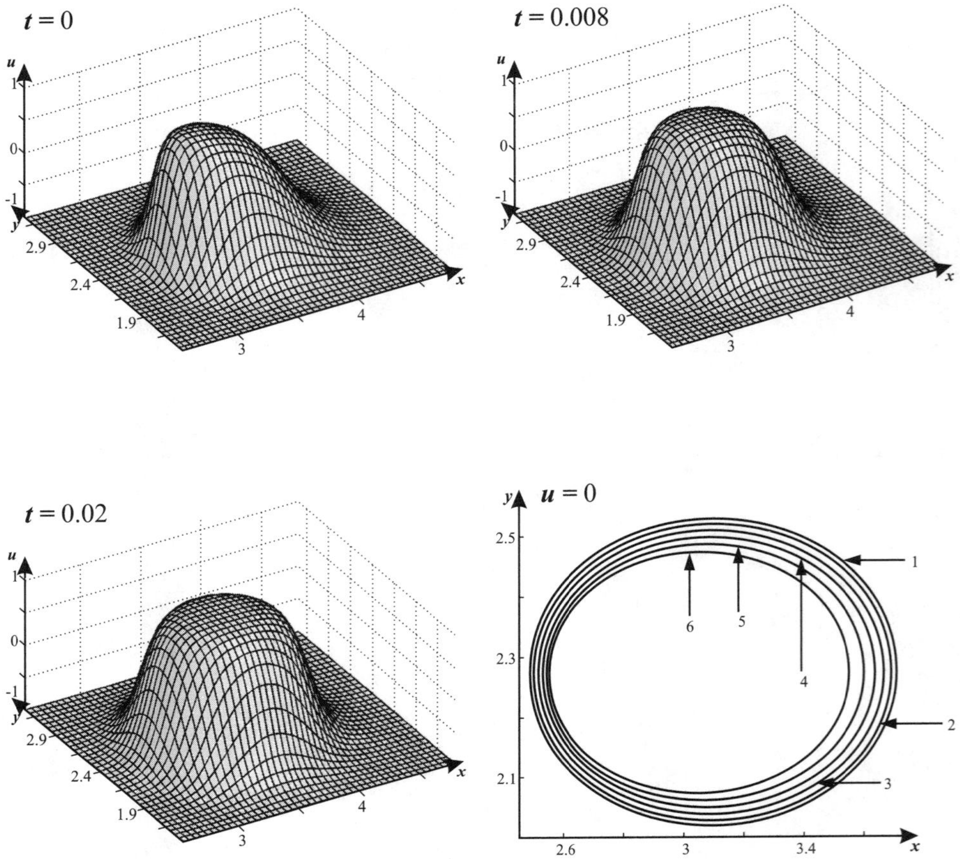

FIGURE 14. Dynamics of the free boundary in the case of positive curvature. The zero level lines are given for: 1: $t = 0$, 2: $t = 0.004$, 3: $t = 0.008$, 4: $t = 0.012$, 5: $t = 0.016$, 6: $t = 0.02$.

of Γ_t of positive curvature. As a result, the concave segments of the boundary expand practically instantly (for a time $t \sim 0.004$), while the variations in the convex segments are insignificant. The tendency for further evolution of Γ_t can be easily observed in the plots in Figure 15. Namely, with increasing t, first, the boundary segments of negative curvature disappear and then the evolution of Γ_t is the same as in case a).

PROOF OF THEOREM 4.3. Suppose that $\omega_0 = \omega_0(\eta)$ is a solution of Eq. (4.58) and $\omega_1(x/\varepsilon)$ is an odd regularization of the function $\operatorname{sign}(x)$, which need not coincide with $\omega_0(x/\varepsilon)$. We introduce a function $\omega(\eta)$ such that

$$\omega_0(\eta) = -1 + 2\int_{-\infty}^{\eta} \omega(\eta')\,d\eta', \qquad \int_{-\infty}^{\infty} \omega(\eta)\,d\eta = 1.$$

It easy to see that the function $\omega_0(\eta)$ belongs to $C^\infty(\mathbb{R}^1)$, is odd, and tends to ± 1 at an exponential rate as $\eta \to \pm\infty$. Respectively, the function $\omega(\eta) \in C^\infty(\mathbb{R}^1)$ is even and tends to zero at an exponential rate as $\eta \to \pm\infty$.

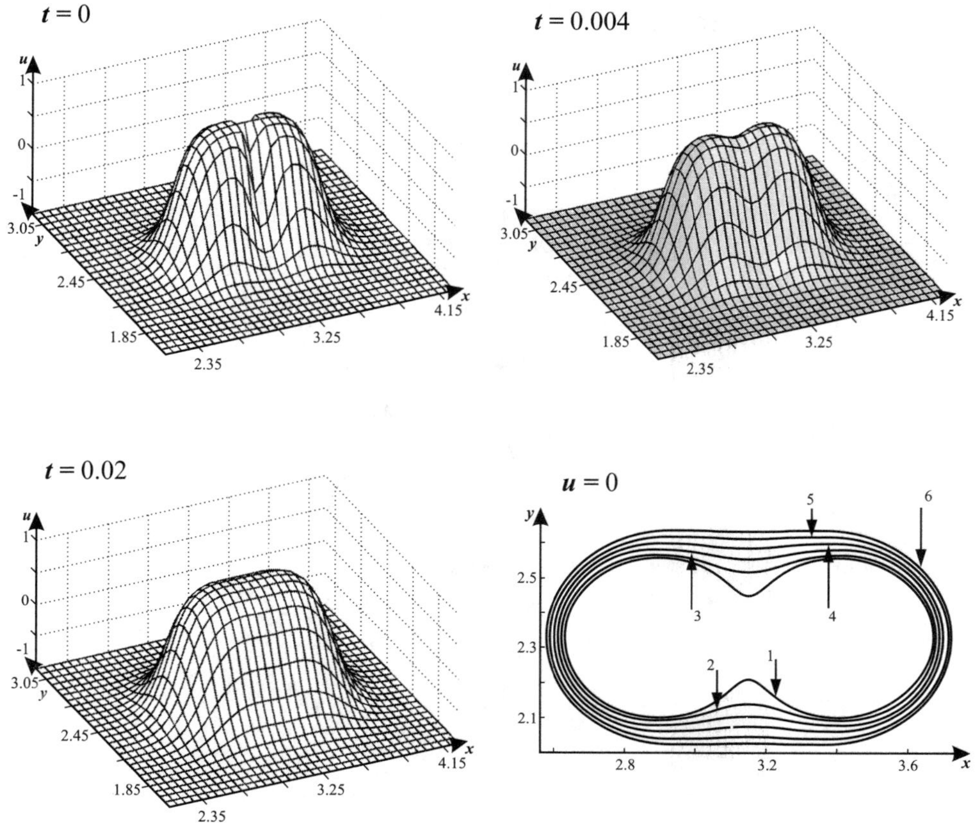

FIGURE 15. Dynamics of the free boundary in the case of segments of negative curvature. The zero level lines are given for: 1: $t = 0$, 2: $t = 0.004$, 3: $t = 0.008$, 4: $t = 0.012$, 5: $t = 0.016$, 6: $t = 0.02$.

We represent the leading term of the asymptotic solution of the phase field system as

$$u = \omega_0\left(\beta\frac{S}{\varepsilon}\right), \qquad \theta = \gamma^+ + \gamma^-\omega_1\left(\beta\frac{S}{\varepsilon}\right), \tag{4.61}$$

where $S = \Phi + t$ and $\Phi = \Phi(x)$, $\beta = \beta(x) > 0$, and $\gamma^\pm = \gamma^\pm(x,t)$ are the desired smooth functions such that

$$\gamma^-\big|_{\Phi+t=0} = 0. \tag{4.62}$$

By $\omega_0(\eta)$ we denote the solution of Eq. (4.58) and by $\omega(z/\varepsilon)$ an arbitrary odd regularization of the function $\operatorname{sign}(z)$.

Similarly to the constructions in Section 1, it is easy to show that the following relations holds for any $\omega_i = \omega_i(\eta)$, $i = 0, 1$, and even $\omega = \omega(\eta)$:

$$\frac{1}{\varepsilon}\omega\left(\beta\frac{S}{\varepsilon}\right) = \frac{1}{\beta}\delta(S) + O_{\mathcal{D}'}(\varepsilon^2), \qquad \omega^l\left(\beta\frac{S}{\varepsilon}\right) = \varepsilon\frac{a_l}{\beta}\delta(S) + O_{\mathcal{D}'}(\varepsilon^3),$$

$$\omega_0^{2k+1}\Big(\beta\frac{S}{\varepsilon}\Big) = \operatorname{sign}(S) + O_{\mathcal{D}'}(\varepsilon^2), \qquad \omega_0^{2k}\Big(\beta\frac{S}{\varepsilon}\Big) = 1 + O_{\mathcal{D}'}(\varepsilon),$$

where $k, l \geq 1$, $a_l = \int \omega^l(\eta)\,d\eta$, the functional $\delta(S)$ acts as

$$(\delta(S), \psi) = \int_{S=0} \psi\sigma,$$

and the 1-form σ is a solution of the equation $dS \wedge \sigma = dx_1 \wedge dx_2$. Moreover, for any smooth even function $F(z)$, $F(\pm 1) = 0$, we have

$$F\Big(\omega_0\Big(\beta\frac{S}{\varepsilon}\Big)\Big) = \varepsilon\frac{a_F}{\beta}\delta(S) + O_{\mathcal{D}'}(\varepsilon^3), \qquad a_F = \int F(\omega_0(\eta))\,d\eta.$$

Furthermore, taking into account the fact that ω is even, we obtain

$$\begin{aligned}\varepsilon\Big(\frac{\partial}{\partial t}\omega_0\Big(\beta\frac{S}{\varepsilon}\Big)\Big)\Big(\frac{\partial}{\partial x_i}\omega_0\Big(\beta\frac{S}{\varepsilon}\Big)\Big) &= 4\Big(\frac{\beta^2}{\varepsilon}\frac{\partial\Phi}{\partial x_i} + \beta\frac{S}{\varepsilon}\frac{\partial\beta}{\partial x_i}\Big)\omega^2\Big(\beta\frac{S}{\varepsilon}\Big)\\ &= 4a_2\beta\frac{\partial\Phi}{\partial x_i}\delta(S) + O_{\mathcal{D}'}(\varepsilon^2).\end{aligned}$$

Similar calculations imply the formulas

$$\begin{aligned}\varepsilon\Big(\frac{\partial}{\partial x_i}\omega_0\Big(\beta\frac{S}{\varepsilon}\Big)\Big)\Big(\frac{\partial}{\partial x_j}\omega_0\Big(\beta\frac{S}{\varepsilon}\Big)\Big) &= 4a_2\beta\frac{\partial\Phi}{\partial x_i}\frac{\partial\Phi}{\partial x_j}\delta(S) + O_{\mathcal{D}'}(\varepsilon^2),\\ \theta\frac{\partial}{\partial x_i}\Big(\omega_0^l\Big(\beta\frac{S}{\varepsilon}\Big)\Big) &= 2\theta\Big|_{\Gamma_t}\frac{\partial\Phi}{\partial x_i}\delta(S) + O_{\mathcal{D}'}(\varepsilon^2),\end{aligned}$$

where $l = 1, 3$ and $\Gamma_t = \{x, \Phi(x) + t = 0\}$. It should be noted that in the last relation we used the fact that the function θ is indeterminate.

Let us consider relation (4.56). It is easy to see that the following vector relation is necessary and sufficient for a vector-function $\vec{\psi} \in \mathcal{D}(\Omega)$ to satisfy relation (4.56):

$$\begin{aligned}2\varepsilon\frac{\partial u}{\partial t}\nabla u - \varepsilon\widehat{L}_1\Big(\Big(\frac{\partial u}{\partial x_1}\Big)^2 - \Big(\frac{\partial u}{\partial x_2}\Big)^2\Big) - 2\varepsilon\widehat{L}_2\frac{\partial}{\partial x_1}\frac{\partial u}{\partial x_2}\\ + \frac{1}{\varepsilon(m+1)}\nabla F(u) - 2\varkappa\theta\nabla\Big(u - \frac{1}{3}u^3\Big) = O_{\mathcal{D}'}(\varepsilon),\end{aligned} \tag{4.63}$$

where the operators $\widehat{L}_i$ are column vectors and

$$\widehat{L}_1 = \begin{pmatrix}\partial/\partial x_1\\ -\partial/\partial x_2\end{pmatrix}, \qquad \widehat{L}_2 = \begin{pmatrix}\partial/\partial x_2\\ \partial/\partial x_1\end{pmatrix}.$$

Relation (4.63) and the above formulas imply the relation

$$\begin{aligned}\delta(S)2\Big(\beta - \frac{\varkappa}{3a_2}\theta\Big|_{\Gamma_t}\Big)\nabla\Phi - \widehat{L}_1\Big\{\beta\Big(\Big(\frac{\partial\Phi}{\partial x_1}\Big)^2 - \Big(\frac{\partial\Phi}{\partial x_2}\Big)^2\Big)\delta(S)\Big\}\\ 2\widehat{L}_2\Big\{\beta\frac{\partial\Phi}{\partial x_1}\frac{\partial\Phi}{\partial x_2}\delta(S)\Big\} + \frac{a_F}{4a_2(m+1)}\nabla\Big\{\frac{1}{\beta}\delta(S)\Big\} = 0.\end{aligned} \tag{4.64}$$

Applying the operators $\widehat{L}_j$ and equating the coefficients of $\delta'(S)$ and $\delta(S)$ with zero, we see that (4.64) implies

$$\frac{a_F}{4a_2(m+1)}\frac{1}{\beta^2}\nabla\Phi = \Big(\Big(\frac{\partial\Phi}{\partial x_1}\Big)^2 - \Big(\frac{\partial\Phi}{\partial x_2}\Big)^2\Big)\widehat{L}_1\Phi + 2\frac{\partial\Phi}{\partial x_1}\frac{\partial\Phi}{\partial x_2}\widehat{L}_2\Phi, \tag{4.65}$$

$$\frac{a_F}{4a_2(m+1)}\nabla\frac{1}{\beta}+2\Big(\beta-\frac{\varkappa}{3a_2}\theta\Big|_{\Gamma_t}\Big)\nabla\Phi \\ -\widehat{L}_1\Big\{\beta\Big(\Big(\frac{\partial\Phi}{\partial x_1}\Big)^2-\Big(\frac{\partial\Phi}{\partial x_2}\Big)^2\Big)\Big\}-2\widehat{L}_2\Big\{\beta\frac{\partial\Phi}{\partial x_1}\frac{\partial\Phi}{\partial x_2}\Big\}=0. \tag{4.66}$$

Then easy calculations show that

$$\Big(\Big(\frac{\partial\Phi}{\partial x_1}\Big)^2-\Big(\frac{\partial\Phi}{\partial x_2}\Big)^2\Big)\widehat{L}_1\Phi+2\frac{\partial\Phi}{\partial x_1}\frac{\partial\Phi}{\partial x_2}\widehat{L}_2\Phi=|\nabla\Phi|^2\nabla\Phi.$$

Hence (4.65) implies the following relation between β and $|\nabla\Phi|$:

$$\beta^2|\nabla\Phi|^2=\frac{a_F}{4a_2(m+1)}. \tag{4.67}$$

We also point out that, introducing ω_0 as a solution of (4.58), we obtain

$$4a_2(m+1)=a_F.$$

Therefore, it follows from (4.67) that $\beta=1/|\nabla\Phi|$.

Furthermore, to reduce relation (4.66), we perform some algebraic calculations and thus obtain the identity

$$\widehat{L}_1\Big(\Big(\frac{\partial\Phi}{\partial x_1}\Big)^2-\Big(\frac{\partial\Phi}{\partial x_2}\Big)^2\Big)+2\widehat{L}_2\Big(\frac{\partial\Phi}{\partial x_1}\frac{\partial\Phi}{\partial x_2}\Big)=2\nabla\Phi\nabla\Phi. \tag{4.68}$$

Next, expressing β in terms of $|\nabla\Phi|$, we obtain the other identity

$$\Big(\Big(\frac{\partial\Phi}{\partial x_1}\Big)^2-\Big(\frac{\partial\Phi}{\partial x_2}\Big)^2\Big)\widehat{L}_1\beta+2\frac{\partial\Phi}{\partial x_1}\frac{\partial\Phi}{\partial x_2}\widehat{L}_2\beta+|\nabla\Phi|^2\nabla\beta=2\langle\nabla\Phi,\nabla\beta\rangle\nabla\Phi. \tag{4.69}$$

Using identities (4.68), (4.69), we readily see that the vector relation (4.66) can be reduced to the scalar relation

$$\beta-\frac{\varkappa}{3a_2}\theta\Big|_{\Gamma_t}=\langle\nabla,\beta\nabla\Phi\rangle. \tag{4.70}$$

It remains to note that the vector $\nu\overset{\text{def}}{=}-\nabla\Phi/|\nabla\Phi|$ is the unit normal on the curve $\Gamma_t=\{x,\Phi(x)+t=0\}$ and the velocity V_ν of Γ_t along ν is equal to $1/|\nabla\Phi|$. Hence (4.70) is exactly the Gibbs–Thompson condition in (4.51) with the mean curvature $\mathcal{K}=\operatorname{div}\nu$ and the constant $\varkappa_1=\varkappa/3a_2$.

To conclude the analysis of (4.56), we note that relation (4.67) can be interpreted as follows. We write explicit formulas for the constants a_2 and a_F as integrals of the corresponding functions. Then (4.67) can be rewritten as

$$\int\Big\{(m+1)\beta^2|\nabla\Phi|^2\Big(\frac{d\omega_0}{d\eta}\Big)^2-F\big(\omega_0(\eta)\big)\Big\}\,d\eta=0. \tag{4.71}$$

Obviously, a variation in the "constant" $\beta^2|\nabla\Phi|^2$ only changes the scale of η. Hence, setting $\beta|\nabla\Phi|=1$, we see that the integrand in (4.71) becomes exactly the left-hand side of the model equation (4.58). So relation (4.58) is a sufficient condition for (4.71) to hold. In turn, (4.71) follows from the definition of the weak asymptotics of the solution.

Now let us discuss relation (4.55). Taking into account condition (4.62) and performing simple calculations, we obtain

$$k(u)\nabla\theta=r^++r^-\operatorname{sign}(S)+O_{\mathcal{D}'}(\varepsilon^2),$$

where

$$r^+ = \frac{1}{2}k^+\nabla(\gamma^+ + \gamma^-) + \frac{1}{2}k^-\nabla(\gamma^+ - \gamma^-),$$
$$r^- = \frac{1}{2}k^+\nabla(\gamma^+ + \gamma^-) - \frac{1}{2}k^-\nabla(\gamma^+ - \gamma^-). \tag{4.72}$$

These relations and (4.55) imply

$$2\{1 - \langle r^-, \nabla\Phi\rangle\}\delta(S) + \left\{\frac{\partial\gamma^-}{\partial t} - \langle\nabla, r^-\rangle\right\}\operatorname{sign}(S) + \frac{\partial\gamma^+}{\partial t} - \langle\nabla, r^-\rangle = 0. \tag{4.73}$$

We denote $\Omega_t^+ = \{x \in \Omega, \Phi(x) + t > 0\}$ and $\Omega_t^+ = \{x \in \Omega, \Phi(x) + t < 0\}$. Then we have

$$\frac{\partial}{\partial t}(\gamma^+ + \gamma^-) - k^+\Delta(\gamma^+ + \gamma^-) = 0 \qquad \text{for } x \in \Omega_t^+,$$
$$\frac{\partial}{\partial t}(\gamma^+ - \gamma^-) - k^+\Delta(\gamma^+ - \gamma^-) = 0 \qquad \text{for } x \in \Omega_t^-, \tag{4.74}$$

$$\langle r^-, \nabla\Phi\rangle = 1 \qquad \text{for } x \in \Gamma_t. \tag{4.75}$$

We introduce the notation

$$\gamma^\pm = \frac{1}{2}(\theta_c^+ \pm \theta_c^-),$$

where $\theta_c^\pm$ are some smooth functions coinciding on Γ_t. Then is becomes clear that (4.74) is exactly the heat equation contained in the modified Stefan problem (4.51). Moreover, relation (4.75) can be readily rewritten as

$$k^+\frac{\partial\theta_c^+}{\partial\nu}\bigg|_{\Gamma_t} - k^-\frac{\partial\theta_c^-}{\partial\nu}\bigg|_{\Gamma_t} = -2V_\nu, \tag{4.76}$$

where, as usual, $\partial/\partial\nu = \langle\nu, \nabla\rangle$. Obviously, (4.76) is exactly the Stefan condition in problem (4.51).

Supplementing (4.74) and (4.76) with conditions (4.70) and (4.62), we see that Definition 4.1 (as well as Definition 4.2) implies the modified Stefan problem (4.51) for the functions $\theta^\pm$. In turn, defining $\theta_c^\pm$ as smooth continuations of $\theta^\pm$ to the entire domain Ω, we complete our construction of the asymptotic solution. The proof of the theorem is complete. □

In conclusion, it remains to note that, in the case $m = 1$, the function ω_0 can be readily calculated: $\omega_0(\eta) = \tanh(\eta/\sqrt{2})$. Then $a_2 \to 1/3\sqrt{2}$ and, hence, $\varkappa_1 = \sqrt{2}\varkappa$, which coincides with the well-known formula (e.g., see [**22, 20**]).

5. Confluence of shock waves

In this section we consider the first nontrivial example of the use of the weak asymptotics method, namely, we obtain uniform-in-time formulas describing the propagation and the interaction of two shock waves in scalar conservation laws.

5.1. A model example. The problem of interaction of shock waves is significantly more interesting than the case of a solitary wave. The time-uniform analytical description of the shock wave confluence has been solved only in the simplest statement, i.e., for the Hopf equation with piecewise smooth initial data,

$$u_t + (u^2)_x = 0, \tag{5.1}$$

$$u\big|_{t=0} = u^0 + e_1^0 H(-x + x_1^0) + e_2^0 H(-x + x_2^0). \tag{5.2}$$

In this case, after the small viscosity regularization, the problem can be reduced to the linear problem (by the Hopf–Cole substitution (see [**128**])). For nonlinearities more general than the quadratic nonlinearity, until quite recently, this problem has not been solved even in the case e_1^0, $e_2^0 = \text{const}$.

The difficulty arising here can be explained as follows. If we set

$$e_1^0 > 0, \qquad e_2^0 > 0, \qquad x_1^0 < x_2^0, \tag{5.3}$$

then, at the first stage, the solution of problem (5.1), (5.2) is the superposition of shock waves

$$u = u^0 + \sum_{k=1}^{2} e_k^0 H\big(-x + \varphi_{k0}(t)\big) \tag{5.4}$$

moving independently of each other (the formulas are given for the Hopf equation (5.1)):

$$\varphi_{20} = x_2^0 + (2u^0 + e_2^0)t, \qquad \varphi_{10} = x_1^0 + \big(2(u^0 + e_2^0) + e_1^0\big)t. \tag{5.5}$$

However, at time t^*,

$$\varphi_{10}(t^*) = \varphi_{20}(t^*), \tag{5.6}$$

the trajectories $x = \varphi_{k0}(t)$ of the shock waves intersect at the point $x^* = \varphi_{k0}(t^*)$ and there does not exist any formula answering the question: What shall occur for $t > t^*$? Of course, it has long been known to physicists that the waves will merge and for $t > t^*$ only one shock wave will move with the total amplitude. Nevertheless, from a formal mathematical viewpoint, another scenario is also possible here. Namely, the second shock wave passes through the first wave (or reflects from the first wave) so that for $t > t^*$ the solution can again be represented in the form (5.4) but with different velocities φ_{k0t} and $\varphi_{01}(t) > \varphi_{20}(t)$. Moreover, in the standard (in the $\mathcal{D}'(\mathbb{R}^2_{x,t})$ sense) approach to (5.1), the time instant of interaction is, in fact, not considered and both these scenarios are equally possible (if we omit stability considerations).

In the case of interaction of two shock waves, the standard definition of a generalized solution readily implies that Eqs. (5.1) (or equations with any other nonlinearity) must hold outside the trajectories of discontinuities, while the Rankine–Hugoniot conditions must hold on the trajectories of discontinuities at the points where the normal exists. But the normal to the singularity trajectory does not exist at the point (x^*, t^*) at which the trajectories of discontinuities merge. Thus the integration by parts does not give any information about the point (x^*, t^*).

For example, the condition of weak continuity with respect to t at time $t = t^*$ is not sufficient for constructing a unique solution (see an example at the end of this section).

We overcome this obstacle by using the weak asymptotics method. Here the most significant point is that we consider the solutions of (5.1) as functions with values in $\mathcal{D}'(\mathbb{R}^1_x)$. By regularizing the initial data and then calculating the weak asymptotics of the solution in the $\mathcal{D}'(\mathbb{R}^1_x)$ sense, we see that, for each chosen ε, the solution is a mapping of $\mathbb{R}^1_{+,t}$ into $\mathcal{D}'(\mathbb{R}^1_x)$, which is smooth in x and continuous and piecewise smooth in t. This, in turn, allows us to give a continuous-in-t description of the interaction of shock waves. Of course, after passing to the limit as $\varepsilon \to 0$, we again obtain a solution of (5.1) in the $\mathcal{D}'(\mathbb{R}^2_{x,t})$ sense. However, the uniform formulas obtained (for $\varepsilon \neq 0$) uniquely show that only the physically meaningful scenario of interaction of shock waves is realized.

Let us state the main result of this subsection.

THEOREM 5.1. *Suppose that conditions* (5.3) *are satisfied. Then the weak asymptotics* mod $O_{\mathcal{D}'}(\varepsilon)$ *of the solution of problem* (5.1)–(5.2) *has the form*

$$u(x,t,\varepsilon) = u^0 + \sum_{k=1}^{2} e_k^0 H_k\Big(-x + \varphi_k\Big(\frac{\psi_0(t)}{\varepsilon}, t\Big), \varepsilon\Big), \tag{5.7}$$

where $H_k(x,\varepsilon)$ *is a regularization of the Heaviside function* $H(x)$, $k = 1, 2$, $\psi_0(t) = \varphi_{20}(t) - \varphi_{10}(t)$ *and* $\varphi_{k0}(t)$ *is determined by formulas* (5.5) *for all* $t \geq 0$, $\varphi_k(\tau, t) \in C^\infty(\mathbb{R}^1 \times \mathbb{R}^1_+)$. *Independently of the choice of the regularization* $H_k(x,\varepsilon)$,

$$\begin{aligned} &\varphi_k(\tau,t) \to \varphi_{k0}(t) && \text{as } \tau \to +\infty \text{ uniformly in } t,\\ &\varphi_k(\tau,t) \to \varphi(t) && \text{as } \tau \to -\infty \text{ uniformly in } t, \end{aligned} \tag{5.8}$$

where $k = 1, 2$ *and*

$$\varphi(t) = x^* + (2u^0 + e_1^0 + e_2^0)(t - t^*), \qquad x^* = \varphi_{k0}(t^*). \tag{5.9}$$

REMARK 5.1. Since $\psi_0(t) > 0$ for $t < t^*$ and $\psi_0(t) < 0$ for $t > t^*$, the limit relations (5.8) correspond as $\varepsilon \to 0$ to the trajectories of shock waves before ($\tau \to +\infty$) and after ($\tau \to -\infty$) the interaction. Clearly, the second relation in (5.8) and (5.9) means that the shock waves merge.

PROOF OF THEOREM 4.3. Let us prove Theorem 5.1 and derive formulas for the phases $\varphi_k(\tau, t)$. We regularize the initial data and write the leading term of the asymptotic expansion in the form (5.7), where the functions $\varphi_k(\tau, t)$ are assumed to be smooth and stabilizing as $\tau \to \pm\infty$,

$$\varphi_k(\tau,t) \to \varphi_k^{\pm}(t) \qquad \text{as } \tau \to \pm\infty. \tag{5.10}$$

To simplify the calculations, we, from the very beginning, assume that the background and the amplitudes of shock waves are constant and the same as in the initial condition. Moreover, we set

$$\varphi_k(\tau,t) = \varphi_{k0}(t) + \psi_0(t)\varphi_{k1}(\tau), \tag{5.11}$$

where φ_{k0} are defined in (5.5) and $\psi_0 = \varphi_{20} - \varphi_{10}$. We also make condition (5.10) sharper by assuming that

$$\varphi_{k1}(\tau) \to 0, \qquad \tau\frac{d}{d\tau}\varphi_{k1}(\tau) \to 0, \qquad \text{as } \tau \to +\infty. \tag{5.12}$$

Obviously, (5.10) and (5.12) mean that as $\varepsilon \to 0$ the function (5.11) turns into a piecewise linear function with a weak singularity (of the type of τ_+) at the point $t = t^*$.

We use formulas from Subsection 1.4 and calculate the weak asymptotics of u and u^2:

$$\begin{aligned} u(x,t,\varepsilon) &= u^0 + \sum_{k=1}^{2} e_k^0 H\big(-x + \varphi_k(\tau,t)\big) + O_{\mathcal{D}'}(\varepsilon),\\ u^2(x,t,\varepsilon) &= {u^0}^2 + \sum_{k=1}^{2} G_k(\tau) H\big(-x + \varphi_k(\tau,t)\big) + O_{\mathcal{D}'}(\varepsilon), \end{aligned} \tag{5.13}$$

where

$$G_k(\tau) = 2u^0 e_k^0 + e_k^{0\,2} + 2e_1^0 e_2^0 B_k\big((-1)^{k-1}\rho(\tau)\big),$$

$$\rho(\tau) \stackrel{\text{def}}{=} \big(\varphi_2(\tau,t) - \varphi_1(\tau,t)\big)/\varepsilon \equiv \tau\big(1 + \varphi_{21}(\tau) - \varphi_{11}(\tau)\big).$$

Recall that $B_k(\rho)$ are smooth functions satisfying the following identity, which is uniform in $\rho \in \mathbb{R}^1$:

$$B_1(\rho) + B_2(-\rho) = 1. \tag{5.14}$$

Moreover, there exists a number $N \gg 1$ depending on the choice of the regularization $H_k(x,\varepsilon)$ and satisfying the condition

$$\begin{aligned} &B_k(\rho) = O(|\rho|^{-N}) && \text{as } \rho \to -\infty \quad \text{and} \\ &B_k(\rho) = 1 + O(\rho^{-N}) && \text{as } \rho \to +\infty, \quad k = 1,2. \end{aligned} \tag{5.15}$$

The explicit formulas for $B_k(\rho)$ were written and discussed in detail in Section 1.

Now let us substitute the asymptotics (5.13) into the Hopf expansion. Obviously, we obtain

$$\sum_{k=1}^{2} \Big\{ e_k^0 \frac{d}{dt}\varphi_k\big(\tau(t,\varepsilon),t\big) - G_k(\tau)\Big\}\delta\big(-x - \varphi_k(\tau,t)\big) = O_{\mathcal{D}'}(\varepsilon).$$

This relation and (5.11) readily imply the equations

$$\frac{d\varphi_{k0}(t)}{dt} + \frac{d\psi_0(t)}{dt}\frac{d}{d\tau}\big(\tau\varphi_{k1}(\tau)\big) = \frac{G_k(\tau)}{e_k^0}, \qquad k = 1,2. \tag{5.16}$$

Let us consider (5.16) as $\tau \to +\infty$. In addition to (5.10), (5.12), we assume that

$$\frac{\rho}{\tau}\bigg|_{\tau\to+\infty} = 1. \tag{5.17}$$

Then, by (5.12) and (5.15), as $\tau \to +\infty$ we obtain

$$\frac{d}{d\tau}(\tau\varphi_{k1}) \to 0, \qquad G_1 \to 2u^0 e_1^0 + e_1^{0\,2} + 2e_1^0 e_2^0, \qquad G_2 \to 2u^0 e_2^0 + e_2^{0\,2}.$$

Hence we obtain the standard Rankine–Hugoniot conditions

$$\frac{d\varphi_{10}}{dt} = 2(u^0 + e_2^0) + e_1^0, \qquad \frac{d\varphi_{20}}{dt} = 2u^0 + e_2^0 \tag{5.18}$$

for the fronts of noninteracting shock waves moving one after another, which allows us to find φ_{k0}. The initial conditions (5.2) and relations (5.18) imply formulas (5.5).

Equations (5.18) were obtained as $\tau \to +\infty$, i.e., in fact, only till the time instant of interaction: for $t < t^*$. But we shall define $\varphi_{k0}(t)$ by formulas (5.5) for all $t \geq 0$.

Now let us find $\varphi_{k1}(\tau)$. We take into account the fact that

$$\tau(\varphi_{21} - \varphi_{11}) = \rho - \tau, \qquad \frac{d\psi_0}{dt} = -(e_1^0 + e_2^0) < 0.$$

Hence, by subtracting one equation in (5.16) from the other and taking account of (5.14) and (5.18), we reduce this system to the equation

$$\frac{d\rho}{d\tau} = 2B_1(\rho) - 1. \tag{5.19}$$

This autonomous ordinary differential equation is typical of our approach. For $t < t^*$ (i.e., as $\tau \to \infty$), the behavior of the function ρ is known: $\rho/\tau \to 0$. But we need to know what happens as $\tau \to -\infty$. Because the right-hand side $2B_1(\rho) - 1$ is monotone with respect to ρ, we have

$$\rho(\tau) \to \rho^* \qquad \text{as } \tau \to -\infty, \tag{5.20}$$

where ρ^* is the maximal root of the equation $B_1(\rho) = 1/2$.

Now from (5.16), (5.18), and the condition that φ_{k1} is bounded at the point $\tau = 0$, we readily derive the desired formulas

$$\varphi_{k1}(\tau) = (-1)^{k-1} 2 \frac{e^0_{3-k}}{e^0_1 + e^0_2} \frac{1}{\tau} \int_0^\tau B_2\big(-\rho(\tau')\big)\, d\tau', \qquad k = 1, 2. \tag{5.21}$$

The fact that the function $B_2(-\rho)$ decreases sufficiently rapidly as $\rho \to +\infty$ (see formula (5.15)) and relation (5.17) guarantee that the integral in the right-hand side of (5.21) converges as $\tau \to +\infty$. Hence

$$\tau\varphi_{k1}(\tau) \to \text{const}(k) \qquad \text{as } \tau \to +\infty,$$

which confirms the *a priori* assumption (5.12). Furthermore, since $B_2(-\rho) \to 1/2$ as $\tau \to -\infty$, the integral in the right-hand side of (5.21) diverges as $O(\tau)$. Hence the limit values $\varphi^-_{k1} = \lim_{\tau\to-\infty} \varphi_{k1}(\tau)$ exist. By using the L'Hospital rule, it is easy to find

$$\varphi^-_{11} = \frac{e^0_2}{e^0_1 + e^0_2}, \qquad \varphi^-_{21} = -\frac{e^0_1}{e^0_1 + e^0_2}.$$

Now elementary calculations show that as $\tau \to -\infty$, both of the functions $\varphi_k(\tau, t)$ have the same limit (5.9). The proof of Theorem 5.1 is complete. □

In conclusion, we prove the following almost obvious consequence of Theorem 5.1.

THEOREM 5.2. *Suppose that condition* (5.3) *is satisfied. Then the weak in the* $\mathcal{D}'(\mathbb{R}^2)$*-sense solution of problem* (5.1), (5.2) *is unique and has the form*

$$u^*(x,t) = u^0 + \sum_{k=1}^{2} e^0_k H\big(-x + \overline{\varphi}_k(t, t^*)\big), \tag{5.22}$$

where t^* *is defined in* (5.6) *and*

$$\overline{\varphi}_1 = x^0_1 + \big(2(u^0 + e^0_2) + e^0_1\big)t - e^0_2(t - t^*)H(t - t^*),$$
$$\overline{\varphi}_2 = x^0_2 + (2u^0 + e^0_2)t + e^0_1(t - t^*)H(t - t^*).$$

PROOF OF THEOREM 4.3. By Theorem 5.1, the smooth (for $\varepsilon \neq 0$) function $u(x,t,\varepsilon)$ ((5.7)) satisfies Eq. (5.1) up to $O_{\mathcal{D}'}(\varepsilon)$. Hence for any test function $\psi(x,t) \in \mathcal{D}(\mathbb{R}^2)$ we have

$$\iint_{-\infty}^{\infty} \big\{u(x,t,\varepsilon)\psi_t + u^2(x,t,\varepsilon)\psi_x\big\}\, dxdt = O(\varepsilon).$$

We represent the left-hand side of this relation as the sum of integrals with respect to t from $-\infty$ to t^* and from t^* to $+\infty$. Now by passing to the limit as $\varepsilon \to 0$ in each of these integrals, we verify the relation

$$\iint_{-\infty}^{\infty} \big\{u^*(x,t)\psi_t + (u^*(x,t,))^2\psi_x\big\}\, dxdt = 0.$$

Thus u^* is a weak solution of problem (5.1), (5.2). Since (5.3) contains the entropy condition $e_k^0 > 0$, $k = 1, 2$, this solution is unique by virtue of Oleinik's theorem (e.g., see [**104**]). The proof of the theorem is complete. □

REMARK 5.2. The right-hand side $O(\varepsilon)$ in the prelimit integral identity is

$$O(\varepsilon) = \int \langle O_{\mathcal{D}'}(\varepsilon), \psi \rangle \, dt.$$

In our example $\langle O_{\mathcal{D}'}(\varepsilon), \psi \rangle$ is a smooth function of the variable t. However, as is easy to see, to prove the limit integral identity, it suffices that this function be integrable (for example, piecewise continuous).

A remainder $O_{\mathcal{D}'}(\varepsilon)$ with such properties arises in the problem of interaction of shock waves with variable amplitudes ($c_i \neq \text{const}$) (see Subsection 5.3).

5.2. The Hopf equation with convex nonlinearity. We show that to obtain a result similar to Theorem 5.1 for the equation

$$u_t + \big(f(u)\big)_x = 0, \tag{5.23}$$

we only need to modify the construction of Subsection 5.1 As in Subsection 5.1, the function $f(u)$ is assumed to be smooth and convex. Supplementing Eq. (5.23) with the initial condition (5.2), we prove the following analog of Theorem 5.1.

THEOREM 5.3. *Suppose that conditions* (5.3) *are satisfied and* $f''(u) \geq 0$. *Then the weak asymptotics* $\operatorname{mod} O_{\mathcal{D}'}(\varepsilon)$ *of the solution of problem* (5.23), (5.2) *has the form* (5.7), *where* $H_k(x, \varepsilon)$ *is a regularization of the Heaviside function,* $k = 1, 2$, $\phi_0(t) = \varphi_{20}(t) - \varphi_{10}(t)$, $\varphi_{k0}(t)$ *are given by the formulas*

$$\begin{gathered} \varphi_{k0}(t) = x_k^0 + V_k t, \qquad t \geq 0, \\ V_1 = \big(f(u^0 + e_1^0 + e_2^0) - f(u^0 + e_2^0)\big)/e_1^0, \quad V_2 = \big(f(u^0 + e_2^0) - f(u^0)\big)/e_2^0, \end{gathered} \tag{5.24}$$

and $\varphi_k(\tau, t) \in C^\infty(\mathbb{R}^1 \times \mathbb{R}^1_+)$. *Independently of the choice of the regularization* H_k,

$$\begin{aligned} &\varphi_k(\tau, t) \to \varphi_{k0}(t) \qquad &&\text{as } \tau \to +\infty \text{ uniformly in } t, \\ &\varphi_k(\tau, t) \to \varphi(t) \qquad &&\text{as } \tau \to -\infty \text{ uniformly in } t. \end{aligned} \tag{5.25}$$

Here $k = 1, 2$ *and* $\varphi(t) = x^* + V(t - t^*)$, *where* t^* *is the moment of time at which the trajectories of the shock waves intersect, i.e.,* $\varphi_{10}(t^*) = \varphi_{20}(t^*)$, $x^* = \varphi_{k0}(t^*)$, *and*

$$V = \big(f(u^0 + e_1^0 + e_2^0) - f(u^0)\big)/(e_1^0 + e_2^0). \tag{5.26}$$

PROOF OF THEOREM 4.3. We write the leading term of the asymptotics of the solution in the form (5.7). We assume that the desired functions have the form (5.11) and satisfy conditions (5.10), (5.12). We use formulas from Subsection 1.5 and calculate the weak asymptotics $u(x, t, \varepsilon)$ ((5.13)) and the function $f(u)$:

$$f\big(u(x, t, \varepsilon)\big) = f(u^0) + \sum_{k=1}^{2} G_k(\tau) H\big(-x + \varphi_k(\tau, t)\big) + O_{\mathcal{D}'}(\varepsilon),$$

where

$$G_k(\tau) = \delta_k^f + B_k^f\big((-1)^{k-1}\rho(\tau)\big), \qquad \delta_k^f = f(u^0 + e_k^0) - f(u^0), \quad k = 1, 2,$$

$\rho(\tau)=\tau\big(1+\varphi_{21}(\tau)-\varphi_{11}(\tau)\big)$, and $B_k^f(\rho)$ are smooth functions whose explicit form is given in Subsection 1.5. For us it is important that

$$B_1^f(\rho)+B_2^f(-\rho)=\delta_3^f, \tag{5.27}$$

$$B_k^f(\rho)=\delta_3^f+O(\rho^{-N}) \quad \text{as } \rho\to\infty, \qquad B_k^f(\rho)=O(|\rho|^{-N}) \quad \text{as } \rho\to-\infty$$

where

$$\delta_3^f=f(u^0+e_1^0+e_2^0)-f(u^0+e_1^0)-f(u^0+e_2^0)+f(u^0).$$

Repeating the constructions of Subsection 5.1, we obtain equations of the form (5.16) from which, using conditions (5.17), we find the functions φ_{k0} ((5.24)). Now, after simple calculations, we transform (5.16) to the form

$$\begin{aligned}\frac{d}{d\tau}(\tau\varphi_{11})&=-\frac{1}{e_1^0(V_2-V_1)}\big(\delta_3^f-B_1^f(\rho)\big),\\ \frac{d}{d\tau}(\tau\varphi_{21})&=\frac{1}{e_2^0(V_2-V_1)}\big(\delta_3^f-B_1^f(\rho)\big).\end{aligned} \tag{5.28}$$

Obviously, (5.28) implies the following analog of Eq. (5.19):

$$\frac{d\rho}{d\tau}=F(\rho), \quad F=c^{-1}\big(\delta_4^f-B_1(\rho)\big), \quad c=\frac{e_1^0e_2^0(V_2-V_1)}{e_1^0+e_2^0}, \quad \delta_4^f=\delta_3^f+c. \tag{5.29}$$

Taking into account the inequality $c<0$ and the limit properties of the function B_1, we readily obtain that $F(\rho)\to 1$ as $\rho\to+\infty$, $F(\rho)\to -F^-$ as $\rho\to-\infty$, and $\delta_4^f\in(0,\delta_3^f)$, where

$$F^-=\frac{f(u^0+e_1^0+e_2^0)e_1^0-f(u^0+e_1^0)(e_1^0+e_2^0)+f(u^0)e_2^0}{f(u^0+e_1^0+e_2^0)e_2^0-f(u^0+e_2^0)(e_1^0+e_2^0)+f(u^0)e_1^0}.$$

By using the inequality

$$(x_2-x)f(x_1)+(x_1-x_2)f(x)+(x-x_1)f(x_2)>0, \qquad x\in(x_1,x_2),$$

valid for any convex function f, we see that F^- is a positive number for any choice of $e_i^0>0$. Therefore, exactly as in Eq. (5.19) in Subsection 5.1, problem (5.29), (5.17) is uniquely solvable, and we have the limit relation $\rho\to\rho^*$ as $\tau\to-\infty$, where ρ^* is the maximal root of the equation

$$B_1(\rho^*)=\delta_4^f. \tag{5.30}$$

Now from (5.28) we obtain the following formulas that are similar to (5.21):

$$\varphi_{k1}=(-1)^k\frac{e_{3-k}^0}{e_1^0e_2^0(V_2-V_1)}\frac{1}{\tau}\int_0^\tau B_2\big(-\rho(\tau')\big)\,d\tau'. \tag{5.31}$$

The analysis of (5.31), which, in fact, does not differ from that in Subsection 5.1, confirms assumptions (5.10) and (5.12) and thus proves the first relation in (5.25). Moreover, it is easy to see that

$$\lim_{\tau\to-\infty}\varphi_{k1}=(-1)^{k+1}e_{3-k}^0/(e_1^0+e_2^0), \qquad k=1,2.$$

This formula readily implies the second relation in (5.25), which completes the proof of the theorem. □

It remains to point out that, repeating the proof of Theorem 5.2 almost word for word, we obtain the following assertion.

THEOREM 5.4. *Suppose that the conditions of Theorem* 5.3 *are satisfied. Then the weak in the* $\mathcal{D}'(\mathbb{R}^2)$*-sense solution of the Cauchy problem* (5.23), (5.2) *is unique and has the form* (5.22), *where* t^* *is defined in* (5.26),

$$\overline{\varphi}_k = x_k^0 + V_k t + (V - V_k)(t - t^*)H(t - t^*), \qquad k = 1, 2,$$

and the velocities are defined in (5.24), (5.26).

5.3. Shock waves with variable amplitude. The above restrictions $u^0 =$ const, $e_k^0 =$ const are purely methodological, since assertions similar to Theorems 5.1–5.4 are also valid when u^0 and e_k^0 are variable. In fact, the main distinction is that in the case of variable initial data we must define the solution outside the fronts of the shock waves, and this procedure is rather cumbersome. Moreover, all our constructions are local with respect to time, i.e., they hold until the next gradient catastrophe occurs.

Let us consider problem (5.1), (5.2). We assume that

$$(5.32) \qquad u^0, e_k^0 \in C^\infty(\mathbb{R}^1), \qquad e_k^0(x) > 0, \quad k = 1, 2, \qquad x_1^0 < x_2^0.$$

In addition, we impose the following restriction on u^0, e_k^0: there exists $T > 0$ such that

$$(5.33) \qquad \begin{gathered} -2 \inf_{\xi > x_2^0} \left(u^0(\xi)\right)'_\xi < T^{-1}, \qquad -2 \inf_{x_1^0 < \xi < x_2^0} \left(u^0(\xi) + e_2^0(\xi)\right)'_\xi < T^{-1}, \\ -2 \inf_{\xi < x_1^0} \left(u^0(\xi) + e_1^0(\xi) + e_2^0(\xi)\right)'_\xi < T^{-1}. \end{gathered}$$

Obviously, under these assumptions a new shock wave cannot arise during the time $(0, T)$.

We introduce some auxiliary functions and the notation, which we shall need to state the result. By $\varphi_{k0} = \varphi_{k0}(t) \in C^\infty([0, t^*])$ and $\varphi = \varphi(t) \in C^\infty([t^*, T])$ we denote functions such that $\varphi_{k0}(0) = x_k^0$, $k = 1, 2$, and $\varphi_{10}(t^*) = \varphi_{20}(t^*) = \varphi(t^*) \overset{\text{def}}{=} x^*$. We assume that

$$(5.34) \qquad t^* < T.$$

Clearly, to the functions φ_{k0}, φ we can assign a graph $\mathcal{T}$ on the plane (x, t) with four vertices $(x_k^0, 0)$, $k = 1, 2$, (x^*, t^*), $(\varphi(T), T)$ and three edges $\Gamma_k = \{(x, t), x = \varphi_{k0}(t), t \in [0, t^*]\}$, $k = 1, 2$, $\Gamma_k = \{(x, t), x = \varphi(t), t \in [t^*, T]\}$. If condition (5.34) is satisfied, then the graph $\mathcal{T}$ divides the strip $\Pi = \mathbb{R}^1 \times [0, T]$ into three domains. Let Ω_2 be the domain bounded by Γ_1, Γ_2, and the straight line $t = 0$, let Ω_1 be the domain lying to the left of the edges $\Gamma_1 \cup \Gamma$, and let Ω_0 be the domain lying to the right of the edges $\Gamma_2 \cup \Gamma$. We denote $\gamma_i = \overline{\Omega}_i \cap \{(x, 0), x \in \mathbb{R}^1\}$, $i = 0, 1, 2$.

We consider the problem

$$(5.35) \qquad \begin{gathered} u_{it} + (u_i^2)_x = 0, \qquad (x, t) \in \Omega_i, \\ u_i\big|_{t=0} = u_i^0(x), \qquad x \in \gamma_i, \quad i = 0, 1, 2, \\ \frac{d\varphi_{k0}}{dt} = \frac{[u^2]\big|_{\Gamma_k}}{[u]\big|_{\Gamma_k}}, \quad k = 1, 2, \qquad \frac{d\varphi}{dt} = \frac{[u^2]\big|_{\Gamma}}{[u]\big|_{\Gamma}}, \end{gathered}$$

where $u_0^0 \equiv u^0$, $u_1^0 = u^0 + e_1^0 + e_2^0$, $u_2^0 = u^0 + e_2^0$, and $[f]\big|_\Gamma$, as usual, denotes the jump of a function f across Γ: $[f]\big|_\Gamma = f\left(x\big|_\Gamma + 0, t\right) - f\left(x\big|_\Gamma - 0, t\right)$. Obviously, in this statement, assumption (5.34) is a condition imposed on the initial data.

Since, as is easy to verify, assumptions (5.32)–(5.34) guarantee the existence of a unique smooth solution of problem (5.35), this problem completely determines both the graph $\mathcal{T}$ and the functions $u_i = u_i(x,t)$, $i = 0, 1, 2$.

The following fact is rather curious. Interpreting the Rankine–Hugoniot conditions as equations on a graph, we at once obtain two variables characterizing the interaction of shock waves, i.e., the jumps of u and u^2 satisfying the Kirchhoff condition

$$[u]\big|_{\Gamma_1} + [u]\big|_{\Gamma_2} = [u]\big|_{\Gamma}, \qquad [u^2]\big|_{\Gamma_1} + [u^2]\big|_{\Gamma_2} = [u^2]\big|_{\Gamma}.$$

We choose a number $\mu > 0$ independent of ε, and by $u_i^c = u_i^c(x,t)$ we denote a smooth continuation of u_i to the entire strip Π such that in the μ-neighborhood of Ω_i and for $t > 0$, u_i^c satisfies the Hopf equation. Constructively, u_i^c can be obtained by using an appropriate continuation u_i^0 in the μ-neighborhood of γ_i.

Finally, we introduce a smooth continuation φ_{k0}^c, φ^c of the functions φ_{k0}, φ for $t > t^*$ and $t < t^*$, respectively. Let

$$\begin{aligned} &\varphi_2^c(t) \le \varphi(t) \le \varphi_1^c(t) && \text{for } t \ge t^*, \\ &\varphi_1(t) \le \varphi^c(t) \le \varphi_2(t) && \text{for } t \le t^*. \end{aligned} \tag{5.36}$$

Moreover, we assume that, in the μ-neighborhood of the point t^*, the continuations φ_{k0}^c, φ^c satisfy the Rankine–Hugoniot conditions whose right-hand sides can be calculated via u_i^c, $i = 0, 1, 2$.

THEOREM 5.5. *Suppose that assumptions* (5.32)–(5.34) *are satisfied. Then the weak asymptotics* mod $O_{\mathcal{D}'}(\varepsilon)$ *of the solution of problem* (5.1), (5.2) *has the form*

$$u(x,t,\varepsilon) = u_0^c(x,t) + \sum_{k=1}^{2} e_k(x,t) H\Big(-x + \varphi_k\Big(\frac{\psi_0(t)}{\varepsilon}, t\Big), \varepsilon\Big), \tag{5.37}$$

where $e_2 = u_2^c - u_0^c$, $e_1 = u_1^c - u_0^c - e_2$, $\psi_0(t) = \varphi_{20}^c(t) - \varphi_{10}^c(t)$, *and* $\varphi_k(\tau,t)$ *is a* $C^\infty(\Pi)$*-function such that*

$$\begin{aligned} &\varphi_k(\tau,t) \to \varphi_{k0}^c(t) && \text{as } \tau \to +\infty, \\ &\varphi_k(\tau,t) \to \varphi^c(t) && \text{as } \tau \to -\infty, \quad k = 1, 2, \end{aligned} \tag{5.38}$$

independently of the choice of the regularization $H_k(x,\varepsilon)$ *and uniformly in* t.

PROOF OF THEOREM 4.3. The formal outline of the construction of the asymptotic solution exactly coincides, up to the change of the constants by functions, with that given in Subsection 5.2. Namely, we represent the asymptotics in the form (5.37), use formulas (5.13), and substitute the asymptotic expansions of u and u^2 into (5.1). As a result, we obtain the relation

$$\begin{aligned} &\sum_{k=1}^{2} \Big\{ e_k \frac{d}{dt} \varphi_k(\tau(t,\varepsilon), t) - G_k(\tau, \varphi_k, t) \Big\} \delta(-x + \varphi_k) \\ &\qquad + \sum_{k=1}^{2} \big\{ e_{kt} + \big(G_k(\tau, x, t)\big)_x \big\} H(-x + \varphi_k) + u_{0t}^c + (u_0^{c2})_x = O_{\mathcal{D}'}(\varepsilon), \end{aligned} \tag{5.39}$$

where

$$G_k = (2u_0^c + e_k)e_k + 2e_1e_2B_k\big((-1)^{k-1}\rho\big), \qquad \rho = (\varphi_2 - \varphi_1)/\varepsilon.$$

We shall use the following representation of φ_k similar to (5.11):

$$\varphi_k(\tau,t) = \varphi_{k0}^c(t) + \psi_0(t)\varphi_{k1}(\tau,t), \qquad \psi_0 = \varphi_{20}^c - \varphi_{10}^c, \tag{5.40}$$

where φ_{k1} are functions satisfying (5.10), (5.12), and (5.17).

Necessary conditions for (5.39) to hold are the relations

$$(5.41)\quad \frac{d}{dt}\varphi_k\big(\tau(t,\varepsilon),t\big) - \left\{2u_0^c + e_k\frac{2B_k((-1)^{k-1}\rho)e_1e_2}{e_k}\right\}\bigg|_{\Gamma_k} = O_{\mathcal{D}'}(\varepsilon), \qquad k=1,2.$$

For $t \in (0, t^*+\mu]$, taking into account (5.35) and (5.40) as well as identities (5.24), we have

$$(5.42)\quad \begin{aligned} \dot{\psi}_0\frac{\partial}{\partial\tau}(\tau\varphi_{11}) + \psi_0\frac{\partial\varphi_{11}}{\partial t} + 2e_2\big|_{\Gamma_1}\big(1-B_1(\rho)\big) &= O_{\mathcal{D}'}(\varepsilon),\\ \dot{\psi}_0\frac{\partial}{\partial\tau}(\tau\varphi_{21}) + \psi_0\frac{\partial\varphi_{21}}{\partial t} - 2e_1\big|_{\Gamma_2}\big(1-B_1(-\rho)\big) &= O_{\mathcal{D}'}(\varepsilon). \end{aligned}$$

Let us consider the equation

$$(5.43)\quad \dot{\psi}_0\frac{\partial}{\partial\tau}\tau(\varphi_{21}-\varphi_{11}) = 2\big(e_1\big|_{\Gamma_2} + e_2\big|_{\Gamma_1}\big)\big(1-B_1(\rho)\big).$$

To analyze this equation, we need the identities

$$\tau(\varphi_{21}-\varphi_{11}) = \rho - \tau, \qquad \dot{\psi}_0 = \alpha - \beta,$$

where

$$(5.44)\quad \beta = e_2\big|_{\Gamma_1} + e_1\big|_{\Gamma_2}, \qquad \alpha = g\big|_{\Gamma_2} - g\big|_{\Gamma_1}, \qquad g = 2u_0^c + e_2 + e_1.$$

Using these identities, we can rewrite (5.43) as

$$(5.45)\quad \frac{\partial\rho}{\partial\tau} = F(\rho,t),$$

with the right-hand side F, which can be written in the two equivalent forms

$$F = \frac{1}{\dot{\psi}_0}\{\dot{\psi}_0 + 2\beta(1-B_1(\rho))\} \equiv \frac{\beta}{\dot{\psi}_0}\left\{1 + \frac{\alpha}{\beta} - 2B_1(\rho)\right\}.$$

The first form of writing F and the properties of B_1 given by (5.15) imply that $F \to 1$ as $\rho \to +\infty$. The smoothness of the functions β and B_1 for $t \in [t^*-\mu, t^*+\mu]$, $\rho \in \mathbb{R}^1$, follows from assumptions (5.32) and (5.33). Moreover, we stress that for $t \in [0, t^*-\mu]$ the inequality $\dot{\psi}_0 = \dot{\varphi}_{20} - \dot{\varphi}_{10} < 0$ is necessarily satisfied, since the velocity of the shock wave increases with its amplitude. For $t \in [t^*-\mu, t^*+\mu]$ and sufficiently small μ, by virtue of (5.44), we have $\dot{\psi}_0 = -\beta\big|_{t=t^*} + O(\mu) < 0$, and hence $F \in C^\infty(\mathbb{R}^1 \times [0, t^*+\mu])$. We continue the function $\alpha(t)$ to be zero, $\alpha(t) \equiv 0$, for $t \geq t^*$. This continuation corresponds to the structure of the graph $\mathcal{T}$: the edges Γ_1 and Γ_2 became the edge Γ for $t \geq t^*$. Then, problem (5.45), (5.17) has a unique (up to a translation) solution $\rho(\tau,t)$, which is smooth in τ, continuous in t, and piecewise smooth in t. Its derivative has a discontinuity at the point t^*. As $\tau \to -\infty$, this solution stabilizes to the value ρ_0, which can be easily found by using the second form of writing F. Namely, ρ_0 is the maximal root of the equation

$$(5.46)\quad 2B_1(\rho_0) = 1 + \alpha/\beta.$$

Next, it follows from (5.44) that $\rho_0(t^*) = \rho_0^*$, where ρ_0^* satisfies the relation $B_1(\rho_0^*) = 1/2$. It follows from (5.46) that

$$(5.47)\quad \rho \to \rho_0 \quad \text{as } \tau \to -\infty, \qquad \tau\rho_t \to 0 \quad \text{as } \tau \to \pm\infty.$$

Let us return to (5.42). Subtracting one relation from the other, we obtain an expression that differs from (5.43) by the additional term $\dot{\psi}_0(\varphi_{21}-\varphi_{11})_t$. However, by (5.47), we have the uniform estimate

$$\dot{\psi}_0(\varphi_{21}-\varphi_{11})_t = \varepsilon\tau\rho_t = O(\varepsilon), \tag{5.48}$$

and hence we can neglect this additional term. Now (5.42) can be rewritten as the integral equation

$$\varphi_{k1}(\tau,t) = -\frac{1}{\tau\dot{\psi}_0}\int_0^\tau \Big\{(-1)^{k-1}2e_{3-k}\big|_{\Gamma_k}\big(1-B_1(\tau',t)\big)+\dot{\psi}_0\varphi_{k1t}(\tau',t)\Big\}\,d\tau'$$

whose solution necessarily exists, since the coefficient $\dot{\psi}_0$ is small in a neighborhood of the point $t=t^*$. Recall that it suffices to consider the $\varepsilon^{1-\delta}$-neighborhood of the point t^*, since, for t from this neighborhood, $\tau\sim(t-t^*)/\varepsilon$ varies from $-\infty$ to $+\infty$ as $\varepsilon\to 0$ and $\delta>0$.

For us it is important that, just as in Subsection 5.1, we again obtain the relations

$$\varphi_{k1}\to 0 \quad \text{as } \tau\to+\infty \quad\text{and}\quad \varphi_{k1}\to(-1)^k e_{3-k}\big|_{\Gamma_k}/\dot{\psi}_0 \quad \text{as } \tau\to-\infty,$$

which confirms the *a priori* assumptions (5.10), (5.12).

Let us consider relations (5.41) for $t\in[t^*+\mu/2,T]$. Such values of t correspond to passing to the limit as $\tau\to-\infty$, and hence, according to the above results,

$$\rho\to\rho_0^*,\qquad \varphi_1-\varphi_2\to 0,\qquad B_k\to 1/2,\qquad \varphi_{k\tau}\to 0.$$

So Eqs. (5.41) take the form

$$\frac{d\varphi_k^-(t)}{dt}-\{2u_0^c+e_1+e_2\}\big|_\Gamma=0,\qquad k=1,2,$$

and we obtain the Rankine–Hugoniot condition on the edge Γ, i.e., the second limit relation in (5.38).

It remains to derive the Hopf equation for problem (5.35). To this end, we return to relation (5.39) and, taking (5.41) into account, rewrite it as

$$\begin{aligned}&\{u_{0t}^c+(u_0^{c2})_x\}H(x-\varphi_2)\\&\quad+\{(u_0^c+e_2)_t+(u_0^{c2})_x+G_{2x}\}H(-x+\varphi_2)H(x-\varphi_1)\\&\quad+\{(u_0^c+e_2+e_1)_t+(u_0^{c2})_x+(G_1+G_2)_x\}H(-x+\varphi_1)=O_{\mathcal{D}'}(\varepsilon).\end{aligned}\tag{5.49}$$

By virtue of the first term in (5.49), the Hopf equation holds in Ω_0. If we introduce the notation $u_1=u_0+e_1+e_2$ and use identity (5.14), the third term implies the Hopf equation in Ω_1. Thus (5.49) can be transformed as

$$\begin{aligned}&\{(u_0^c+e_2)_t+(u_0^c+e_2)_x^2\\&\quad+2(e_2e_1)_xB_2(-\rho)\}H(-x+\varphi_2)H(x-\varphi_1)=O_{\mathcal{D}'}(\varepsilon).\end{aligned}\tag{5.50}$$

Note that for any test function $\psi(x)$ we have

$$\left|\int H(-x+\varphi_2)H(x-\varphi_1)\psi(x)\,dx\right|\le|\varphi_1-\varphi_2|.$$

Hence for any smooth functions e_1 and e_2 we have the estimate

$$(e_1e_2)_xB_2(-\rho)H(-x+\varphi_2)H(x-\varphi_1)=O_{\mathcal{D}'}(\varepsilon).$$

Clearly, the terms remaining in (5.50) imply the Hopf equation for the function $u_0^c + e_2$ in Ω_2. Recall that the smoothness of u_i, and hence of e_1 and e_2, is ensured by condition (5.32). The proof of the theorem is complete. $\square$

To complete this section, we present the following result that is similar to Theorem 5.2.

THEOREM 5.6. *Suppose that the assumptions of Theorem* 5.5 *are satisfied. Then the weak in the* $\mathcal{D}'(\mathbb{R}^1 \times (0,T))$*-sense solution of the Cauchy problem* (5.1), (5.2) *is unique and has the form*

$$u^*(x,t) = u_0(x,t)H(x - \overline{\varphi}_2(t,t^*)) + u_1(x,t)H(-x + \overline{\varphi}_1(t,t^*))$$
$$+ u_2(x,t)H(-x + \overline{\varphi}_2(t,t^*))H(-x + \overline{\varphi}_1(t,t^*)),$$

where

$$\overline{\varphi}_k(t,t^*) = \varphi_{k0}(t) + \big(\varphi(t) - \varphi_{k0}(t)\big)H(t - t^*), \qquad k = 1,2,$$

the functions u_i, $i = 0,1,2$, φ_{k0}, $k = 1,2$, φ *constitute the solution of* (5.35), *and* $t^* : \varphi_{10}(t^*) = \varphi_{20}(t^*) = \varphi(t^*)$.

EXAMPLE 5.1. The solution described in this section is of a rather simple natural form. In fact, it is sewed together from the following parts: two shock waves till the time instant t^* and one shock wave for $t > t^*$. We cannot say anything about the limit equation in the usual sense at the point $x = x^*$, $t = t^*$, but we see that the trajectories of the discontinuities are continuous at this point. Moreover, one can easily see that the weak limits of the solutions described above are continuous with respect to t in the weak sense. Thus we have the following "naive" situation: everywhere outside the curves $x = \phi_k(t)$, $k = 1,2$, the solution can be found as a function satisfying the equation $L[u] = 0$, while the curves $x = \phi_k(t)$, $k = 1,2$, are determined by the Rankine–Hugoniot conditions for $t \neq t^*$ and the condition that they are continuous at the point $t = t^*$.

However, as we shall see now, one can construct a great deal of such points. As an example, we consider the following Cauchy problem for the Hopf equation:

$$u_t + (u^2)_x = 0, \quad u(x,t)\big|_{t=0} = H(-x-1) - H(x-1).$$

One can easily verify that for $t < t^* = 1$ the solution $u(x,t)$ consists of two shock waves approaching each other:

$$u(x,t) = H(-x+t-1) - H(x+t-1).$$

By using the techniques used above for calculating the solution $u^*(x,t)$, we see that the solution satisfying the integral identity for $t \geq t^* = 1$ (this solution is one of "naive" solutions) is the stationary solution of the Hopf equation:

$$u(x,t) = H(-x) - H(x),$$

which consists of two shock waves sticking together.

Another different "naive" solution can be constructed as follows. Let us consider the solution of the Hopf equation with the initial condition

$$W(x,t)\Big|_{t=0} = \begin{cases} H(-x-1), & x \leq -1, \\ ax, & -1 \leq x \leq 1, \\ -H(x-1), & x > 1. \end{cases}$$

The solution $W(x,t)$ of this problem has the form

$$W(x,t) = H(-x-\phi(t)) - H(x-\phi(t)) + \frac{ax}{1+2at}[1 - H(-x-\phi(t)) - H(x-\phi(t))].$$

Its plot consists of two horizontal lines ($y = 1$ to the left of the point $x_- = -\phi$ and $y = 1$ to the right of the point $x_+ = \phi$) and the inclined line $y = ax/(1+2at)$ between the points x_- and x_+.

At the points $x_\pm$ the solution has a jump whose coordinates $x_\pm = \pm\phi(t)$ can be calculated from the Rankine–Hugoniot conditions

$$\phi_t = \frac{(-1)^2 - (ax/(1+2at))^2}{-1 - ax/(1+2at)}\bigg|_{x=\phi} = \frac{a\phi}{1+2at} - 1, \qquad \phi(0) = 1.$$

Assume that $a > 1$. Then, obviously, $\phi_t > 0$ on some interval $[0, t_0]$, $t_0 > 0$.

We consider the limit of the function $W(x,t)$ in $\mathcal{D}'(\mathbb{R}'_x)$ as $t \to -1/(2a) + 0$. For any test function $\varphi(x)$ we have

$$\lim_{t\to -1/(2a)+0} \langle W(x,t), \varphi(x)\rangle$$
$$= \lim_{t\to -1/(2a)+0} \left[\frac{1}{1+2at}\int_{-\phi}^{\phi} x\varphi(x)\,dx + \int_{-\phi}^{-\infty} \varphi(x)\,dx - \int_{\infty}^{\phi} \varphi(x)\,dx\right].$$

By using the L'Hospital rule, we obtain

$$\lim_{t\to -1/(2a)+0} \langle W(x,t), \varphi(x)\rangle$$
$$= \lim_{t\to -1/(2a)+0} \left[\frac{\phi_t(\phi\varphi(\phi)) - \phi_t(\phi\varphi(-\phi))}{2a}\right]$$
$$+ \int_0^{-\infty} \varphi(x)\,dx - \int_{\infty}^{0} \varphi(x)\,dx = \int_{\infty}^{-\infty} (H(-x) - H(x))\varphi(x)\,dx.$$

Consider the function

$$U(x,t) = \begin{cases} u(x,t), & 0 \le t \le t^*, \\ W(x, t - 1/(2a) - t^*), & t > t^*. \end{cases}$$

This function is continuous in t in $\mathcal{D}'_x$ and satisfies the classical Hopf equation everywhere except for the discontinuity trajectories

$$x_\pm = \begin{cases} \pm(t-1), & t \le t^* = 1, \\ \pm\phi(x, t - 1/(2a) - t^*), & t > 1, \end{cases}$$

which are continuous curves. Thus this is the second "naive" solution of the problem of merging discontinuities. In this solution, the discontinuities seem to interact elastically and then to run away after the interaction.

Thus our constructions are not useless. This theorem allows one to choose a "true naive" solution among all possible solutions.

6. Soliton interaction in the KdV model

In this and in the following sections, by using the ideas considered above, we qualitatively describe the interaction of ε–δ-solitons (or narrow solitary waves) in models governed by KdV type equations. From methodological considerations, we first study the simplest case, i.e., the KdV equation with small dispersion

$$u_t + (u^2)_x + \varepsilon^2 u_{xxx} = 0 \tag{6.1}$$

supplemented with the initial data, which are the superposition of two solitary waves,

$$u\big|_{t=0} = u^0(x,\varepsilon), \qquad u^0 = A_1\omega\Big(\beta_1\frac{x-x_1^0}{\varepsilon}\Big) + A_2\omega\Big(\beta_2\frac{x-x_2^0}{\varepsilon}\Big). \tag{6.2}$$

We set $\beta_i = \sqrt{A_i/12}$, $\omega(\eta) = 1/2\cosh^2(\eta)$, and assume that

$$x_2^0 < x_1^0, \qquad A_2 > A_1 > 0$$

so that there exists a time moment t^* at which the trajectories of the solitary waves corresponding to (6.2) intersect: $V_1t^* + x_1^0 = V_2t^* + x_2^0$, $V_i = 4\beta_i^2$.

Needless to say, for the KdV equation, as well as for the modified KdV (MKdV) equation, the behavior of the solution is well known: the solitons interact passing through each other without changing their shapes and, after the interaction, they are shifted by some distance. This assertion can be obtained directly, for instance, from the formula for the exact two-soliton solution of problem (6.1), (6.2) derived by the inverse scattering transform method (e.g., see [**7, 129**])

$$u_{\mathrm{sol}} = 6\beta_1^2(\beta_2^2-\beta_1^2)v_1(\eta_1,\rho) + 6\beta_2^2(\beta_2^2-\beta_1^2)v_2(\eta_2,\rho), \tag{6.3}$$

$$v_1 = \Big\{\beta_1\sinh(\beta_1\eta_1-\mu) - \beta_2\coth\big(\beta_2(\eta_1-\rho)+\mu\big)\cosh(\beta_1\eta-\mu)\Big\}^{-2},$$

$$v_2 = \Big\{\beta_2\cosh(\beta_2\eta_2+\mu) - \beta_1\tanh\big(\beta_1(\eta_2+\rho)-\mu\big)\sinh(\beta_2\eta_2+\mu)\Big\}^{-2},$$

where $\eta_i = \big(x-\varphi_{i0}(t)\big)/\varepsilon$, $\varphi_{i0} = V_it + x_i^0$, $\rho = \big(\varphi_{20}(t)-\varphi_{10}(t)\big)/\varepsilon$, and

$$\mu = \frac{1}{2}\ln\frac{\beta_2-\beta_1}{\beta_2+\beta_1} \tag{6.4}$$

is the displacement of the soliton trajectories after the interaction.

It is well known that as $\rho\to-\infty$ (that is, for $\varphi_{20}-\varphi_{10}<0$, $\varepsilon\to0$, i.e., prior to the interaction) and as $\rho\to\infty$ (that is, for $\varphi_{20}-\varphi_{10}>0$, $\varepsilon\to0$), formula (6.3) is the sum of solitary solitons with accuracy up to terms admitting the estimate $O(\varepsilon^N)$ for any $N>0$. However, after the interaction, some constants are added to the arguments of the solutions.

In fact, the problem of describing the soliton interaction (as soon as the qualitative mechanism of interaction becomes clear) is just the calculation of these constants.

As mentioned above, the weak asymptotics method permits us to obtain a description of the single soliton dynamics in an extremely simple way. There is a natural question of whether we can describe the interaction of solitons passing to the limit in the weak sense as $\varepsilon\to0$.

In what follows, we present some results obtained in this way.

According to our approach, for problem (6.1), (6.2) we present the two-soliton weak asymptotic solution mod $O_{\mathcal{D}'}(\varepsilon^2)$ in the form

$$u = g_1(\tau)\omega\Big(\beta_1\frac{x-\phi_1(t,\tau,\varepsilon)}{\varepsilon}\Big) + g_2(\tau)\omega\Big(\beta_2\frac{x-\phi_2(t,\tau,\varepsilon)}{\varepsilon}\Big), \tag{6.5}$$

$$g_i = A_i + S_i(\tau), \qquad \phi_i = \varphi_{i0}(t) + \varepsilon\varphi_{i1}(\tau), \qquad \tau = \beta_1\big(\varphi_{20}(t)-\varphi_{10}(t)\big)/\varepsilon,$$

where $\varphi_{i0}(t) = V_it + x_i^0$ so that $x = \varphi_{i0}(t)$ is the trajectory of motion of the solitary wave with amplitude A_i, the function $\tau = \tau(t,\varepsilon)$ has the meaning of "fast" time,

$\tau(t^*, \varepsilon) = 0$, S_i and φ_{i1} are corrections to the amplitude and the phases rapidly varying during the time of interaction with exponential rates:

$$S_i(\tau) \to 0 \quad \text{as } \tau \to \pm\infty, \tag{6.6}$$

$$\varphi_{i1}(\tau) \to 0 \quad \text{as } \tau \to -\infty, \qquad \varphi_{i1}(\tau) \to \varphi_{i1}^{\infty} = \text{const} \quad \text{as } \tau \to +\infty.$$

It turned out that (in contrast to the inverse transform method) the construction of the solution significantly depends on the ratio of the parameters β_1 and β_2, i.e., on $\theta = \beta_1/\beta_2$, and this ratio can be easily recalculated as the ratio of soliton amplitudes. Namely, it turned out that the solution in the form of (6.5) can be constructed for $0 < \theta < 1/2$. For $\theta = 1/2$ the formulas degenerate, and this case must be studied separately. For $\theta \in (\frac{1}{2}, 1)$, the functions contained in the expression in the right-hand side of (6.5) become complex-valued, and, in order to deal with real functions, we must seek the solution in the form

$$u = g_1(\tau)\omega\Big(\beta_1 \frac{x - \phi_1(t,\tau,\varepsilon)}{\varepsilon}\Big) + g_2(\tau)\omega\Big(\beta_2 \frac{x - \phi_2(t,\tau,\varepsilon)}{\varepsilon}\Big) + \text{c.c.}$$

It is clear that these additional terms make our calculations much more complicated. So, in what follows, we restrict our study of interaction to the case

$$0 < \theta < 1/2.$$

Strictly speaking, we can prove that there exists a $\theta^* < 1/2$ such that our statements are true for $0 < \theta < \theta^*$. But numerical calculations show that $\theta^* = 1/2$.

Under this restriction on the initial amplitudes of solitons, we construct an asymptotic solution of the KdV equation and, in the next section, for the KdV type equations.

Our aim is to reconstruct all the qualitative properties of soliton interaction obtained by the inverse transform method:

1) solitons pass through each other without changing their structure,
2) the result of interaction is a phase shift.

In what follows, such a mechanism of interaction of solitary waves will be called the *soliton scenario of interaction.* The description of interaction is reduced to the study of autonomous first-order differential equations. Moreover, these equations are of similar form for both the KdV equation and the nonintegrable KdV type equations.

One cannot solve these equations explicitly. However, for small values of the parameter θ, it is possible to obtain enough information about the solution and to calculate the phase shifts. The formulas thus obtained differ from (6.4). The matter is that, as will be shown below, the KdV type equations with small dispersion are not well defined in the asymptotic sense. Namely, an arbitrarily small perturbation of the solution can lead to a contribution of order $O_{\mathcal{D}'}(\varepsilon^2)$ to the solution and thus change the phase shift.

The main result of this section, which we obtain without using the inverse transform method, is the following theorem.

THEOREM 6.1. *Let $\theta = \beta_1/\beta_2 \in (0, 1/2)$. Then the interaction of ε-δ-solitons of the KdV equation follows the soliton scenario. The weak asymptotics of the solution of problem* (6.1), (6.2) *is asymptotically not unique in the terms of order $O_{\mathcal{D}'}(\varepsilon^2)$.*

We prove this assertion in two steps. First, we construct the weak asymptotics modulo $O_{\mathcal{D}'}(\varepsilon^2)$ of the solution of problem (6.1), (6.2) in the form (6.5), (6.6).

At the second stage of the proof, we show that a perturbation (arbitrarily small in the sense of $\mathcal{D}'$) of the leading term of the asymptotics of (6.5) changes the constants φ_{i1}^{∞}.

Prior to proving the theorem, let us discuss the leading term of the weak asymptotics of (6.5) and compare it with the exact solution (6.3). First, recall that we understand the solution of problem (6.1), (6.2) in the weak sense, that is, by a weak asymptotics modulo $O_{\mathcal{D}'}(\varepsilon^2)$ of the solution we mean a function $u = u(x,t,\varepsilon)$ such that for any test function $\psi = \psi(x)$ the following relations are satisfied:

$$\begin{aligned} &\frac{d}{dt}\int u\psi\,dx - \int u^2\frac{\partial\psi}{\partial x}\,dx = O(\varepsilon^2),\\ \text{(6.7)}\qquad &\frac{d}{dt}\int u^2\psi\,dx - \frac{4}{3}\int u^3\frac{\partial\psi}{\partial x}\,dx + 3\int\Big(\varepsilon\frac{\partial u}{\partial x}\Big)^2\frac{\partial\psi}{\partial x}\,dx = O(\varepsilon^2),\\ &\int\big(u\big|_{t=0} - u^0(x,\varepsilon)\big)\psi\,dx = O(\varepsilon^2). \end{aligned}$$

Here and in the following, the notation $\int\cdots dx$ means the integration over $\mathbb{R}^1$.

The most unexpected result obtained while proving the theorem is the following fact.

THEOREM 6.2. *The condition* $\theta\in(0,1/2)$, *the conservation laws*

$$\frac{d}{dt}\int u\,dx = 0, \tag{6.8}$$

$$\frac{d}{dt}\int u^2\,dx = 0, \tag{6.9}$$

and the energy relations

$$\frac{d}{dt}\int xu\,dx - \int u^2\,dx = 0, \tag{6.10}$$

$$\frac{d}{dt}\int xu^2\,dx - \frac{4}{3}\int u^3\,dx + 3\int\Big(\varepsilon\frac{\partial u}{\partial x}\Big)^2\,dx = 0 \tag{6.11}$$

are necessary and sufficient conditions for the function (6.5) *to be a weak asymptotics of the solution of problem* (6.1), (6.2) *in the sense of* (6.7).

Let us discuss the structure of the asymptotic solution of (6.5). Obviously, the above formulas imply that, in the pointwise sense, the function (6.5) is the sum of solitary waves

$$u = \sum_{i=1}^{2} A_i\omega\Big(\beta_i\frac{x-\varphi_{i0}(t)}{\varepsilon}\Big) + O(\varepsilon^\infty), \qquad t < t^*,$$

until they interact and the sum of solitary waves with a phase shift

$$u = \sum_{i=1}^{2} A_i\omega\Big(\beta_i\frac{x-\varphi_{i0}(t)}{\varepsilon} - \beta_i\varphi_{i1}^{\infty}\Big) + O(\varepsilon^\infty), \qquad t > t^*,$$

after the interaction. We stress that the resulting phase shifts $\beta_i\varphi_{i1}^{\infty}$ satisfy the same relation

$$\beta_1\varphi_{11}^{\infty} + \beta_2\varphi_{21}^{\infty} = 0 \tag{6.12}$$

as those obtained for the exact two-soliton solution. Moreover, this relation is also valid if the function (6.5) is perturbed by terms of order $O_{\mathcal{D}'}(\varepsilon^n)$, $n \geq 2$.

In the sense of $\mathcal{D}'$, for the function (6.5), uniformly in t we have

$$u = \varepsilon \sum_{i=1}^{2} \frac{g_i(\tau)}{\beta_i} \delta(x - \phi_i) + O_{\mathcal{D}'}(\varepsilon^3). \tag{6.13}$$

However, $\varepsilon\delta(x - \phi_i) = \varepsilon\delta(x - \varphi_{i0}) + O_{\mathcal{D}'}(\varepsilon^2)$ and we see that only the soliton components S_i of the amplitude g_i are essential, while the role of phase shifts $\varepsilon\varphi_{i1}$ seems to be unessential. Moreover, a similar conclusion can readily be obtained by calculating the weak asymptotics of the exact solution (6.3):

$$u_{\text{sol}} = \varepsilon \sum_{i=1}^{2} \frac{P_i(\tau)}{\beta_i} \delta(x - \varphi_{i0}(t)) + O_{\mathcal{D}'}(\varepsilon^2), \tag{6.14}$$

where

$$P_i = 6\beta_i^2(\beta_2^2 - \beta_1^2) \int v_i(\eta, \tau/\beta_1)\, d\eta \to A_i \qquad \text{as } \tau \to \pm\infty.$$

Nevertheless, the phase shifts play an important role. To verify this, we calculate the weak asymptotics for the derivatives of the function (6.5). Taking into account the fact that S_i and φ_{i1} vary rapidly with time, we obtain

$$\frac{\partial u}{\partial t} = \sum_{i=1}^{2} \left\{ \frac{c}{\beta_i} \frac{\partial S_i}{\partial \tau} \delta(x - x^*) - \varepsilon \frac{A_i}{\beta_i} V_i \delta'(x - \varphi_{i0}) \right.$$
$$\left. - c\varepsilon \frac{\partial}{\partial \tau} \left[V_i \frac{\tau}{c} \frac{S_i}{\beta_i} + \frac{g_i}{\beta_i} \varphi_{i1} \right] \delta'(x - x^*) \right\} + O_{\mathcal{D}'}(\varepsilon^2),$$

where $c = \beta_1(V_2 - V_1)$ and $x^* = \varphi_{i0}(t^*)$ is the point of intersection of the soliton trajectories. Similar calculations for the exact solution (6.3) lead to the expansion

$$\frac{\partial u_{\text{sol}}}{\partial t} = \sum_{i=1}^{2} \left\{ \frac{c}{\beta_i} \frac{\partial P_i^s}{\partial \tau} \delta(x - x^*) - \varepsilon \frac{A_i}{\beta_i} V_i \delta'(x - \varphi_{i0}) \right.$$
$$\left. - c\varepsilon \frac{\partial}{\partial \tau} \left[V_i \frac{\tau}{c} \frac{P_i^s}{\beta_i} + \frac{D_i}{\beta_i} \right] \delta'(x - x^*) \right\} + O_{\mathcal{D}'}(\varepsilon^2),$$

where $D_i = 6\beta_i^2(\beta_2^2 - \beta_1^2) \int \eta v_i(\eta, \tau/\beta_1)\, d\eta$, and we set $P_i = A_i + P_i^s$ so that $P_i^s \to 0$ as $\tau \to \pm\infty$. Comparing these two formulas, as well as (6.13) and (6.14), we readily see that the varying phase shifts of the function (6.5) play the same role as the odd (with respect to η) parts of the functions $v_i(\eta, \tau/\beta_i)$ in the exact solution.

If we discuss the interaction process and compare the functions (6.3) and (6.5) in the pointwise sense, then we see that these functions are, of course, different. Nevertheless, they possess an important common property: the soliton type (with respect to time) components of the amplitudes compensate one another. We mean that, uniformly in $\tau \in \mathbb{R}^1$, the relation

$$S_1(\tau)/\beta_1 + S_2(\tau)/\beta_2 = 0 \tag{6.15}$$

holds for the weak asymptotics (6.5) (we prove this later). This coincides with the identity

$$P_1(\tau)/\beta_1 + P_2(\tau)/\beta_2 = A_1/\beta_1 + A_2/\beta_2,$$

which can be derived directly for the exact solution by rewriting formula (6.3) in the form usual in the inverse transform method or by using the conservation law (6.8).

Prior to proving Theorem 6.1, we introduce some auxiliary functions needed to describe the weak asymptotics (6.5) exactly. By $\lambda_j = \lambda_j(\sigma)$ and $L_i = L_i(\sigma)$ we denote the following convolutions depending on the parameter $\theta = \beta_1/\beta_2$:

$$\begin{aligned} \lambda_0 &= \frac{1}{a_2}\int \omega(\eta)\omega(\theta\eta+\sigma)\,d\eta, \qquad \lambda_1 = \frac{1}{a_2}\int \eta\omega(\eta)\omega(\theta\eta+\sigma)\,d\eta, \\ \lambda_2 &= \frac{1}{a_2'}\int \omega'(\eta)\omega'(\theta\eta+\sigma)\,d\eta, \qquad \omega'(z) = \partial\omega(z)/\partial z, \end{aligned} \tag{6.16}$$

$$\begin{aligned} L_1 &= \frac{1}{a_3}\int \Big\{\theta(\theta+k)\omega(\theta\eta+\sigma) + (1-k)\omega(\eta)\Big\}^3 d\eta - (1+\theta^5), \\ L_2 &= \theta^3 k(2\theta+k) + k^2 - 2k + 2\theta^2(\theta+k)(1-k)\lambda_2, \end{aligned} \tag{6.17}$$

where a_2, a_3, a_2' are the numbers

$$\begin{aligned} a_2 &= \int \omega^2(\eta)\,d\eta = \frac{1}{3}, \qquad a_3 = \int \omega^3(\eta)\,d\eta = \frac{2}{15}, \\ a_2' &= \int \big(\omega'(\eta)\big)^2 d\eta = \frac{4}{15}, \end{aligned} \tag{6.18}$$

and $k = k(\sigma)$ is the function determined for $\theta \in (0, 1/2)$ by the expressions

$$\begin{aligned} k &= \Big\{(1-\theta)(1+\theta-\lambda_0\theta) - \sqrt{I}\Big\}(1+\theta-2\lambda_0\theta)^{-1}, \\ I &= (1+\theta)^2(1-\theta-\lambda_0\theta)^2 + 2\lambda_0\theta^2(1+\theta)(1-2\theta). \end{aligned} \tag{6.19}$$

For $\theta = 1/2$ we set

$$k = \lambda_0/(3-2\lambda_0). \tag{6.20}$$

Next, we introduce a function $\mathcal{F} = \mathcal{F}(\sigma)$ by the formula

$$\mathcal{F} = F/(1-\theta+\mathcal{J}'), \tag{6.21}$$

where

$$\begin{aligned} F &= \Big(\frac{4}{3}\frac{a_3}{a_2^2}(L_1-L_2) + L_2\Big)\big(\theta(1-\theta)^2\big)^{-1} - \mathcal{J}', \\ \mathcal{J}' &= k\Big(\frac{1-\theta-\theta^2}{\theta} - k\Big) + \Big\{\Big(\frac{1-\theta-\theta^2}{\theta} - 2k\Big)\sigma + 2\theta(1-\theta-2k)\lambda_1\Big\}\frac{\partial k}{\partial\sigma} \\ &\quad + 2\theta(\theta+k)(1-k)\frac{\partial\lambda_1}{\partial\sigma}. \end{aligned}$$

Let us define a function $\sigma_0 = \sigma_0(\tau)$ depending also on the parameter θ as a solution of the problem

$$\frac{d\sigma_0}{d\tau} = \mathcal{F}(\tau+\sigma_0), \qquad \sigma_0 \to 0 \qquad \text{as } \tau \to -\infty. \tag{6.22}$$

In what follows, we show that problem (6.22) has a solution for all $\tau \in \mathbb{R}^1$. Moreover,

$$\sigma_0 \to \sigma_0^\infty(\theta) \qquad \text{as } \tau \to +\infty \text{ like a function of exponential type.} \tag{6.23}$$

THEOREM 6.3. *Let $\theta \in (0, 1/2)$. Then the weak asymptotic solution of problem* (6.1), (6.2) *has the form* (6.5), *where*

$$S_1 = \sqrt{A_1A_2}\,k(\tau+\sigma_0), \qquad S_2 = -A_2 k(\tau+\sigma_0), \tag{6.24}$$

(6.25)
$$\varphi_{11} = \frac{1}{\beta_1(1+\theta)}\{\tau k(\tau+\sigma_0) - \sigma_0(1-k(\tau+\sigma_0))\},$$
$$\varphi_{21} = \frac{1}{\beta_2\theta(1+\theta)}\{\theta\sigma_0 + (\tau+\sigma_0)k(\tau+\sigma_0)\}.$$

REMARK 6.1. The above formulas readily imply that

$$k(\sigma)\to 0 \quad \text{as } \sigma\to\pm\infty, \qquad \varphi_{i1}(\tau)\to 0 \quad \text{as } \tau\to-\infty, \quad i=1,2,$$
$$\beta_1\varphi_{11}(\tau)\to -\sigma_0^\infty/(1+\theta), \quad \beta_2\varphi_{21}(\tau)\to \sigma_0^\infty/(1+\theta) \quad \text{as } \tau\to+\infty$$

like functions of exponential type.

Proof of Theorem 6.3. We present the leading term of the weak asymptotics of the solution of problem (6.1), (6.2) in the form (6.5), where $S_i = S_i(\tau)$ and $\phi_i = \varphi_{i0}(t) + \varepsilon\varphi_{i1}(\tau)$ are the desired smooth functions. We assume that

$$S_i(\tau)\to 0 \quad \text{as } \tau\to\pm\infty, \quad i=1,2,$$
$$\varphi_{i1}(\tau)\to 0 \quad \text{as } \tau\to-\infty, \qquad \varphi_{i1}(\tau)\to\varphi_{i1}^\infty \quad \text{as } \tau\to+\infty$$

like functions of exponential type. Now we apply formulas from Section 1 and calculate the weak asymptotics

(6.26)
$$u^n = \varepsilon a_n\sum_{i=1}^2 K_{i0}^{(n)}\delta(x-\phi_i) + \varepsilon a_n\Big\{\sum_{i=1}^2 K_{i1}^{(n)} + R_n\Big\}\delta(x-x^*) + O_{\mathcal{D}'}(\varepsilon^2),$$
$$(\varepsilon u_x)^2 = \varepsilon a_2'\sum_{i=1}^2 \beta_i^2 K_{i0}^{(2)}\delta(x-\phi_i) + \varepsilon a_2'\Big\{\sum_{i=1}^2 \beta_i^2 K_{i1}^{(2)} + Q_2\Big\}\delta(x-x^*) + O_{\mathcal{D}'}(\varepsilon^2),$$

(6.27)
$$\frac{\partial u}{\partial t} = c\frac{\partial}{\partial\tau}\sum_{i=1}^2 K_{i1}^{(1)}\delta(x-x^*) - \varepsilon\sum_{i=1}^2 V_iK_{i0}^{(1)}\delta'(x-\phi_i)$$
$$- \varepsilon c\frac{\partial}{\partial\tau}\Big\{\sum_{i=1}^2 K_i^{(1)}\varphi_{i1} + \frac{V_i}{c}\tau K_{i1}^{(1)}\Big\}\delta'(x-x^*) + O_{\mathcal{D}'}(\varepsilon^2),$$
$$\frac{\partial u^2}{\partial t} = ca_2\frac{\partial}{\partial\tau}\Big\{\sum_{i=1}^2 K_{i1}^{(2)} + R_2\Big\}\delta(x-x^*) - \varepsilon a_2\sum_{i=1}^2 V_iK_{i0}^{(2)}\delta'(x-\phi_i)$$
$$- \varepsilon ca_2\frac{\partial}{\partial\tau}\Big\{\sum_{i=1}^2\Big(K_i^{(2)}\varphi_{i1} + \frac{V_i}{c}\tau K_{i1}^{(2)}\Big) + R_2^{(1)}\Big\}\delta'(x-x^*) + O_{\mathcal{D}'}(\varepsilon^2).$$

Here $K_i^{(n)} = g_i^n/\beta_i$, $K_{i0}^{(n)} = A_i^n/\beta_i$, $K_{i1}^{(n)} = K_i^{(n)} - K_{i0}^{(n)}$, and the numbers a_n for $n=2,3$ and a_2' are calculated by formulas (6.18), while $a_1 = 1$. We also set $V_i = \varphi_{i0t}'(t^*)$, $c = \beta_1(V_2 - V_1)$, and

$$R_n = \frac{1}{\beta_2 a_n}\int \big(g_1\omega(\theta\eta+\sigma) + g_2\omega(\eta)\big)^n\,d\eta - \sum_{i=1}^2 K_i^{(n)}.$$

Moreover, using the notation (6.16), we obtain

(6.28)
$$R_2 = 2\frac{g_1g_2}{\beta_2}\lambda_0, \qquad R_2^{(1)} = 2\frac{g_1g_2}{\beta_2}\Big\{\Big(\frac{V_2}{c}\tau + \varphi_{21}\Big)\lambda_0 + \frac{1}{\beta_2}\lambda_1\Big\},$$
$$Q_2 = 2g_1g_2\beta_1\lambda_6.$$

By definition, the weak asymptotics of the solution must satisfy the relations

$$(6.29)\qquad \frac{\partial u}{\partial t}+\frac{\partial u^2}{\partial x}=O_{\mathcal{D}'}(\varepsilon^2),\qquad \frac{\partial u^2}{\partial t}+\frac{4}{3}\frac{\partial u^3}{\partial x}-3\frac{\partial}{\partial x}(\varepsilon u_x)^2=O_{\mathcal{D}'}(\varepsilon^2).$$

We substitute expansions (6.26), (6.27) into (6.29) and collect the terms containing $\delta(x-x^*)$, $\varepsilon\delta'(x-\phi_i)$, and $\varepsilon\delta'(x-x^*)$. By equating the obtained relations with zero, we arrive at the relations

$$(6.30)\qquad \frac{d}{d\tau}\sum_{i=1}^{2}K_{i1}^{(1)}=0,$$

$$(6.31)\qquad \frac{d}{d\tau}\Big\{\sum_{i=1}^{2}K_{i1}^{(2)}+R_2\Big\}=0,$$

$$(6.32)\qquad \frac{d}{d\tau}\sum_{i=1}^{2}\Big\{K_i^{(1)}\varphi_{i1}+\tau\frac{V_i}{c}K_{i1}^{(1)}\Big\}=0,$$

$$(6.33)\qquad c\frac{d}{d\tau}\Big\{\sum_{i=1}^{2}\Big(K_i^{(2)}\varphi_{i1}+\tau\frac{V_i}{c}K_{i1}^{(2)}\Big)+R_2^{(1)}\Big\}$$
$$=\frac{4}{3}\frac{a_3}{a_2}\Big\{\sum_{i=1}^{2}K_{i1}^{(3)}+R_3\Big\}-3\frac{a_2'}{a_2}\Big\{\sum_{i=1}^{2}\beta_i^2K_{i1}^{(2)}+Q_2\Big\},$$

as well as the relations

$$(6.34)\qquad \begin{gathered}-V_iK_{i0}^{(1)}+a_2K_{i0}^{(2)}=0,\\ -a_2V_iK_{i0}^{(2)}+\frac{4}{3}a_3K_{i0}^{(3)}-3a_2'\beta_i^2K_{i0}^{(2)}=0,\qquad i=1,2,\end{gathered}$$

describing the motion of solitary waves without taking into account their interaction. Relations (6.34) readily imply

$$(6.35)\qquad V_i=a_2A_i,\qquad \beta_i^2=\gamma A_i,\qquad \gamma=(\tfrac{4}{3}a_3-a_2^2)/3a_6'.$$

Let us transform system (6.30)–(6.33). Integrating the first three relations and performing simple transformations, we obtain

$$(6.36)\qquad K_{11}^{(1)}+K_{21}^{(1)}=0,$$

$$(6.37)\qquad \sum_{i=1}^{2}K_{i1}^{(2)}+R_2=0,$$

$$(6.38)\qquad \frac{\sigma}{\beta_1}K_{11}^{(1)}-\sum_{i=1}^{2}K_{i0}^{(1)}\varphi_{i1}=0.$$

Next, we use (6.28), (6.36) and rewrite (6.37) as follows:

$$(6.39)\quad -2(A_2-A_1)K_{11}^{(1)}+(\beta_1+\beta_2)(K_{11}^{(1)})^2+\frac{2}{\beta_2}(A_1+\beta_1K_{11}^{(1)})(A_2-\beta_2K_{11}^{(1)})\lambda_0=0.$$

Solving (6.39) for $K_{11}^{(1)}$ and choosing the root from the condition that $K_{11}^{(1)}\to 0$ as $\tau\to-\infty$, we obtain

$$(6.40)\qquad K_{11}^{(1)}=\frac{\beta_2}{\gamma}k(\sigma),$$

where the function $k(\sigma)$ is defined in (6.19) for $\theta\neq 1/2$ and in (6.20) for $\theta=1/2$.

Let us analyze formulas (6.19), (6.20). The obvious inequality $\lambda_0(\sigma,\theta) \le 1/\sqrt{\theta}$ for $\theta < 1$ implies

$$1+\theta-2\lambda_0(\sigma,\theta)\theta \ge (1-\sqrt{\theta})^2 \ge \text{const} > 0.$$

Next, let us consider the expression I defined in (6.19). It is easy to see that the maximal value (with respect to σ) of λ_0, which we denote by $\lambda_0(0,\theta)$, is a function of $\theta \in [0,1]$ monotonically decreasing from 3/2 to 1. Therefore, for $\theta \ge 2/5$, there exists $\sigma^* = \sigma^*(\theta)$ for which the first term in the formula for I is zero. Nevertheless, the second term in this formula is positive for $\theta < 1/2$ and negative for $\theta > 1/2$. Hence for $\theta < 1/2$ we have the following estimate uniform with respect to σ:

$$I \ge \text{const} > 0, \tag{6.41}$$

and we see that for $\theta > 1/2$ this inequality does not hold.

For $\theta = 1/2$ we define k by formula (6.20). Since $3 - 2\lambda_0\big|_{\theta=1/2} \ge \text{const} > 0$, this formula determines the smooth function $k(\sigma)$.

Finally, for $\theta \in (c_1, 1/2]$, $c_1 > 0$, the function $k(\sigma)$ is smooth. It is also easy to see that this function is positive, even with respect to σ, and tends to zero as $\sigma \to \pm\infty$ like an exponential function.

Now let us consider relation (6.38) together with the relation

$$\sigma = \tau + \beta_1(\varphi_{21} - \varphi_{11}). \tag{6.42}$$

By solving (6.38), (6.42) for φ_{i1}, we obtain

$$\begin{aligned} \varphi_{11} &= \left(\beta_1 \sum_{i=1}^{2} K_{i0}^{(1)}\right)^{-1} (\tau K_{20}^{(1)} - \sigma K_2^{(1)}), \\ \varphi_{21} &= \left(\beta_1 \sum_{i=1}^{2} K_{i0}^{(1)}\right)^{-1} (-\tau K_{10}^{(1)} + \sigma K_1^{(1)}). \end{aligned} \tag{6.43}$$

Let us transform Eq. (6.33). After simple calculations, taking into account (6.28), (6.36), (6.40), and (6.43), we obtain

$$\sum_{i=1}^{2}\left(K_i^{(2)}\varphi_{i1} + \tau\frac{V_i}{c}K_{i1}^{(2)}\right) + R_2^{(1)} = \frac{\beta_2^2}{\gamma^2}\{-\tau(1-\theta)+\sigma(1-\theta)+\mathcal{J}(\sigma)\},$$

where

$$\mathcal{J}(\sigma) = \sigma\big(\theta^{-1}(1-\theta-\theta^2)k - k^2\big) + 2\theta(\theta+k)(1-k)\lambda_1. \tag{6.44}$$

Now we note that the following relations hold:

$$\sum_{i=1}^{2} K_{i1}^{(3)} + R_3 = \frac{\beta_2^5}{\gamma^3}L_1(\sigma), \qquad \sum_{i=1}^{2} \beta_i^2 K_{i1}^{(2)} + Q_2 = \frac{\beta_2^5}{\gamma^2}L_2(\sigma),$$

$$c = a_2\frac{\beta_2^3}{\gamma}\theta(1-\theta^2), \qquad 3\gamma a_2' = \frac{4}{3}a_3 - a_2^2,$$

where L_1, L_2 are defined in (6.17). Hence we can rewrite Eq. (6.33) in the autonomous form

$$(1-\theta+\mathcal{J}')\frac{d\sigma}{d\tau} = 1-\theta+\frac{F_1(\sigma)}{\theta(1-\theta^2)}, \tag{6.45}$$

where

$$F_1 = \frac{4}{3}\frac{a_3}{a_2^2}(L_1 - L_2) + L_2, \qquad \mathcal{J}' = \frac{\partial \mathcal{J}}{\partial \sigma}.$$

Formulas (6.16), (6.44) and the properties of $k(\sigma)$ readily imply that $\mathcal{J}(\sigma) \in C^\infty(\mathbb{R}^1)$ for $0 < \theta \le 1/2$. Moreover, $\mathcal{J}(\sigma)$ is odd and tends to zero as $\sigma \to \pm\infty$ like an exponential function.

Consider $\mathcal{J}(\sigma)$ for small θ. Expanding $\omega(\theta\eta + \sigma)$ in the Taylor series around the point $\theta = 0$, we obtain

$$\lambda_0 = \frac{1}{a_2}\omega(\sigma) + O(\theta^2), \qquad \lambda_1 = O(\theta).$$

Thus we have

$$k(\sigma) = \frac{\theta^2}{a_2}\omega(\sigma) + O(\theta^3), \qquad \mathcal{J}(\sigma) = \frac{\theta(1-\theta)}{a_2}\sigma\omega(\sigma) + O(\theta^3), \tag{6.46}$$

and there exists $\theta_1^* \in (0, 1/2]$ such that uniformly in σ

$$1 - \theta + \mathcal{J}'(\sigma) \ge \text{const} > 0 \qquad \text{for } \theta < \theta_1^*. \tag{6.47}$$

It follows from this relation and the fact that $F_1(\sigma)$ is smooth that Eq. (6.45) has a solution for all $\tau \in \mathbb{R}^1$. Next, for small θ we find

$$\lambda_2 = O(\theta), \qquad L_1 = -\frac{3}{2}\theta^2\omega(\sigma) + O(\theta^3), \qquad L_2 = -6\theta^2\omega(\sigma) + O(\theta^4).$$

Now we readily calculate that for small θ

$$1 - \theta + \frac{F_1(\sigma)}{\theta(1-\theta^2)} = 1 - \theta + \frac{6}{5}\theta\omega(\sigma) + O(\theta^2) > 0. \tag{6.48}$$

Hence there exists $\theta_2^* \in (0, 1/2]$ such that uniformly in σ

$$1 - \theta + \frac{F_1(\sigma)}{\theta(1-\theta^2)} \ge \text{const} > 0 \qquad \text{for } \theta < \theta_2^*. \tag{6.49}$$

Inequalities (6.47) and (6.49) and the fact that the function $F_1(\sigma)$ is even in σ imply that $\sigma(\tau)$ tends to $\pm\infty$ as $\tau \to \pm\infty$ and is odd in τ up to an additive constant.

We set $\sigma_0 = \sigma - \tau$ and

$$\theta^* = \min\{\theta_1^*, \theta_2^*\}.$$

As mentioned above, we cannot rigorously prove a more precise estimate for θ^*, but computer calculations give $\theta^* = 1/2$.

Obviously, it follows from the above that problem (6.22) has a solution smooth in τ and it is odd up to an additive constant $\sigma_0(0)$. The exponential stabilization rate of σ_0 as $\tau \to \infty$ is a direct consequence of the fact that the function $\mathcal{F}(\sigma)$ decreases like an exponential function as $\tau \to \infty$.

To complete the proof of Theorem 6.3, it remains to note that formulas (6.21), (6.22) readily follow from (6.33), (6.37), and (6.40). The third relation in (6.7) for functions of the form (6.5) is obvious. □

Proof of Theorem 6.2. Let us consider (6.27). Taking into account the normalization of the function ω, we obtain

$$\sum_{i=1}^{2} K_{i1}^{(1)} = \sum_{i=1}^{2} \frac{g_i - A_i}{\beta_i}\int \omega(\eta)\,d\eta = \sum_{i=1}^{2} g_i \int \omega(\beta_i z)\,dz + \text{const}$$

$$= \frac{1}{\varepsilon} \int \sum_{i=1}^{2} g_i \omega\Big(\beta_i \frac{x-\phi_i}{\varepsilon}\Big)\, dx + \text{const}\,.$$

Since $cd/d\tau = \varepsilon d/dt$, we see that (6.27) is exactly the conservation law (6.8) calculated for functions u of the form (6.5).

Consider relation (6.30). With the help of the identity

$$\frac{V_i}{c}\tau = \frac{\varphi_{i0}}{\varepsilon} + q, \qquad q = \frac{V_1\varphi_{20} - V_2\varphi_{10}}{\varepsilon(V_2 - V_1)}, \qquad i = 1, 2,$$

by setting $\xi = \theta\eta + \sigma$, we calculate

$$\begin{aligned} M &\overset{\text{def}}{=} a_2 \sum_{i=1}^{2} \Big\{ K_i^{(2)} \varphi_{i1} + \frac{V_i}{c}\tau K_{i1}^{(2)} \Big\} + a_2 R_2^{(1)} \\ &= \frac{g_1^2}{\beta_1}\frac{\phi_1}{\varepsilon} \int \omega^2(\xi)\, d\xi + \frac{g_2^2}{\beta_2}\frac{\phi_2}{\varepsilon} \int \omega^2(\eta)\, d\eta \\ &\quad + 2\frac{g_1 g_2}{\beta_2} \Big\{ \Big(\frac{\phi_2}{\varepsilon} + q\Big) \int \omega(\eta)\omega(\xi)\, d\eta + \frac{1}{\beta_2} \int \eta\omega(\eta)\omega(\xi)\, d\eta \Big\} \\ &\quad + a_2 \sum_{i=1}^{2} \Big(qK_{i1}^{(2)} - \frac{\varphi_{i0}}{\varepsilon} K_{i0}^{(2)} \Big). \end{aligned}$$

Performing the change $\eta = \beta_2(x - \phi_2)/\varepsilon$, we prove the relation

$$\begin{aligned} &\frac{\phi_2}{\varepsilon} \int \omega(\eta)\omega(\xi)\, d\eta + \frac{1}{\beta_2} \int \eta\omega(\eta)\omega(\xi)\, d\eta \\ &\quad = \frac{\beta_2}{\varepsilon^2} \int x\omega\Big(\beta_1 \frac{x-\phi_1}{\varepsilon}\Big) \omega\Big(\beta_2 \frac{x-\phi_2}{\varepsilon}\Big)\, dx. \end{aligned}$$

Now it is easy to see that the even functions $\omega(\eta)$ satisfy the relation

$$\begin{aligned} M = &\frac{1}{\varepsilon^2} \int x\Big(g_1\omega\Big(\beta_1\frac{x-\phi_1}{\varepsilon}\Big) + g_2\omega\Big(\beta_2\frac{x-\phi_2}{\varepsilon}\Big)\Big)^2 dx \\ &+ a_2 q\Big(\sum_{i=1}^{2} K_{i1}^{(2)} + R_2 \Big) - \frac{a_2}{\varepsilon} \sum_{i=1}^{2} \varphi_{i0} K_{i0}^{(2)}. \end{aligned}$$

In a similar way, we obtain

$$a_3\Big\{ \sum_{i=1}^{2} K_{i1}^{(3)} + R_3 \Big\} = \frac{1}{\varepsilon} \int \Big(\sum_{i=1}^{2} g_i\omega\Big(\beta_i \frac{x-\phi_i}{\varepsilon}\Big)\Big)^3 dx - a_3 \sum_{i=1}^{2} K_{i0}^{(3)},$$

$$a_2'\Big\{ \sum_{i=1}^{2} \beta_i K_{i1}^{(2)} + Q_2 \Big\} = \frac{1}{\varepsilon} \int \Big\{ \varepsilon\frac{\partial}{\partial x}\Big(\sum_{i=1}^{2} g_i\omega\Big(\beta_i \frac{x-\phi_i}{\varepsilon}\Big)\Big)\Big\}^2 dx - a_2' \sum_{i=1}^{2} \beta_i^2 K_{i0}^{(2)}.$$

We substitute the relations obtained into (6.33) and take into account (6.29). Then (6.33) takes the form

$$\begin{aligned} &\frac{1}{\varepsilon}\Big\{ \frac{d}{dt} \int xu^2\, dx - \frac{4}{3} \int u^3\, dx + 3 \int (\varepsilon u_x)^2\, dx \Big\} \\ &\quad = \sum_{i=1}^{2} \Big\{ a_2 \frac{d}{dt} \varphi_{i0} K_{i0}^{(2)} - \frac{4}{3} a_3 K_{i0}^{(3)} + 3a_2' \beta_i^2 K_{i0}^{(2)} \Big\}. \end{aligned} \tag{6.50}$$

However, we have $d\varphi_{i0}/dt = V_i$, and thus the right-hand side in (6.50) is zero by virtue of the second relation in (6.34). Hence Eq. (6.33) is exactly the energy relation (6.11) calculated for the function u of the form (6.5).

In a similar way we can prove that (6.31) and (6.32) coincide with the equalities (6.9) and (6.10) calculated for the function u of the form (6.5).

To complete the proof of Theorem 6.2, it remains to recall that relations (6.30) and (6.33) are necessary and sufficient conditions for the function (6.5) to satisfy (6.7) with any test function ψ.

It should be noted that relations (6.34) also coincide exactly with the energy relations (6.10) and (6.11) whose left-hand sides, however, are calculated for the solitary wave $A_i\omega(\beta_i(x-\varphi_{i0})/\varepsilon)$. □

Proof of Theorem 6.1. Let $W_i(\eta)$ be some even functions from $C^\infty(\mathbb{R}^1)$ decreasing faster than $|\eta|^{-2}$ as $\eta \to \pm\infty$. By $W_{i,\ell}(\eta)$ we denote their $2\ell+1$-order derivatives with respect to η for arbitrary $\ell \geq 0$ and consider the function

$$U = \sum_{i=1}^{2} \Big\{ g_i(\tau)\omega\Big(\beta_i \frac{x-\phi_i}{\varepsilon}\Big) + A_i G_i(\tau) W_{i,\ell}\Big(\frac{\beta_i}{\nu_i(\theta)}\,\frac{x-\phi_i}{\varepsilon}\Big)\Big\}, \tag{6.51}$$

where ω, g_i, β_i, ϕ_i are the same as in (6.5), $G_i(\tau) \in C^\infty$ are arbitrary functions exponentially decreasing as $\tau \to \pm\infty$, and $\nu_i > 0$ are sufficiently small variables depending on the parameter θ.

It is easy to prove that in the sense of $\mathcal{D}'$ the function (6.51) differs from the function (6.5) by a value of order $O(\varepsilon^{2\ell+2})$, while their first derivatives differ by a value of order $O(\varepsilon^{2\ell+1})$. Nevertheless, we have

$$\begin{aligned} U^n &= (u^n)_0 + \varepsilon a_n M_{(n)}(\tau,\nu,G)\delta(x-x^*) + O_{\mathcal{D}'}(\varepsilon^2),\\ (\varepsilon U_x)^2 &= (\varepsilon u_x)_0^2 + \varepsilon a_2' P(\tau,\nu,G)\delta(x-x^*) + O_{\mathcal{D}'}(\varepsilon^2),\\ \frac{\partial U^2}{\partial t} &= \Big(\frac{\partial u^2}{\partial t}\Big)_0 + a_2 C(\tau,\nu,G)\delta(x-x^*)\\ &\quad + \varepsilon a_2 C^{(1)}(\tau,\nu,G)\delta'(x-x^*) + O_{\mathcal{D}'}(\varepsilon^2), \end{aligned}$$

where $(u^n)_0$, $(\varepsilon u_x)_0$, and $(\partial u^2/\partial t)_0$ are distributions defined in formulas (6.26), (6.27), the functions $M_{(n)}$, P, C, and $C^{(1)}$ tend to zero as $\tau \to \pm\infty$ like exponential functions, and

$$M_{(n)} = O(\nu_1+\nu_2), \qquad P = O(1/\nu_1 + 1/\nu_2), \qquad C = O(\nu_1+\nu_2).$$

We require that the function U is the weak asymptotics modulo $O_{\mathcal{D}'}(\varepsilon^2)$ of the solution of problem (6.1), (6.2). Just as above, we obtain a system of model equations from (6.7). Four of these equations coincide with (6.30), (6.32), and (6.34), and the other two have the form

$$\frac{d}{d\tau}\Big\{\sum_{i=1}^{2} K_{i1}^{(2)} + R_2 + M_{(2)}\Big\} = 0, \tag{6.52}$$

$$\begin{aligned} c\frac{d}{d\tau}\Big\{\sum_{i=1}^{2}\Big(K_i^{(2)}\varphi_{i1} + \tau\frac{V_i}{c}K_{i1}^{(2)}\Big) + R_2^{(1)} + C^{(1)}\Big\} \\ = \frac{4}{3}\frac{a_3}{a_2}\Big\{\sum_{i=1}^{2} K_{i1}^{(2)} + R_3 + M_{(3)}\Big\} - 3\frac{a_2'}{a_2}\Big\{\sum_{i=1}^{2}\beta_i^2 K_{i1}^{(2)} + Q_2 + P\Big\}. \end{aligned} \tag{6.53}$$

Repeating the above constructions and using the notation (6.40), we obtain the auxiliary function k from (6.30) and (6.52). However, in contrast to (6.19), we now have $k = k(\sigma, \theta, \nu, G)$. Here it is important that we can choose parameters ν_i arbitrarily small so that the corresponding quadratic equation has a real root for $\theta < \theta_1^*$.

Furthermore, formulas (6.52) and (6.42) yield

$$\varphi_{11} = \Big(\beta_1 \sum_{i=1}^{2} K_{i0}^{(1)}\Big)^{-1} (\tau K_{20}^{(1)} - \sigma K_2^{(1)} + D_1),$$

$$\varphi_{21} = \Big(\beta_1 \sum_{i=1}^{2} K_{i0}^{(1)}\Big)^{-1} (-\tau K_{10}^{(1)} + \sigma K_1^{(1)} + D_2),$$

which differ from (6.43) by additional functions $D_i = D_i(\tau, \sigma, \nu, G)$ such that $D_i \to 0$ as $\tau \to \pm\infty$. Here it should be pointed out that, since $M_{(2)}$ vanishes as $\tau \to \pm\infty$, the symmetry law for the resulting phase shifts of trajectories (6.12) remains valid for all perturbations of the function u contained in (6.51).

Obviously, we now can transform (6.53) to the equation similar to (6.45):

$$\frac{d}{d\tau}\{(\sigma - \tau)(1-\theta) + \mathcal{J}(\sigma) + \mathcal{J}_1(\tau, \sigma, \nu, G)\} = \frac{F_2}{\theta(1-\theta^2)}, \tag{6.54}$$

where

$$F_2 = F_1(\sigma) + \widetilde{F}_1(\tau, \sigma, \nu, G), \qquad \widetilde{F}_1 = \frac{4}{3}\frac{a_3}{a_2^2}(M_{(3)} - P) + P,$$

$\mathcal{J}$ and F_1 are the same functions as in (6.45), and $\mathcal{J}_1$ is a smooth function tending to zero as $\tau \to \pm\infty$. Let us integrate (6.54) with respect to τ, and let $\tau \to \infty$. Since $\sigma = \tau + \sigma_0$, we obtain the relation

$$\sigma_0^\infty(1-\theta) = \Big(\int_{-\infty}^{\infty} F_1\, d\tau + \int_{-\infty}^{\infty} \widetilde{F}_1\, d\tau\Big) \Big/ \theta(1-\theta^2). \tag{6.55}$$

The first integral in the right-hand side of (6.55) corresponds to the fixed value of $\sigma_0^\infty(\theta)$ obtained in the proof of Theorem 6.3. We see that the perturbations of u contained in (6.51) yield a function of soliton type (with respect to τ), namely, the function $\widetilde{F}_1 = (1 - 4a_3/3a_2^2)P + O(\nu_1 + \nu_2)$. Hence, by changing G_i and ν_i, we can make the limit σ_0^∞ be an arbitrary finite number. This fact indicates asymptotic nonuniqueness. The proof of Theorem 6.1 is complete. □

In conclusion, we analyze Eq. (6.45) for small θ. As shown in (6.48), $F_1(\sigma) = \frac{6}{5}\theta^2\omega(\sigma) + O(\theta^3)$. Hence we have $(1-\theta)\sigma_0^\infty = \frac{6}{5}\theta + O(\theta^2)$. Thus the trajectory of the first soliton is shifted by $-\varepsilon\beta_1\varphi_{11}^\infty = \frac{6}{5}\varepsilon\theta + O(\varepsilon\theta^2)$, while, according to the exact formula, this shift is equal to $-2\varepsilon\mu = \varepsilon\theta + O(\varepsilon\theta^2)$. In view of the fact that the weak asymptotics (6.5) is asymptotically nonunique, this coincidence is astonishingly good.

7. Soliton interaction in KdV type models

In this section we discuss a natural generalization of the KdV equation

$$u_t + (u^m)_x + \varepsilon u_{xxx} = 0. \tag{7.1}$$

Under the assumption that $m \geq 2$ is an integer, we study the problem of interaction of two solitary waves. Suppose that

$$u\Big|_{t=0} = u^0(x,\varepsilon), \qquad u^0 = A_1\omega\Big(\beta_1\frac{x-x_1^0}{\varepsilon}\Big) + A_2\omega\Big(\beta_2\frac{x-x_2^0}{\varepsilon}\Big), \tag{7.2}$$

where $\omega(\eta)$ is defined by (3.24) via the exact solution of the model equation (3.23) corresponding to (7.1) and

$$\beta_i^2 = \gamma^{m-1}A_i^{m-1}, \qquad \gamma = \Big(\frac{m-1}{m+3}\frac{a_m a_2}{a_2'}\Big)^{1/(m-1)}. \tag{7.3}$$

We assume that $A_i > 0$ for even m and that the sign of A_i can be arbitrary for odd m. In both cases, we assume that

$$|A_2| > |A_1|, \qquad x_1^0 < x_1^0,$$

which ensures the interaction of solitons at time t^*.

Precisely as in Section 6, a weak asymptotics modulo $O_{\mathcal{D}'}(\varepsilon^2)$ of the solution of problem (7.1) (7.2) is a function u satisfying for all $\psi = \psi(x) \in \mathcal{D}(\mathbb{R}^1)$ the relations

$$\begin{aligned} &\frac{d}{dt}\int u\psi\,dz - \int u^m \frac{\partial\psi}{\partial x}\,dx = O(\varepsilon^2),\\ &\frac{d}{dt}\int u^2\psi\,dz - \frac{2m}{m+1}\int u^{m+1}\frac{\partial\psi}{\partial x}\,dx + 3\int\Big(\varepsilon\frac{\partial u}{\partial x}\Big)^2\frac{\partial\psi}{\partial x}\,dx = O(\varepsilon^2),\\ &\int\big(u\big|_{t=0} - u^0\big)\psi\,dx = O(\varepsilon^2). \end{aligned} \tag{7.4}$$

The main result of this section is the proof of the following assertion.

THEOREM 7.1. *There exists $\theta^*(m) \in (0,1)$ such that the interaction of solitary waves in the KdV type equation* (7.1) *follows the soliton scenario for $\theta \stackrel{\text{def}}{=} \beta_1/\beta_2 \in \big(0, \theta^*(m)\big)$. The weak asymptotics of the solution of problem* (7.1), (7.2) *is asymptotically nonunique in the terms of order $O_{\mathcal{D}'}(\varepsilon^2)$.*

The proof of this theorem differs only in detail from the proof of a similar statement in Section 6. Namely, we represent the leading term of the weak asymptotics in the form

$$u = \sum_{i=1}^{2}(A_i + S_i(\tau))\omega\Big(\beta_i\frac{x-\varphi_{i0}(t)-\varepsilon\varphi_{i1}(\tau)}{\varepsilon}\Big), \tag{7.5}$$

where S_i and φ_{i1} are assumed to satisfy the same properties as in Section 5, while the function $\omega(\eta)$, precisely as in (7.2), depends on the parameter m. As in Section 5, we establish the following assertion.

THEOREM 7.2. *The condition $\theta \in (0, \theta^*(m))$, the conservation laws*

$$\frac{d}{dt}\int u\,dx = 0, \qquad \frac{d}{dt}\int u^2\,dx = 0,$$

and the energy relations

$$\frac{d}{dt}\int xu\,dx - \int u^m\,dx = 0,$$

$$\frac{d}{dt}\int xu^2\,dx - \frac{2m}{m+1}\int u^{m+1}\,dx + 3\int\Big(\varepsilon\frac{\partial u}{\partial x}\Big)^2\,dx = 0$$

are necessary and sufficient for the function (7.5) *to be a weak asymptotics modulo* $O_{\mathcal{D}'}(\varepsilon^2)$ *of the solution of problem* (7.1), (7.2).

The key point in the proof of these statements is the derivation of model equations for S_i, ϕ_i and their analysis. To state the result, we introduce some auxiliary notation and functions. We write

$$s_i = \operatorname{sign} A_i, \qquad s = s_1 s_2, \qquad \varkappa = \frac{2}{m-1},$$

and the convolutions $\lambda_l = \lambda_l(\sigma)$, $l = 0, 1, 2$, are calculated by formulas of the form (6.16).

By $\theta_0^* \in (0,1)$ we denote the number determined by the relation

$$1 - \theta_0^* = 2(\theta_0^*)^{\varkappa}$$

and introduce a function $k = k(\sigma)$ determined in the case $s = 1$, $\theta = \theta_0^*$ by the formula

$$k = \frac{\lambda_0(1-\theta)}{1+\theta-2\lambda_0\theta}\bigg|_{\theta=\theta_0^*} \tag{7.6a}$$

and in all other cases by the formula

$$k = \frac{1-\theta\lambda_0 + s(\lambda_0 - 1)\theta^{\varkappa} - \sqrt{I}}{1+\theta-2\theta\lambda_0}, \tag{7.6}$$

where

$$I = \big(1 - s\theta^{\varkappa} - \lambda_0(\theta^{\varkappa}+\theta)\big)^2 + 2\lambda_0\theta^{\varkappa}\big(1 - s\theta - \theta^{\varkappa}(1+s)\big). \tag{7.7}$$

Furthermore, we introduce the convolutions $\lambda_{n,j} = \lambda_{n,j}(\sigma)$:

$$\lambda_{n,j} = \frac{1}{a_n}\int \omega^{n-j}(\theta\eta+\sigma)\omega^j(\eta)\,d\eta \tag{7.8}$$

and the functions $L_{i,n} = L_{i,n}(\sigma)$:

$$\begin{aligned} L_{1,n} &= \theta^{n-1}\big[(k+s\theta^{\varkappa-1})^n - (s\theta^{\varkappa-1})^n\big] + (1-k)^n - 1 \\ &\quad + \sum_{j=1}^{n-1} c_n^j \theta^{n-j}(k+s\theta^{\varkappa-1})^{n-j}(1-k)^j\lambda_{n,j}, \end{aligned} \tag{7.9}$$

$$L_{2,n} = (1+\theta^3)k^2 - 2k(1-s\theta^{\varkappa+2}) + 2\theta^2(k+s\theta^{\varkappa-1})(1-k)\lambda_2, \tag{7.10}$$

where c_n^j are binomial coefficients.

Now for all $m \ge 2$ and any sign of s except the case

$$s = -1, \qquad m = 3, \tag{7.11}$$

we define the function $\sigma_0 = \sigma_0(\tau)$ as a solution of the problem

$$\frac{d\sigma_0}{d\tau} = \frac{\mathcal{F}_m(\tau+\sigma_0) - \mathcal{J}'_m(\tau+\sigma_0)}{r_m + \mathcal{J}'_m(\tau+\sigma_0)}, \qquad \sigma_0 \to 0 \qquad \text{as } \tau \to -\infty. \tag{7.12}$$

Here $\mathcal{J}'_m(\sigma) = \partial\mathcal{J}_m(\sigma)/\partial\sigma$,

$$\mathcal{J}_m(\sigma) = \sigma k\left[\frac{1-2s\theta^{\varkappa}-\theta^{2\varkappa-1}}{1+s\theta^{\varkappa-1}} - \theta k\right] + 2\theta(\theta k + s\theta^{\varkappa})(1-k)\lambda_1, \tag{7.13}$$

$$r_m = \theta^{\varkappa-1}\frac{s-\theta^{\varkappa}}{1+s\theta^{\varkappa-1}}, \tag{7.14}$$

$$(7.15)\qquad \mathcal{F}_m(\sigma) = (1-\theta^2)^{-1}\left\{\frac{2m}{m+1}\cdot\frac{a_{m+1}}{a_2 a_m}(L_{1,m+1}-L_{2,m})+L_{2,m}\right\} - (1-r_m)L_{1,m}.$$

In the exceptional case (7.11) we find the function $\sigma_0 = \sigma_0(\tau)$ as the solution of the problem

$$(7.16)\qquad \frac{d\sigma_0}{d\tau} = \frac{L^-_{1,3}(\tau+\sigma_0) - \mathcal{J}_3'(\tau+\sigma_0)}{1+\mathcal{J}_3'(\tau+\sigma_0)}, \qquad \sigma_0 \to 0 \qquad \text{as } \tau \to -\infty,$$

where $\mathcal{J}_3'(\sigma) = \partial\mathcal{J}_3(\sigma)/\partial\sigma$, $\mathcal{J}_3(\sigma) = -\sigma k(\sigma)$, and $L^-_{1,3} = L_{1,3}\big|_{s=-1}$.

By using the functions $\sigma_0(\tau)$ and setting $\sigma = \tau + \sigma_0(\tau)$, we define $\psi_1 = \psi_1(\tau)$ in all the cases except the case (7.11) by the formula

$$(7.17)\qquad \psi_1(\tau) = (1+s\theta^{\varkappa-1})^{-1}\left\{(\sigma_0+\tau)k(\sigma_0+\tau) - \sigma_0 + \int_{-\infty}^{\tau} L_{1,m}\big(\tau'+\sigma_0(\tau')\big)\,d\tau'\right\}.$$

In the exceptional case (7.11) we define the function $\psi_1(\tau)$ as follows:

$$(7.18)\qquad \psi_1(\tau) = (1+\theta)^{-1}\left\{\theta\sigma(k^2-2k) - \sigma_0 + 2\theta^2(1-k)^2\lambda_1 + \int_{-\infty}^{\tau}\big(\mathcal{F}_3 + (1-r_3)L^-_{1,3}\big)\,d\tau'\right\}.$$

THEOREM 7.3. *There exists a number $0 < \theta^* < 1$ depending on m and s such that for $0 < c_1 \le \theta \le \theta^*$ problems (7.12) and (7.15) have unique solutions. For such θ the weak asymptotics of the solution of problem (7.1), (7.2) has the form (7.5), where*

$$(7.19)\qquad \begin{gathered} S_1 = \theta A_2 k(\tau+\sigma_0), \qquad S_2 = -A_2 k(\tau+\sigma_0), \\ \varphi_{i0} = V_i t + x_i^0, \qquad V_i = a_m A_i^{m-1}, \qquad \tau = \beta_1\frac{\varphi_{20}(t)-\varphi_{10}(t)}{\varepsilon}, \\ \psi_2(\tau) = \big(\psi_1(\tau)+\sigma_0(\tau)\big)/\theta, \qquad \varphi_{i1} = \psi_i/\beta_i, \quad i=1,2. \end{gathered}$$

Proof of Theorem 7.3. We represent the weak asymptotics of the solution in the form (7.5) and apply formulas from Section 1. By substituting these expansions into (7.4) and equating the coefficients of $\delta'(x-x^*)$, $\varepsilon\delta'(x-\phi_i)$, and $\varepsilon\delta'(x-x^*)$ with zero, we obtain the following system of equations similar to system (6.30)–(6.34):

$$(7.20)\qquad \frac{d}{d\tau}\sum_{i=1}^{2} K_{i1}^{(1)} = 0,$$

$$(7.21)\qquad \frac{d}{d\tau}\left\{\sum_{i=1}^{2} K_{i1}^{(2)} + R_2\right\} = 0,$$

$$(7.22)\qquad c\frac{d}{d\tau}\sum_{i=1}^{2}\left\{K_i^{(1)}\varphi_{i1} + \frac{V_i}{c}\tau K_{i1}^{(1)}\right\} = f_m,$$

$$(7.23)\qquad c\frac{d}{d\tau}\left\{\sum_{i=1}^{2}\left(K_i^{(2)}\varphi_{i1} + \frac{V_i}{c}\tau K_{i1}^{(2)}\right) + R_2^{(1)}\right\} = F_m,$$

$$V_i K_{i0}^{(1)} = a_m K_{i0}^{(m)}, \tag{7.24}$$

$$V_i K_{i0}^{(2)} = \frac{2m}{m+1}\frac{a_{m+1}}{a_2} K_{i0}^{(m+1)} - 3\frac{a_2'}{a_2}\beta_i^2 K_{i0}^{(2)}, \tag{7.25}$$

where we use the notation

$$f_m = a_m\Big(\sum_{i=1}^{2} K_{i1}^{(m)} + R_m\Big), \tag{7.26}$$

$$F_m = \frac{2m}{m+1}\frac{1}{a_2} f_{m+1} - 3\frac{a_2'}{a_2}\Big(\sum_{i=1}^{2}\beta_i^2 K_{i1}^{(2)} + Q_2\Big). \tag{7.27}$$

Clearly, expressions (7.24), (7.25) imply the same relations between the amplitude A_i, the velocity V_i, and the coefficient β_i as in the problem of motion of a single solitary wave:

(7.28)

$$V_i = a_m A_i^{m-1}, \qquad \beta_i^2 = \gamma^{m-1} A_i^{m-1}, \qquad \gamma^{m-1} = \frac{(2m/(m+1))a_{m+1} - a_2 a_m}{3a_2'}.$$

Since $\omega(\eta)$ is given in terms of the solution of Eq. (3.23), using the properties of the coefficients a_1 and a_2', we can rewrite the coefficient γ in the form (7.3).

Let us consider Eqs. (7.20) and (7.21). Integrating by τ, we arrive at an equation of the form (6.39). We set

$$A_i = s_i \frac{\beta_i^{\varkappa}}{\gamma}, \qquad K_{11}^{(1)} = s_2 \frac{\beta_2^{\varkappa-1}}{\gamma} k, \qquad \varkappa = \frac{2}{m-1}. \tag{7.29}$$

Then, solving this equation for $k = k(\sigma)$, we obtain formulas (7.6), (7.7). Since $\lambda_0 \le 1/\sqrt{\theta}$, we have $1 + \theta - 2\theta\lambda_0 \ge \text{const} > 0$ for $\theta < 1$. Next, relation (7.7) readily implies that for $s = -1$ the inequality $I \ge \text{const} > 0$ holds uniformly in $\sigma \in \mathbb{R}^1$ and $\theta \in (0,1)$. Let us consider the case $s = 1$. Obviously, the second term in (7.7) is positive only for $\theta < \theta_0^*$. For $\theta = \theta_0^*$, the first term takes the form $(1+\theta)^2(1-\lambda_0)^2/4$. Since $\lambda_0(0)$ is a monotonically decreasing function of θ and $\lambda_0(0)\big|_{\theta=1} = 1$, for all $\theta \ge \theta_0^*$, there exists a value $\sigma = \sigma^*(m,\theta)$ such that the first term is zero at this point. Thus we obtain the condition $\theta < \theta_0^*$ for the estimate

$$I \ge \text{const} > 0$$

to be uniform with respect to τ.

Finally, the function k defined in (7.6a), (7.6) is real and smooth only if

$$\theta \le \theta_1^*, \quad \theta_1^* = \theta_0^* \ \text{ for } s = 1 \qquad \text{and} \qquad \theta_1^* = 1 - \mu, \quad \mu > 0, \ \text{ for } s = -1.$$

Let us consider Eqs. (7.22) and (7.23). Using formulas (7.9), (7.10), and (7.29), we transform these equations to the form

$$\frac{d}{d\tau}\big\{-\sigma k + s\theta^{\varkappa-1}\psi_1 + \theta\psi_2\big\} = L_{1,m}, \tag{7.30}$$

$$\begin{aligned}(1-\theta^2)\frac{d}{d\tau}\big\{&-\sigma k(2s\theta^{\varkappa} + \theta k) + \theta^{2\varkappa-1}\psi_1 + \theta\psi_2\\ &+ 2\theta(s\theta^{\varkappa} + \theta k)(1-k)\lambda_1\big\} = \frac{2m}{m+1}\frac{a_{m+1}}{a_2 a_m}(L_{1,m+1} - L_{2,m}) + L_{2,m}.\end{aligned} \tag{7.31}$$

Here we have used the notation

$$\psi_i = \beta_i \varphi_{i1}, \qquad i = 1, 2. \tag{7.32}$$

We supplement relations (7.30) and (7.31) by the identity

$$\sigma = \tau + \sigma_0, \qquad \sigma_0 = \theta\psi_2 - \psi_1, \tag{7.33}$$

and consider this system in the exceptional case (7.11). Since $\varkappa = 1$ in this case, Eq. (7.30) takes the form

$$\{1 - (\sigma k)'\}\frac{d\sigma}{d\tau} = 1 + L^-_{1,3}(\sigma), \tag{7.34}$$

where the prime denotes the derivative with respect to σ and the superscript "$-$" means that we set $s = -1$ in the formula for $L_{1,3}(\sigma)$. Now, by using the explicit form of the function k, we can verify that

$$1 - \frac{\sqrt{1+\theta}}{1-\sqrt{\theta}} \le k \le 0, \qquad 1 - \frac{c\theta}{(1-\sqrt{\theta})^2} \le 1 - (\sigma k)' \le \frac{\sqrt{1+\theta}}{1-\sqrt{\theta}},$$

where the constant c independent of θ is the maximal value of the function $-\sigma\lambda_0'(\sigma)$.

The above formulas readily imply that there exists $\theta_2^* > 0$ such that, for $0 < \theta \le \theta_2^*$, Eq. (7.34) can be solved for all $\tau \in \mathbb{R}^1$.

Obviously, problem (7.16) follows from relations (7.33) and (7.34) and the condition $\varphi_{i1} \to 0$ as $\tau \to -\infty$.

Let us study the general case. We set $\theta\psi_2 = \psi_1 + \sigma - \tau$ and, by using (7.30), define $d\psi_1/d\tau$ as a function of σ and τ. Then (7.31) can be transformed to the form

$$\big(r_m + \mathcal{J}_m'(\sigma)\big)\frac{d\sigma}{d\tau} = \mathcal{F}_m(\sigma) + r_m, \tag{7.35}$$

where the notation (7.13), (7.15) is used.

Let us prove that Eq. (7.35) is globally solvable for, at least, sufficiently small θ. For this purpose, we first note that for $\theta \ll 1$

$$\begin{aligned} \lambda_0 &= \lambda_0^0 + O(\theta^2), & \lambda_0^0 &= \frac{1}{a_2}\omega(\sigma), \\ \lambda_1 &= \theta\lambda_1^0 + O(\theta^3), & \lambda_1^0 &= \frac{1}{a_2}\omega'(\sigma)\int \eta^2\omega(\eta)\,d\eta. \end{aligned} \tag{7.36}$$

Therefore,

$$k = s_1\theta^{\varkappa}\lambda_0^0 + O(\theta^{\varkappa+1}),$$

and we calculate

$$\begin{aligned} \mathcal{J}_3 &= \frac{1}{2}\sigma\lambda_0^0\theta + O(\theta^2), & \mathcal{J}_4 &= \sigma\lambda_0^0\theta + O(\theta^{4/3}), \\ \mathcal{J}_5 &= -2s_1\sigma\lambda_0^0\theta^{3/2} + O(\theta^2), & \mathcal{J}_m &= -s_2\sigma\lambda_0^0\theta^{2\varkappa} + O(\theta) \quad \text{for } m \ge 6. \end{aligned}$$

At the same time, we have

$$r_3 = \frac{1}{2} + O(\theta), \qquad r_4 = 1 + O(\theta^{1/3}), \qquad r_m = 1 + O(\theta^{\varkappa}) \quad \text{for } m \ge 7.$$

Hence there exists θ_m^* such that

$$r_m + \mathcal{J}_m'(\sigma) \ge \text{const} > 0 \qquad \text{for } \theta < \theta_m^*.$$

It remains to note that $\mathcal{J}_m'(\sigma)$ and $\mathcal{F}_m(\sigma)$ are bounded by a constant and tend to zero both as $\sigma \to \pm\infty$ and $\theta \to 0$. This implies that Eq. (7.35) is solvable for $\theta \le \theta^* = \min\{\theta_1^*, \theta_m^*\}$, and hence problem (7.12) is solvable.

The last step in deriving formulas (7.19) is the calculation of the function ψ_1. In the exceptional case (7.11) formula (7.18) is obtained by integrating (7.33) for

$m = 3$ and $s = -1$. In the general case we obtain (7.17) by integrating (7.32). The proof of Theorem 7.3 is complete. □

The proof of Theorem 7.2 and of the asymptotic nonuniqueness of the weak asymptotics is carried out in the same way as the proofs in Section 6.

8. Confluence of free boundaries in the modified Stefan problem

8.1. Statement of the problem and the qualitative description of the results. Suppose that at the initial instant of time, in the domain $\Omega \subset \mathbb{R}^n$ filled with a substance with two possible phase states, there are two subdomains Ω_{0i}^+, $i = 1, 2$, filled with the phase "+" and the subdomain $\Omega_0^- \stackrel{\text{def}}{=} \overline{\Omega} \setminus \bigcup_i \overline{\Omega}_{0i}$ is filled with the phase "−." Suppose that the subdomains Ω_{01}^+ and Ω_{02}^+ do not intersect each other and do not have common points with the external boundary

$$\overline{\Omega}_{01}^+ \cap \overline{\Omega}_{02}^+ = \emptyset, \qquad \partial\Omega_{0i}^+ \cap \partial\Omega = \emptyset. \tag{8.1}$$

Then to describe the temperature evolution of the phases "+" and "−" and that of the interface Γ_{ti} in the case of media with surface tension, we have the modified Stefan problem

$$\frac{\partial \theta_i^\pm}{\partial t} = k^\pm \Delta \theta_i^\pm, \qquad x \in \Omega_{ti}^\pm, \quad t > 0, \tag{8.2}$$

$$\left[k^\pm \frac{\partial \theta^\pm}{\partial \nu_i}\right]\Bigg|_{\Gamma_{ti}} = -2V_{\nu_i}, \qquad [\theta_i^\pm]\Big|_{\Gamma_{ti}} = 0, \tag{8.3}$$

$$\varkappa_1 \theta_i^\pm\Big|_{\Gamma_{ti}} = V_{\nu_i} + \mathcal{K}_i, \qquad i = 1, 2, \tag{8.4}$$

which is a natural generalization of problem (4.51). Here Ω_{ti}^+, $i = 1, 2$, are the domains filled with the phase "+" for $t \geq 0$, and it is assumed that

$$\partial\Omega_{ti}^+ \cap \partial\Omega = \emptyset, \qquad i = 1, 2, \quad t > 0, \tag{8.5}$$

and θ_i^+ is the temperature of the substance in Ω_{ti}^+. For brevity, we write $\Omega_{t1}^- \stackrel{\text{def}}{=} \Omega_{t2}^- \stackrel{\text{def}}{=} \Omega_t^-$ for the domain filled with the solid phase "−" at time $t \geq 0$ and $\theta_1^- \stackrel{\text{def}}{=} \theta_2^- \stackrel{\text{def}}{=} \theta^-$ for the temperature in this domain. We assume that

$$\overline{\Omega}_t^- \cap \overline{\Omega}_{t1}^+ \cap \overline{\Omega}_{t2}^+ = \overline{\Omega} \tag{8.6}$$

for all $t \geq 0$. Just as above (see Subsection 4.3), $\varkappa_1 = \text{const}$, $k^\pm > 0$ are the thermal conductivity coefficients of the phases "+" and "−," $\Gamma_{ti} = \partial\Omega_{ti}^+$, ν_i is the normal to Γ_{ti}, V_{ν_i} is the normal velocity of Γ_{ti}, and $\mathcal{K}_i$ is the mean curvature of Γ_{ti}. Let

$$\mathcal{K}_i\big|_{t=0} > 0, \quad i = 1, 2, \qquad \operatorname{dist}(\Gamma_{01}, \Gamma_{02}) \geq \text{const} > 0. \tag{8.7}$$

We assume that the initial temperature $\theta_i^\pm\big|_{t=0}$ is positive and sufficiently large in a sufficiently large domain $\widetilde{\Omega} \supset \Omega_{10}^+ \cap \Omega_{20}^+$. Then, as follows from the Gibbs–Thompson laws (8.4), the domains Ω_{ti}^+ will expand with t. Obviously, until the boundaries Γ_{ti} contact each other, i.e., under the condition

$$\operatorname{dist}(\Gamma_{t1}, \Gamma_{t2}) \geq \text{const} > 0, \tag{8.8}$$

problem (8.2)–(8.4) is, in fact, the "superposition" of the standard modified Stefan problems (4.51). Therefore, under reasonable restrictions (sufficient smoothness of the initial data and the matching conditions on Γ_{0i}, $i = 1, 2$, [**117**]) and under conditions (8.5), (8.8), this problem has a classical solution.

Suppose that at time t^* condition (8.8) is violated. Then as $t \to t^*$ the statement of the Stefan problems in the form (8.2)–(8.4) is no longer meaningful at least because for $x \in \mathbb{R}^n$, $n > 1$, the curve $\Gamma_{t^*1} \cup \Gamma_{t^*2}$ has a self-intersection and, for $x \in \mathbb{R}^1$, the points Γ_{t1} and Γ_{t2} become a single point at time $t = t^*$.

The goal of this section is to describe the confluence of the free boundaries. Here the key point is the passage from the modified Stefan problem to its regularization, i.e., to the phase field system

$$\frac{\partial \theta}{\partial t} + \frac{\partial u}{\partial t} = \langle \nabla, k(u) \nabla \theta \rangle, \qquad x \in \Omega, \quad t > 0, \tag{8.9}$$

$$\varepsilon \frac{\partial u}{\partial t} = \varepsilon \Delta u + \frac{1}{\varepsilon}(u - u^{2m+1}) + \varkappa(1 - u^2)\theta, \tag{8.10}$$

where $m \geq 1$, $k(u) = (k^+ + k^-)/2 + u(k^+ - k^-)/2$, and the notation has the same meaning as in (4.52), (4.53).

It follows from general considerations that the confluence of boundaries in the one-dimensional and multidimensional cases occurs in different ways. Therefore, we shall study these cases separately.

For $n = 1$, as it follows from the physical considerations, the confluence of the free boundaries must result in the disappearance of the domain Ω_t^- filled with the phase "−." Hence we must describe the bifurcation of the two-kink solution into the leading solution. To achieve this goal, we shall use the same technique as above (see Sections 6 and 7). Again the key point is the definition of the weak solution of the phase field system, which admits passage to the limit as $\varepsilon \to 0$. For the convenience of the reader, we present this definition here (see also Subsection 3.3).

DEFINITION 8.1. A sequence of the pairs $(u = u(x, t, \varepsilon), \theta = \theta(x, t, \varepsilon))$ that for $\varepsilon > 0$ belong to $C^\infty(0, T; C^\infty(\Omega))$ is called the weak asymptotic solution of the phase field system (8.9), (8.10) mod $O_{\mathcal{D}'}(\varepsilon)$ in the one-dimensional case $\Omega \subset \mathbb{R}^1_x$ if, for any test functions $\psi, \psi_1 \in \mathcal{D}(\Omega)$, the following conditions are satisfied:

$$\frac{d}{dt} \int (\theta + u)\psi \, dx + \int k(u) \frac{\partial \theta}{\partial x} \frac{\partial \psi}{\partial x} \, dx = O(\varepsilon), \tag{8.11}$$

$$\begin{aligned} \varepsilon \int \left(\frac{\partial u}{\partial x} \right)^2 \frac{\partial \psi_1}{\partial x} \, dx &- \frac{1}{\varepsilon(m+1)} \int F(u) \frac{\partial \psi_1}{\partial x} \, dx \\ &+ 2\varepsilon \int \frac{\partial u}{\partial t} \frac{\partial u}{\partial x} \psi_1 \, dx + 2\varkappa \int \left(u - \frac{1}{3} u^3 \right) \frac{\partial}{\partial x} (\theta \psi_1) \, dx = O(\varepsilon), \end{aligned} \tag{8.12}$$

where

$$F(u) = u^{2m+2} - (m+1)u^2 + m. \tag{8.13}$$

The exact statement of the result requires very cumbersome formulas. Therefore, this statement will be given below (see Theorem 8.1). Here we restrict ourselves to its qualitative description.

We write the asymptotic solution of the phase field system in the form that potentially describes different scenarios of the processes that occur in the confluence of free boundaries $\Gamma_{ti} = \{x = \varphi_i(t, \varepsilon)\}$, $i = 1, 2$. Here the main characteristics are the normed distance $\rho = (\varphi_2(t, \varepsilon) - \varphi_1(t, \varepsilon))/\varepsilon$ between the boundaries Γ_{t2} and Γ_{t1} and the "fast" time $\tau \sim (t^* - t)/\varepsilon$. Till the moment of interaction, i.e., in the limit as $\tau \to +\infty$ and for $\rho \to +\infty$ as $\varepsilon \to 0$, our solution describes the situation (8.8), i.e., is the superposition of the noninteracting waves. After the interaction $\tau \to -\infty$ as $\varepsilon \to 0$, the realization of some scenario is determined by the behavior of the function ρ. If it turns out that $\rho \to 0$, then the scenario of the confluence of the points Γ_{t1} and Γ_{t2} is realized and the domain Ω_t^- disappears. If it turns out that $\rho \to -\infty$, then the global reconstruction occurs: after the interaction the merging kinks become diverging kinks, and the phase structure is simultaneously changed: Ω_{ti}^+ for $t > t^*$ turns into Ω_{ti}^- and Ω_t^- into Ω_t^+. Finally, if $\rho \to \text{const} \neq 0$ as $\tau \to -\infty$, then the mushy region is being formed: assumption (8.6) is violated and a region appears in which the state of the medium differs from both solid and liquid (from the phases "+" and "−").

Our main result is to establish the fact that for the positive temperature (in the region of interaction) the scenario of confluence of the boundaries is realized. In this case we obtain the following formulas uniform in t:

$$\frac{d\varphi_1}{dt} = \Phi_1(\theta, t, \varepsilon), \qquad \frac{d\varphi_2}{dt} = -\Phi_2(\theta, t, \varepsilon), \tag{8.14}$$

generalizing the one-dimensional Gibbs–Thompson law (8.4) to the case of interaction. Here the functions Φ_i are such that $\Phi_i \to \varkappa_1 \theta^\pm\big|_{\Gamma_{ti}}$ as $\tau \to +\infty$. So, for the time till the interaction, (8.14) turns into the standard Gibbs–Thompson condition (8.4). Simultaneously, $\Phi_1, \Phi_2 \to \Phi^* = \text{const}$ as $\tau \to -\infty$, which means the confluence of the boundaries. Simultaneously, we show that as $\tau \to -\infty$ the Stefan conditions (8.3) turn into the following conditions:

$$k\left(\frac{\partial\theta^+}{\partial x}\bigg|_{\Gamma_{t2}} - \frac{\partial\theta^+}{\partial x}\bigg|_{\Gamma_{t1}}\right) = \Phi_3(t, \varepsilon) \to 0 \qquad \text{as } \tau \to -\infty. \tag{8.15}$$

Hence the temperature is smoothed. The explicit formulas for the functions Φ_i, $i = 1, 2$, are given below (see (8.58), (8.59), and (8.69)).

In the multidimensional case (we restrict ourselves to $\Omega \subset \mathbb{R}^2$), the interaction of free boundaries is much more complicated. The main reason is that here, together with the "functional" bifurcation (two kinks turn into a single kink), a geometric bifurcation takes place. Namely, as the result of interaction, two curves Γ_{ti} turn into one curve with self-interaction at $t = t^*$. It still remains unknown how to describe this process uniformly in t. So we reinforce the notion of weakness of the solution, i.e., we weaken the solution replacing the multidimensional analog of Definition 3.2, (i.e., Definition 4.2 in Subsection 4.3) by the following definition.

Definition 8.2. A sequence of the pairs $(u = u(x, t, \varepsilon), \theta = \theta(x, t, \varepsilon))$ that for $\varepsilon > 0$ belong to $C^\infty(0, T; C^\infty(\Omega))$ is called the weak asymptotic solution of the phase field system (8.9), (8.10) $\operatorname{mod} O_{\mathcal{D}'(Q)}(\varepsilon^\mu)$, $Q = (0, T) \times \Omega$, in the two-dimensional case $\Omega \subset \mathbb{R}_x^2$ if, for any test functions ψ and $\vec{\psi} = (\psi_1, \psi_2) \in \mathcal{D}(Q)$, the following conditions are satisfied:

$$-\iint (\theta + u)\frac{\partial\psi}{\partial t}\, dx dt + \iint k(u)\langle\nabla\theta, \nabla\psi\rangle\, dx dt = O(\varepsilon^\mu),$$

$$\begin{aligned} &2\varepsilon \iint \frac{\partial u}{\partial t}\langle\vec{\psi}, \nabla u\rangle\, dx dt + \varepsilon \iint \left(\frac{\partial\psi_1}{\partial x_1} - \frac{\partial\psi_2}{\partial x_2}\right)\left(\left(\frac{\partial u}{\partial x_1}\right)^2 - \left(\frac{\partial u}{\partial x_2}\right)^2\right) dx dt \\ &+ 2\varepsilon \iint \left(\frac{\partial\psi_1}{\partial x_2} + \frac{\partial\psi_2}{\partial x_1}\right)\frac{\partial u}{\partial x_1}\frac{\partial u}{\partial x_2}\, dx dt - \frac{1}{\varepsilon(m+1)}\iint F(u)\langle\nabla, \vec{\psi}\rangle\, dx dt \end{aligned} \tag{8.16}$$

$$+ 2\varkappa \int \left(u - \frac{1}{3}u^3\right) \langle \nabla, \theta \vec{\psi} \rangle \, dxdt = O(\varepsilon^\mu),$$

where the function $F(u)$ is the same as in (8.13).

We show that (see Subsection 8.3) under sufficiently general assumptions on the geometry of Γ_{ti}, $t < t^*$, it is possible to find solutions of system (8.9), (8.10) such that for $t < t^* + \varepsilon^{\mu_1}$ and $t > t^* - \varepsilon^{\mu_1}$, $\mu_1 \in (0, 1/2)$, their sewing together satisfies relations (8.16) with exponent $\mu_1 \in (0, 1/2)$ whose value depends on the geometric properties of Γ_{ti} as $t \to t^*$. The main result of Subsection 8.3 is the conclusion that the temperature is "instantly" smoothed at the moment of contact of the boundaries in the sense that for $t < t^*$ the modified Stefan problem (8.2)–(8.4) for three domains Ω^+_{t1}, Ω^+_{t2}, and Ω^-_t becomes the problem for two domains Ω^+_t and Ω^-_t separated by a smooth boundary Γ_t for $t > t^*$.

Finally, in Subsection 8.3 we present rather curious (from our viewpoint) results of numerical modeling of the solution of the boundary value problem for the phase field system. The most unexpected (for us) fact is that the qualitative conclusions, which we draw on the basis of the asymptotic analysis as $\varepsilon \to 0$, were extremely clear in the numerical computations carried out for $\varepsilon = 0.1$.

8.2. Construction of the asymptotic solution ($\Omega \subset \mathbb{R}^1$). We introduce the functions $\varphi_{i0} = \varphi_{i0}(t)$ describing the dynamics of the free boundaries $\Gamma^0_{ti} = \{x = \varphi_{i0}(t)\}$ without interaction as follows. We set

$$\varphi_{10}(0) < \varphi_{20}(0) \tag{8.17}$$

and assume that the temperature $\theta^-\big|_{t=0}(x)$ is bounded below by a sufficiently large positive constant for $x \in [\varphi_{10}(0), \varphi_{20}(0)]$.

We consider the modified Stefan problem (8.2)–(8.4) and recall (see Subsection 4.3) that in the one-dimensional case the Stefan conditions (8.3) and the Gibbs–Thompson conditions (8.4) have the form

$$\begin{aligned} \dot{\varphi}_{10} &= \varkappa_1 \theta^\pm_1\Big|_{x=\varphi_{10}}, \qquad k^+ \frac{\partial \theta^+_1}{\partial x}\Big|_{x=\varphi_{10}-0} - k^- \frac{\partial \theta^-}{\partial x}\Big|_{x=\varphi_{10}+0} = -2\dot{\varphi}_{10}, \\ \dot{\varphi}_{20} &= -\varkappa_1 \theta^\pm_2\Big|_{x=\varphi_{20}}, \qquad k^+ \frac{\partial \theta^+_2}{\partial x}\Big|_{x=\varphi_{20}+0} - k^- \frac{\partial \theta^-}{\partial x}\Big|_{x=\varphi_{20}-0} = -2\dot{\varphi}_{20}, \end{aligned} \tag{8.18}$$

where, as usual, the point on a function denotes differentiation with respect to t, and $\varkappa_1$ is related to the constant $\varkappa$ in the phase field system as $\varkappa_1 = \varkappa/2a_2$. We also recall (see Subsection 4.3) that a_2 is the integral of the squared special function $\omega(\eta)$ defined as follows. By $\omega_0 = \omega_0(\eta)$ we denote the solution of the problem

$$\left(\frac{d\omega_0(\eta)}{d\eta}\right)^2 = \frac{1}{m+1} F(\omega_0), \qquad \omega_0 \to \pm 1 \qquad \text{as } \eta \to \pm\infty. \tag{8.19}$$

Then we set

$$\omega(\eta) = \frac{1}{2} \frac{d\omega_0(\eta)}{d\eta}. \tag{8.20}$$

It is easy to show that $\omega_0(\eta)$ is an odd function such that $\omega_0 \to \pm 1$, while $\omega \to 0$ as $\eta \to \pm\infty$ at an exponential rate and

$$\int \omega(\eta) \, d\eta = 1.$$

By $(\theta_i^\pm, \varphi_{i0})$, $i = 1, 2$, we denote the classical solution of the modified Stefan problem (8.2), (8.18). Obviously, we have

$$\dot\varphi_{10} > 0, \qquad \dot\varphi_{20} < 0 \tag{8.21}$$

because of (8.18) and our assumption that the temperature is positive. Therefore, the boundaries Γ_{ti}^0 will move towards each other with increasing t. We choose a number $\mu_1 \in (0,1)$ and let t_1 denote a moment of time such that

$$\operatorname{dist}(\Gamma_{t1}^0, \Gamma_{t2}^0)\big|_{t=t1} = \varepsilon^{\mu_1}. \tag{8.22}$$

For $t > t_1$, we continue the functions $\varphi_{i0}(t)$ so that these functions remain continuous and satisfy inequalities (8.21). We preserve the notation $\varphi_{i0}(t)$ for these functions continued so that they are globally defined (for $t \geq 0$).

Let

$$\sigma_0(t) = \varphi_{20}(t) - \varphi_{10}(t). \tag{8.23}$$

Then the function

$$\tau = \sigma_0(t)/\varepsilon \tag{8.24}$$

has the meaning of "fast" time. If by $t = t^*$ we denote the conditional moment of time at which the free boundaries contact each other, i.e., if $\sigma_0(t_0^*) = 0$, then as $\varepsilon \to 0$ we have for any $\mu_2 \in (0,1)$:

$$\tau \to \infty \quad \text{as } t \leq t_0^* - \varepsilon^{\mu_2}, \qquad \tau \to -\infty \quad \text{as } t \geq t_0^* + \varepsilon^{\mu_2}. \tag{8.25}$$

We write the asymptotic solution of the phase field system (8.9), (8.10) in the form

$$u = \omega_0\Big(\beta\frac{\varphi_1 - x}{\varepsilon}\Big) + \omega_0\Big(\beta\frac{x - \varphi_2}{\varepsilon}\Big) + \omega_\mu(\rho), \tag{8.26}$$

$$\theta = \gamma^+ + \gamma_1^-\omega_1\Big(\beta\frac{\varphi_1 - x}{\varepsilon}\Big) + \gamma_2^-\omega_1\Big(\beta\frac{x - \varphi_2}{\varepsilon}\Big). \tag{8.27}$$

Here $\omega_0(\eta)$ is a special function determined by the solution of problem (8.19), $\omega_1(x/\varepsilon)$ is an odd regularization of the discontinuous function $\operatorname{sign}(x)$, $\beta = \beta(\tau,t) > 0$, $\varphi_i = \varphi_i(\tau, t)$, $\gamma^+ = \gamma^+(x,t)$ and $\gamma_i^- = \gamma_i^-(x,t)$ are the desired smooth functions, $\rho = (\varphi_2 - \varphi_1)/\varepsilon$, $\omega_\mu(\rho) = \omega_0(\rho + \varepsilon^{-\mu})$, $\mu \in (0,1)$ is some chosen number, and τ is the "fast" time defined in (8.24). We write the function φ_i as

$$\varphi_i(\tau, t) = \varphi_{i0}(t) + \sigma_0(t)\varphi_{i1}(\tau, t) \tag{8.28}$$

and assume that

$$|\varphi_{i1}(\tau,t)| \leq \text{const}, \qquad \varphi_{i1} \to 0 \quad \text{as } \tau \to +\infty. \tag{8.29}$$

Moreover, we assume that

$$\beta > 0 \quad \text{and} \quad \beta \to 1 \quad \text{as } \tau \to +\infty. \tag{8.30}$$

Finally, we assume that

$$\gamma_i^-\big|_{x=\varphi_{i0}(t)} = 0, \qquad i = 1, 2. \tag{8.31}$$

Let us clarify the meaning of the ansatz (8.26). Before the interaction of free boundaries starts, i.e., for $t \leq t_0^* - \varepsilon^{\mu_2}$ and $\mu_2 < \mu_1$, formulas (8.28) and (8.29) imply

$$\tau \to +\infty, \qquad \varphi_{i1} \to \varphi_{i0}, \qquad \rho \to +\infty \qquad \text{as } \varepsilon \to 0.$$

Therefore, for such t, with accuracy $O(\varepsilon^\infty)$, we have the formula

$$u = \omega_0\left(\frac{\varphi_{10} - x}{\varepsilon}\right) + \omega_0\left(\frac{x - \varphi_{20}}{\varepsilon}\right) + 1,$$

which, in the limit as $\varepsilon \to 0$, describes the function $\overline{u}$ equal to 1 in $\Omega_{t1}^+ = \Omega \cap \{x < \varphi_{10}\}$ and $\Omega_{t2}^+ = \Omega \cap \{x > \varphi_{20}\}$ and equal to -1 in $\Omega_t^- = \Omega \cap \{\varphi_{10} < x < \varphi_{20}\}$.

After the confluence of the boundaries, the plot of the function u significantly depends on the limit behavior of ρ. If $\rho \to 0$ as $\tau \to -\infty$, then

$$\varphi_1 \to \varphi_2, \qquad u \to 1,$$

i.e., the domain Ω_t^- disappears.

Conversely, if $\rho \to -\infty$ as $\tau \to -\infty$, then, for $t \geq t_0^* + \varepsilon^{\mu_2}$ and $\mu_2 < 1 - \mu$, we obtain $\omega_\mu(\rho) \to -1$ and $\varphi_1 > \varphi_2$. Respectively, as $\varepsilon \to 0$ we obtain the function $\overline{u}$ equal to -1 for $x < \varphi_2$ and $x < \varphi_1$ and equal to $+1$ for $\varphi_2 < x < \varphi_1$. Therefore, this means that, as the result of the collision of the free boundaries, a domain Ω_t^+ increasing (with increasing $t > t^*$) appears instead of the domain Ω_t^- decreasing (as $t \to t_0^*$) and domains Ω_{ti}^- appear instead of the domain Ω_{ti}^+. Finally, if $\rho \to \text{const} \neq 0$ as $\tau \to -\infty$, then the distance between the boundaries Γ_{t1} and Γ_{t2} stabilizes at a value $O(\varepsilon)$, i.e., an analog of the "mushy region" arises.

An important property of the ansatz (8.27) is that it implies the continuity of the function θ (see conditions (8.31)).

Now we shall construct the asymptotic solution. For simplicity, we consider only the time such that

$$t \leq t_0^* + \varepsilon^{\mu_2}, \qquad 0 < 1 - \mu < \mu_2 < 1. \tag{8.32}$$

The assumption that $\mu_2 < 1$ ensures the possibility to pass to the limit as $\varepsilon \to 0$. At the same time, since $\mu_2 > 1 - \mu$ and we assume that conditions (8.29) are satisfied, the inequality $\rho < \varepsilon^{-\mu}$ is uniform in t determined by (8.32). Therefore, if conditions (8.29) are satisfied, then we have $\omega_\mu(\rho) = 1 + O(\varepsilon^\infty)$ and, instead of (8.32), we obtain the simplified formula

$$u = 1 + \omega_0\left(\beta\frac{\varphi_1 - x}{\varepsilon}\right) + \omega_0\left(\beta\frac{x - \varphi_2}{\varepsilon}\right). \tag{8.33}$$

Now we need to calculate some weak asymptotic expansions. As is easy to calculate, by (8.20) and (8.26), we have

$$\varepsilon\left(\frac{\partial u}{\partial x}\right)^2 = \frac{4}{\varepsilon}\beta^2\left\{\omega^2\left(\beta\frac{\varphi_1 - x}{\varepsilon}\right) + \omega^2\left(\beta\frac{x - \varphi_2}{\varepsilon}\right) - 2\omega\left(\beta\frac{\varphi_1 - x}{\varepsilon}\right)\omega\left(\beta\frac{x - \varphi_2}{\varepsilon}\right)\right\}.$$

However, since $\omega(\eta)$ is even,

$$\beta\omega^2\left(\beta\frac{x - \varphi_i}{\varepsilon}\right) = \varepsilon a_2\delta(x - \varphi_i) + O_{\mathcal{D}'}(\varepsilon^3),$$

$$\beta\omega\left(\beta\frac{x - \varphi_1}{\varepsilon}\right)\omega\left(\beta\frac{x - \varphi_2}{\varepsilon}\right) = \frac{1}{2}\varepsilon a_2\lambda(\beta\rho)\big(\delta(x - \varphi_1) + \delta(x - \varphi_2)\big) + O_{\mathcal{D}'}(\varepsilon^2),$$

where a_2 is the integral of $\omega^2(\eta)$ and λ is the convolution of the functions ω:

$$\lambda(z) = \frac{1}{a_2}\int \omega(\eta)\omega(\eta - z)\,d\eta. \tag{8.34}$$

Therefore,

$$\varepsilon\left(\frac{\partial u}{\partial x}\right)^2 = 4\beta a_2\big(1-\lambda(\beta\rho)\big)\big(\delta(x-\varphi_1)+\delta(x-\varphi_2)\big)+O_{\mathcal{D}'}(\varepsilon). \tag{8.35}$$

Next, we use the following relation, which is easy to verify and which holds for any even functions ω:

$$\int \eta\omega(\eta+z)\omega(\eta-z)\,d\eta = 0.$$

Then, after some calculations, under condition (8.32), we obtain the relation

$$\begin{aligned}\varepsilon\frac{\partial u}{\partial x}\frac{\partial u}{\partial t} &= -2a_2\beta\Big\{\big(2\dot\varphi_1-\lambda(\dot\varphi_1+\dot\varphi_2)\big)\delta(x-\varphi_1)+\big(2\dot\varphi_2-\lambda(\dot\varphi_1+\dot\varphi_2)\big)\delta(x-\varphi_1)\Big\}\\ &\quad+2a_2\dot\beta\rho\lambda\big(\delta(x-\varphi_2)-\delta(x-\varphi_1)\big)+O_{\mathcal{D}'}(\varepsilon),\end{aligned}$$

where $\lambda=\lambda(\beta\rho)$ and the point on a function denotes the total derivative with respect to t. In particular,

$$\dot\beta=\frac{\partial\beta(\tau,t)}{\partial t}+\frac{1}{\varepsilon}\frac{d\sigma_0}{dt}\frac{\partial\beta(\tau,t)}{\partial\tau}.$$

Next, we take into account the fact that $\lambda(\beta\rho)$ disappears at an exponential rate as $\rho\to\pm\infty$, which implies the following relation that is uniform in $\rho\in\mathbb{R}^1$:

$$\lambda\big(\delta(x-\varphi_2)-\delta(x-\varphi_1)\big)=\frac{1}{2}\varepsilon\rho\lambda\big(\delta'(x-\varphi_2)+\delta'(x-\varphi_1)\big)+O_{\mathcal{D}'}(\varepsilon^2).$$

Hence we have

$$\begin{aligned}\varepsilon\frac{\partial u}{\partial t}\frac{\partial u}{\partial x} &= -2a_2\beta\Big\{\big(2\dot\varphi_1-\lambda(\dot\varphi_1+\dot\varphi_2)\big)\delta(x-\varphi_1)\\ &\qquad+\big(2\dot\varphi_2-\lambda(\dot\varphi_1+\dot\varphi_2)\big)\delta(x-\varphi_1)\Big\}\\ &\quad+a_2\dot\sigma_0\beta_\tau\rho^2\lambda\big(\delta'(x-\varphi_2)+\delta'(x-\varphi_1)\big)+O_{\mathcal{D}'}(\varepsilon).\end{aligned} \tag{8.36}$$

Similar simple calculations lead to the relation

$$\frac{\partial}{\partial x}\left(u-\frac{1}{3}u^3\right)=2A(\beta\rho)\big(\delta(x-\varphi_2)-\delta(x-\varphi_1)\big)+O_{\mathcal{D}'}(\varepsilon), \tag{8.37}$$

where, under condition (8.32),

$$A(z)=\int\Big\{1-\big(1+\omega_0(\eta)-\omega_0(\eta+z)\big)^2\Big\}\omega(\eta)\,d\eta. \tag{8.38}$$

But if condition (8.32) is violated, then the coefficient A is calculated by the formula

$$A(z)=\int\Big\{1-\big(1+\omega_\mu(\rho)+\omega_0(\eta)-\omega_0(\eta+z)\big)^2\Big\}\omega(\eta)\,d\eta. \tag{8.39}$$

A straightforward verification shows that the following relations hold for both formulas:

$$A(z)\to\frac{2}{3}\quad\text{as } z\to+\infty,\qquad A(z)\to 4a_2z+O(z^2)\quad\text{as } z\to 0. \tag{8.40}$$

To calculate the limit as $\rho\to-\infty$, we use formula (8.39) and also obtain $A(\beta\rho)\to 2/3$ in this case.

The next step is to calculate the weak asymptotics of $F(u)$. In the general case, this was done in Subsection 1.5. If $u\to\pm1$ as $x\to\pm\infty$ and $F(\pm1)=0$, then the general formula can be made more precise.

LEMMA 8.1. *Let $F(z) \in C^\infty$, and let $F(\pm 1) = 0$. Then*

$$\begin{aligned}(8.41)\qquad &\frac{1}{\varepsilon}F\Big(1+\omega_0\Big(\beta\frac{\varphi_1-x}{\varepsilon}\Big)+\omega_0\Big(\beta\frac{x-\varphi_2}{\varepsilon}\Big)\Big)\\ &\quad=\frac{1}{\beta}\big(a_F^-\delta(x-\varphi_1)+a_F^+\delta(x-\varphi_2)\big)\\ &\quad+\frac{1}{2\beta}B_\Delta(\beta\rho)\big(\delta(x-\varphi_1)+\delta(x-\varphi_2)\big)+O_{\mathcal{D}'}(\varepsilon),\end{aligned}$$

where

$$a_F^\pm = \int F(\pm\,\omega_0(\eta))\,d\eta,$$

$$(8.42)\qquad B_\Delta(z) = \int\Big\{F\big(1+\omega_0(\eta)-\omega_0(\eta+z)\big)-F\big(\omega_0(\eta)\big)-F\big(-\omega_0(\eta+z)\big)\Big\}\,d\eta,$$

$$(8.43)\qquad B_\Delta(z)\to 0 \qquad \textit{as } z\to\infty.$$

PROOF OF THEOREM 4.3. The proof is based on the remark that, for u of the form (8.33), the function $F(u)$ is concentrated in the ε-neighborhoods of the points $x=\varphi_1$ and $x=\varphi_2$. □

In our case F is an odd function and ω_0 satisfies Eq. (8.19). Hence we have

$$a_F^+ = a_F^- = 4(m+1)a_2.$$

Now we are ready to analyze relations (8.12). Combining formulas (8.35)–(8.37) and (8.41), we see that the following relations are necessary and sufficient for (8.12) to hold:

$$\begin{aligned}&\frac{1}{\beta}\Big\{1-\beta^2(1-\lambda)+\frac{B_\Delta}{8(m+1)a_2}+\frac{1}{2}\dot\sigma_0\lambda\rho^2\beta\beta_\tau\Big\}\big(\delta'(x-\varphi_1)+\delta'(x-\varphi_2)\big)\\ &\quad-\beta\Big\{\big(2\dot\varphi_1-\lambda(\dot\varphi_1+\dot\varphi_2)\big)\delta(x-\varphi_1)+\big(2\dot\varphi_2-\lambda(\dot\varphi_1+\dot\varphi_2)\big)\delta(x-\varphi_2)\Big\}\\ &\quad-\frac{\varkappa}{a_2}A\big(\theta\delta(x-\varphi_2)-\theta\delta(x-\varphi_1)\big)=0,\end{aligned}$$

where $\beta\rho$ is the argument of the functions λ, A, and B_Δ.

This fact readily implies three equations

$$(8.44)\qquad 2\dot\varphi_1-\lambda(\dot\varphi_1+\dot\varphi_2)=\frac{\varkappa}{a_2}\frac{A}{\beta}\theta\Big|_{x=\varphi_1},$$

$$(8.45)\qquad 2\dot\varphi_2-\lambda(\dot\varphi_1+\dot\varphi_2)=-\frac{\varkappa}{a_2}\frac{A}{\beta}\theta\Big|_{x=\varphi_2},$$

$$(8.46)\qquad -\dot\sigma_0\frac{\partial\beta}{\partial\tau}=G(\beta,\rho\beta),$$

where

$$(8.47)\qquad G=\frac{2}{\lambda\rho^2\beta}\Big\{1-\beta^2+\lambda\beta^2+\frac{B_\Delta}{8(m+1)a_2}\Big\}.$$

Now we analyze system (8.44)–(8.46). Recall that $\rho=(\varphi_2-\varphi_1)/\varepsilon$ and $\varphi_i=\varphi_{i0}+\sigma_0\varphi_{i1}$, $\sigma_0=\varphi_{20}-\varphi_{10}$. Hence we have

$$\dot\varphi_2-\dot\varphi_1=\dot\sigma_0\frac{d}{d\tau}\tau(1+\varphi_{21}-\varphi_{11})\overset{\text{def}}{=}\dot\sigma_0\frac{\partial\rho}{\partial\tau}.$$

We also take into account the fact that our definition of the functions $\varphi_{i0}(t)$ implies

$$\dot{\sigma}_0 = -\frac{\varkappa}{3a_2}\Big(\theta\Big|_{x=\varphi_{20}} + \theta\Big|_{x=\varphi_{10}}\Big).$$

Now, subtracting one equation from the other in (8.44), (8.45), we obtain

$$\frac{\partial \rho}{\partial \tau} = \frac{3}{2}\frac{A}{\beta}\frac{\theta|_{x=\varphi_2} + \theta|_{x=\varphi_1}}{\theta|_{x=\varphi_{20}} + \theta|_{x=\varphi_{10}}}. \tag{8.48}$$

Recall that, according to our assumption, the temperature is continuous and positive. Therefore, taking into account (8.29), (8.30), and (8.40), we see that

$$\frac{\partial \rho}{\partial \tau} \to 1 \qquad \text{as } \tau \to \infty. \tag{8.49}$$

Next, the second estimate in (8.40) implies that $A(\beta\rho)\beta^{-1}\big|_{\rho=0} = 0$ uniformly in all bounded β. Moreover, A is positive for $\rho > 0$. This follows from the fact that, for $z > 0$, the function $(1 + \omega_0(\eta) - \omega_0(\eta + z))^2$ is close to 1 everywhere except the neighborhoods of the points $\eta = 0$ and $\eta = -z$ where this function is less than 1. So $\rho = 0$ is the maximal value of ρ at which the right-hand side of Eq. (8.48) is zero. This implies that there exists a solution of Eq. (8.48) such that

$$\rho/\tau \to 1 \qquad \text{as } \tau \to +\infty \quad \text{and} \quad \rho \to 0 \qquad \text{as } \tau \to -\infty. \tag{8.50}$$

Next, we use the identity

$$\dot{\varphi}_i = \dot{\varphi}_{i0} + \dot{\sigma}_0 \frac{\partial}{\partial \tau}(\tau \dot{\varphi}_{i1}),$$

which can be easily verified. Now Eqs. (8.44), (8.45) can be rewritten as

$$2\dot{\sigma}_0 \frac{\partial}{\partial \tau}(\tau \varphi_{i1}) = \lambda(\dot{\varphi}_1 + \dot{\varphi}_2) \pm \frac{\varkappa}{a_2}\Big(\frac{A}{\beta}\theta\Big|_{x=\varphi_i} - \frac{2}{3}\theta\Big|_{x=\varphi_{i0}}\Big), \qquad i = 1, 2. \tag{8.51}$$

By letting τ tend to $+\infty$ and using relations (8.30), (8.40) and the fact that λ disappears for large τ, we see that the right-hand side of Eq. (8.51) is zero. This implies that $\varphi_{i1} \to 0$ as $\tau \to +\infty$.

Now we return to Eq. (8.46). Using relations (8.43), we examine the limit of the right-hand side as $\tau \to +\infty$ and see that the assumption that $\beta \to 1$ as $\tau \to \infty$ is a necessary condition for G to be bounded.

Let us determine the sign of G for $\tau \gg 1$. We note that the integrand in (8.42) is an even function with respect to the point $\eta = -z/2$. Hence we can rewrite the function $B_\Delta(z)$ as

$$\begin{aligned} B_\Delta(z) = 2\int_0^\infty \Big\{ & F\big(\omega_0(\eta - z/2) + 1 - \omega_0(\eta + z/2)\big) \\ & - F\big(\omega_0(\eta - z/2)\big) - F\big(1 - (1 - \omega_0(\eta + z/2))\big)\Big\}\, d\eta. \end{aligned} \tag{8.52}$$

For sufficiently large z and positive η, the function $W = 1 - \omega_0(\eta + z/2)$ is an infinitesimal of exponential type with respect to $\eta + z/2$. Using the Taylor expansion, we see that the integrand in (8.52) can be written as

$$F'_{\omega_0}\big(\omega_0(\eta - z/2)\big)W(\eta, z) + O\big(W^2(\eta, z)\big).$$

It follows from (8.51) that

$$F'_{\omega_0}\big(\omega_0(\eta - z/2)\big) = 2(m+1)\omega''_{0\eta}(\eta - z/2).$$

Then, after simple calculations, we obtain the relation

$$\int_0^\infty F'_{\omega_0}\big(\omega_0(\eta - z/2)\big)W(\eta, z)\,d\eta = 4(m+1)\Big\{a_2\lambda(z) - \omega(z/2)\big(1-\omega_0(z/2)\big)\Big\}. \tag{8.53}$$

The second term in the right-hand side of (8.53) is $O(\omega^2(z))$, i.e., it is of the same order as the integral of $W^2(\eta, z)$. Hence we have

$$B_\Delta(z) = 8(m+1)a_2\lambda(z) + O(\omega^2(z)).$$

This and (8.47) imply the relation

$$G(\beta, z) = \frac{2}{\lambda\beta^2\rho}\Big\{1 - \beta^2 + \lambda(\beta^2+1) + O(\omega^2(z))\Big\} \qquad \text{for } z \gg 1$$

whose obvious consequence is the conclusion that

$$G(\beta, z) > 0 \quad \text{for } \beta < 1 \qquad \text{and} \qquad G(\beta, z) < 0 \quad \text{for } \beta > 1. \tag{8.54}$$

Let us consider the right-hand side of Eq. (8.46) as $\tau \to -\infty$. We take into account conditions (8.50) and the fact that a function $F(z)$ of the form (8.13) satisfies the relations $F'(1) = 0$ and $F''(1) = 4m(m+1)$. Then it follows from the Taylor expansions that

$$B_\Delta = -8(m+1)a_2(1-mz^2) + O(z^3), \qquad \lambda(z) = 1 - a_2' z^2/2a_2 + O(z^4). \tag{8.55}$$

Here a_2' is the integral of $(\omega'(\eta))^2$ related to a_2 as

$$a_2' + (2m+1)\int \omega_0^{2m}(\eta)\omega^2(\eta)\,d\eta = a_2. \tag{8.56}$$

Relation (8.55) implies

$$G(\beta, \rho\beta) = \beta\frac{a_2'}{a_2}\bigg(2m\frac{a_2'}{a_2} - \beta^2\bigg) + O\big((\rho\beta)^2\big) \qquad \text{as } \rho \to 0,$$

and (8.56) implies the inequality $a_2/a_2' > 1$.

So, as $\tau \to -\infty$, the function β tends either to zero or to the number $\beta^* = \sqrt{2ma_2/a_2'} > 1$. However, accurate calculations show that all the derivatives of the function $G(\beta(\tau), \rho(\tau)\beta(\tau))$ with respect to τ are zero for $\beta = 0$. Hence $\beta = 0$ is an isolated solution of Eq. (8.46). It follows from this and the boundary condition that $\beta \to 1$ as $\tau \to +\infty$, that is, assumptions (8.30) are satisfied and, moreover, the limit relation holds:

$$\beta \to \beta^* \qquad \text{as } \tau \to -\infty. \tag{8.57}$$

Summarizing the above constructions, we conclude that relations (8.12) imply the equations

$$\dot\varphi_1 = \frac{\varkappa}{4a_2}\frac{A}{\beta}\frac{2\theta|_{x=\varphi_1} - \lambda(\theta|_{x=\varphi_1} + \theta|_{x=\varphi_2})}{1-\lambda} \stackrel{\text{def}}{=} \Phi_1(\theta, t, \varepsilon), \tag{8.58}$$

$$\dot\varphi_2 = \frac{\varkappa}{4a_2}\frac{A}{\beta}\frac{2\theta|_{x=\varphi_2} - \lambda(\theta|_{x=\varphi_1} + \theta|_{x=\varphi_2})}{1-\lambda} \stackrel{\text{def}}{=} \Phi_2(\theta, t, \varepsilon), \tag{8.59}$$

$$\frac{\partial\beta}{\partial\tau} = \frac{3a_2}{\varkappa}\frac{G}{\theta|_{x=\varphi_{10}} + \theta|_{x=\varphi_{20}}}, \tag{8.60}$$

where $A = A(\rho, \beta)$, $\lambda = \lambda(\rho\beta)$, and $G = G(\beta, \rho\beta)$ are defined in (8.34), (8.39), and (8.47), and $\rho = (\varphi_2 - \varphi_1)/\varepsilon$.

Equations (8.58) and (8.59) are an analog of the one-dimensional version of the Gibbs–Thompson condition, more precisely, a generalization of this condition to the case of the confluence of the boundaries. As $\tau \to +\infty$, i.e., prior to the interaction, they turn into the standard conditions (8.18). As $\tau \to -\infty$, i.e., after the interaction, these equations determine some functions φ_i such that $\varphi_1 - \varphi_2 \to 0$. Calculating the indeterminacy in the right-hand sides of (8.58) and (8.59), we see that the functions φ_i are bounded as $\tau \to -\infty$. Namely, using formulas (8.40), (8.55), and (8.57), we obtain

$$\dot{\varphi}_i = 2\frac{a_2}{a_2'}\frac{\varkappa}{\beta^{*2}}\frac{\theta|_{x=\varphi_1} - \theta|_{x=\varphi_2}}{\rho} + O(\rho) \qquad \text{as } \rho \to 0, \quad i = 1, 2. \tag{8.61}$$

Since the temperature θ is continuous, the right-hand side in formula (8.61) remains bounded as $\rho \to 0$, i.e., as $\tau \to -\infty$.

Now let us consider relation (8.11). Easy calculations show that under assumption (8.31) we have the relation

$$\begin{aligned}\frac{\partial\theta}{\partial t} &= \frac{\partial\gamma^+}{\partial t} - \frac{\partial\gamma_1^-}{\partial t}\operatorname{sign}(x-\varphi_1) + \frac{\partial\gamma_2^-}{\partial t}\operatorname{sign}(x-\varphi_2) + O_{\mathcal{D}'}(\varepsilon),\\ k(u)\frac{\partial\theta}{\partial x} &= k^+\Big\{2\frac{\partial\gamma^+}{\partial x} - \operatorname{sign}(x-\varphi_1)\frac{\partial}{\partial x}(\gamma^+ + \gamma_1^- - \gamma_2^-)\\ &\qquad + \operatorname{sign}(x-\varphi_2)\frac{\partial}{\partial x}(\gamma^+ - \gamma_1^- + \gamma_2^-)\Big\}\\ &\quad + \frac{1}{2}k^-\big(\operatorname{sign}(x-\varphi_1) - \operatorname{sign}(x-\varphi_2)\big)\frac{\partial}{\partial x}(\gamma^+ - \gamma_1^- - \gamma_2^-) + O_{\mathcal{D}'}(\varepsilon).\end{aligned} \tag{8.62}$$

Moreover, we have

$$u = 1 - \operatorname{sign}(x-\varphi_1) + \operatorname{sign}(x-\varphi_2) + O_{\mathcal{D}'}(\varepsilon^2).$$

Therefore, for (8.11) to hold, we need the relations

$$\begin{aligned}&\frac{\partial\gamma^+}{\partial t} - k^+\frac{\partial^2\gamma^+}{\partial x^2} + \Big\{-\frac{\partial\gamma_1^-}{\partial t} + \frac{k^+}{2}\frac{\partial^2}{\partial x^2}(\gamma^+ + \gamma_1^- - \gamma_2^-)\\ &\qquad\qquad + \frac{k^-}{2}\frac{\partial^2}{\partial x^2}(\gamma^+ - \gamma_1^- - \gamma_2^-)\Big\}\operatorname{sign}(x-\varphi_1)\\ &+ \Big\{-\frac{\partial\gamma_2^-}{\partial t} - \frac{k^+}{2}\frac{\partial^2}{\partial x^2}(\gamma^+ - \gamma_1^- + \gamma_2^-) - \frac{k^-}{2}\frac{\partial^2}{\partial x^2}(\gamma^+ - \gamma_1^- - \gamma_2^-)\Big\}\operatorname{sign}(x-\varphi_2)\\ &+ \Big\{2\dot{\varphi}_1 - k^+\frac{\partial}{\partial x}(\gamma^+ + \gamma_1^- - \gamma_2^-) - k^-\frac{\partial}{\partial x}(\gamma^+ - \gamma_1^- - \gamma_2^-)\Big\}\delta(x-\varphi_1)\\ &- \Big\{2\dot{\varphi}_2 - k^+\frac{\partial}{\partial x}(\gamma^+ - \gamma_1^- + \gamma_2^-) + k^-\frac{\partial}{\partial x}(\gamma^+ - \gamma_1^- - \gamma_2^-)\Big\}\delta(x-\varphi_2) = O_{\mathcal{D}'}(\varepsilon).\end{aligned} \tag{8.63}$$

Let us consider the case $x < \varphi_1$. Then $\operatorname{sign}(x-\varphi_1) = \operatorname{sign}(x-\varphi_2) = -1$ and it follows from (8.63) that

$$\frac{\partial}{\partial t}(\gamma^+ + \gamma_1^- - \gamma_2^-) = k^+\frac{\partial^2}{\partial x^2}(\gamma^+ + \gamma_1^- - \gamma_2^-). \tag{8.64}$$

Obviously, for $\varphi_1 < x < \varphi_2$, $\operatorname{sign}(x-\varphi_1) = 1$ and $\operatorname{sign}(x-\varphi_2) = -1$. Therefore, for such x we obtain

$$\frac{\partial}{\partial t}(\gamma^+ - \gamma_1^- - \gamma_2^-) = k^- \frac{\partial^2}{\partial x^2}(\gamma^+ - \gamma_1^- - \gamma_2^-). \tag{8.65}$$

Similarly, for $x > \varphi_2$, relation (8.63) implies

$$\frac{\partial}{\partial t}(\gamma^+ - \gamma_1^- + \gamma_2^-) = k^+ \frac{\partial^2}{\partial x^2}(\gamma^+ - \gamma_1^- + \gamma_2^-). \tag{8.66}$$

As is easy to see, we have obtained the heat equations (8.2) in the domains Ω_{ti}^+ and Ω_t^-. If (8.64)–(8.66) are satisfied, then (8.63) takes the form

$$\begin{aligned}\Big\{2\dot\varphi_1 - k^+\frac{\partial}{\partial x}(\gamma^+ + \gamma_1^- - \gamma_2^-) - k^-\frac{\partial}{\partial x}(\gamma^+ - \gamma_1^- - \gamma_2^-)\Big\}\delta(x-\varphi_1)\\ - \Big\{2\dot\varphi_2 - k^+\frac{\partial}{\partial x}(\gamma^+ - \gamma_1^- + \gamma_2^-) + k^-\frac{\partial}{\partial x}(\gamma^+ - \gamma_1^- - \gamma_2^-)\Big\}\delta(x-\varphi_2) = O_{\mathcal{D}'}(\varepsilon).\end{aligned} \tag{8.67}$$

Obviously, this implies the Stefan conditions

$$\begin{aligned} k^-\frac{\partial}{\partial x}(\gamma^+ - \gamma_1^- - \gamma_2^-)\Big|_{x=\varphi_1} - k^+\frac{\partial}{\partial x}(\gamma^+ + \gamma_1^- - \gamma_2^-)\Big|_{x=\varphi_1} &= -2\dot\varphi_1,\\ k^+\frac{\partial}{\partial x}(\gamma^+ - \gamma_1^- + \gamma_2^-)\Big|_{x=\varphi_2} - k^-\frac{\partial}{\partial x}(\gamma^+ - \gamma_1^- - \gamma_2^-)\Big|_{x=\varphi_2} &= 2\dot\varphi_2.\end{aligned} \tag{8.68}$$

We consider (8.58), (8.59), (8.64)–(8.66), and (8.68) as $\tau \to +\infty$, i.e., prior to the interaction of the boundaries. For this time we have

$$\varphi_i \to \varphi_{i0}, \qquad \dot\varphi_i \to \dot\varphi_{i0}, \qquad \Phi_i \to \pm\varkappa_1\theta\big|_{x=\varphi_{i0}}, \quad i=1,2, \qquad \text{as} \quad \varepsilon \to 0.$$

Thus we obtain the standard modified Stefan problem (8.2) (the one-dimensional case) and (8.18). Recall that we define the time $t_1 = t_0^* - \varepsilon^{\mu_1}$, $\mu_1 \in (0,1)$, to be the upper bound for the admissibility of the Stefan and Gibbs–Thompson conditions (8.18). If we choose a number $\mu_2 \in (0,\mu)$ and consider the time interval $t_0^* - \varepsilon^{\mu_2} \le t \le t_0^* - \varepsilon^{\mu_1}$, then on this interval both statements of the boundary conditions i.e., (8.18) and (8.58), (8.59), (8.68), are simultaneously well defined and, moreover, the "fast" time τ is not too large, $\tau \sim \varepsilon^{-1+\mu}$, $\mu \in [\mu_2, \mu_1]$.

Let us consider the boundary conditions (8.58), (8.59), (8.68) for finite time τ, (i.e., for $|t - t_0^*| = O(\varepsilon)$) and as $\tau \to -\infty$. All these boundary conditions are meaningful for such time. Moreover, it should be noted that for this time, with accuracy $O_{\mathcal{D}'}(\varepsilon)$, relation (8.67) implies, together with (8.58), the relation

$$k^+\Big\{\frac{\partial}{\partial x}(\gamma^+ - \gamma_1^- + \gamma_2^-)\Big|_{x=\varphi_2} - \frac{\partial}{\partial x}(\gamma^+ + \gamma_1^- - \gamma_2^-)\Big|_{x=\varphi_1}\Big\} = 2\dot\sigma_0\frac{\partial\rho}{\partial\tau}. \tag{8.69}$$

Since $\rho_\tau \to 0$ as $\tau \to -\infty$, relation (8.69) means the smoothing of the temperature when the free boundaries merge.

It remains to establish the correspondence between the functions γ^+ and γ^- and the solution of the free boundary problem. We assume that $\theta_i^+ = \theta_i^+(x,t,\varepsilon)$ and $\theta_i^- = \theta_i^-(x,t,\varepsilon)$ satisfy the problem

$$\begin{aligned}\frac{\partial\theta_i^+}{\partial t} &= k^+\frac{\partial^2\theta_i^+}{\partial x^2}, \qquad x < \varphi_1 \quad \text{for } i=1 \quad \text{and} \quad x > \varphi_2 \quad \text{for } i=2,\\ \frac{\partial\theta^-}{\partial t} &= k^-\frac{\partial^2\theta^-}{\partial x^2}, \qquad \varphi_1 < x < \varphi_2,\end{aligned} \tag{8.70}$$

$$k^{-}\frac{\partial\theta^{-}}{\partial x}\bigg|_{x=\varphi_1} - k^{+}\frac{\partial\theta_1^{+}}{\partial x}\bigg|_{x=\varphi_1} = -2\dot{\varphi}_1, \quad k^{+}\frac{\partial\theta_2^{+}}{\partial x}\bigg|_{x=\varphi_2} - k^{-}\frac{\partial\theta^{-}}{\partial x}\bigg|_{x=\varphi_2} = 2\dot{\varphi}_2,$$
$$\dot{\varphi}_1 = \Phi_1, \qquad \dot{\varphi}_2 = -\Phi_2.$$

Here φ_i and Φ_i are defined in (8.58)–(8.60) and it is assumed that problem (8.70) is supplemented with an initial condition satisfying (8.17) and a boundary condition on the boundary $\partial\Omega$.

By θ_{ic}^{+} and θ_c^{-} we denote smooth continuations of θ_i^{+} and θ^{-} to the entire domain Ω and set

$$\gamma^{+} = \frac{1}{2}(\theta_{1c}^{+} + \theta_{2c}^{+}), \qquad \gamma_i^{-} = \frac{1}{2}(\theta_{ic}^{+} - \theta_c^{-}), \quad i = 1, 2. \tag{8.71}$$

Obviously, under this choice of γ^{+}, γ_i^{-}, Eqs. (8.64)–(8.66), and (8.68) exactly coincide with the corresponding equations in (8.70). The continuity of the functions θ_i^{+} and θ^{-} on the free boundaries ensures that conditions (8.31) hold.

Now we state the result obtained.

THEOREM 8.1. *Suppose that conditions* (8.1), (8.17) *are satisfied and the initial temperature is positive and sufficiently large for* $x \in [\varphi_1(0), \varphi_2(0)]$. *Suppose also that, for* $t \le t1$, *the classical solution* θ_i^{+}, θ^{-}, $\varphi_{i0}(t)$, $i = 1, 2$, *of the modified Stefan problem* (8.2), (8.18) *exists and* $t1 = t_0^* - \varepsilon^{\mu_1}$, *where* $\mu_1 \in (0, 1)$ *and* t_0^* *is the time of intersection of the curves* $\Gamma_{t1}^0 = \{x = \varphi_{10}(t)\}$ *and* $\Gamma_{t2}^0 = \{x = \varphi_{20}(t)\}$. *Then the evolution of the temperature and the confluence of the free boundaries* $\Gamma_{t1} = \{x = \varphi_1(t, \varepsilon)\}$ *and* $\Gamma_{t2} = \{x = \varphi_2(t, \varepsilon)\}$ *are described by problem* (8.70). *As the result of the confluence of the free boundaries, the phase "–" disappears, and the free boundary problem turns into the heat equation for the phase "+" considered in the entire domain* Ω.

If we avoid the detailed description of the confluence of boundaries and pass to the limit as $\varepsilon \to 0$, then we can only state that there exists a moment of time t^* such that for $t < t^*$ the temperature evolution is described by the modified Stefan problem for two phases and for $t > t^*$ by the heat equation for a single phase. The Cauchy condition for this equation is posed at $t = t^*$ with the initial data at least from $C^1(\Omega)$ (see (8.69)).

8.3. Confluence of free boundaries. In the multidimensional case, the confluence of free boundaries is much more complicated and interesting than that in the one-dimensional case. The main reason is that the dynamics of free boundaries significantly depends on their geometry, and hence the interaction is not of local character.

As already noted, it is still not clear how to describe such processes uniformly in t. Hence we weaken the asymptotic solution so that it is understood in the sense of Definition 8.2. Using additional integration with respect to t, we see that taking into account the detailed description of the process for small time yields a change of order $O(\varepsilon^{\mu})$, $\mu > 0$, in the correction, which does not give any other information. Hence we use the possibility to construct weak asymptotics prior to the moment of contact and after the moment of contact. Then we sew them together. However, at the moment of contact $t = t^*$, a geometric bifurcation occurs, namely, at $t = t^*$ two curves Γ_{t1} and Γ_{t2} became a single curve with self-intersection, and, for $t > t^*$, this curve transforms into a smooth curve. So the key point of the whole construction is to answer the question: Is it possible to choose the initial data at the moment of contact in a reasonable way so that they correspond to the result of the preceding evolution and to pose a well-defined initial condition?

Efforts to answer this question are the main contents of this section. So let us consider the modified Stefan problem (8.2)–(8.4) for the temperature in three domains Ω_{ti}^+, $i = 1, 2$, Ω_t^- and, respectively, for two interfaces Γ_{ti}, $i = 1, 2$. We assume that $\Omega \subset \mathbb{R}^2$ and the initial data satisfy conditions (8.1), (8.7). Among all possible scenarios of the medium evolution for $t > 0$, we choose a scenario for which the domains Ω_{ti}^+ expand with increasing t. To this end, we assume that the curvature of Γ_{0i} is positive and the initial temperature is sufficiently large. We also assume that, for the entire interval of time under study, the front is rigid (condition (8.6)) and the domains Ω_{ti}^+ do not contact the external boundary (condition (8.5)).

Regularizing the problem, we pass to the phase field system (8.9), (8.10).

For the time prior to the bifurcation, we write the asymptotic solution of this system as the simple superposition of two noninteracting solutions (see Subsection 5.3):

$$(8.72)\qquad \begin{aligned} \theta &= \theta_1 \overset{\text{def}}{=} \gamma^+ + \gamma_1^- \omega_1\Big(\beta_1 \frac{S_1}{\varepsilon}\Big) + \gamma_2^- \omega_1\Big(\beta_2 \frac{S_2}{\varepsilon}\Big), \\ u &= u_1 \overset{\text{def}}{=} 1 + \omega_0\Big(\beta_1 \frac{S_1}{\varepsilon}\Big) + \omega_0\Big(\beta_2 \frac{S_2}{\varepsilon}\Big). \end{aligned}$$

Here

$$S_i = t + \Phi_i(x), \qquad i = 1, 2,$$

and $\varphi_i(x)$, $\beta_i = \beta_i(x) > 0$, $\gamma^+ = \gamma^+(x,t)$, $\gamma_j^- = \gamma_j^-(x,t)$, $j = 1, 2$, are the desired smooth functions. We assume that for each $i = 1, 2$

$$(8.73)\qquad \Phi_i(x) > 0 \quad \text{for } x \in \Omega_{0i}^+ \qquad \text{and} \qquad \Phi_i(x) < 0 \quad \text{for } x \in \Omega \setminus \overline{\Omega}_{0i}^+.$$

Precisely as in Subsection 8.2, $\omega_0(z/\varepsilon)$ is a special regularization of the function $\operatorname{sign}(z)$ determined by the solution of Eq. (8.19), and $\omega_1(z/\varepsilon)$ is an arbitrary odd regularization of $\operatorname{sign}(z)$.

Repeating the constructions of Subsection 4.3 practically word for word, we see that $\beta_i = 1/|\nabla\Phi_i|$ and the functions φ_i and γ^+, γ_j^- are determined by the solution of the modified Stefan problem (8.2)–(8.4). Namely, at each moment of time t the free boundaries Γ_{ti} are determined by the relations $\Phi_i(x) + t = 0$, while the normals ν_i, the normal velocities V_{ν_i}, and the mean curvatures $\mathcal{K}_i$ are calculated according to the formulas

$$(8.74)\qquad \nu_i = -\nabla\Phi_i/|\nabla\Phi_i|, \qquad V_{\nu_i} = 1/|\nabla\Phi_i|, \qquad \mathcal{K}_i = \operatorname{div}\nu_i, \quad i = 1, 2.$$

Finally, the functions γ^+ and γ_j^- are calculated in terms of θ_i^+, θ^- by formulas (8.71).

Similarly to the one-dimensional case, it is easy to see that problem (8.2)–(8.4) gives a correct description of the process only till the moment of time t_1 at which the boundaries Γ_{t1} and Γ_{t2} approach each other sufficiently nearly so that relation (8.22) holds for some $\mu_1 \in (0, 1)$. Nevertheless, till the moment of time at which the free boundaries contact each other, i.e., till $t^* = 0$, problem (8.2)–(8.4) remains well defined, because the curves Γ_{t_i} remain smooth for such time. We define the temperature and the curves Γ_{ti} continuously at the moment of time $t = t^*$ at which the free boundaries contact each other forming the curve $\Gamma^* = \Gamma_{t_1^*} \cup \Gamma_{t_2^*}$ with a self-intersection. Next, we continue the temperatures θ_i^+, θ^- to the time interval

$[t^*, t^* + \varepsilon^{\mu_1}]$. For this time, we define the curves Γ_{ti}, $i = 1, 2$, from the modified Stefan problems (8.2)–(8.4) considered independently. Obviously, the curves Γ_{ti} move at a finite velocity, so that at $t = t^* + \varepsilon^{\mu_1}$ they intersect cutting out the domain $\Omega^+_{t1} \cap \Omega^+_{t2}$ of area $\sim \varepsilon^{2\mu_1}$.

Let us consider the state of the medium at time t^* and, simplifying the situation, assume that there exists a local system of coordinates (ξ, ζ) such that the curve Γ^* can be describes as

$$\zeta = \pm|\xi|^{1/p}, \qquad p > 1,$$

in a neighborhood of the point of contact $(0, 0)$.

Now we regularize this curve by setting

$$\Gamma^*_\varepsilon = \Big\{(\xi, \zeta), \zeta = \pm(\xi^2 + \varepsilon^{2q})^{1/2p}\Big\}, \qquad q = \frac{p}{4(2p-1)}.$$

Clearly, under this regularization, significant changes occur only in a small neighborhood of the point $(0, 0)$, while the changes that occur at finite distances from this point are negligibly small.

Next, we take into account the fact that, under this regularization, the point of contact $(0, 0)$ splits into two points

$$z^0_\pm = (0, \pm\zeta^*_\varepsilon)$$

at a distance $\sim \varepsilon^{\lambda_1}$, $\lambda_1 = q/p$, from each other.

It is easy to calculate the mean curvatures $\mathcal{K}(z^0_\pm)$ of the curve Γ^*_ε at the points $z^0_\pm$ and to see that they are negative and take the value $|\mathcal{K}(z^0_\pm)| \sim \varepsilon^{-\alpha_1}$ with exponent $\alpha_1 = q(2p-1)/p = 1/4$.

Next, we move the curve Γ^*_ε in reverse time by $\Delta t \sim \varepsilon^{\mu_1}$ according to the Gibbs–Thompson law. The curve obtained is denoted by $\Gamma^*_{\varepsilon,0}$, and the moment of time obtained is denoted by t^*_0. Let $\mu_1 \in (1/2, 1)$. Then, after this transformation, the points z^0_+ and z^0_- move towards each other through a distance of $\sim \varepsilon^{\lambda_2}$, where $\lambda_2 > \lambda_1$. Hence $\Gamma^*_{\varepsilon,0}$ is a smooth curve without self-intersection (for $\varepsilon > 0$). Moreover, easy calculations show that the curvatures at the critical points $\Gamma^*_{\varepsilon,0}$ remain negative and their absolute values are the same as those of $\mathcal{K}(z^0_\pm)$. This is because of the fact that, by the Gibbs–Thompson law, the motion of the segments of the curve of negative curvature in reverse time is equivalent to the motion in direct time of the segments of positive curvature, and this motion is at a finite velocity.

By Ω^+_{30} we denote the domain bounded by the curve $\Gamma^*_{\varepsilon,0}$. Let $\Omega^-_{30} = \Omega \setminus \overline{\Omega}^+_{30}$. Let $\Phi_3 = \Phi_3(x, \varepsilon)$ be a smooth (for $\varepsilon > 0$) function such that $\Phi_3(x, \varepsilon) > 0$ for $x \in \Omega^+_{30}$ and $\Phi_3(x, \varepsilon) < 0$ for $x \in \Omega^-_{30}$. Then, for any time after the bifurcation, we can write the asymptotic solution of the phase field system in the form

$$\tag{8.75} \begin{aligned} \theta &= \theta_2 \overset{\text{def}}{=} \widetilde{\gamma}^+ + \widetilde{\gamma}^- \omega_1\Big(\beta_3 \frac{t + \Phi_3}{\varepsilon}\Big), \\ u &= u_2 \overset{\text{def}}{=} \omega_o\Big(\beta_3 \frac{t + \Phi_3}{\varepsilon}\Big), \qquad \beta_3 = \frac{1}{|\nabla \Phi_3|}, \end{aligned}$$

which is standard for the situation with a single interface. It follows from the results of Subsection 5.3 that, to the asymptotic solution (8.75) of the phase field system, there corresponds the solution of the modified Stefan problem with a single interface. The initial condition for this problem is posed at $t = t^*$. Namely, we take the curve $\Gamma^*_{\varepsilon,0}$ to be the initial position of the interface. At each point $x \in \overline{\Omega}^\pm_{30}$ we take the temperature $\theta^{\pm}\big|_{t=t_0^*}$ obtained calculating solution (8.72) (i.e., prior to the moment of bifurcation) to be the initial value of the temperature, independently of whether this point previously belonged to $\overline{\Omega}_i^+$ or $\overline{\Omega}^-$.

To analyze the solution of the modified Stefan problem with the above initial conditions $\theta^{\pm}\big|_{t=t_0^*}$ and $\Gamma^*_{\varepsilon,0}$ for any small time, we shall use the coordinates (ξ,ζ). Using formulas of the form (8.73), similarly to [**20**], we rewrite the Gibbs–Thompson condition as a quasilinear parabolic equation in which the variable ζ plays the role of time and the variable ξ that of the spatial coordinate. Since the curvature of the curve $\Gamma^*_{\varepsilon,0}$ is negative in the domain under study, the sign of the dissipative term in this equation is negative. Respectively, as in the reverse heat equation, the function Φ_3 increases extremely rapidly. Moreover, straightforward Cauchy–Kowalewski type calculations show that the points $z_+(t)$ lying on the upper part of the curve $\Gamma^*_{\varepsilon,t}$ (in the coordinates (ξ,ζ)) move upwards at a velocity increasing with $|\mathcal{K}(z_+(t))|$. A similar behavior is demonstrated by the points $z_-(t)$ lying on the lower part of the curve $\Gamma^*_{\varepsilon,t}$, but they move downwards. Taking into account the fact that $\Gamma^*_{\varepsilon,t}=\{x,\Phi_3(x,\varepsilon)+t=0\}$ and the curvatures at the critical points of $\Gamma^*_{\varepsilon,0}$ are $\sim\varepsilon^{-1/4}$, we see that these points move away from one another through a distance $\sim\Delta t\varepsilon^{-1/4}$ for a small time. So the curve $\Gamma^*_{\varepsilon,t}$ becomes smoother with increasing time t. However, $|\mathcal{K}|$ decreases simultaneously, and hence the rate of smoothing also decreases. Nevertheless, since the character of parabolic equations is not local, it is clear that the fast motion of the critical points of $\Gamma^*_{\varepsilon,t}$ reconstructs the curve, in general, at distances $O(1)$.

Now we can sew together the solutions (8.72) and (8.75), by setting

$$\theta=k(\tau)\theta_1+\big(1-k(\tau)\big)\theta_2,\qquad u=k(\tau)u_1+\big(1-k(\tau)\big)u_2, \tag{8.76}$$

where $\tau=(t-t^*)/\varepsilon$ and $k(\tau)\in C^\infty$ is a function such that $k(\tau)\to 1$ as $\tau\to-\infty$ and $k(\tau)\to 0$ as $\tau\to+\infty$ at an exponential rate. The estimate of the error caused by the substitution of (8.76) into the equations, in the sense of Definition 8.2, is based on the fact that for $t<t_1$ and $t>t^*$, respectively, each of the pairs (θ_1,u_1) and (θ_2,u_2) is a weak solution uniformly in t, i.e., in the sense of Definition 8.1. Passing from Definition 8.1 to Definition 8.2, we weaken the solution, which allows us to overcome the difficulties caused by the roughness in the matching relations (8.75). Moreover, it turns out that the behavior of the solution (8.75) in the domain bounded by the curve $\Gamma^*_{\varepsilon,t}$ and the nonsmooth curve $\Gamma_{t1}\cup\Gamma_{t2}\setminus\Gamma_{\Delta t}$, where $\Gamma_{\Delta t}=\partial(\Omega^+_{t1}\cap\Omega^+_{t2})$, makes the largest contribution to the error. Nevertheless, choosing an appropriate μ, one can show that the contribution of this domain to the error is of order $O(\varepsilon^\mu)$, where the exponent $\mu=\mu(p)>0$ does not exceed $1/2$.

Here we present the ideas and the main steps of the constructions rather than concrete calculations. The first step (to pose the initial data for (8.74) and to analyze the solution for small time) is actually performed by using the Taylor expansions. The second step (to estimate the error) requires ideologically simple, but technically very cumbersome, integral estimates. All these calculations are, in our opinion, not of great interest, since they are in the channel of the ideas developed here.

8.4. Numerical simulation. We describe the results of the numerical simulation[4] of the solution of the phase field system (8.9), (8.10) supplemented with the

[4]The authors are grateful to V. Rudnev who performed all the numerical computations.

Neumann conditions on the external boundary

$$\frac{\partial\theta}{\partial\nu_b}\Big|_{\partial\Omega}=0,\qquad \frac{\partial u}{\partial\nu_b}\Big|_{\partial\Omega}=0, \tag{8.77}$$

where ν_b is the normal on $\partial\Omega$.

In the one-dimensional case we choose the initial data

$$\theta\Big|_{t=0}=5,\qquad u\Big|_{t=0}=1+\omega_0\Big(\frac{x-x_0^+}{\varepsilon}\Big)+\omega_0\Big(\frac{x_0^- - x}{\varepsilon}\Big), \tag{8.78}$$

where $\omega_0(\eta)$ is the solution of Eq. (8.19) for $m=1$, i.e.,

$$\omega_0(\eta)=\tanh(\eta/\sqrt{2}), \tag{8.79}$$

$x_0^+ < x_0^-$, and $\theta\big|_{t=0}=5$ is a "sufficiently" high temperature. This solution ensures the motion of interfaces towards each other for the whole time interval we are interested in. The parameter ε was set to be 0.1.

The evolution of the order function u and the temperature θ in the case $m=1$ is represented in Figures 16 and 17. The plots show that the kinks $\omega_0((x-\varphi_1)/\varepsilon)$ and $\omega_0((\varphi_2-x)/\varepsilon)$ annihilate sufficiently uniformly in t on the time interval $\Delta t \sim 0.08 \sim O(\varepsilon)$.

The plots of the temperature (Figure 17) clearly show the smoothing effect in the interaction domain. At the first stage of the confluence of the boundaries, when the normed distance $\rho=(\varphi_2(t,\varepsilon)-\varphi_1(t,\varepsilon))/\varepsilon$ between the boundaries $\Gamma_{ti}=\{x=\varphi_i(t,\varepsilon)\}$ is sufficiently large, the temperature in neighborhoods of Γ_{ti} decreases and forms a W-profile. So a tendency to the formation of singularities at lower points of this profile is clearly demonstrated: in Γ_{ti}, as $\varepsilon\to 0$, we would obtain breaks corresponding to the Stefan conditions (8.3) (see the curves in the upper plots in Figure 16 and curve 1 in the lower plots of this figure). Nevertheless, simultaneously with decreasing ρ, the central part of the W-profile moves downwards. As a result, the W-profile turns into a U-profile as $\rho\to 0$ (see curve 1 and curves 2, 3 in the lower plots in Figure 17). At the final stage of the confluence of the boundaries, the derivatives in the lower part of the U-profile become smoother and smoother (see curves 3, 4 and 5 in the lower plots in Figure 17). Namely, the tendency of the lower part to move downwards and to the formation of a break ceases at $t\approx 0.14$, and the lower part of the U-profile begins to move upwards. We also note that the result of the existence of the domain Ω_0^- with the phase "$-$" at $t=0$ ($\overline{u}=-1$ in Ω_0^-) is the formation of a cooled region. Computations show that, starting from the final stage of the process of confluence, i.e., starting from $t\sim 0.14$, this region is being heated, while the temperature outside this region decreases simultaneously.

To visualize the nonlinearity of the confluence of interfaces, we calculated the evolution of the differences (see Figure 18)

$$\delta_u=u-(u^+ + u^- + 1),\qquad \delta_\theta=\theta-(\theta^+ + \theta^-),$$

where (θ,u) is the solution of problem (8.9), (8.10), (8.76) with the initial condition (8.77) and $(\theta^\pm,u^\pm)$ are the waves calculated without interaction, i.e., the solutions of problems (8.9), (8.10), (8.76) with the initial conditions

$$\theta^\pm\Big|_{t=0}=5,\qquad u^\pm\Big|_{t=0}=\omega_0\Big(\pm\frac{x-x_0^\pm}{\varepsilon}\Big). \tag{8.80}$$

The plots clearly show that, in the process of interaction, the solution of problem (8.9), (8.10), (8.76), (8.77) cannot be described as the simple superposition of

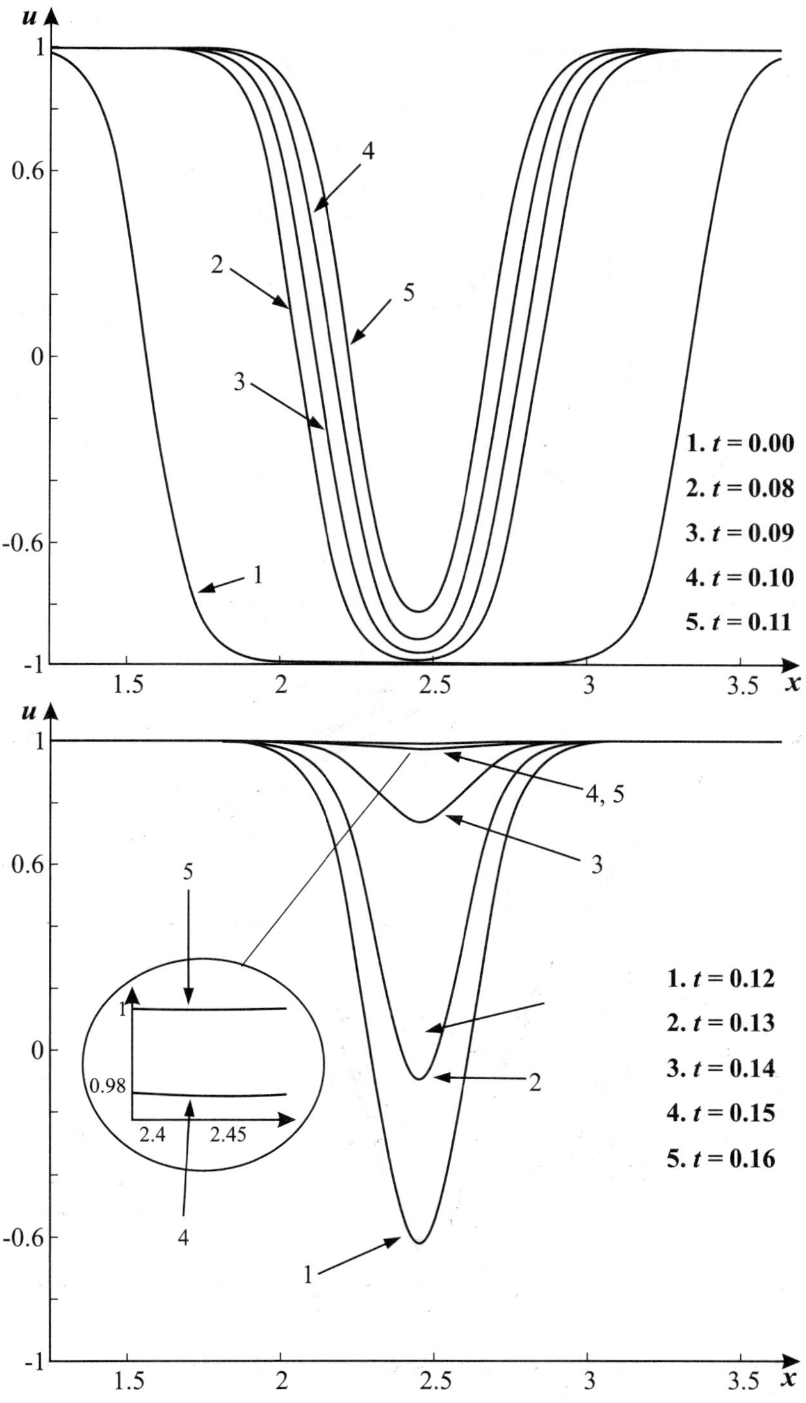

FIGURE 16. Dynamics and annihilation of kinks in the one-dimensional case, $m = 1$.

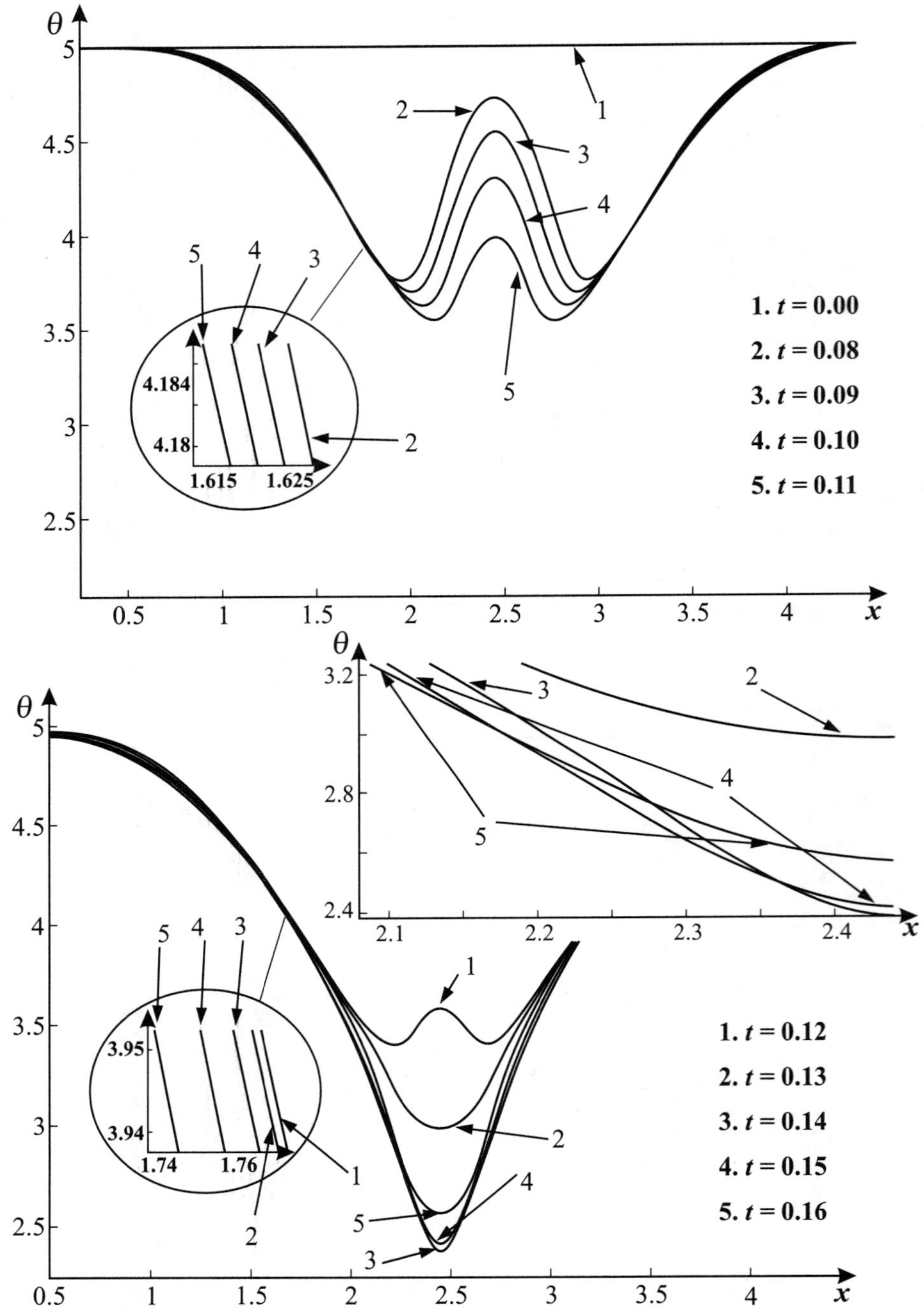

FIGURE 17. Evolution of the temperature in the process of confluence of boundaries, $m = 1$.

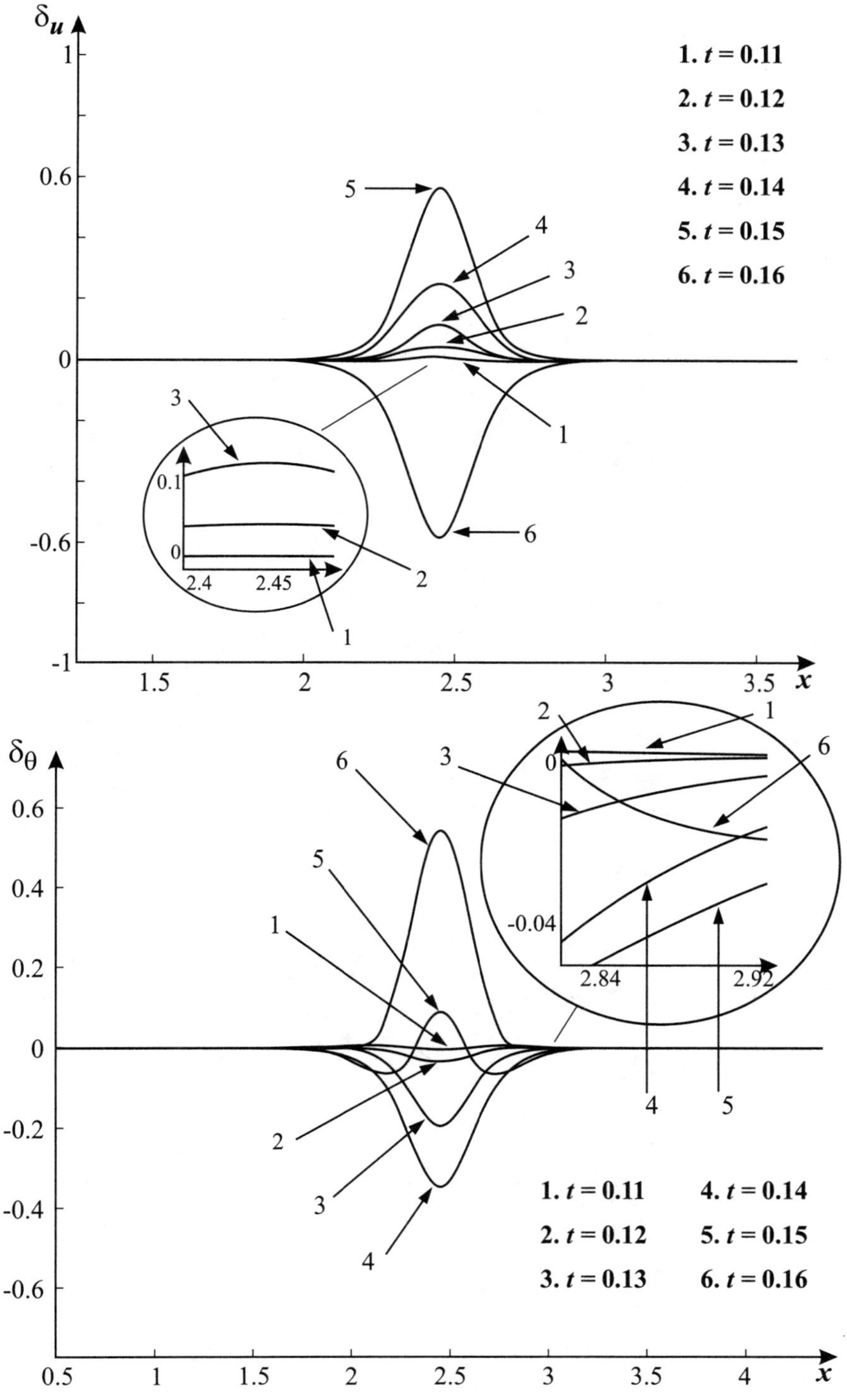

FIGURE 18. Nonlinear interaction effects in the one-dimensional case.

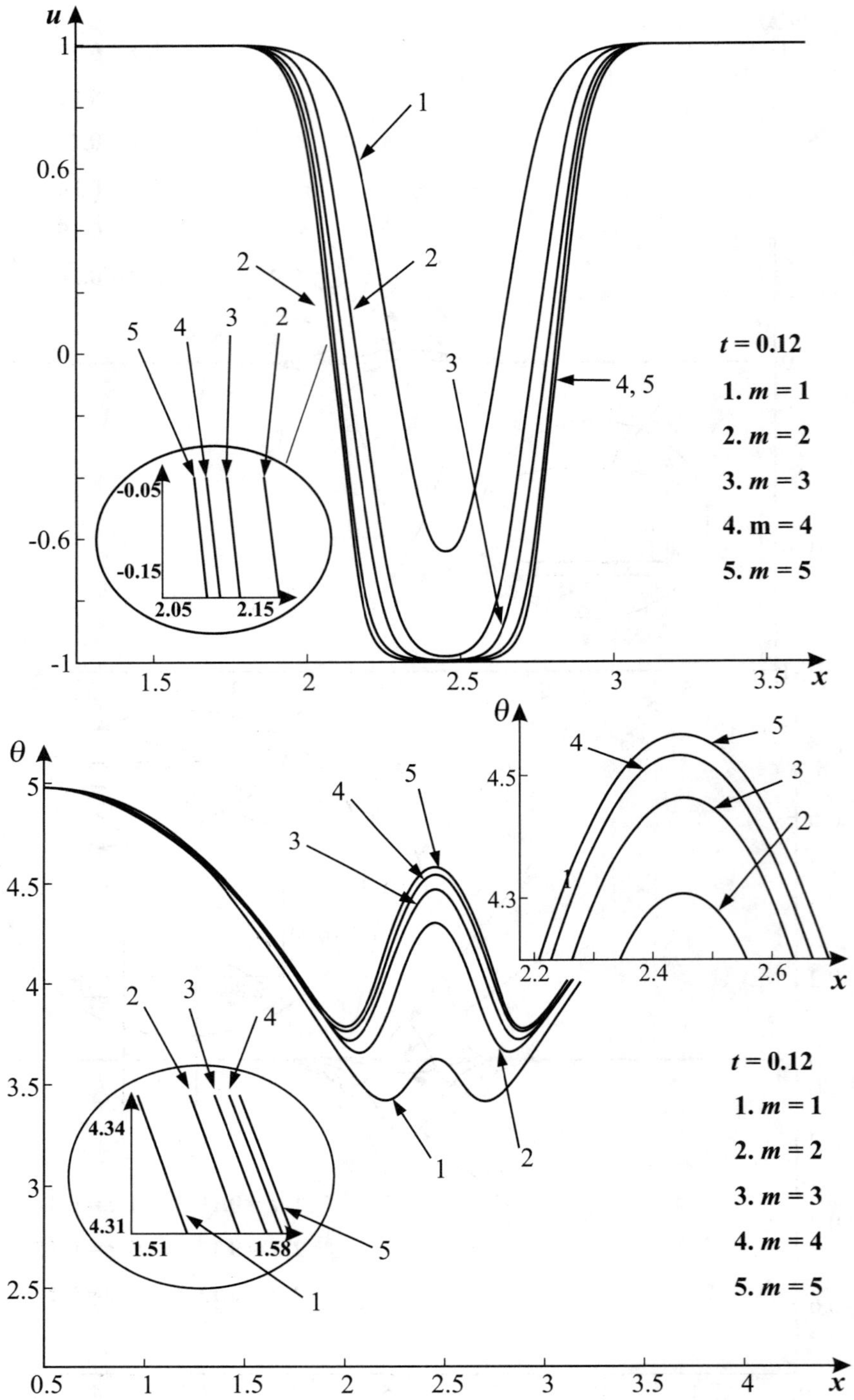

FIGURE 19. Variation of the kink profile and its velocity with increasing m and the corresponding plots of the temperature.

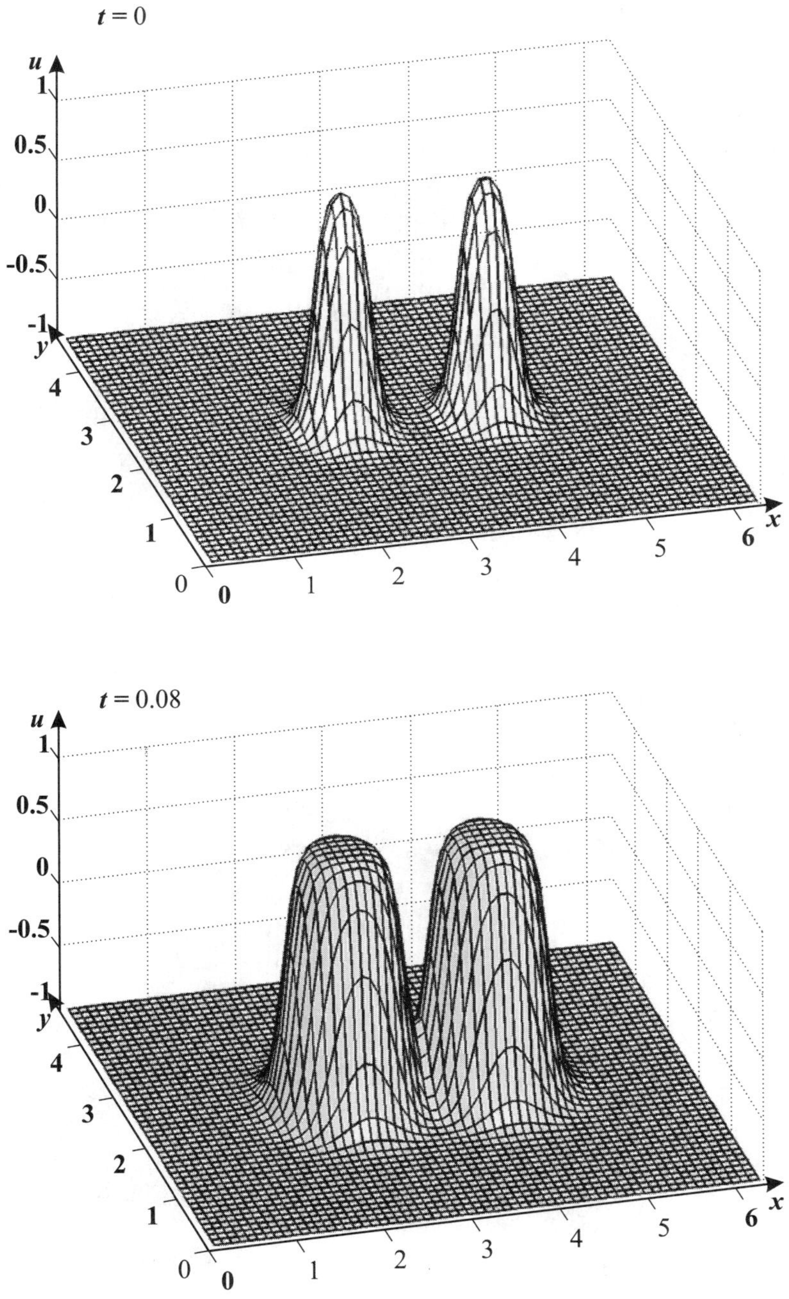

FIGURE 20. Evolution and confluence of two kinks in the two-dimensional case, $m = 1$ ($t = 0$ and $t = 0.08$).

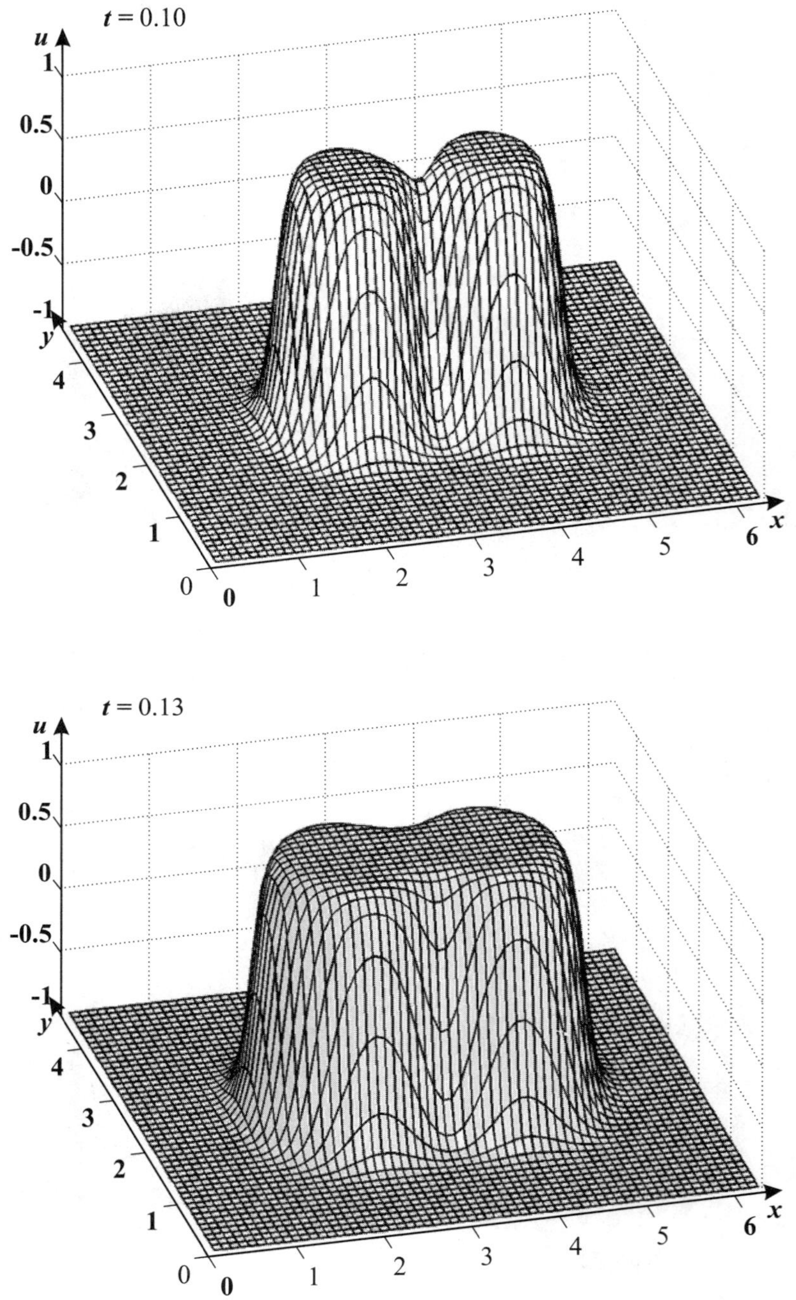

FIGURE 21. Evolution and confluence of two kinks in the two-dimensional case, $m = 1$ ($t = 0.10$ and $t = 0.13$).

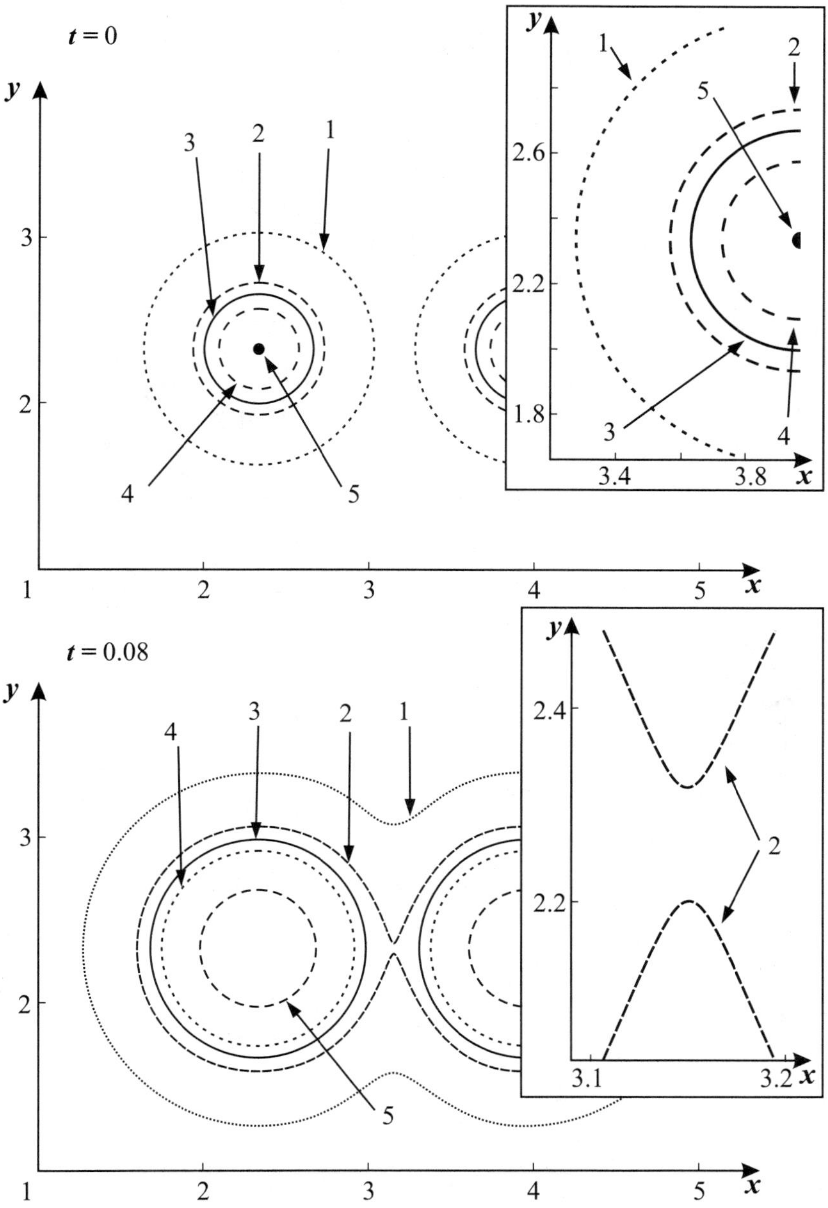

FIGURE 22. Level lines of the order function ($m = 1$) at different moments of time ($t = 0$ and $t = 0.08$). 1: $u = -1$, 2: $u = -0.5$, 3: $u = 0$, 4: $u = 0.5$, 5: $u = 1$.

solitary waves (θ^+, u^+) and (θ^-, u^-). The plots in Figure 18 also show that the interaction occurs on a spatial interval of the same length as that of the domain of the "fast" variation of the regularization (recall that we set $\varepsilon = 0.1$ in our computations). So, in the one-dimensional case, the interaction of kinks of the order function u is localized in x and t in a domain of width $O(\varepsilon)$ in each direction.

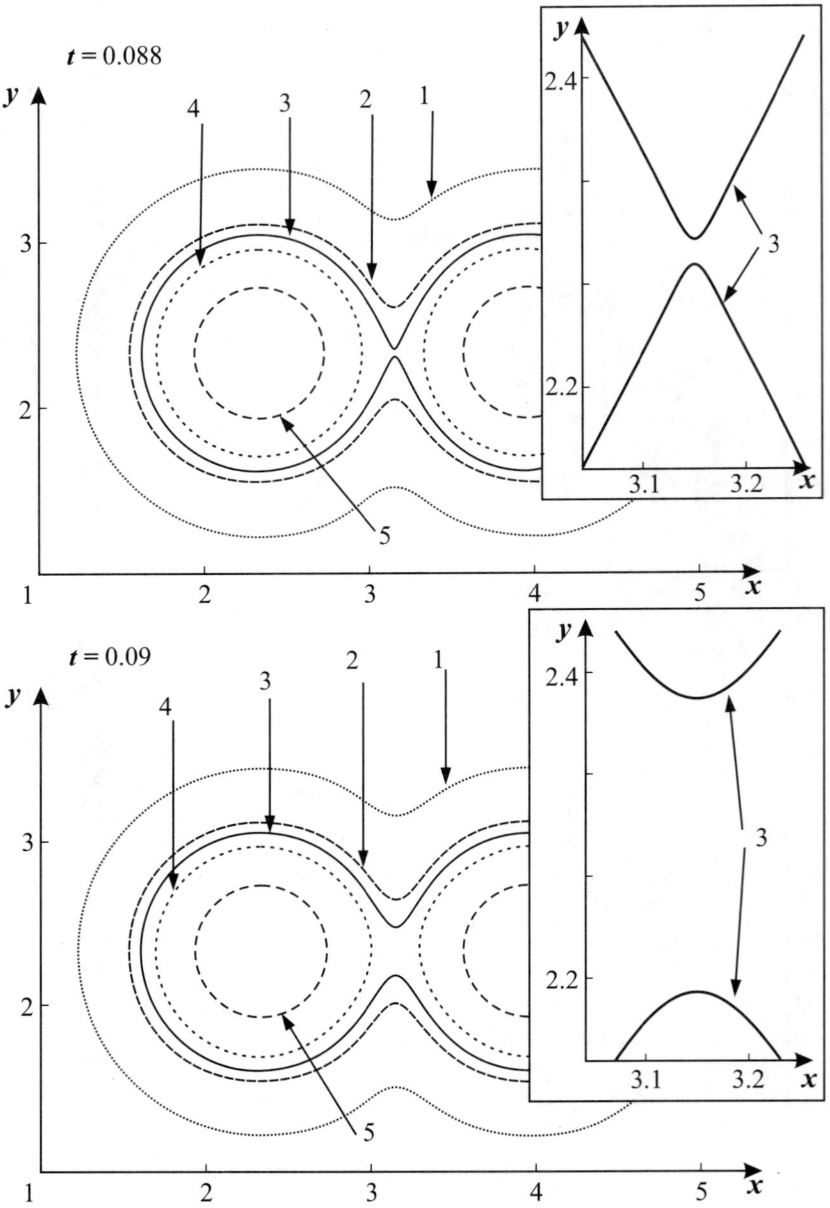

FIGURE 23. Level lines of the order function ($m = 1$) at different moments of time ($t = 0.088$ and $t = 0.09$). 1: $u = -1$, 2: $u = -0.5$, 3: $u = 0$, 4: $u = 0.5$, 5: $u = 1$.

For $m > 1$, the qualitative picture of the confluence of free boundaries turns out to be the same as for $m = 1$. There are only qualitative distinctions related to an increase in the steepness of the kink profile and to a decrease in its velocity with increasing m. This effect is illustrated by the plots in Figure 19, which present the solution of problem (8.9), (8.10), (8.76), (8.77) for different m at a fixed moment of time.

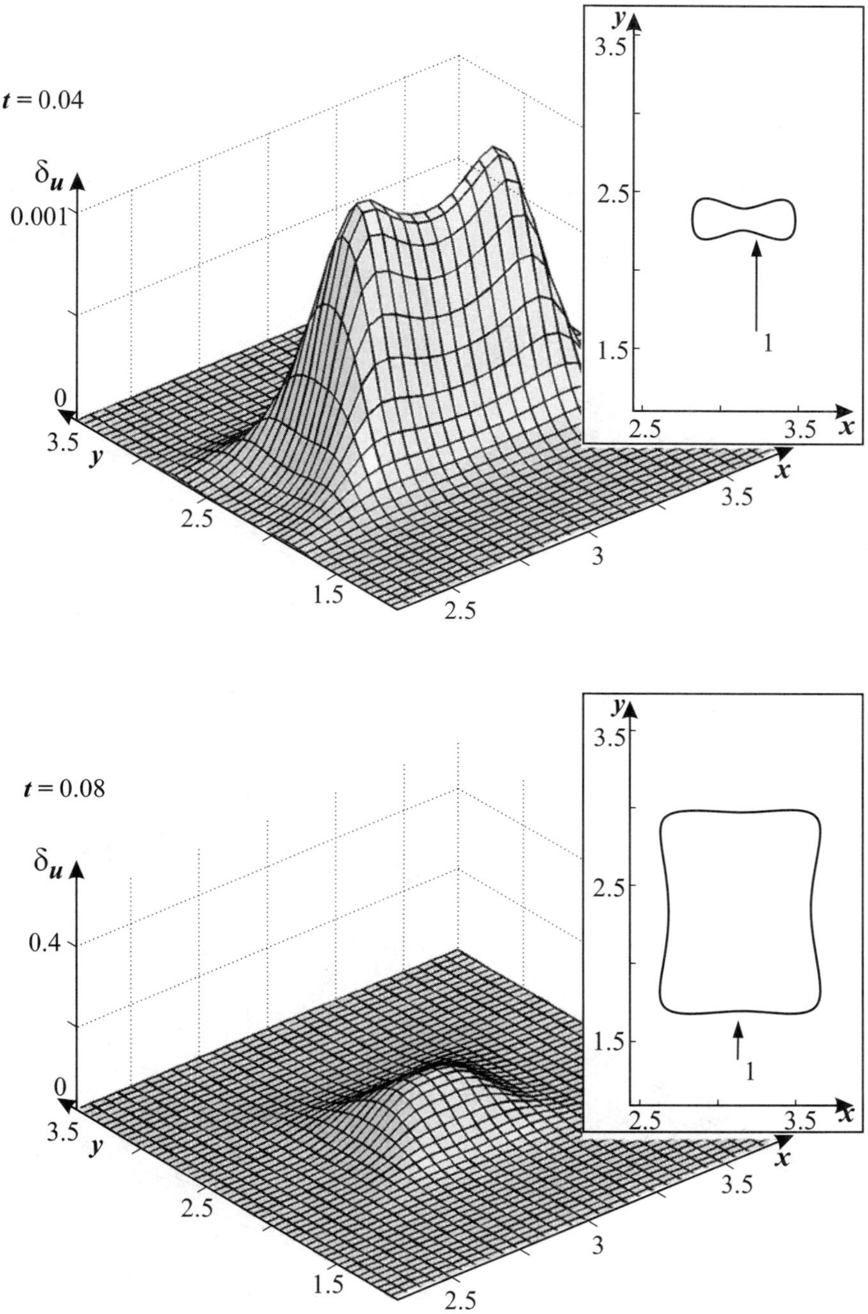

FIGURE 24. Nonlinear effects of kink interaction ($m = 1$). Level lines of the difference $\delta_u = u - (\sum_{\pm} u^{\pm} + 1)$. 1: $u = 0$, 2: $u = 0.2$, 3: $u = 0.4$ ($t = 0.04$ and $t = 0.08$).

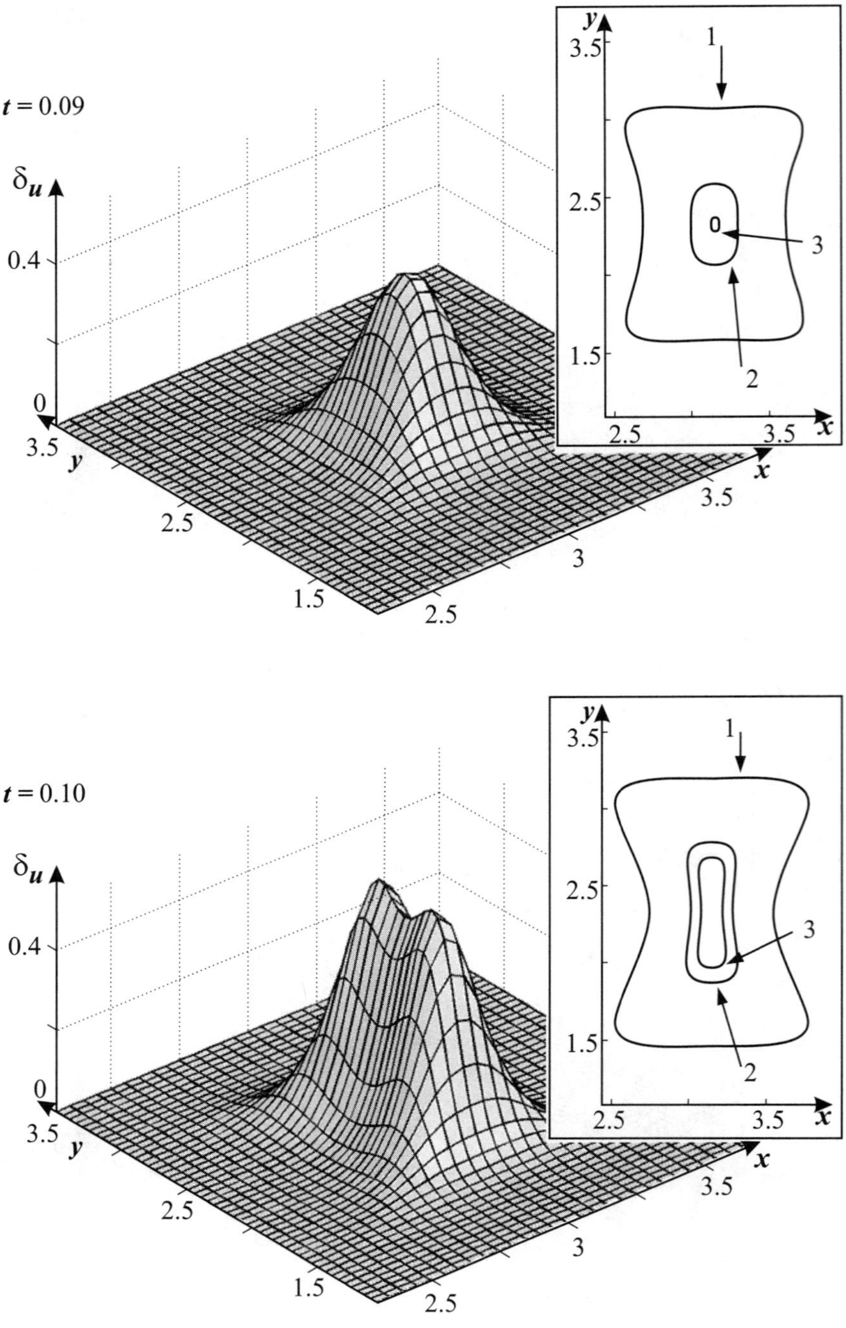

FIGURE 25. Nonlinear effects of kink interaction ($m = 1$). Level lines of the difference $\delta_u = u - (\sum_{\pm} u^{\pm} + 1)$. 1: $u = 0$, 2: $u = 0.2$, 3: $u = 0.4$ ($t = 0.09$ and $t = 0.10$).

Figures 20 and 21 show the evolution of the order function in the two-dimensional case calculated as a solution of problem (8.9), (8.10), (8.76) with $m = 1$ and the initial data

$$\theta\Big|_{t=0} = 5, \qquad u\Big|_{t=0} = 1 + \omega_0\Big(\frac{S_1^0}{\varepsilon}\Big) + \omega_0\Big(\frac{S_2^0}{\varepsilon}\Big). \tag{8.81}$$

Here $\omega_0(\eta)$ has the form (8.78), $S_i^0 = R_i - |x - x_0^i|$, and R_i, x_i^0 are constants. As in the one-dimensional case, the parameters in (8.81) were chosen by using the Gibbs–Thompson conditions (8.4) so that the free boundaries move towards each other.

For a detailed analysis of the process of interaction, we consider the level lines of the function u. The plots in Figures 22 and 23 show that, prior to the moment of contact of the free boundaries Γ_{ti}, the noninteracting kinks evolve preserving the radial symmetry. At time $t^* \sim 0.088$, there is a contact of the free boundaries Γ_{ti} (which are the zero level lines of the function u; see Figures 22 and 23 for $t = 0.088$). After this, a new free boundary is formed. This free boundary is instantly smoothed. As a result, already for $t \sim 0.09$, the distance between the nearest points of Γ_t (coinciding at $t = t^*$) becomes approximately equal to 0.2 and then increases approximately at a rate such that at $t = 0.1$ this distance is ≈ 1. Additional information characterizing the process of interaction can be found in Figures 24 and 25 presenting the difference between the "exact" solution of problem (8.9), (8.10), (8.76), (8.80) and the sum of noninteracting kinks u^+ and u^- calculated for some initial data similar to (8.79). It is clearly seen that now, in contrast to the one-dimensional case, the spatial domain of interaction has dimensions $\sim O(1)$. Moreover, this domain has a typical form of a butterfly whose wings are stretched along the boundary Γ_t.

There are small quantitative differences between the process of interaction of two-dimensional waves for $m > 1$ and the case $m = 1$ described above, caused by the differences in the steepness of the profile of u and in the velocity of the motion of kinks.

The character of evolution of the confluence of interfaces, as well as the principal difference between the one-dimensional and multidimensional cases, were studied in Subsections 8.2 and 8.3. It should be noted that numerical experiments were realized after the mathematical analysis of this problem [**106**]. We were struck to what extent the effects discovered by the asymptotic analysis as $\varepsilon \to 0$ were visualized in numerical simulation with $\varepsilon = 0.1$.

Acknowledgement. The authors wish to thank J.-A. Marti, M. Oberguggenberger, S. Pilipović, and M. Nedelkov for friendly discussions which were very useful in developing the approach presented here. The authors are grateful to D. Konopkin for his help in preparing the figures.

References

[1] M. J. Ablowitz and H. Segur, *Solitons and inverse scattering transform*, SIAM, Philadilphia, 1981.
[2] M. J. Ablowitz, H. Segur, and X. P. Wang, *Wave collapse and instability of solitary waves of a generalized Kadomtsev–Petviashvili equation*, Physica D **78** (1994), nos. 3–4, 241–265.
[3] H. W. Alt and I. Pawlow, *A mathematical model of dynamics of non-isothermal phase separation*, Physica D **59** (1992), 389–416.
[4] H. A. Biagioni, *A nonlinear theory of generalized functions*, Springer-Verlag, Berlin, 1992.

[5] H. A. Biagioni and M. Oberguggenberger, *Generalized solutions to Burgers' equation*, J. Diff. Equations **97** (1992), 263–287.
[6] ———, *Generalized solutions to the Korteweg–de Vries and the regularized long-wave equations*, SIAM J. Math. Anal. **23** (1992), 923–940.
[7] B. K. Bullough and P. J. Caudrey (eds.), *Solitons*, Topics in Current Physics, vol. 17, Springer-Verlag, Berlin, 1980,
[8] G. Caginalp, *An analysis of a phase field model of a free boundary*, Arch. Rat. Mech. Anal. **92** (1986), 205–245.
[9] ———, *A conserved phase system: implication for kinetic undercooling*, Phys. Rev. B **38** (1988), 789–891.
[10] ———, *The dynamics of a conserved phase field system: Stefan like, Hele–Shaw, and Cahn–Hillard models as asymptotic limits*, IMA J. Appl. Math. **44** (1990), 77–94.
[11] F. Calogero and A. Degasperis, *Spectral transform and solitons*, vol. 1, *Tools to solve and investigate nonlinear evolution equations*, North-Holland, Amsterdam, 1982.
[12] S. Chapman and T. G. Cowling, *The mathematical theory of non-uniform gases*, Cambridge Univ. Press, Cambridge, 1952.
[13] J. F. Colombeau, *Elementary introduction to new generalized functions*, North Holland, 1985.
[14] J. F. Colombeau and A. Y. Le Roux, *Multiplication of distributions in elasticity and hydrodynamics*, J. Math. Phys. **29** (1988), 315–319.
[15] C. M. Dafermos, *Hyperbolic conservation laws in continuum physics*, Springer-Verlag, Berlin–New York, 2000.
[16] V. G. Danilov, *Generalized solutions describing singularity interaction*, Int. J. of Math. and Math. Sci. **29** (2002), no. 8, 481–494.
[17] V. G. Danilov, V. P. Maslov, and V. M. Shelkovich, *Algebra of singularities of singular solutions to first-order quasilinear strictly hyperbolic systems*, Theor. Math. Phys. **114** (1998), no. 1, 1–42.
[18] V. G. Danilov, V. P. Maslov, and K. A. Volosov, *Mathematical modeling of heat and mass transfer processes*, Kluwer, Dordrecht, 1995.
[19] V. G. Danilov and G. A. Omel′yanov, *Propagation of singularities and related problems of solidification*, Matematichki Vestnik (Belgrade) **49** (1997), 143–150.
[20] V. G. Danilov, G. A. Omel′yanov, and E. V. Radkevich, *Asymptotics of the solution to the phase field system and the modified Stefan problem*, Differentsial′nye Uravneniya **31** (1995), 446–454.
[21] ———, *Weak solutions to the phase field system*, Integral Transforms and Special Functions **6** (1997), 27–35.
[22] ———, *Hugoniot-type conditions and weak solutions to the phase-field system*, European J. Appl. Math. **10** (1999), 55–77.
[23] V. G. Danilov and V. M. Shelkovich, *Generalized solutions of nonlinear differential equations and the Maslov algebras of distributions*, Integral Transformations and Special Functions **6** (1997), 137–146.
[24] ———, *Propagation and interaction of nonlinear waves*, Eighth International Conference on Hyperbolic Problems, Theory–Numerics–Applications, Abstracts, Magdeburg, Germany, February 28–March 3, 2000, pp. 326–328.
[25] ———, *Propagation and interaction of shock waves of quasilinear equation*, Nonlinear Studies **8** (2001), no. 1, 211–245.
[26] ———, *Propagation of infinitely narrow δ-solitons*, http://arXiv.org/abs/math-ph/0012002.
[27] ———, *Propagation and interaction of delta-shock waves*, Ninth International Conference on Hyperbolic Problems, Theory–Numerics–Applications, Abstracts, California Institute of Technology, California USA, March 25–29, 2002, pp. 106–110.
[28] J. Dieudonné and L. Schwartz, *La dualité dans les espaces $(\mathcal{F})$ et $(\mathcal{LF})$*, Ann. Inst. Fourier (Grenoble) **1** (1949), 61–101.
[29] N. Djapić, N. Kunziger, and M. Pilipović, *Symmetry group analysis of weak solutions*, Proc. London Math. Soc. **84** (2002), 686–710.
[30] S. Yu. Dobrokhotov, *Hugoniot–Maslov chains for solitary vortices of the shallow water equations*, I. *Derivation of the chains for the case of variable Coriolis forces and reduction to the Hill equation*, Russian J. Math. Phys. **6** (1999), no. 2.

[31] B. A. Dubrovin, V. B. Matveev, and S. P. Novikov, *Nonlinear equations of Korteweg–de Vries type, finite-zone linear operators, and Abelian varieties*, Uspekhi Mat. Nauk **31** (1976), no. 1, 55–136; English transl., Russian Math. Surveys **31** (1976), no. 1, 59–146.
[32] Yu. V. Egorov, *A contribution to the theory of generalized functions*, Uspekhi Mat. Nauk. **45** (1990), no. 5, 3–40; English transl., Russian Math. Surveys **45** (1990), no. 5, 1–49.
[33] G. Ercole, *Delta-shock waves as self-similar viscosity limits*, Ouart. Appl. Math. **LVIII** (2000), no. 1, 177–199.
[34] H. Flaschka, M. Forest, and D. McLaughlin, *Multiphase averaging and the inverse spectral solution of the Korteweg–de Vries equation*, Comm. Pure Appl. Math. **33** (1980), 739–784.
[35] G. R. Fowles and A. F. P. Houwing, *Instabilities of shock and deformation waves*, Phys. Fields **24** (1984), 1982–1990.
[36] C. S. Gardner, J. M. Green, M. D. Kruskal, and R. M. Miura, *A method for solving the Korteweg–de Vries equation*, Phys. Rev. Lett. **19** (1967), 1095–1097.
[37] I. M. Gelfand and G. E. Shilov, *Generalized functions*, vol. 1, *Properties and operations*, Academic Press, New York, 1964.
[38] J. Glimm, *Solutions in the large for nonlinear systems of equations*, Comm. Pure Appl. Math. **18** (1965), 697–715.
[39] K. A. Gorshkov and L. A. Ostrovsky, *Interactions of solitons in nonintegrable systems: direct perturbation method and applications*, Physica D **182** (1981), 428–438.
[40] R. Grimshaw, *Slowly varying solitary waves.* I. *Korteweg–de Vries equation*, Proc. Roy. Soc. London Ser. A **368** (1979), 359–375.
[41] H. Grosse, M. Oberguggenberger, I. T. Todorov, *Generalized functions for quantum fields obeying quadratic exchange relations*, Proc. Steklov Inst. Math. **228** (2000), 90–100.
[42] L. Hsiao, *Quasilinear Hyperbolic Systems and Dissipative Mechanics*, World Scientific, Singapore, 1997.
[43] A. M. Il′in. *Matching of asymptotic expansions of solutions of boundary value problems*, Nauka, Moscow, 1989; English transl., Amer. Math. Soc., Providence, RI, 1992.
[44] A. M. Il′in and L. A. Kalyakin, *Perturbation of finite-soliton solutions to the Korteweg–de Vries equation*, Dokl. Ros. Akad. Nauk **336** (1994), 595–598: English transl., Phys. Dokl. **40** (1995), 603–606.
[45] V. K. Ivanov, *An associate algebra of the simplest generalized functions*, Siberian Math. J. **20** (1979), 509–516.
[46] B. B. Kadomtsev, *Collective effects in plasma*, Nauka, Moscow, 1976; French. transl., Mir, Moscow, 1979.
[47] B. B. Kadomtsev and V. I. Petviashvili, *On the stability of solitary waves in weakly dispersing media*, Dokl. Akad. Nauk SSSR **192** (1970), 753–756; English transl., Soviet Phys. Dokl. **15** (1970), 539–541.
[48] L. A. Kalyakin, *Longwave asymptotics. Integrable equations as asymptotic limits of nonlinear systems*, Uspekhi Mat. Nauk **44** (1989) no. 1, 5–34; English transl., Russian Math. Surveys **44** (1989) no. 1, 3–42.
[49] ———, *Perturbation of the Korteweg–de Vries soliton*, Teoret. Mat. Fiz. **92** (1992), 62–76; English transl. Theoret. and Math. Phys. **92** (1992), 736–747.
[50] ———, *On the problem of first correction in the theory of soliton perturbation*, Mat. Sb. **186** (1995), no. 7, 51–76; English transl., Sb. Math. **186** (1995), 977–1002.
[51] ———, *Weak interaction of the KdV solitons under perturbation*, Physica D **87** (1995), 193–200.
[52] ———, *Asymptotic well-posedness for the problem of soliton perturbation*, Russian J. Math. Phys. **4** (1998), no. 3, 3–27.
[53] ———, *Weak interaction of perturbed KdV solitons*, Fundam. i Prikl. Mat. **4** (1998), 1179–1205. (Russian).
[54] ———, *Perturbation of a singular solution to the Liouville equation*, Teoret. Mat. Fiz. **118** (1999), 390–397; English transl., Theoret. and Math. Phys. **118** (1999), 307–314.
[55] M. V. Karasev, *Quantization and intrinsic dynamics*, this volume, pp. 1–32.
[56] V. I. Karpman, *Soliton evolutions in the presence of perturbations*, Physica Scripta **20** (1979), 462–478.
[57] V. I. Karpman and E. M. Maslov, *The structure of tails formed under effects of perturbations on solitons*, Zh. Eksper. Teoret. Fiz. **75** (1978), 504–517; English transl., Soviet Phys. JEPT **48** (1978), 252–259.

[58] D. J. Kaup and A. C. Newell, *Solitons as particles, oscillators and in slowly changing media: a singular perturbation theory*, Proc. Roy. Soc. London Ser. A **361** (1978), 413–446.

[59] D. J. Kaup, *A perturbation theory for inverse scattering transforms*, SIAM J. Appl. Math. **31** (1976), 121–123.

[60] D. J. Kaup and A. C. Newell, *Solitons as particles, oscillators and in slowly changing media: a singular perturbation theory*, Proc. Roy Soc. Lond. A **361** (1978), 413–446.

[61] J. P. Keener and D. W. McLaughlin, *Solitons under perturbations*, Phys. Rev. A **16** (1977), 777–790.

[62] B. L. Keyfitz and H. C. Kranzer, *Spaces of weighted measures for conservation laws with singular shock solutions*, Journal of Differential Equations **118** (1995), 420–451.

[63] B. L. Keyfitz, *Conservation laws, delta-shocks and singular shocks.* In: M. Grosser, G. Hormann and M. Oberguggenberger (eds.), *Nonlinear Theory of Generalized Functions*, Chapman & Hall/CRC, 1999, pp. 99–112.

[64] A. Yu. Khrennikov, V. M. Shelkovich, and O. G. Smolyanov, *A linear space of vector-valued distributions with a multiplicative structure*, Abstracts of the International Conference: Stochastic Analysis and Related Topics, St.Petersburg, June 4–10 2001, pp. 38–39.

[65] ———, *Multiplicative structures in the linear space of vector-valued distributions*, Dokl. Ros. Akad. Nauk **383** (2002), no. 1, 1–4; English transl. in Russian Acad. Sci. Dokl. Math.

[66] ———, *Locally convex spaces of vector-valued distributions with multiplicative structures*, to appear in Infinite-Dimensional Analysis, Quantum Probability and Related Topics.

[67] Yu. S. Kivshar′, *On the perturbation theory for solitons: the sine–Gordon equation*, preprint no. 21-84, Fiz.-Tekhn. Inst. Nizkikh Temperatur Akad. Nauk Ukrain. SSR, Kharkov, 1984.

[68] K. Ko and H. H. Kuehl, *Korteweg–de Vries soliton in a slowly varying medium*, Phys. Rev. Lett. **40** (1978), 233–236.

[69] Y. Kodama and T. Taniuti, *Higher order corrections to the soliton velocity and the linear dispersion relation*, Physica Scripta **20** (1979), 486–489.

[70] I. M. Krichever, *Method of averaging for two-dimensional "integrable" equations*, Funktsional. Anal. i Prilozhen. **22** (1988), no. 3, 37–52; English transl., Functional Anal. Appl. **22** (1988), 200–213.

[71] S. N. Kruzhkov *First-order quasilinear equations in several independent variables*, Math. USSR Sb. **10** (1970), 127–243.

[72] G. E. Kuzmak, *Asymptotic solutions of nonlinear second-order differential equations with variable coefficients*, Prikl. Mat. Mekh. **23** (1959), 515–526; English transl., J. Appl. Math. Mech. **23** (1959), 730–744.

[73] G. L. Lamb, Jr., *Elements of soliton theory*, Wiley, New York, 1980.

[74] L. D. Landau and E. M. Lifshits, *Course of theoretical physics*, vol. 6: *Fluid Mechanics*, Nauka, Moscow, 1986; English transl., Pergamon Press, Oxford, 1987.

[75] P. D. Lax and C. D. Livermore, *The small dispersion limit for the KdV equation*, I, Comm. Pure Appl. Math. **36** (1983), 253–290.

[76] ———, *The small dispersion limit for the KdV equation*, II, Comm. Pure Appl. Math. **36** (1983), 571–594.

[77] Ph. LeFloch, *An existence and uniqueness result for two nonstrictly hyperbolic systems*, Nonlinear Evolution Equations that Change Type, Springer-Verlag, Berlin, 1990, pp. 126–138.

[78] S. Leibovich and A. R. Seebass (eds.), *Nonlinear waves*, Cornell Univ. Press, Ithaca, NY, 1974.

[79] Li Bang-He, *Non-standard analysis and multiplication of distributions*, Acta Scientia Sinica **21** (1978), no. 5, 561–585.

[80] Li Bang-He and Li Yaqing, *On the harmonic and analytic representations of distributions*, Acta Scientia Sinica **28** (1985), no. 9, 923–937.

[81] Ya. B. Livchak, *Towards the theory of generalized functions*, Trudy Rizhskogo Algebr. Seminar, Riga 1969, pp. 98–164 (Russian).

[82] K. Lonngren and A. Scott (eds.), *Solitons in action*, Academic Press, New York, 1978.

[83] J.-A. Marti, *Fundamental structures and asymptotic microlocalization in sheaves of generalized functions*, Integral Transforms and Special Functions **6** (1997), 223–228.

[84] ———, *$(\mathcal{C}, \mathcal{E}, \mathcal{P})$-sheaf structures and applications*, Nonlinear Theory of Generalized Functions, M. Grosser, G. Hormann, and M. Oberguggenberger (eds.), Chapman and Hall/CRC, 1999, pp. 175–186.

[85] E. M. Maslov, *On the theory of perturbations of solitons in second approximation*, Teoret. Mat. Fiz. **42** (1980), 362–373; English transl. Theoret. and Math. Phys. **42** (1980), 237–245.

[86] ———, *Dynamics of rotationally symmetric solitons in near-sg field model with applications to large-area Josephson junctions and ferromagnetics*, Physica D **15** (1985), 433–443.

[87] V. P. Maslov, *Propagation of shock waves in isoentropi nonviscous gas*, Itogi Nauki i Tekhn.: Sovremennye Probl. Mat., vol. 8, VINITI, Moscow, 1977, pp. 199–271; English transl., J. Soviet Math. **13** (1980), 119–163.

[88] ———, *Three algebras corresponding to nonsmooth solutions of systems of quasilinear hyperbolic equations*, Uspekhi Mat. Nauk **35** (1980) no. 2, 252–253. (Russian).

[89] V. P. Maslov and G. A. Omel′yanov *Asymptotic soliton-form solutions of equations with small dispersion*, Uspekhi Mat. Nauk. **36** (1981), no. 3, 63–126; English transl., Russian Math. Surveys **36** (1981), no. 3, 73-149.

[90] ———, *Hugoniot-type conditions for infinitely narrow solutions of equations for simple waves*, Sibirsk. Mat. Zh. **24** (1983), no. 5, 172–182; English transl. **24** (1983), 787–795.

[91] ———, *Geometric asymptotics for nonlinear PDE*. I, Amer. Math. Soc., Providence, RI, 2001.

[92] V. P. Maslov and V. A. Tsupin, *Propagation of a shock wave in an isoentropic gas with small viscosity*, Itogi Nauki i Tekhn.: Sovremennye Probl. Mat., vol. 8, VINITI, Moscow, 1977, pp. 273–308; English transl., J. Soviet Math. **13** (1980), 163–185.

[93] ———, *Necessary conditions for the existence of infinitely narrow solitons in gas dynamics*, Dokl. Akad. Nauk SSSR **246** (1979), 298–300; English transl., Soviet Phys. Dokl. **24** (1979), no. 5, 354–356.

[94] ———, *Delta-shaped Sobolev generalized solutions of quasilinear equations*, Uspekhi Mat. Nauk **34** (1979), no. 1, 235–236; English transl., Russian Math. Surveys **34** (1979), no. 1, 231–232.

[95] D. W. McLaughlin and A. C. Scott, *A multisoliton perturbation theory*, in [**82**], pp. 201–256.

[96] I. A. Molotkov and S. A. Vakulenko, *Concentrated nonlinear waves*, Izd. Leningrad Univ., Leningrad, 1988. (Russian)

[97] M. Nedelkov, S. Pilipović, and D. Scarpalezos, *The linear theory of Colombeau generalized functions*, Longman, Harlow, 1998.

[98] M. Nedelkov and S. Pilipović, *Hypoelliptic differential operators with generalized constant coefficients*, Proc. Edinb. Math. Soc. **41** (1998), 47–60.

[99] G. Nicolas and I. Prigogine, *Self-organization in nonequilibrium systems. From dissipative structure to order through fluctuations*, Wiley, New York, 1977.

[100] M. Oberguggenberger, *Multiplication of distributions and applications to partial differential equations*, Pitman Res. Notes Math., vol. 259, Longman, Harlow, 1992.

[101] M. Oberguggenberger and E. E. Rosinger, *Solution of continuous nonlinear PDEs through order completion*, North-Holland Math. Studies, vol. 181, North-Holland, Amsterdam, 1994.

[102] M. Oberguggenberger and F. Russo, *Nonlinear SPDEs: Colombeau solutions and pathwise limits. Stochastic Analysis and Related Topics*, VI, L. Decreusefond, J. Gjerde, B. Oksendal, and A. S. Ustunel (eds.), Birkhaeuser, Boston, 1998, pp. 319–332.

[103] O. A. Oleinik, *Discontinuous solutions of nonlinear differential equations*, Uspekhi Mat. Nauk **12** (1957), no. 3, 3–73; English transl. Amer. Math. Soc. Transl. (2) **26** (1963), 95–172.

[104] ———, *On uniqueness and stability of the generalized solution to the Cauchy problem for a quasilinear equation*, Uspekhi Mat. Nauk **14** (1959), no. 2, 165–170; English transl. Amer. Math. Soc. Transl. (2) **23** (1963), 285–290.

[105] ———, *On the construction of a generalized solution to the Cauchy problem for a quasilinear equation by introducing the "vanishing viscosity,"* Uspekhi Mat. Nauk **14** (1959), no. 2, 159–164; English transl. Maer. Math. Coc. Transl. (2) **23** (1963), 277–283.

[106] G. A. Omel′yanov, *Weak singularity dynamics in a nonlinear viscous medium*, Nonlinear Phenomena in Complex Systems **5** (2002), no. 1, 21–32.

[107] ———, *Dynamics and interaction of nonlinear waves: multidimensional case*, Int. Conf. "Differential Equations and Related Topics," dedicated to the Centenary Anniversary of I. G. Petrovskii. Book of abstracts. Moscow Univ. Press, 2001, pp. 305–306.

[108] G. A. Omel′yanov and V. V. Trushkov, *Tanh-type asymptotic solution of the conserved phase field system*, Adv. Math. Sci. Appl. **8** (1998), 663–689.

[109] ———, *Dynamics of a free boundary in a binary medium with variable thermal conductivity*, Mat. Zametki **66** (1999), 231–241; English transl. Math. Notes **66** (1999), 181–189.

[110] L. A. Ostrovskii and E. N. Pelinovskii, *Nonlinear evolution of tsunami-type waves.* In: Theoretical and Experimental Study of Tsunami Waves, Nauka, Moscow, 1077, pp. 52–60.

[111] A. Pankov and K. Pflügar, *Periodic and solitary traveling waves for the generalized Kadomtsev–Petviashvili equations*, Math. Mech. Appl. Sci. **22** (1999), 733–752.

[112] S. Pilipović, *Local and microlocal analysis in the space of Colombeau generalized functions.* In: M. Grosser, G. Hormann and M. Oberguggenberger (ed.), *Nonlinear Theory of Generalized Functions*, Chapman and Hall/CRC, 1999, pp. 187–200.

[113] ———, *Colombeau's generalized functions and the pseudo-differential calculus*, Lect. in Math. Sci., Univ. Tokyo, 1994.

[114] S. Pilipović and D. Scarpalézos, *Differential cooperators with generalized coefficients*, Portug. Math. **53** (1996), 305–324.

[115] ———, *Colombeau generalized ultradistributions*, Math. Proc. Camb. Phil. Soc, 2000.

[116] P. I. Plotnikov and V. N. Starovoitov, *Stefan problem as the limit of the phase field system*, Differential Equations **29** (1993), 461–471.

[117] E. V. Radkevich, *On conditions for the existence of a classical solution of the modified Stefan problem, (the Gibbs–Thompson law)*, Mat. Sb. **185** (1992), no. 2, 77–101; English transl. Math. USSR-Sb. **75** (1993), 221–246.

[118] B. L. Rozhdestvenskii and N. N. Yanenko, *Systems of Quasilinear Equations and Their Applications to Gas Dynamics*, Nauka, Moscow, 1978; English transl., Amer. Math. Soc., Providence, RI, 1983.

[119] L. Schwartz, *Sur l'impossibilité de la multiplication des distributions*, Compt. Rend. Acad. Sci. Paris **239** (1954), 847–848.

[120] V. M. Shelkovich, *An associative algebra of distributions and multipliers*, Dokl. Akad. Nauk SSSR **314** (1990), 159–164; English transl. Soviet Math. Dokl. **42** (1991), 409–414.

[121] ———, *An associative-commutative algebra of distributions that includes multiplicators, generalized solutions of nonlinear equations*, Mat. Zametki **57** (1995), no. 5, 765–783; English transl. in Math. Notes.

[122] ———, *Colombeau generalized functions: a theory based on harmonic regularizations*, Mat. Zametki, 63 (1998), no. 2, 313–316; English transl. in Math. Notes. 63 (1998).

[123] V. M. Shelkovich and A. P. Yuzhakov, *The structure of a class of Ivanov asymptotic distributions*, Izv. Vyssh. Uchebn. Zaved. Mat. **4** (1991), 70–73. English transl. in Soviet Math. (Iz. VUZ).

[124] H. M. Soner, *Convergence of the phase field equations to the Mullins–Sekerka problem with kinetic undercooling*, Arch. Rat. Mech. Anal. **131** (1995), 139–197.

[125] J. C. Tolédano and P. Tolédano, *The Landau Theory of Phase Transitions*, World Scientific, Singapore, 1987.

[126] A. L. Velikovich and M. A. Liberman, *Physics of shock waves in gases and plasma*, Nauka, Moscow, 1987; English transl., Springer-Verlag, Berlin, 1986.

[127] V. S. Vladimirov, *Generalized functions in mathematical physics.* Mir, Moscow, 1979.

[128] G. B. Whitham, *Linear and nonlinear waves*, Wiley, New York–London–Toronto, 1974.

[129] V. E. Zakharov, S. V. Manakov, S. P. Novikov, and L. P. Pitaevskii, *Theory of solitons. The inverse problem method*, Nauka, Moscow, 1980; English transl., Plenum Press, New York, 1984.

[130] V. E. Zakharov, S. L. Musher, and A. M. Rubenchik, *On the nonlinear stage of excitation of waves in a plasma*, Psi'ma Zh. Eksper. Teoret. Fiz. **19** (1974), 249–251; English transl., JETP Letters **19** (1974), 151–152.

[131] V. E. Zakharov and A. M. Rubenchik, *Instability of waveguides and solitons in nonlinear media*, Zh. Eksper. Teoret. Fiz. **65** (1973), 997–1011; English transl., Soviet Phys. JETP **38** (1973), 494–500.

[132] V. E. Zakharov and A. B. Shabat, *A scheme for integrating the nonlinear equations of mathematical physics by the method of inverse scattering problem.* I, Funktsional. Anal. i Prilozhen. **8** (1974), no. 3, 43–53; English transl., Functional Anal. Appl. **8** (1974), 226–235.

[133] Ya. B. Zeldovich, *Theory of shock waves and introduction to gas dynamics*, Akademizdat, Moscow, 1946. (Russian).

Moscow Technical University of Communication and Informatics, Russia
E-mail address: danilov@miem.edu.ru

Moscow Institute of Electronics and Mathematics, Russia
Current address: University of Sonora, Mexico
E-mail address: omel@miem.edu.ru

St.-Petersburg State Architecture and Civil Engineering University, Russia
E-mail address: shelkv@svm.abu.spb.ru

Amer. Math. Soc. Transl.
(2) Vol. **208**, 2003

Global Asymptotics and Quantization Rules for Nonlinear Differential Equations

M. V. Karasev and A. V. Pereskokov

ABSTRACT. We present a review of methods for constructing global asymptotic solutions of nonlinear differential equations with a small parameter at the second order derivative. We consider the Kuzmak–Whitham method, the method of matching asymptotic expansions, the isomonodromy deformation method, and the averaging method. The emphasis is on the eigenvalue problems for ordinary differential equations with cubic and saturating nonlinearities, as well as for the Hartree equation. For these equations we calculate the phase shifts at turning points and present "quantization rules" similar to the Bohr–Sommerfeld rule.

CONTENTS

Introduction

Part 1. Local Asymptotic Expansions
1. Kuzmak–Whitham method
2. Kuzmak–Whitham method for an equation with cubic nonlinearity
3. Averaging method for the Painlevé equations. Modulation equation
4. Asymptotic behavior of singular-at-zero solutions of the Painlevé equation of the third kind

Part 2. Global Asymptotic Solutions of Equations with Cubic Nonlinearity
5. Isomonodromy deformation method. Connection formulas for the second Painlevé transcendent
6. Matching the asymptotic solutions for equations with cubic nonlinearity. Quantization rule

---

2000 *Mathematics Subject Classification.* Primary 34E20, 34L30.
Key words and phrases. Semiclassical approximation, Kuzmak–Whitham method, turning point, quantization rule.
The first author was supported in part by RFBR, Grant 02-01-00952.

7. Cauchy problem for the second order nonlinear equation.
Asymptotics near the turning point

Part 3. Asymptotic Solutions of Self-consistent Field Equations
8. Radially symmetric asymptotic solutions of the Hartree equation
9. Asymptotic solution of the Hartree equation concentrated near plane disks
10. One-dimensional self-consistent field equation with additional cubic nonlinearity

Part 4. Global Asymptotic Solution of Equations with a Saturating Nonlinearity
11. Statement of the problem
12. Structure of the asymptotic solution
13. Scattering problem. Calculation of the phase shift.
Final theorems
References

Introduction

The effect of focusing for second order linear differential equations with a small parameter in the derivatives is well known. Namely, the amplitude of the solution is singular with respect to the small parameter on the boundary of the attainable set (turning points, caustics, etc.) of the classical Newton system. Similar but more complicated effects are present in nonlinear equations. To construct global asymptotics of solutions of nonlinear differential equations, i.e., asymptotics including turning points or caustics, is a difficult problem. Such constructions are known only for some special classes of equations.

In the present paper, we review results concerning this problem, but only for stationary equations (i.e., for the eigenvalue problems) and, mainly, in one-dimensional cases. It should be noted that most of these results were obtained during the last 15–20 years when essential progress was made in this field.

The present paper can also serve as an introduction to the theory of asymptotic solutions of nonlinear differential equations, because here some fundamental methods (such as the *Kuzmak–Whitham method, the averaging method, the isomonodromy deformation method, the method of matching asymptotic expansions*, etc.) are discussed and justified in detail from a unified viewpoint. To this end, we consider the most important examples for nonlinear eigenvalue problems.

To calculate the asymptotics of eigenvalues, we introduce and study a "nonlinear" analog of the Bohr–Sommerfeld quantization rule. In some situations, the nonlinearity of the equation leads only to some corrections to the standard "linear" quantization rule; in some other cases, this rule in the nonlinear version is absolutely different.

The paper is organized as follows. In Part 1 we study methods for constructing fast oscillating real asymptotic solutions which can be used only locally, i.e., between the turning points. The solutions obtained are used later for constructing global asymptotics.

First, we consider the Kuzmak–Whitham method. This method generalizes the well-known WKB-method to equations with strong nonlinearity. The asymptotics of the solution is constructed as the two-scale expansion

$$y(x,\varepsilon) = y_0(\tau,x) + \varepsilon y_1(\tau,x) + \cdots + \varepsilon^n y_n(\tau,x) + O(\varepsilon^{n+1}), \qquad \varepsilon \to 0,$$

where $\tau = S(x)/\varepsilon$ is the fast variable, x is the slow variable, and the functions $y_j(\tau,x)$ are 2π periodic in τ. In Part 1 we present sufficient conditions for the applicability of the Kuzmak–Whitham method and write explicit formulas for the leading term of the asymptotics. Moreover, we justify the method, i.e., we prove that the discrepancy between the formal asymptotic and the exact solution is small.

For the special class of Painlevé equations, another approach to the construction of local fast oscillating asymptotics is possible. In Part 1 we study the averaging method based on the existence of matrix commutator representations for the Painlevé equations. Under this approach, the solution can be constructed as an elliptic function modulated in a special way. The parameters determining this elliptic function satisfy the modulation equations. They can be written using a single first order ordinary differential equation.

In Part 2, using the method of matching asymptotic expansions, we construct the global asymptotic solution of the following equation with cubic nonlinearity:

$$\varepsilon^2 \frac{d^2y}{dx^2} + V(x)y + 2y^3 = 0, \qquad x \in \mathbb{R}^1. \tag{1}$$

We assume that the smooth function $V(x)$ attains its maximum at a single point, monotonically decreases away from this point, and

$$\lim_{x\to\pm\infty} V(x) = -\infty.$$

Between the turning points, the solution of equation (1) is determined by the Kuzmak–Whitham expansion. Near the turning points, the solution can be described using the model second Painlevé equation with a parameter

$$\varepsilon^2 \frac{d^2y}{dx^2} - xy + 2y^3 = 0, \qquad x \in \mathbb{R}^1. \tag{2}$$

We are interested in solutions of equation (2) which exponentially decay in one direction away from the turning point and oscillate fast in the other direction away from the turning point.

To obtain formulas connecting the asymptotics on each side of the turning point, we use the isomonodromy deformation method. This method allows us to express the first integral of the Painlevé equation first in terms of the Kuzmak–Whitham expansion and then in terms of the exponentially decaying asymptotics. Comparing these two expressions, we can find the connection formulas for the asymptotics and, in particular, calculate the phase shift. In the case of equation (2), the phase shift at the turning point is $\pi/2$ (modulo π).

As was pointed out in [**1**], the existence of a small parameter ε in equation (2) leads to a problem much more general and difficult than the problem of studying relations between asymptotics at infinity for the Painlevé equations without the parameter ($\varepsilon = 1$). This is related to the fact that the general solution of equation (2) simultaneously depends on two arbitrary functions of the parameter ε with respect to which the asymptotics is calculated.

As a final result, in Part 2 we describe the quantization rule of Bohr–Sommerfeld type for equation (1) with cubic nonlinearity. In the case of the eigenvalue problem, this rule allows us to find the asymptotics of eigenvalues. The quantization rule obtained differs significantly from the "linear" Bohr–Sommerfeld rule. Namely, the position of the turning points, the form of the phase function $S(x)$, and the phase shift at the turning points are different.

In Part 3 we consider the self-consistent field equations. These equations contain the convolution of the squared modulus of the solution and the fundamental solution of the Laplace operator. First, we study the Hartree equation with Coulomb interaction, which, after a change of the variables, can be written as

$$\left(\varepsilon^2\Delta + \frac{\mu}{|x|} + (1-\mu)\int_{\mathbb{R}^3}\frac{|\psi(\tilde{x})|^2}{|x-\tilde{x}|}\,d\tilde{x} - 1\right)\psi(x) = 0, \qquad x \in \mathbb{R}^3,$$
$$\|\psi\|_{L_2} = 1.$$

Here Δ is the Laplace operator and $0 \le \mu \le 1$ is a parameter. First, attention should be paid to the construction of radially symmetric asymptotic solutions. Taking into account that the integrals of fast oscillating functions are small and performing the substitution $\psi(r) = y(r)/(2\sqrt{\pi}r)$, $r = |x|$, we can approximate the Hartree equation with the linear equation

$$\varepsilon^2\frac{d^2y}{dr^2} + V_0(r)y = 0, \tag{3}$$

where the limit self-consistent potential $-V_0(r)$ is a solution of some boundary value problem, which can be solved explicitly. Furthermore, the methods developed for linear equations can be applied to equation (3). In particular, global asymptotic solutions can be constructed using the Fourier method. In the case of the polaron equation ($\mu = 0$), we need to construct an additional local expansion near the zero point.

In Part 3 we also consider a one-dimensional Thomas–Fermi–von Weizäcker type equation, which contains both integral and local cubic nonlinearity.

Moreover, in Part 3 we discuss methods for constructing essentially multidimensional not radially symmetric asymptotic solutions of the Hartree equation, which are concentrated near plane disks. Such solutions are constructed by using new model equations: the polaron equation along the normal to the disk and the Airy-polaron equation along the transversal to the edge of the disk.

Finally, in Part 4 we study equations with a saturating nonlinearity, i.e., with a nonlinearity tending to a constant with increasing amplitude of the solution. A typical example of this situation is offered by the equation

$$\varepsilon^2\frac{d^2y}{dx^2} + \big(U(x) + a(x)f(y^2)\big)y = 0, \qquad x \in \mathbb{R}^1. \tag{4}$$

Here $-U(x)$ is a smooth potential well, $f(\sigma)$ is a smooth monotone function exponentially decreasing as $\sigma \to \infty$, and $a(x)$ is a bounded function. This is the second example of local nonlinearity of general form (in addition to the case with the Painlevé equations as model equations near the turning points) where the global asymptotics can be constructed explicitly. The asymptotic solution decaying at infinity is constructed using the method of matching asymptotic expansions and has a rather complicated structure. The real axis is divided into seven zones in each of which the asymptotics is determined by its own model equation.

Here the main idea is the following. Since the amplitude of the solution near the turning point is large but the function f is fast decreasing, the contributions of the nonlinearity can be neglected outside small neighborhoods of the zeros of the solution. So here the asymptotics is expressed in terms of the Airy functions.

However, near the zeros of the solution, one may assume that the functions $U(x)$ and $a(x)$ in equation (4) are approximately constant. As a result, one obtains an equation that can be integrated in quadratures. It remains to sew together the local asymptotics thus constructed. Calculations show that, for equation (4), in the leading term the phase shift at the turning points is equal to the linear shift $\pi/4$, and then an essentially nonlinear ε-depending correction to the shift appears. Moreover, the nonlinearity in equation (4) implies a change in the form of the curve in the quantization rule and an increase in the support of the solution compared to the linear case.

Part 1. Local Asymptotic Expansions

1. Kuzmak–Whitham method

1.1. Outline of the Kuzmak–Whitham method. We consider the nonlinear equation

$$\varepsilon^2 \frac{d^2 y}{dx^2} + f(x, y) = 0, \qquad \alpha \leq x \leq \beta, \tag{1.1}$$

where f is a smooth real function and $\varepsilon > 0$ is a small parameter. We introduce the primitive of the function $2f$ with respect to the variables y:

$$\frac{\partial H(x, y)}{\partial y} = 2f(x, y).$$

On the function H we impose the following condition under which equation (1.1) has a fast oscillating real asymptotic solution.

CONDITION 1. There exist smooth real functions $E_1(x)$ and $E_2(x)$ such that for each fixed $x \in [\alpha, \beta]$ on the interval $E_1 < E < E_2$, the equation $H(x, z) = E$ has at least two solutions $z^{\pm}(x, E)$, $z^- < z^+$, satisfying the condition $H(x, z) < E$ for $z^- < z < z^+$. Moreover, let $z^{\pm}$ smoothly depend on x and E, and let

$$\frac{\partial H}{\partial z}\big(x, z^{\pm}(x, E)\big) \neq 0.$$

We present a method for constructing asymptotic solutions of equation (1.1) as $\varepsilon \to 0$. For nonlinear ordinary differential equations, such a method was first introduce by Kuzmak in [**2**], where the ideas introduced in [**3**] were used. Later, a similar method was developed for partial differential equations by Whitham [**4**] and Luke [**5**]. Namely, the method became widely known just because of these works. An intimate connection between the method and variational principles and conservation laws was found (see [**4**], [**6**]).

We look for an asymptotic solution of equation (1.1) in the form

$$y(x, \varepsilon) = y_0(\tau, x) + \varepsilon y_1(\tau, x) + \cdots + \varepsilon^n y_n(\tau, x) + O(\varepsilon^{n+1}), \tag{1.2}$$

where $\tau = S(x)/\varepsilon$ is the fast variable, $S(x)$ and $y_j(\tau, x)$ are unknown smooth real functions, and $y_j(\tau, x)$ are 2π-periodic in τ. Since (1.2) contains two variables, the fast variable τ and the slow variable x, such an expansion is said to be *two-scale*.

Differentiating (1.2), we obtain

$$\varepsilon^2 \frac{d^2 y}{dx^2} = \sum_{j=0}^{n} \varepsilon^j \Big\{ (S'(x))^2 \frac{\partial^2 y_j}{\partial \tau^2} + \varepsilon \Big[2S'(x) \frac{\partial^2 y_j}{\partial \tau \partial x} + S''(x) \frac{\partial y_j}{\partial \tau} \Big] + \varepsilon^2 \frac{\partial^2 y_j}{\partial x^2} \Big\} + O(\varepsilon^{n+1}). \tag{1.3}$$

We substitute expansions (1.2), (1.3) in (1.1). Then, expanding the function f into the Taylor series and comparing the coefficients of powers of ε, we obtain the recurrent system of equations

$$(S'(x))^2 \frac{\partial^2 y_0}{\partial \tau^2} + f(x, y_0) = 0, \tag{1.4}$$

$$L y_j = F_j, \qquad j = 1, \dots, n. \tag{1.5}$$

Here the operator L has the form

$$L = (S'(x))^2 \frac{\partial^2}{\partial \tau^2} + \frac{\partial f}{\partial y}(x, y_0),$$

and the right-hand sides F_j are given by the formulas

$$F_1 = -2S' \frac{\partial^2 y_0}{\partial \tau \partial x} - S'' \frac{\partial y_0}{\partial \tau}, \tag{1.6}$$

$$F_2 = -2S' \frac{\partial^2 y_1}{\partial \tau \partial x} - S'' \frac{\partial y_1}{\partial \tau} - \frac{\partial^2 y_0}{\partial x^2} - \frac{\partial^2 f}{\partial y^2}(x, y_0) \frac{y_1^2}{2}, \tag{1.7}$$

$$F_j = -2S' \frac{\partial^2 y_{j-1}}{\partial \tau \partial x} - S'' \frac{\partial y_{j-1}}{\partial \tau} - \frac{\partial^2 y_{j-2}}{\partial x^2} + P_j(x, y_0, y_1, \dots, y_{j-1}),$$

where $j = 3, \dots, n$ and P_j are polynomials in the variables $y_1, \dots, y_{j-1}$.

If the functions S and y_j satisfy equations (1.4), (1.5), then, obviously, the function (1.2) is a formal asymptotic solution of (1.1) with accuracy $O(\varepsilon^{n+1})$. Equations (1.4), (1.5) are ordinary differential equations with respect to τ with coefficients depending on x as a parameter. The first equation is nonlinear; the others are linear nonhomogeneous equations.

Let us find a solution of equation (1.4) that is 2π-periodic in τ. Multiplying (1.4) by $2\partial y_0/\partial \tau$ and then integrating, we obtain

$$(S')^2 \Big(\frac{\partial y_0}{\partial \tau} \Big)^2 + H(x, y_0) = E(x),$$

where $E(x)$ is the "integration constant" with respect to the variable τ. So we have

$$\frac{\partial y_0}{\partial \tau} = \pm \frac{1}{S'} \sqrt{E(x) - H(x, y_0)}. \tag{1.8}$$

To integrate equation (1.8), we define a function $Y(\tau, E, x)$ for $E_1 < E < E_2$. First, we define the period $T = T(x, E)$ as follows:

$$T = 2S'(x) \int_{z^-(x,E)}^{z^+(x,E)} \frac{dz}{\sqrt{E - H(x, z)}}. \tag{1.9}$$

On the interval $\tau \in [0, T/2]$, we define the function Y by the implicit equation

$$\tau = S' \int_{Y}^{z^+(x,E)} \frac{dz}{\sqrt{E - H(x, z)}}, \tag{1.10}$$

and then extend it to the interval $\tau \in [-T/2, 0]$ as an even function: $Y(\tau, E, x) \equiv Y(-\tau, E, x)$. By Condition 1, we obtain a smooth function that is T-periodic in τ and ranges between z^- and z^+.

LEMMA 1.1. *Let $x \in [\alpha, \beta]$. We choose smooth functions $S(x)$ and $E(x)$ (where $E_1(x) < E(x) < E_2(x)$) so that the condition $T = 2\pi$ holds, i.e.,*

$$\frac{\pi}{S'(x)} = \int_{z^-(x,E(x))}^{z^+(x,E(x))} \frac{dz}{\sqrt{E(x) - H(x,z)}}. \tag{1.11}$$

Suppose also that $\eta(x)$ is any smooth real function. Then the function

$$y_0(\tau, x) = Y(\tau + \eta(x), E(x), x) \tag{1.12}$$

is a 2π-periodic in τ solution of equation (1.4) *smoothly depending on x.*

DEFINITION 1.1. The values of x such that $S'(x) = 0$ will be called *turning points* for equation (1.1)

It follows from formulas (1.11) that by Condition 1, there are no turning points on the interval $[\alpha, \beta]$. Obviously, expansion (1.3), and hence (1.2), cannot be used near the turning points.

The periodicity condition (1.11) allows us to express $S'(x)$ in terms of $E(x)$. However, formula (1.12) for y_0 still contains two arbitrary functions $E(x)$ and $\eta(x)$. We can find these functions only studying the equations for the subsequent terms of the asymptotic expansion (1.2). Such a situation is typical of nonlinear problems.

1.2. Solvability conditions. Let us analyze equations (1.5). First, we consider the homogeneous equation

$$L\omega = (S'(x))^2 \frac{\partial^2 \omega}{\partial \tau^2} + \frac{\partial f}{\partial y}(x, y_0)\omega = 0. \tag{1.13}$$

By (1.11), the coefficients of the linear operator L are functions 2π-periodic in τ. Equation (1.13) has a 2π-periodic in τ solution $\omega_1(\tau, x) = \partial y_0/\partial \tau$. To prove this fact, it suffices to differentiate equation (1.4) with respect to τ.

Suppose that $\omega_2(\tau, x)$ is the second independent solution of equation (1.13). Then

$$\omega_2(\tau + 2\pi, x) = \omega_2(\tau, x) + A\omega_1(\tau, x), \tag{1.14}$$

where $A = A(x)$ is a function constant in τ. This identity follows from the Floquet–Lyapunov theory for equations with periodic coefficients (e.g., see [**7**]). Recall the following condition.

CONDITION 2. For any $x \in [\alpha, \beta]$ not all solutions of equation (1.13) are 2π-periodic in τ.

Hence any solution of equation (1.13) that is independent of $\omega_1 = \partial y_0/\partial \tau$ is not 2π-periodic. Thus $A(x) \neq 0$ in formula (1.14). Identity (1.14) allows us to write ω_2 as

$$\omega_2(\tau, x) = g(\tau, x) + \frac{A(x)\tau}{2\pi}\omega_1(\tau, x),$$

where the secular term proportional to τ is written explicitly. Here the function $g(\tau, x)$ is 2π-periodic in τ.

REMARK 1.1. The situation in which the coefficient $q(\tau)$ and all the solutions of the equation

$$\frac{d^2\omega}{d\tau^2} + q(\tau)\omega = 0 \tag{1.15}$$

are simultaneously 2π-periodic functions is, of course, an exceptional situation. Nevertheless, as was pointed out in [**8**], there exist infinitely many 2π-periodic functions $q(\tau) \not\equiv \text{const}$ for which this situation takes place.

REMARK 1.2. The following formula obtained in [**9**] and [**10**] can be used to verify Condition 2:

$$\omega_2 = \omega_1 \int_0^\tau \left((S')^2 \frac{\partial f}{\partial y}(x, y_0) - 1\right) \frac{\omega_1^2 - (\partial\omega_1/\partial\tau)^2}{\left(\omega_1^2 + (\partial\omega_1/\partial\tau)^2\right)^2}\, d\tau + \frac{\partial\omega_1/\partial\tau}{\omega_1^2 + (\partial\omega_1/\partial\tau)^2}.$$

This formula expresses the second independent solution of equation (1.13) in terms of $\omega_1 = \partial y_0/\partial\tau$. As compared to the well-known expression for the second solution $\omega_2 = \omega_1 \int \omega_1^{-2}\, d\tau$, this formula is more cumbersome, but, in turn, this formula can also be used (without any additional regularizations) in the case where ω_1 is zero at some points.

Now let us consider the nonhomogeneous equation

$$L\omega = F(\tau, x), \tag{1.16}$$

where $F(\tau, x)$ is a smooth function 2π-periodic in τ. The following assertion [**8**], [**11**] holds.

LEMMA 1.2. *Suppose that Condition* 2 *is satisfied. A necessary and sufficient condition for the solvability of equation* (1.15) *in the class of smooth functions 2π-periodic in τ is the relation*

$$\int_0^{2\pi} F(\tau, x) \frac{\partial y_0}{\partial\tau}(\tau, x)\, d\tau = 0. \tag{1.17}$$

If (1.17) *holds, then the 2π-periodic in τ solution of equation* (1.16) *has the form*

$$\begin{aligned} \omega(\tau) = a_0\omega_1(\tau) + \frac{1}{\Delta(S')^2}\Big\{\omega_1(\tau) \int_0^\tau F(\tau')\omega_2(\tau')\, d\tau' \\ - \omega_2(\tau)\Big(A^{-1}\int_0^{2\pi} F(\tau')\omega_2(\tau')\, d\tau' + \int_0^\tau F(\tau')\omega_1(\tau')\, d\tau'\Big)\Big\}. \end{aligned} \tag{1.18}$$

Here $a_0 = a_0(x)$ is an arbitrary smooth function, and Δ denotes the Wronskian

$$\Delta(x) = \omega_2 \frac{\partial\omega_1}{\partial\tau} - \omega_1 \frac{\partial\omega_2}{\partial\tau}.$$

The dependence on x of all the functions in (1.18) *is omitted for brevity.*

So the condition for the existence of a 2π-periodic solution of the jth equation in (1.5) has the form

$$\int_0^{2\pi} F_j \frac{\partial y_0}{\partial\tau}\, d\tau = 0. \tag{1.19}$$

Moreover,

$$y_j(\tau, x) = y_j^0(\tau, x) + a_j(x)\frac{\partial y_0}{\partial\tau}(\tau, x), \tag{1.20}$$

where the "constant" of integration $a_j(x)$ is a smooth function and $y_j^0(\tau, x)$ is some particular 2π-periodic in τ solution of equation (1.5).

Let us study the solvability conditions (1.19). For $j = 1$, where F_1 is determined by formula (1.6), we have

$$\int_0^{2\pi} \left(2S' \frac{\partial^2 y_0}{\partial \tau \partial x} + S'' \frac{\partial y_0}{\partial \tau} \right) \frac{\partial y_0}{\partial \tau} \, d\tau = 0.$$

This relation can be rewritten as

$$\frac{d}{dx} \left(S' \int_0^{2\pi} \left(\frac{\partial y_0}{\partial \tau} \right)^2 d\tau \right) = 0,$$

whence

$$S' \int_0^{2\pi} \left(\frac{\partial y_0}{\partial \tau} \right)^2 d\tau = 4c. \tag{1.21}$$

Here c is a positive constant.

Passing in (1.21) from the integration variable τ to the integration variable y_0 and using expression (1.8) for $\partial y_0 / \partial \tau$, we write the solvability condition for $j = 1$ in the form

$$\int_{z^-(x,E(x))}^{z^+(x,E(x))} \sqrt{E(x) - H(x, z)} \, dz = 2c. \tag{1.22}$$

We obtain an equation for the function $E(x)$. Let us impose the following condition.

CONDITION 3. Suppose that on $[a, b]$ there exists a smooth solution $E(x)$ of equation (1.22) satisfying the inequality $E_1(x) < E(x) < E_2(x)$.

Furthermore, if the function $E(x)$ is known, then we can find the phase $S(x)$ from the periodicity condition (1.11). In formula (1.12) for y_0, it remains to determine only the function $\eta(x)$. To find this function, we consider the solvability condition (1.19) for $j = 2$. By (1.7), it has the form

$$\int_0^{2\pi} \left(2S' \frac{\partial^2 y_1}{\partial \tau \partial x} + S'' \frac{\partial y_1}{\partial \tau} + \frac{\partial^2 y_0}{\partial x^2} + \frac{\partial^2 f}{\partial y^2}(x, y_0) \frac{y_1^2}{2} \right) \frac{\partial y_0}{\partial \tau} \, d\tau = 0. \tag{1.23}$$

To transform (1.23), we use the fact that the function

$$\Phi(\tau, x) \stackrel{\text{def}}{=} (\tau + \eta) \frac{\partial y_0}{\partial \tau} + 2\pi \left(\frac{\partial T}{\partial E}(x, E(x)) \right)^{-1} \frac{\partial Y}{\partial E}(\tau + \eta, E(x), x),$$

where $T(x, E)$ is determined by formula (1.9), is a 2π-periodic in τ solution of the equation

$$L\Phi = 2(S')^2 \frac{\partial^2 y_0}{\partial \tau^2}.$$

The following assertion holds.

LEMMA 1.3 ([**11**]). *For $j = 2$ the solvability condition (1.19) is equivalent to the equation*

$$\eta'(x) = \frac{2c_1}{\pi^2} \big(S'(x) \big)^2 \frac{\partial T}{\partial E}(x, E) \bigg|_{E=E(x)}, \tag{1.24}$$

where c_1 is a constant.

REMARK 1.3. The constant c_1 in equation (1.24) determines a correction of order ε to the parameter c in (1.22). This follows from the relation

$$\frac{S'(x,c)}{\varepsilon} + \eta'(x,c,c_1) = \frac{S'(x,c-\varepsilon c_1)}{\varepsilon} + O(\sqrt{\varepsilon}).$$

Hence in the study of the leading term of the asymptotic expansion, we can set $\eta' = 0$. However, in this case the parameter in (1.22) is assumed to be $\tilde{c} = c - \varepsilon c_1$ rather than c.

The solvability conditions for $j \geq 3$ are studied similarly. These conditions determine the "constants" of integration $a_k(x)$ contained in y_k (see (1.20)). Moreover, the function $a_k(x)$ is found from condition (1.19) for $j = k+2$. This is possible, because for $j = k+1$ the integral in the left-hand side of the solvability condition (1.19) is independent of a_k (although a_k is contained in F_{k+1}).

Combining the above results, we arrive at the following concluding result.

THEOREM 1.1. *Suppose that Conditions 1–3 are satisfied. If the smooth functions $E(x)$, $S(x)$, and $\eta(x)$ satisfy equations* (1.11), (1.22), *and* (1.24), *then the function*

$$y_0 = Y\left(\frac{S(x)}{\varepsilon} + \eta(x), E(x), x\right)$$

is the leading term of some formal asymptotic solution of equation (1.1). *The next terms of expansion* (1.2) *can be found from equations* (1.5) *and the solvability conditions* (1.19). *If*

$$Y_N(x,\varepsilon) \stackrel{\text{def}}{=} \sum_{i=0}^{N} \varepsilon^i y_i\left(\frac{S(x)}{\varepsilon}, x\right) \tag{1.25}$$

is a partial sum of the asymptotic series (1.2), *then for any $N \geq 1$ we have*

$$\varepsilon^2 \frac{d^2 Y_N}{dx^2} + f(x, Y_N) = \varepsilon^{N+1} R_N(\tau, x, \varepsilon)\Big|_{\tau = S(x)/\varepsilon}, \tag{1.26}$$

where the remainder $R_N(\tau,x,\varepsilon)$ belongs to $C^\infty(\mathbb{R}^1 \times [\alpha,\beta] \times [0,1])$ and is 2π-periodic in τ.

We note that in the framework of the Kuzmak–Whitham method, the constants of integration in the first order equations (1.11) and (1.24) remain undetermined. They are found only in the process of constructing global asymptotic solutions belonging to some functional space.

DEFINITION 1.2. The number

$$\delta(x_*) = -\frac{S(x_*)}{\varepsilon} - \eta(x_*)$$

is called the *phase shift* (or phase increment) at the turning point x_*.

In the next sections we write down the phase shifts for solutions of several nonlinear equations.

1.3. Justification of the Kuzmak–Whitham method. We shall justify asymptotic formulas, i.e., prove that the constructed formal asymptotic solution approximates the exact solution of (1.1). This was analyzed by Fedoryuk [**8**].

Let $Y_N(x, \varepsilon)$ be the function (1.25), and let $y(x, \varepsilon)$ be a solution of equation (1.1) with the same initial data, i.e.,

$$\varepsilon^2 \frac{d^2 y}{dx^2} + f(x, y) = 0, \qquad y(\alpha, \varepsilon) = \tilde{y}_0, \qquad \varepsilon \frac{dy}{dx}(\alpha, \varepsilon) = \tilde{y}_1, \tag{1.27}$$

where $\tilde{y}_0 = Y_N(\alpha, \varepsilon)$ and $\tilde{y}_1 = \varepsilon dY_N(\alpha, \varepsilon)/dx$. Let the conditions of Theorem 1.1 be satisfied. Furthermore, we impose the following condition.

CONDITION 4. There exist positive constants δ, ε_0, and k, and a constant $\beta > \alpha$, such that if $|\tilde{y}_0 - y_0^*| \le \delta$, $|\tilde{y}_1 - y_1^*| \le \delta$, $0 < \varepsilon \le \varepsilon_0$, then on $[\alpha, \beta]$ the solution of equation (1.1) with the initial conditions

$$y(\alpha, \varepsilon) = y_0^*, \qquad \varepsilon \frac{dy}{dx}(\alpha, \varepsilon) = y_1^*$$

exists, is unique, and belongs to the class C^∞. Moreover, the estimates

$$|y(x, \varepsilon)| \le k, \qquad \left|\varepsilon \frac{dy}{dx}(x, \varepsilon)\right| \le k$$

hold on the set $D_0 = (0, \varepsilon_0] \times [\alpha, \beta]$.

Note that if $\partial f/\partial x \le 0$ for $\alpha \le x \le \beta$ and $|y| \le \max |z_\pm(x, E(x))|$, then Condition 4 is satisfied because the function

$$\Lambda\left(x, y, \frac{dy}{dx}\right) = \frac{\varepsilon^2}{2}\left(\frac{dy}{dx}\right)^2 + \int_\alpha^y f(t, z)\, dz$$

is the Lyapunov function of equation (1.1).

We consider the difference $u(x, \varepsilon) = y(x, \varepsilon) - Y_N(x, \varepsilon)$ for which we have the Cauchy problem

$$\varepsilon^2 \frac{d^2 u}{dx^2} + f(x, Y_N + u) - f(x, Y_N) = -\varepsilon^{N+1} R_N, \qquad u(\alpha, \varepsilon) = 0, \qquad \frac{du}{dx}(\alpha, \varepsilon) = 0. \tag{1.28}$$

By Condition 4 and Theorem 1.1, the function $u(x, \varepsilon)$ is bounded on the set D_0. Hence the Cauchy problem (1.28) can be represented as

$$Mu = g(x, \varepsilon) u^2 - \varepsilon^{N+1} R_N, \qquad u(\alpha, \varepsilon) = 0, \qquad \frac{du}{dx}(\alpha, \varepsilon) = 0. \tag{1.29}$$

Here

$$Mu \overset{\text{def}}{=} \varepsilon^2 \frac{d^2 u}{dx^2} + \frac{\partial f}{\partial y}(x, Y_N) u,$$

and g is a function bounded on D_0.

Let us find the asymptotic solutions of the equation $Mv = 0$. (We shall need these solutions later in constructing the solution of problem (1.29).) We introduce the class K of smooth functions of the variables $\tau \in \mathbb{R}^1$, $x \in [\alpha, \beta]$, and $\varepsilon \in [0, \varepsilon_0]$, which are 2π-periodic in τ. We set

$$v_1(x,\varepsilon) = v_1^*\left(\frac{S(x)}{\varepsilon}, x, \varepsilon\right), \qquad \text{where } v_1^*(\tau, x, \varepsilon) = \sum_{i=0}^{N} \varepsilon^i \frac{\partial y_i}{\partial \tau}.$$

Differentiating relation (1.26) with respect to the variable τ, we see that the function $v_1^*(\tau, x, \varepsilon)$ belongs to K and

$$Mv_1 = \varepsilon^{N+1} h_1\left(\frac{S(x)}{\varepsilon}, x, \varepsilon\right), \qquad \text{where } h_1 \in K.$$

We shall look for the second "almost solution" of the equation $Mv = 0$ in the form

$$v_2(x,\varepsilon) = v_2^*\left(\frac{S(x)}{\varepsilon}, x, \varepsilon\right), \qquad \text{where } v_2^*(\tau, x, \varepsilon) = \tau v_1^*(\tau, x, \varepsilon) + \tilde{v}(\tau, x, \varepsilon).$$

Let $\Delta_2(x, \varepsilon)$ be the Wronskian of the functions v_1, v_2. We have the following assertion.

LEMMA 1.4 ([**8**]). *If the number $\beta - \alpha > 0$ is sufficiently small, then there exists a function $\tilde{v} \in K$ such that*

$$Mv_2 = \varepsilon h_2\left(\frac{S(x)}{\varepsilon}, x, \varepsilon\right), \qquad \textit{where } h_2 \in K,$$

and $\Delta_2(x, \varepsilon) = \varepsilon^{-1} S'(x)[1 + O(\varepsilon)]$.

Now we study problem (1.28). Let $u(x, \varepsilon)$ be a solution of this problem. We choose functions $c_1(x, \varepsilon)$ and $c_2(x, \varepsilon)$ so that

$$u = c_1 u_1 + c_2 u_2, \qquad \frac{du}{dx} = c_1 \frac{du_1}{dx} + c_2 \frac{du_2}{dx}, \qquad c_1(\alpha, \varepsilon) = 0, \qquad c_2(\alpha, \varepsilon) = 0. \tag{1.30}$$

Here

$$u_1 \stackrel{\text{def}}{=} \varepsilon^{(N+1)/2} v_1, \qquad u_2 \stackrel{\text{def}}{=} \varepsilon^{(N+3)/2} v_2. \tag{1.31}$$

Using representation (1.29), we can prove the following assertion by straightforward calculations.

LEMMA 1.5. *The Cauchy problem* (1.28) *is equivalent to the system of integral equations*

$$c(x, \varepsilon) = \int_\alpha^x A(t, \varepsilon) c(t, \varepsilon)\, dt + \int_\alpha^x B(t, \varepsilon)\, dt, \tag{1.32}$$

where

$$A = \frac{1}{\varepsilon^2 \Delta_1} \begin{pmatrix} u_2 M u_1 & u_2 M u_2 \\ -u_1 M u_1 & -u_1 M u_2 \end{pmatrix}, \qquad c = \begin{pmatrix} c_1 \\ c_2 \end{pmatrix},$$

$$B = \frac{1}{\varepsilon^2 \Delta_1} [g(c_1 u_1 + c_2 u_2)^2 - \varepsilon^{N+1} R_N] \begin{pmatrix} -u_2 \\ u_1 \end{pmatrix}.$$

Here $\Delta_1(x, \varepsilon)$ is the Wronskian of the functions u_1, u_2.

Next we estimate the solution of system (1.32) using the Gronwall lemma, which allows us to prove the following assertion.

LEMMA 1.6 ([**8**]). *Let $N \geq 3$. Then the solution of system* (1.32) *exists, is unique, and belongs to the class C^∞ on the set D_0. Moreover, the following estimate holds*:

$$\sqrt{c_1^2(x,\varepsilon) + c_2^2(x,\varepsilon)} \leq k_0(x-\alpha)\varepsilon^{(N-3)/2},$$

where k_0 is a positive constant.

Finally, the following theorem holds.

THEOREM 1.2. *Suppose that Conditions* 1–4 *are satisfied, $y(x,\varepsilon)$ is a solution of the Cauchy problem* (1.27), *and the number $\beta - \alpha > 0$ is sufficiently small. Then the estimates*

$$|y(x,\varepsilon) - Y_N(x,\varepsilon)| \leq k_N(x-\alpha)\varepsilon^{N-1}, \tag{1.33}$$

$$\left|\frac{dy}{dx}(x,\varepsilon) - \frac{dY_N}{dx}(x,\varepsilon)\right| \leq k_N(x-\alpha)\varepsilon^{N-2} \tag{1.34}$$

hold for any $N \geq 3$. Here $\alpha \leq x \leq \beta$, $0 < \varepsilon \leq \varepsilon_0$, and k_N are constants.

To prove this theorem, we use the fact that the functions v_1 and εv_2 are bounded. Then Lemma 1.6 and formulas (1.29)–(1.32) imply the estimate (1.33):

$$|u(x,\varepsilon)| \leq \varepsilon^{(N+1)/2}\sqrt{c_1^2 + c_2^2}\big(|v_1| + \varepsilon|v_2|\big) \leq k_N(x-\alpha)\varepsilon^{N-1}.$$

The estimate (1.34) can be proved similarly.

REMARK 1.4. In [**12**], for equations with saturating nonlinearity, it was proved that the asymptotic and exact solutions are close to each other not only on an interval of a sufficiently small length, but on the entire domain of applicability of the Kuzmak–Whitham expansion.

2. Kuzmak–Whitham method for an equation with cubic nonlinearity

Let us consider the equation with cubic nonlinearity

$$\varepsilon^2\frac{d^2y}{dx^2} + V(x)y + 2y^3 = 0, \qquad x \in \mathbb{R}^1, \tag{2.1}$$

where ε is a small parameter. This equation arises in the study of nonlinear waves, in particular, in nonlinear optics. We study the case where a smooth function $V(x)$ attains its maximum only at a single point $\overline{x}$ and monotonically decreases with x moving away from $\overline{x}$ in such a way that $\lim_{x\to\pm\infty} V(x) = -\infty$. (The behavior of the function $V(x)$ is shown in Figure 1.)

For such functions $V(x)$, equation (2.1) has exactly two turning points $x_- < x_+$. Expansion (1.2) holds on the interval between these points. Let us write explicitly the first two terms of this expansion:

$$y(x,\varepsilon) = y_0\left(\frac{S(x)}{\varepsilon}, x\right) + \varepsilon y_1\left(\frac{S(x)}{\varepsilon}, x\right) + O(\varepsilon^2). \tag{2.2}$$

Substituting (2.2) into (2.1), we obtain

$$(S')^2\frac{\partial^2 y_0}{\partial\tau^2} + V(x)y_0 + 2y_0^3 = 0, \tag{2.3}$$

$$(S')^2\frac{\partial^2 y_1}{\partial\tau^2} + V(x)y_1 + 6y_0^2y_1 = -2S'\frac{\partial^2 y_0}{\partial\tau\partial x} - S''\frac{\partial y_0}{\partial\tau}.$$

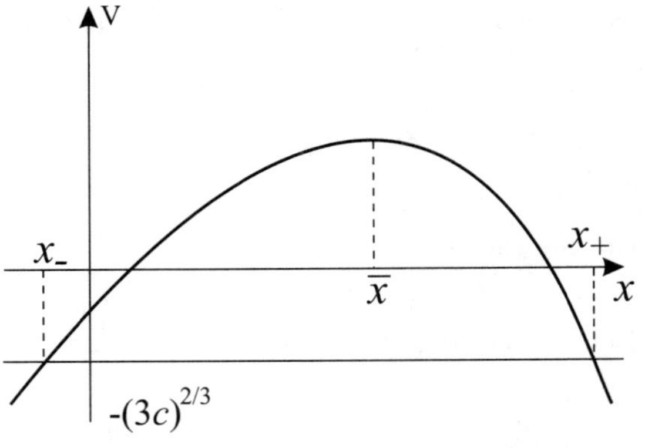

FIGURE 1

According to Condition 2 under which the Kuzmak–Whitham method can be used to study (2.1), the kernel of the operator

$$L = (S')^2 \frac{\partial^2}{\partial\tau^2} + V + 6y_0^2$$

in the class of 2π-periodic in τ functions must be one-dimensional. This fact is proved in [**8**].

Condition 1 of the Kuzmak–Whitham method allows us to find the domain where expansion (2.2) holds. In the case of equation (2.1), the function H has the form $H(x,z) = V(x)z^2 + z^4$. Let $E_1(x) \equiv 0$, and let $E_2(x) \equiv +\infty$. Then for $E_1 < E < E_2$, equation $E - H(x,z) = 0$ has exactly two real roots

$$z^{\pm}(x,E) = \pm\big(-V + (V^2+4E)^{1/2}\big)^{1/2}/\sqrt{2};$$

moreover, $H(x,z) < E$ for $z^- < z < z^+$ and $\partial H(x,z^{\pm})/\partial z = 2z^{\pm}\sqrt{V^2+4E} \neq 0$. The Kuzmak–Whitham method can be used on any interval $x \in [\alpha,\beta]$, where the solution $E(x)$ of (1.22) is a smooth positive function. We show that such intervals lie strictly between turning points.

We write $p(x) \overset{\text{def}}{=} z^+(x,E(x))$. Instead of the function $E(x)$, we shall look for a function $p(x)$ in terms of which $E(x)$ can be expressed as follows:

$$E(x) = V(x)p^2(x) + p^4(x). \tag{2.4}$$

Since $E(x) - H(x,z) = (p^2 - z^2)(V + p^2 + z^2)$, in the case of (2.1), equations (1.22), (1.11) take the form

$$\int_0^p \sqrt{(p^2-z^2)(V+p^2+z^2)}\,dz = c, \tag{2.5}$$

$$S' = \frac{\pi}{2}\left(\int_0^p \frac{dz}{\sqrt{(p^2-z^2)(V+p^2+z^2)}}\right)^{-1}, \tag{2.6}$$

and the turning points x_-, x_+ (where $S'(x_\pm) = 0$) satisfy the relations

$$V(x_\pm) + p^2(x_\pm) = 0, \tag{2.7}$$

$$\int_0^{p(x_\pm)} z\sqrt{p^2(x_\pm) - z^2}\,dz = c. \tag{2.8}$$

By (2.4) and (2.7), the function $E(x)$ vanishes precisely at the turning points and remains positive between these points.

Calculating the integral in (2.8), we obtain $p(x_\pm) = \sqrt[3]{3c}$ and, hence, relation (2.7) implies the following relation for the turning points:

$$V(x_\pm) = -(3c)^{2/3}. \tag{2.9}$$

We assume that if $V(\overline{x}) > 0$, then $c > 0$ and if $V(\overline{x}) \le 0$, then $c > |V(\overline{x})|^{3/2}/3 + h$, where $h > 0$ is an arbitrary number. Then the conditions imposed on $V(x)$ guarantee that equation (2.9) has exactly two roots $x_- < x_+$, which are just the desired turning points for equation (2.1). We emphasize that the position of the turning points depends on the parameter c.

We write the formulas for y_0. First, we find the function $p = p(x)$ as a solution of equation (2.5) (the plots of numerical solutions can be found in [**2**]). Next, from (2.6) we find the phase $S(x)$. The equation for the correction to the phase is obtained in the following lemma.

LEMMA 2.1. *The function $\eta(x)$ is determined by the equation*

$$\eta'(x) = -\frac{4c_1(S')^3}{\pi^2 p^2(V+2p^2)} \int_0^p \frac{(z^2+p^2)\,dz}{\sqrt{(p^2-z^2)(V+p^2+z^2)^3}}, \tag{2.10}$$

where c_1 is a constant.

PROOF. In (1.24) we calculate $\partial T/\partial E$. In the case of equation (2.1), after z is replaced by $u = z/z^+$, formula (1.9) for the period T takes the form

$$T(x,E) = 4S'(x)z^+ \int_0^1 \frac{du}{\sqrt{E - V(x)(z^+u)^2 - (z^+u)^4}}.$$

Hence we have

$$\begin{aligned} \frac{\partial T}{\partial E}(x,E) &= 4S'\frac{\partial z^+}{\partial E}\int_0^1 \frac{du}{\sqrt{E - V(x)(z^+u)^2 - (z^+u)^4}} \\ &\quad - \frac{4S'z^+}{2}\int_0^1 \frac{\left(1 - 2V(x)u^2 z^+\partial z^+/\partial E - 4u^4(z^+)^3\partial z^+/\partial E\right)du}{\sqrt{\left(E - V(x)(z^+u)^2 - (z^+u)^4\right)^3}}. \end{aligned} \tag{2.11}$$

Differentiating the relation

$$E = V(x)(z^+)^2 + (z^+)^4 \tag{2.12}$$

with respect to E, we obtain

$$\frac{\partial z^+}{\partial E} = \left(2V(x)z^+ + 4(z^+)^3\right)^{-1}. \tag{2.13}$$

It remains to substitute (2.12) and (2.13) into the right-hand side of (2.11), to collect similar terms, and to multiply the expression obtained by $2c_1(S')^2\pi^{-2}$. Replacing z^+ by p, we obtain (2.10). The proof of the lemma is complete. □

In the case of equation (2.1), the terms of the asymptotic expansion are expressed in terms of the Jacobi elliptic functions [**13**]. For example, equation (2.6) can be written as

$$S' = \frac{\pi\sqrt{V+p^2}}{2K(k)}, \tag{2.14}$$

where

$$K(k) = \int_0^1 \frac{dx}{\sqrt{(1-x^2)(1-k^2x^2)}}$$

is the complete elliptic integral of the first kind whose module is $k = ip/\sqrt{V+p^2}$. Equations (2.5) and (2.10) can also be written in terms of $K(k)$ and the complete elliptic integral of the second kind

$$E(k) = \int_0^1 \sqrt{\frac{1-k^2x^2}{1-x^2}}\,dx$$

(see [**14**]).

Let us express $Y(\tau, E(x), x)$ in terms of the elliptic sine $\operatorname{sn}(\lambda, k)$. From relations (1.10), (2.6), and (2.12), we have

$$\tau = S' \int_Y^p \frac{dz}{\sqrt{(p^2-z^2)(V+p^2+z^2)}} = \frac{\pi}{2} - \frac{S'}{\sqrt{V+p^2}} \int_0^{Y/p} \frac{dz}{\sqrt{(1-z^2)(1+k^2z^2)}},$$

where $k = ip/\sqrt{V+p^2}$. Hence, by (2.14), we have

$$K(k)\left(1 - \frac{2\tau}{\pi}\right) = \int_0^{Y/p} \frac{dz}{\sqrt{(1-z^2)(1+k^2z^2)}},$$

which implies

$$Y(\tau, E(x), x) = p(x)\operatorname{sn}\left(K(k)\left(1-\frac{2\tau}{\pi}\right), k\right). \tag{2.15}$$

REMARK 2.1. Formula (2.15) shows that the function $p(x)$ is the amplitude of the solution. Since $p(x)$ also depends on the parameter c, to define this function, we need some additional condition. For example, in Section 6 we shall use the additional condition posed to normalize the global asymptotic solution.

So we have shown how to calculate all the functions determining the leading term of the asymptotic expansion

$$y_0\left(\frac{S(x)}{\varepsilon}, x\right) = Y\left(\frac{S(x)}{\varepsilon} + \eta(x), E(x), x\right). \tag{2.16}$$

The next term of the asymptotic expansion

$$y_1\left(\frac{S(x)}{\varepsilon}, x\right) = Y_1\left(\frac{S(x)}{\varepsilon} + \eta(x), x\right) \tag{2.17}$$

is also expressed in terms of elliptic functions. We define

$$\operatorname{dn}(\lambda, k) \stackrel{\text{def}}{=} \sqrt{1-k^2\operatorname{sn}^2(\lambda,k)}, \qquad \operatorname{cn}(\lambda,k) \stackrel{\text{def}}{=} \cos\varphi(\lambda,k).$$

Here $\varphi = \varphi(\lambda, k)$ is determined by the equation $\operatorname{sn}(\lambda, k) = \sin\varphi$; in this case $\varphi = 0$ for $\lambda = 0$ and φ increases for $0 \le \lambda \le K(k)$. Moreover, we introduce the Jacobi zeta function

$$Z(\lambda, k) \stackrel{\text{def}}{=} E(\lambda, k) - \lambda\frac{E(k)}{K(k)},$$

where $E(\lambda, k) = \int_0^\lambda \operatorname{dn}^2(\omega, k)\,d\omega$ and $K(k)$, $E(k)$ are complete elliptic integrals of the first and the second kind. We have the following lemma.

LEMMA 2.2 ([**14**]). *The function* $Y_1(\tau, x)$ *has the form*

$$\begin{aligned}Y_1(\tau,x) &= a_1(x)\operatorname{cn}\operatorname{dn}\\ &-\frac{(V+p^2)^{3/2}}{p(V+2p^2)^2}\biggl\{\frac{c_1}{K(k)}\biggl[-\frac{VZ\operatorname{cn}\operatorname{dn}}{(V+p^2)}+\operatorname{sn}\operatorname{dn}^2+\frac{p^4\operatorname{sn}\operatorname{cn}^2}{(V+p^2)^2}\biggr]\\ &+\frac{V'}{2}\biggl[-\frac{VZ^2\operatorname{cn}\operatorname{dn}}{2(V+p^2)}+Z\operatorname{sn}\operatorname{dn}^2+\frac{Zp^4\operatorname{sn}\operatorname{cn}^2}{(V+p^2)^2}+\frac{p^2V\operatorname{sn}^2\operatorname{cn}\operatorname{dn}}{2(V+p^2)^2}\biggr]\biggr\}.\end{aligned}\tag{2.18}$$

Here $a_1(x)$ *is a smooth function. In the notation for* Z, sn, cn, *and* dn, *the argument* $K(k)\bigl(1-2\tau/\pi\bigr)$ *and the module* $k = ip/\sqrt{V+p^2}$ *are omitted.*

The results obtained above imply the following assertion.

THEOREM 2.1. *Suppose that the interval* $x \in [\alpha, \beta]$ *lies strictly between the turning points. We define the amplitude* $p(x)$ *as a solution of equation* (2.5) *and then find the phase* $S(x)$ *and the correction to the phase* $\eta(x)$ *integrating the first order differential equations* (2.14) *and* (2.10). *Then the function*

$$y(x,\varepsilon) = y_0\biggl(\frac{S(x)}{\varepsilon}, x\biggr) + \varepsilon y_1\biggl(\frac{S(x)}{\varepsilon}, x\biggr),$$

where y_0 *and* y_1 *are given by formulas* (2.15)–(2.18), *is a formal asymptotic solution of equation* (2.1) *with accuracy* $O(\varepsilon^2)$.

3. Averaging method for the Painlevé equations. Modulation equation

3.1. The Painlevé equations. The second order equations

$$\frac{d^2y}{dz^2} = R\biggl(z, y, \frac{dy}{dz}\biggr),\tag{3.1}$$

without moving branching points and with R a rational function, are called the *Painlevé equations.* This means that in the complex plane there are finitely many fixed points $z_1, \dots, z_n$ and only these points can be branching points for any solution of equation (3.1).

At the beginning of the 20th century such equations were classified by Painlevé. It turned out that there exist exactly six irreducible types of equations for which not all solutions can be expressed in terms of elementary or known special functions. Moreover, these six types of equations cannot be transformed into one another using the change of the independent variable or of the unknown function.

The Painlevé equations of the first type have the form

$$\frac{d^2y}{dz^2} = 3y^2 + z;\tag{3.2}$$

of the second type they have the form

$$\frac{d^2y}{dz^2} = zy - 2y^3 + \alpha,\tag{3.3}$$

where α is a constant. Since these equations cannot be reduced to simpler ordinary differential equations, they determine some new special functions, the Painlevé transcendents. The Painlevé equations are usually denoted by P_j, $j = 1, 2, \dots, 6$.

Interest in the Painlevé equations has increased recently, since these equations are closely related to nonlinear equations of mathematical physics, which can be integrated by the inverse scattering problem method [**15**]. For example, the Korteweg–de Vries equation

$$\frac{\partial u}{\partial t} - 6u^2\frac{\partial u}{\partial x} + \frac{\partial^3 u}{\partial x^3} = 0 \tag{3.4}$$

has a self-similar solution of the form

$$u(x,t) = \frac{y(z)}{(3t)^{1/3}}, \qquad z = \frac{x}{(3t)^{1/3}}, \tag{3.5}$$

where $y(z)$ is the second Painlevé transcendent. Indeed, if we substitute (3.5) into (3.4), we obtain the equation

$$\frac{d^3y}{dz^3} - 6y^2\frac{dy}{dz} - \frac{d}{dz}(zy) = 0,$$

which coincides with (3.3) after a single integration.

The Painlevé equations, which can be obtained by exact reductions from partial differential equations integrable by the inverse scattering problem method, inherit the extremely rich structure of such equations. Hence, using the methods developed for partial differential equations, it is possible to solve several problems for nonlinear ordinary differential equations, which have been unsolved for a long time.

So, one of the most important problems in the study of the Painlevé equations is to analyze their solutions asymptotically as the independent variable tends to infinity and to obtain connection formulas for the asymptotics at $+\infty$ and $-\infty$ or in sectors of the complex plane as $z \to \infty$. Remarkable progress in the asymptotic study of the Painlevé transcendents is related to the isomonodromy deformation method [**16**]–[**18**], [**1**].

3.2. Averaging of the Painlevé equations. In several papers, the ideas of the Whitham averaging were successfully used in the construction of expansions for the Painlevé equations. For instance, in [**19**], where models of the string theory were studied, the equation

$$\frac{d^2y}{dx^2} = 3y^2 + \varepsilon x, \qquad \varepsilon \to +0, \tag{3.6}$$

was asymptotically studied using the Whitham approach. The character of the solution of this problem is determined by the initial data: $y(0,\varepsilon)$, $dy(0,\varepsilon)/dx$. For a special class of initial data fast oscillating as $\varepsilon \to 0$, by using the scale transformation, one can easily recalculate problem (3.6) supplemented with the condition $\varepsilon x = O(1)$ into the problem of constructing the asymptotics as $z \to \infty$ for the first Painlevé equation (3.2).

The Whitham approach allows one to represent the leading term of the asymptotic expansion of the first transcendent (if the modulus of the independent variable is large) as a specially modulated elliptic function (more precisely, the Weierstrass $\wp$-function for equation P_1). In this case the parameters determining this elliptic function depend on the variable x according to the special differential equation usually called the *modulation equation* or the *Whitham equation.* In [**20, 21**] this method for constructing asymptotics was generalized to the other Painlevé equations.

The Painlevé equations are known as integrable nonlinear ordinary differential equations. Apparently, this fact is best illustrated by the existence of the matrix commutator representation (the Lax pair)

$$\frac{\partial L_j}{\partial \lambda} - \frac{\partial A_j}{\partial x} + [L_j, A_j] = 0, \qquad j = 1, 2, \dots, 6, \tag{3.7}$$

where $L_j = L_j(y, y', x, \lambda)$ and $A_j = A_j(y, y', x, \lambda)$ are 2×2 matrices rationally depending on the spectral parameter λ so that the jth Painlevé equation $y'' = P_j(y, y', x)$ is equivalent to system (3.7). From now on, the prime denotes differentiation with respect to x. The explicit form of the matrices L_j and A_j is known and is presented, e.g., in [**16**].

The statements obtained below are of the same character for all six Painlevé equations and will be illustrated in detail for equation P_1, which technically is the simplest case. Hence the matrices L_1 and A_1 have the form

$$L_1 = \begin{pmatrix} 0 & 1 \\ y - \lambda & 0 \end{pmatrix}, \qquad A_1 = \begin{pmatrix} -y' & 2y + 4\lambda \\ -x - y^2 + 2y\lambda - 4\lambda^2 & y' \end{pmatrix}. \tag{3.8}$$

Now let us construct asymptotic solutions of the Painlevé equations. We introduce an additional variable $X = X(x)$ and replace the variable x, which is explicitly contained in the matrices L_j and A_j, by X:

$$L_j = L_j(y, y', X, \lambda), \qquad A_j = A_j(y, y', X, \lambda).$$

LEMMA 3.1 ([**20**]). *Suppose that ε is some positive number and j is one of the numbers $1, \dots, 6$. Then the relation*

$$\varepsilon \frac{\partial L_j(y, y', X, \lambda)}{\partial \lambda} - \frac{\partial A_j(y, y', X, \lambda)}{\partial x} + \big[L_j(y, y', X, \lambda), A_j(y, y', X, \lambda)\big] = 0 \tag{3.9}$$

is equivalent to the system

$$\frac{\partial X}{\partial x} = \varepsilon, \tag{3.10}$$

$$y'' = P_j(y, y', X), \tag{3.11}$$

where $P_j(y, y', x)$ is the right-hand side of the jth Painlevé equation.

The lemma is proved by direct calculations. So for P_1, by (3.8), we obtain

$$\varepsilon \frac{\partial L_1}{\partial \lambda} = \begin{pmatrix} 0 & 0 \\ -\varepsilon & 0 \end{pmatrix}, \qquad \frac{\partial A_1}{\partial x} = \begin{pmatrix} -y'' & 2y' \\ -\partial X/\partial x - 2yy' + 2y'\lambda & y'' \end{pmatrix},$$

$$L_1 A_1 - A_1 L_1 = \begin{pmatrix} -X - 3y^2 & 2y' \\ -2yy' + 2\lambda y' & X + 3y^2 \end{pmatrix}.$$

Hence, for $j = 1$, system (3.10), (3.11) becomes

$$\frac{\partial X}{\partial x} = \varepsilon, \qquad y'' = 3y^2 + X. \tag{3.12}$$

If in (3.9) we set $\varepsilon = 0$ and $X = \text{const}$, then (3.9) becomes equivalent to (3.11), where $X = \text{const}$. So we have the following assertion.

LEMMA 3.2 ([**20**]). *For $X = \text{const}$ the solution of equation (3.11) can be represented by the formula*

$$y_0(x) = f_j(xU + \varphi; \vec{g}), \tag{3.13}$$

where f_j are periodic functions, which can be expressed in terms of the elliptic Weierstrass or Jacobi functions (see [**13**]*) for each of the six Painlevé equations; $U = U(X)$, $\varphi = \varphi(X)$; and $\vec{g} = \vec{g}(X)$ are parameters determining the elliptic function f_j.*

The statement of the lemma follows from the theory of elliptic functions [**13**]. For the first Painlevé equation, multiplying (3.12) by $2y'$ and integrating the obtained relation, we have

$$\left(\frac{dy}{dx}\right)^2 = 2y^3 + 2Xy + F_1. \tag{3.14}$$

Here $F_1 = F_1(X)$. The solution of equation (3.14) is expressed in terms of the Weierstrass $\wp$-function as follows:

$$y_0(x) = 2\wp(x + \varphi; g_2, g_3), \tag{3.15}$$

where $g_2 = -X$ and $g_3 = -F_1/4$. Indeed, the $\wp$-function, which can be determined by inverting the integral

$$z = \int_{\wp(z;g_2,g_3)}^{\infty} (4t^3 - g_2 t - g_3)^{-1/2}\, dt, \tag{3.16}$$

satisfies the differential equation

$$\left(\frac{dy}{dz}\right)^2 = 4y^3 - g_2 y - g_3.$$

Note that in formula (3.16) any curve that does not pass through the roots of the Weierstrass polynomial $R_3(t) = 4t^3 - g_2 t - g_3$ can be taken for the integration path.

Now we assume that the variable x tends to infinity. We introduce the fast variable $\tau = xU(X)$, which, according to (3.10), can be written as $\tau = S(X)/\varepsilon$, where ε is a small positive number and $S(X) = XU(X)$. So the function y_0 in (3.13) takes the form

$$y_0(\tau, X) = f_j\left(\frac{S(X)}{\varepsilon} + \varphi(X); \vec{g}(X)\right). \tag{3.17}$$

Furthermore, assuming that ε is positive and small, we seek the formal asymptotic solution of the system (3.10), (3.11) in the form

$$y = y_0(x) + \varepsilon y_1(x) + O(\varepsilon^2), \qquad \varepsilon \to +0. \tag{3.18}$$

In the expansion (3.18), the leading term y_0 is determined by (3.17), where the parameters $\vec{g}(X)$, which determine the elliptic functions f_j, are subject to some special ordinary differential equations. These equations are called the *modulation equations.* We have the following assertion.

LEMMA 3.3. *The modulation equations for the Painlevé equations have the form*

$$\frac{\partial}{\partial X}\det A_j = a_{22}\overline{\frac{\partial l_{11}}{\partial \lambda}} + a_{11}\overline{\frac{\partial l_{22}}{\partial \lambda}} - a_{12}\overline{\frac{\partial l_{21}}{\partial \lambda}} - a_{21}\overline{\frac{\partial l_{12}}{\partial \lambda}}, \qquad j = 1, 2, \dots, 6, \tag{3.19}$$

where $A_j = (a_{mn})$ and $L_j = (l_{mn})$; $m, n = 1, 2$. The bar over a symbol in (3.19) denotes the result of averaging over the period of the elliptic function $f_j(\tau + \varphi; \vec{g})$ (over the fast variable τ).

PROOF. It is easy to see that equation (3.9) with $\varepsilon = 0$ implies that the spectral characteristics of the matrix A_j are independent of the variable x. In particular, the condition $\partial \det A_j/\partial x = 0$ holds.

For the small parameter $\varepsilon > 0$, equation (3.9) ensures the relation

$$(a_{nm})' = \varepsilon \frac{\partial l_{nm}}{\partial \lambda} + [L_j, A_j]_{nm}, \tag{3.20}$$

where $a_{nm} = (A_j)_{mn}$ and $l_{nm} = (L_j)_{mn}$; $n, m = 1, 2$. Next, substituting relations (3.20) in the identity

$$\frac{\partial}{\partial X} \det A_j = a'_{11}a_{22} + a_{11}a'_{12} - a'_{12}a_{21} - a_{12}a'_{21}$$

and using the rule of differentiation of functions of the form (3.17):

$$\frac{\partial f}{\partial x} = \frac{\partial S}{\partial X}\frac{\partial f}{\partial \tau} + \varepsilon \frac{\partial f}{\partial X},$$

we obtain the equation

$$\begin{aligned} &\left(\frac{\partial S}{\partial X}\frac{\partial}{\partial \tau} + \varepsilon \frac{\partial}{\partial X}\right) \det A_j \\ &\quad = \varepsilon\left(a_{22}\frac{\partial l_{11}}{\partial \lambda} + a_{11}\frac{\partial l_{22}}{\partial \lambda} - a_{12}\frac{\partial l_{21}}{\partial \lambda} - a_{21}\frac{\partial l_{12}}{\partial \lambda}\right) + O(\varepsilon^2). \end{aligned} \tag{3.21}$$

It remains to perform the averaging in (3.21), i.e., to integrate over the period (with respect to the fast variable τ). This procedure annihilates the terms that are the total derivatives in τ. The proof of the lemma is complete. □

It follows from the modulation equations that there exists only one coefficient of the polynomial (in λ) $\det A_j$ whose dynamics with respect to the variable X is not trivial (the sixth Painlevé equation is an exception). We denote this coefficient by F_j. So the modulation equations can be written by using the single first order ordinary differential equation for $F_j = F_j(X)$. This can be shown by straightforward calculations in each of the cases $j = 1, 2, \ldots, 5$.

So for P_1, by (3.8) and (3.14), we have

$$\det A_1 = 16\lambda^3 + 4X\lambda - F_1, \tag{3.22}$$

where $F_1 = (y')^2 - 2y^3 - 2yX$. In this case the modulation equation (3.19) takes the form

$$\frac{\partial}{\partial X} \det A_1 = 4\lambda + 2\overline{y}. \tag{3.23}$$

Because relation (3.22) implies

$$\frac{\partial}{\partial X} \det A_1 = 4\lambda - \frac{\partial F_1}{\partial X},$$

equation (3.23) can be rewritten as

$$\frac{\partial F_1}{\partial X} = -2\overline{y}. \tag{3.24}$$

For the solution given by (3.15), after the averaging over the period of the $\wp$-function, from (3.24) we obtain the explicit form of the modulation equation:

$$\frac{\partial F_1}{\partial X} = -\frac{2\eta}{\omega} = 2e_1 + 2(e_3 - e_1)\frac{E(k)}{K(k)}. \tag{3.25}$$

Here $K(k)$, $E(k)$ are complete elliptic integrals, $k^2 = (e_2 - e_3)/(e_1 - e_3)$; e_1, e_2, e_3 are the roots of the Weierstrass polynomial $R_3(t) = 4t^3 - g_2 t - g_3$; $g_2 = -X$ and $g_3 = -F_1/4$; and 2ω is the period of the Weierstrass function $\wp$. Moreover, $\eta = \zeta(\omega)$, where ζ is the Weierstrass zeta function [**13**].

So, in the case of P_1, the modulation equation (3.25) determines the function $F_1(X)$.

REMARK 3.1. The simplest method for calculating the elliptic ansatz is to solve the equations

$$F_j = c_1, \qquad X = c_2,$$

where c_1, c_2 are constants.

Finally, we have the following theorem.

THEOREM 3.1 ([**20**]). *The function y_0 given by* (3.13), (3.19) *describes the leading term of the asymptotic expansion for the solution of the corresponding Painlevé equation as the variable x tends to infinity.*

Note that in the proof of the theorem we use the scale transformation under which the expansion (3.18) with respect to the small parameter ε becomes a series in inverse powers of the large variable x.

REMARK 3.2. In [**22**] the Cauchy problem for a nonlinear ordinary differential equation of the form

$$y'' + A(y, y', x) = \varepsilon B(y, y', x), \qquad x > 0, \quad 0 < \varepsilon \ll 1,$$
$$y\big|_{x=0} = y_0, \qquad y'\big|_{x=0} = y_1$$

was considered. The case in which the exact solution has a finite order singularity and the singular point depends on a small parameter ε was studied. A method similar to the Kuzmak–Whitham method was proposed. Using this method, one can construct an asymptotics on the interval from zero to the first moving singular point. On the basis of the expansions thus obtained, it is possible to construct an asymptotic approximation of the solution with arbitrary accuracy outside an arbitrarily narrow (with respect to ε) neighborhood of the singular point. This method can be applied to the Painlevé equations, as well as to the Liouville equation.

4. Asymptotic behavior of singular-at-zero solutions of the Painlevé equation of the third kind

Now we apply the above results to the equation

$$\frac{d^2y}{dx^2} + \frac{1}{x}\frac{dy}{dx} + \sin y = 0. \tag{4.1}$$

This is a special case of the third Painlevé equation in the most frequently used form. The following problem can be considered as an example.

It is well known that the elliptic sine-Gordon equation

$$\frac{\partial^2 y}{\partial t^2} + \frac{\partial^2 y}{\partial z^2} + \sin y + f = 0 \tag{4.2}$$

has a class of solutions singular at zero with singularities of the form

$$y = \alpha \log\frac{1}{x} + \beta + o(1), \qquad x^2 = z^2 + t^2 \to 0, \tag{4.3}$$

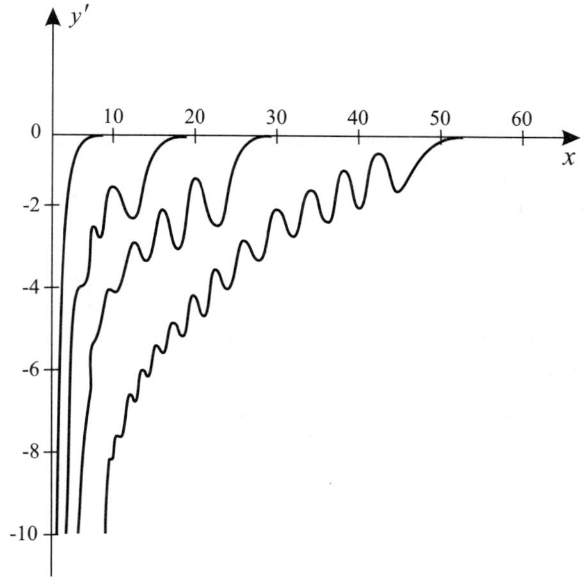

FIGURE 2

where α, β are parameters. This class is distinguished among other solutions, since it is important for applications [**23**]. For examples, solutions with singularities (4.3) can be used to describe bound states in field theory.

In [**23**] the isomonodromy deformation method was used for studying the case of the radially symmetric equation (4.2):

$$\frac{d^2y}{dx^2} + \frac{1}{x}\frac{dy}{dx} + \sin y + f = 0, \tag{4.4}$$

which coincides with (4.1) for $f = 0$. It turned out that solutions of equation (4.4) with singularities of the form (4.3) at the origin tend to the asymptotic limit at infinity:

$$y(x) = 2\pi k + \gamma\sqrt{\frac{\pi}{2x}}e^{-x}\Big[1 + O\Big(\frac{1}{x}\Big)\Big], \qquad x \to +\infty,$$

where k is an integer and $\gamma = \text{const}$. The behavior of these solutions in finite domains is of special interest, namely, these solutions form a wave packet of soliton series type at distances that are not very large. Numerical experiments [**23**] show that these oscillations have the form shown in Figure 2. In this figure, different curves correspond to different values of the parameter α in formula (4.3). One can see that the solution forms the typical "Whitham" zone of oscillations with increasing α. We shall construct the asymptotics of singular solutions of equation (4.1) as $\alpha \to \infty$ in a domain of finite values of the variable $X = x/|\alpha|$ in the form of a modulated elliptic function determining the above oscillations.

Let ε be a small positive number. We shall seek the solution of system (3.10), (3.11) in the form of the expansion

$$y = \varepsilon^{-1}y_{-1}(X) + y_0(x) + \varepsilon y_1(x) + O(\varepsilon^2) \tag{4.5}$$

such that the parameters $\vec{g}(X)$ determining the function $y_0 = f_3$ satisfy the modulation equation (3.19). The leading term of the asymptotics can be easily found by substituting expansion (4.5) into equation (4.1). Integrating the equation

$$y''_{-1} + \frac{1}{x} y'_{-1} = 0,$$

we find $y_{-1} = \log X$, where $X = \varepsilon x$. From now on, the prime denotes differentiation with respect to x.

Then we shall construct the next term of expansion (4.5) in the form

$$y_0(\tau, X) = f_3\Big(\frac{S(X)}{\varepsilon} + \varphi(X); \vec{g}(X)\Big).$$

In this case we restrict ourselves only to stating the results obtained in [**21**].

Since for the third Painlevé equation (4.1) we have

$$A_3 = i\Big(\frac{\cos y}{\lambda^2} - \frac{x^2}{16}\Big)\sigma_3 - \frac{ixy'}{4\lambda}\sigma_1 + \frac{i \sin y}{\lambda^2}\sigma_2, \qquad L_3 = -\frac{ix\lambda}{8}\sigma_3 - \frac{iy'}{2}\sigma_1,$$

where

$$\sigma_1 = \begin{pmatrix} 0 & 1 \\ 1 & 0 \end{pmatrix}, \qquad \sigma_2 = \begin{pmatrix} 0 & i \\ -i & 0 \end{pmatrix}, \qquad \sigma_3 = \begin{pmatrix} 1 & 0 \\ 0 & -1 \end{pmatrix},$$

equation (3.9) is equivalent to the system

$$X' = \varepsilon, \qquad X(y'' + \sin y) + X'y' = 0. \tag{4.6}$$

A direct calculation of the determinant of the matrix A_3 implies

$$\lambda^4 \det A_3 = \frac{\lambda^4 X^4}{256} - \frac{\lambda^2 X^2}{8} F_3 + 1, \tag{4.7}$$

where the parameter F_3 appears to be the constant of integration for the second equation in system (4.6) with $\varepsilon = 0$:

$$F_3 = \cos y - \frac{1}{2}(y')^2. \tag{4.8}$$

Substituting concrete matrix entries into relation (3.19), we obtain

$$\frac{\partial}{\partial X} \det A_3 = \frac{X^3}{64} - \frac{X \overline{\cos y}}{4\lambda^2}.$$

On the other hand, differentiating relation (4.7) with respect to X, we derive the modulation equation explicitly. So we have the following statement.

THEOREM 4.1 ([**21**]). *The modulation equation* (3.19) *ensures the condition*

$$(\varepsilon^{-1} y_{-1} + y_0)'' + \frac{1}{x}(\varepsilon^{-1} y_{-1} + y_0)' + \sin(\varepsilon^{-1} y_{-1} + y_0) = O(\varepsilon), \qquad \varepsilon \to +0,$$

and can be rewritten as

$$\frac{\partial F_3}{\partial X} = -\frac{2}{X}(\overline{\cos y} - F_3) = \frac{1}{X}\Big[\frac{17}{16}(F_3 + \sqrt{F_3^2 - 1})\frac{E(k)}{K(k)} - 2F_3\Big], \tag{4.9}$$

where $E(k), K(k)$ are complete elliptic integrals,

$$k^2 = k^2(F_3) = \frac{2\sqrt{F_3^2 - 1}}{F_3 + \sqrt{F_3^2 - 1}}.$$

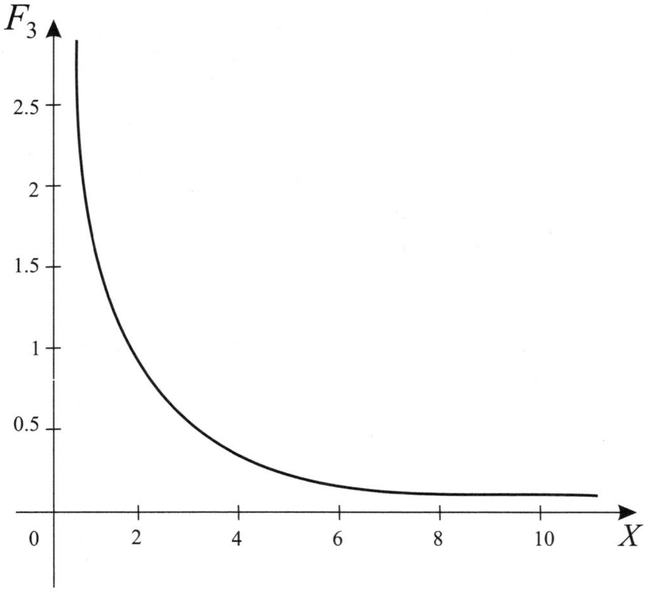

FIGURE 3

REMARK 4.1. The general solution of the differential equation (4.9) is a solution of the equation

$$\int_{F_{3,0}}^{F_3(X)} \frac{df}{\dfrac{17}{16}(f+\sqrt{f^2-1})\dfrac{E(k(f))}{K(k(f))}-2f} = \log X,$$

where $F_{3,0}$ is a constant. The typical behavior of the parameter F_3 as a function of X is shown in Figure 3.

Finally, the elliptic ansatz (3.13) is determined by the condition (4.8). We have

$$y_0 = \pi(2n+1) - i\log\{4\wp(x+\varphi; g_2, g_3)\}, \qquad n \in \mathbb{Z}, \tag{4.10}$$

where $\wp$ is the Weierstrass function determined by the parameters

$$g_2 = \frac{1}{3}F_3^2 - \frac{1}{4}, \qquad g_3 = \frac{1}{24}F_3 - \frac{1}{27}F_3^3.$$

Part 2. Global Asymptotic Solutions of Equations with Cubic Nonlinearity

5. Isomonodromy deformation method. Connection formulas for the second Painlevé transcendent

5.1. Monodromy data. We consider the second Painlevé equation with a parameter

$$\varepsilon^2 \frac{d^2y}{dx^2} - xy + 2y^3 = 0, \qquad x \in \mathbb{R}^1. \tag{5.1}$$

Such an equation describes a nonlinear analog of the Airy functions and is the Painlevé equation most frequently encountered in physical applications (e.g., see [**24**]–[**27**]).

Analyzing the results of numerical integration, Miles [**28**], [**29**] conjectured that the solution of the Painlevé equation exponentially decreasing on the right of the turning point has the Kuzmak–Whitham asymptotics on the left of the turning point and the phase shift in this asymptotics is equal to $\pi/2$ at the turning point. We shall prove this conjecture using the isomonodromy deformation method. In what follows, the connection formulas for the asymptotics obtained here will help us to construct the global asymptotic solution of equation (2.1).

In the framework of the isomonodromy deformation method, the following linear system is associated with (5.1):

$$\varepsilon\frac{d\Psi}{d\lambda} = A(\lambda)\Psi, \qquad \lambda \in \mathbb{C}, \tag{5.2}$$

where

$$A(\lambda) = \begin{pmatrix} -4i\lambda^2 - ix + 2iy^2 & 4y\lambda + 2i\varepsilon y' \\ -4y\lambda + 2i\varepsilon y' & 4i\lambda^2 + ix - 2iy^2 \end{pmatrix}. \tag{5.3}$$

This is a system of matrix dimension 2×2 with a single irregular singular point at infinity. The exact sense of the relation between the nonlinear equation (5.1) and the linear system (5.2) was explained in [**16**].

THEOREM 5.1. *The monodromy data for system* (5.2) *are independent of x if and only if $y(x)$ is a smooth solution of equation* (5.1).

The monodromy data for system (5.2) are exhausted by the set of the Stokes matrices $S_k = \Psi_k^{-1}\Psi_{k+1}$, $k = 1, \dots, 6$. They relate the canonical solutions of (5.2) $\Psi_k(\lambda)$, $k = 1, \dots, 7$, which as $\lambda \to \infty$,

$$\lambda \in \widehat{\Omega}_k = \left\{\lambda : \frac{\pi(k-2)}{3} < \arg\lambda < \frac{\pi k}{3}\right\}, \qquad k = 1, \dots, 7,$$

have the asymptotics

$$\Psi_k(\lambda) = \left(I + O\left(\frac{1}{\lambda}\right)\right)\begin{pmatrix} \exp(-4i\lambda^3/(3\varepsilon) - ix\lambda/\varepsilon) & 0 \\ 0 & \exp(4i\lambda^3/(3\varepsilon) + ix\lambda/\varepsilon) \end{pmatrix}. \tag{5.4}$$

Here I is the unit matrix. All $\Psi_k(\lambda)$ are entire functions of the variable λ. Therefore, in particular, $\Psi_1(\lambda) \equiv \Psi_7(\lambda)$.

We shall describe the monodromy data for system (5.2) in detail. We shall consider only the real solutions of the Painlevé equation (5.1), which exponentially decrease on the right of the turning point. Then it follows from the structure of $A(\lambda)$ that the Stokes matrices are parametrized by one real parameter g and have the form [**30**]

$$\begin{gathered} S_1 = \begin{pmatrix} 1 & 0 \\ g & 1 \end{pmatrix}, \quad S_3 = \begin{pmatrix} 1 & 0 \\ -g & 1 \end{pmatrix}, \quad S_4 = \begin{pmatrix} 1 & -g \\ 0 & 1 \end{pmatrix}, \quad S_6 = \begin{pmatrix} 1 & g \\ 0 & 1 \end{pmatrix}, \\ S_2 = I, \qquad S_5 = I. \end{gathered} \tag{5.5}$$

This description of the monodromy data allows us to specify the statement of Theorem 5.1. Namely, the real value $g = g(x, y, \varepsilon y')$ parametrizing the monodromy data is the first integral of the Painlevé equation (5.1).

In what follows, we construct the asymptotics of solutions of system (5.2). This allows us to express the function g first in terms of the Kuzmak–Whitham asymptotics for equation (5.1), which holds on the left of the turning point, and then in terms of the asymptotics exponentially decreasing on the right of the turning point. Since the function g is the first integral of equation (5.1), comparing the expressions for g calculated by two methods, we can obtain connection formulas for the asymptotics on both sides of the turning point.

5.2. Calculation of the first integral for the Kuzmak–Whitham expansion. Let us consider expansion (2.2) for the Painlevé equation (5.1). The formulas for this expansion follow from formulas (2.5), (2.6), (2.10), (2.15)–(2.18) obtained in Section 2 for equation (2.1) if we set $V(x) = -x$ in these formulas. In this case, in contrast to equation (2.1), equation (5.1) will have only one turning point $x_* = (3c)^{2/3}$ determined by relation (2.9). Here c is the constant contained in the equation

$$\int_0^p \sqrt{(p^2 - z^2)(-x + p^2 + z^2)}\, dz = c. \tag{5.6}$$

The Kuzmak–Whitham expansion holds on the left of the turning point x_* whose position depends on the value of c.

Now let us calculate the first integral g. To this end, we construct the asymptotics of the solution of system (5.2). We begin by calculating the eigenvalues of the matrix A, which satisfy the equation

$$\det(A - \mu I) = 0. \tag{5.7}$$

Solving this equation, we obtain the following statement.

LEMMA 5.1. *The eigenvalues of the matrix A given by* (5.3) *have the form* $\mu_1 = \mu$, $\mu_2 = -\mu$, *where*

$$\mu = 4i\left(\left(\lambda^2 + \frac{x}{4}\right)^2 + \frac{\varepsilon^2 (y')^2 - xy^2 + y^4}{4}\right)^{1/2}. \tag{5.8}$$

The turning points of the linear system (5.2) are determined by the condition $\mu_1 = \mu_2$, which implies the relation $\mu = 0$. The WKB-approximation does not hold near these points.

If the Kuzmak–Whitham expansion (2.2) holds for $y(x, \varepsilon)$, then integrating (2.3) (where $V(x) = -x$) once with the initial conditions $y_0 = p(x)$ and $\partial y_0/\partial\tau = 0$ for $\tau = -\eta(x)$, we have

$$\varepsilon^2 (y')^2 - xy^2 + y^4 = (S')^2 \left(\frac{\partial y_0}{\partial \tau}\right)^2 - xy_0^2 + y_0^4 + O(\varepsilon) = p^2(-x + p^2) + O(\varepsilon).$$

Here the function $p = p(x)$ satisfies equation (5.6) and hence, in the domain where the Kuzmak–Whitham expansion holds, the equation $\mu = 0$ takes the form

$$(\lambda^2 + x/4)^2 + p^2(-x + p^2)/4 + O(\varepsilon) = 0.$$

Its solutions with respect to λ determine four turning points $\lambda_1, \lambda_2, \lambda_3, \lambda_4$, where $\lambda_1 = (\sqrt{-x + p^2} + ip)/2 + O(\varepsilon)$, $\lambda_2 = \overline{\lambda}_1$, $\lambda_3 = -\lambda_1$, and $\lambda_4 = -\overline{\lambda}_1$. (They are shown in Figure 4.) If x is finite and $x < x_* - h$, where $h > 0$, then the turning points are simple.

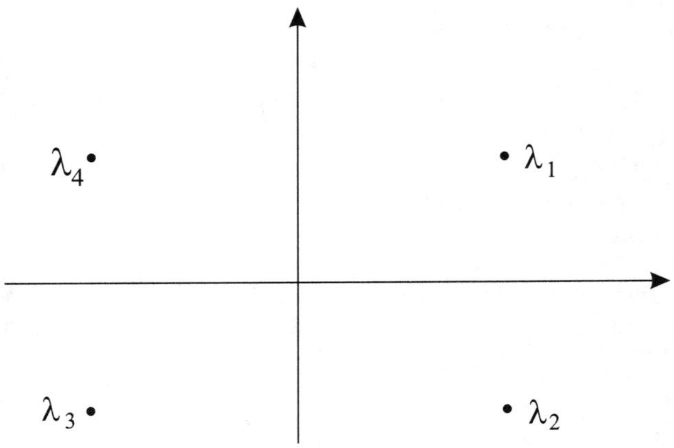

FIGURE 4

We shall construct the WKB-approximation of system (5.2). To this end, we reduce this system to the diagonal form [**31**]. Namely, we define $\Lambda_0 = \begin{pmatrix} -\mu & 0 \\ 0 & \mu \end{pmatrix}$ and assume that the matrix T_0 reduces the matrix A to the diagonal form: $T_0^{-1}AT_0 = \Lambda_0$. If we set $\Psi = T_0(\lambda)\tilde{\varphi}$, then

$$\varepsilon\frac{d\tilde{\varphi}}{d\lambda} = (\Lambda_0 + \varepsilon B_1)\tilde{\varphi}, \qquad \text{where } B_1 = -T_0^{-1}\frac{dT_0}{d\lambda}.$$

Next, we seek $\tilde{\varphi}$ in the form $\tilde{\varphi} = \big(I + \varepsilon T_1 + \varepsilon^2 T_2 + O(\varepsilon^3)\big)\varphi$, where T_1, T_2 are the desired matrices. Then φ satisfies the equation

$$\begin{aligned} \varepsilon\frac{d\varphi}{d\lambda} &= \big(I + \varepsilon T_1 + \varepsilon^2 T_2 + O(\varepsilon^3)\big)^{-1}(\Lambda_0 + \varepsilon B_1)\big(I + \varepsilon T_1 + \varepsilon^2 T_2 + O(\varepsilon^3)\big)\varphi \\ &\quad - \varepsilon\big(I + O(\varepsilon)\big)^{-1}\Big(\varepsilon\frac{dT_1}{d\lambda} + O(\varepsilon^2)\Big)\varphi. \end{aligned} \tag{5.9}$$

We choose T_1, T_2 so that

$$\varepsilon\frac{d\varphi}{d\lambda} = \big(\Lambda_0 + \varepsilon\Lambda_1 + \varepsilon^2\Lambda_2 + O(\varepsilon^3)\big)\varphi, \tag{5.10}$$

where $\Lambda_0, \Lambda_1, \Lambda_2$ are diagonal matrices. Comparing equations (5.9) and (5.10), we obtain the following equation for T_1:

$$\Lambda_0 T_1 - T_1\Lambda_0 = \Lambda_1 - B_1. \tag{5.11}$$

Since the matrix on the left-hand side of (5.11) has zeros on its main diagonal, for equation (5.11) to be solvable, it is necessary to set $\Lambda_1 = \operatorname{diag} B_1$, where diag is an operator that reduces matrices so that all entries except those lying on the main diagonal are zero. Then the solution of equation (5.11) is given by the formula

$$T_1 = \frac{1}{2\mu}\begin{pmatrix} 0 & (B_1)_{12} \\ -(B_1)_{21} & 0 \end{pmatrix}.$$

The equation for T_2 has the form

$$\Lambda_0 T_2 - T_2\Lambda_0 = \Lambda_2 + \frac{dT_1}{d\lambda} + T_1(\Lambda_1 - B_1),$$

and thus we have the matrix

$$\Lambda_2 = -\operatorname{diag}\left(\frac{dT_1}{d\lambda} + T_1(\Lambda_1 - B_1)\right) = \frac{1}{2\mu}(B_1)_{12}(B_1)_{21}\begin{pmatrix}1 & 0\\ 0 & -1\end{pmatrix}.$$

So we have $\Psi = T_0(I + \varepsilon T_1 + O(\varepsilon^2))\varphi$, where φ can be found from system (5.10), which can be split. Integrating this system, we obtain the following assertion.

LEMMA 5.2. *The WKB-approximation for system* (5.2) *has the form*

$$\Psi = T_0\Big(I + O(\varepsilon T_1) + O\Big(\varepsilon\int_\gamma^\lambda \Lambda_2\,d\lambda\Big)\Big)\exp\left(\frac{1}{\varepsilon}\int_\gamma^\lambda \Lambda_0\,d\lambda + \int_\gamma^\lambda \Lambda_1\,d\lambda\right)C_0, \tag{5.12}$$

where γ is independent of λ and C_0 is a matrix constant with respect to λ.

Let us write the matrices contained in (5.12) in more detail via the elements of the matrix $A = \begin{pmatrix}a_{11} & a_{12}\\ a_{21} & a_{22}\end{pmatrix}$ and μ. We choose T_0 in the form

$$T_0 = \begin{pmatrix}(a_{11}-\mu)/a_{21} & 1\\ 1 & a_{21}/(a_{11}+\mu)\end{pmatrix}. \tag{5.13}$$

Then

$$\begin{aligned}B_1 &= -T_0^{-1}\frac{dT_0}{d\lambda}\\ &= \begin{pmatrix}\dfrac{a_{21}}{2\mu}\dfrac{d}{d\lambda}\left(\dfrac{a_{11}}{a_{21}}\right) - \dfrac{1}{2}\dfrac{d}{d\lambda}\ln\left(\dfrac{\mu}{a_{21}}\right) & -\dfrac{(a_{11}+\mu)}{2\mu}\dfrac{d}{d\lambda}\left(\dfrac{a_{21}}{a_{11}+\mu}\right)\\ -\dfrac{(a_{11}+\mu)}{2\mu}\dfrac{d}{d\lambda}\left(\dfrac{a_{11}-\mu}{a_{21}}\right) & \dfrac{a_{21}}{2\mu}\dfrac{d}{d\lambda}\left(\dfrac{a_{11}}{a_{12}}\right) - \dfrac{1}{2}\dfrac{d}{d\lambda}\ln\left(\dfrac{\mu}{a_{12}}\right)\end{pmatrix}.\end{aligned}$$

Thus, taking (5.7) into account, we have

$$\Lambda_1 = \operatorname{diag} B_1 = \begin{pmatrix}-b + \dfrac{d}{d\lambda}\ln\left(\dfrac{\sqrt{a_{12}}a_{21}}{\sqrt{\mu}(a_{11}-\mu)}\right) & 0\\ 0 & b + \dfrac{d}{d\lambda}\ln\sqrt{\dfrac{a_{12}}{\mu}}\end{pmatrix}, \tag{5.14}$$

where

$$\begin{aligned}b &\overset{\text{def}}{=} \frac{a_{12}}{2\mu}\frac{d}{d\lambda}\frac{a_{11}}{a_{12}}\\ &= -\frac{\lambda^2 - x/4 + y^2/2 + i\varepsilon y^{-1}y'\lambda}{(2\lambda + i\varepsilon y^{-1}y')((\lambda^2 + x/4)^2 + (\varepsilon^2(y')^2 - xy^2 + y^4)/4)^{1/2}}.\end{aligned} \tag{5.15}$$

The matrices T_1 and Λ_2 can be written similarly (see [**14**]). Estimating the singularities arising at the turning points of T_1 and Λ_2, as well as relations (5.13) and (5.14), we can write down formula (5.12) in more detail.

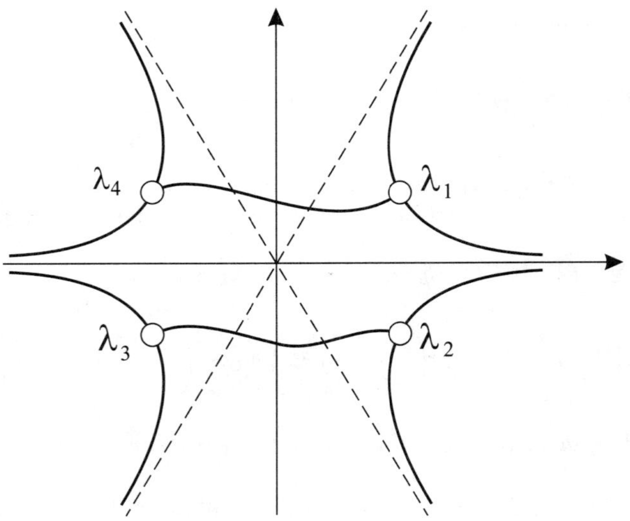

FIGURE 5

LEMMA 5.3. *Suppose that the Kuzmak–Whitham expansion is valid for $y(x)$. Then the WKB-approximation for system* (5.2) *has the form*

$$\Psi = \frac{1}{\sqrt{\mu a_{12}}}\begin{pmatrix} a_{12} & a_{12} \\ -(\mu + a_{11}) & \mu - a_{11}\end{pmatrix}$$

$$\times \begin{pmatrix} \exp\Big(-\frac{1}{\varepsilon}\int_{\lambda_1}^{\lambda}\mu\,d\lambda - \int_{\lambda_1}^{\lambda} b\,d\lambda\Big) & 0 \\ 0 & \exp\Big(\frac{1}{\varepsilon}\int_{\lambda_1}^{\lambda}\mu\,d\lambda + \int_{\lambda_1}^{\lambda} b\,d\lambda\Big)\end{pmatrix} \tag{5.16}$$

$$\times\, C_0\Big(I + O(\varepsilon) + \sum_{j=1}^{4} O(\varepsilon(\lambda-\lambda_j)^{-3/2})\Big).$$

Here C_0 is a matrix constant with respect to λ, a_{ij} are the entries of the matrix A given by (5.3), *and the functions μ, b are given by formulas* (5.8), (5.15).

Now let us determine the domains where the WKB-asymptotics constructed above can be applied. To this end, it is necessary to study the structure of the Stokes lines determined by the relation $\operatorname{Re}\big(\varepsilon^{-1}\int_{\lambda_j}^{\lambda}\mu\,d\lambda\big) = 0$. Since

$$\int_{\lambda_j}^{\lambda}\mu\,d\lambda = \int_{\lambda_j}^{\lambda}\big(4i\lambda^2 + ix + O(\lambda^{-2})\big)\,d\lambda = 4i\frac{\lambda^3}{3} + ix\lambda + O(1)$$

as $\lambda \to \infty$ and $x = O(1)$, for large λ the Stokes lines tend to the rays $\arg\lambda = \pi n/3$, where n are integers. Since λ_j are simple turning points, three Stokes lines are issued from each point λ_j. Moreover, in [**14**] it is shown that

$$\operatorname{Re}\Big(\frac{1}{\varepsilon}\int_{\lambda_1}^{-\overline{\lambda}_1}\mu\,d\lambda\Big) = 0, \qquad \operatorname{Re}\Big(\frac{1}{\varepsilon}\int_{-\lambda_1}^{\overline{\lambda}_1}\mu\,d\lambda\Big) = 0.$$

The global structure of the Stokes lines is shown in Figure 5.

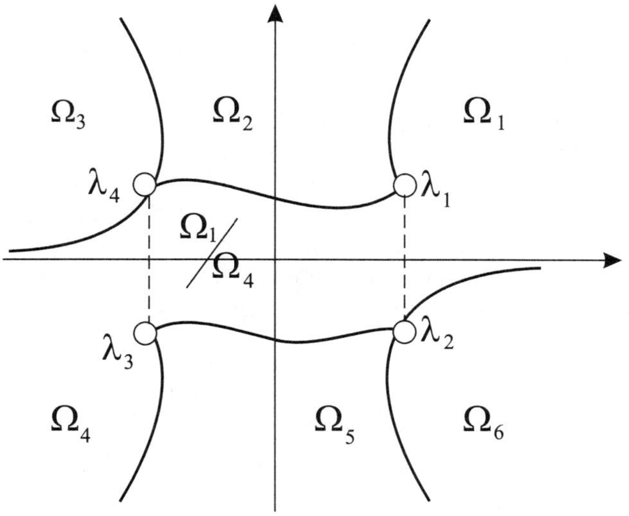

FIGURE 6

It is precisely these Stokes lines that determine the boundaries of the domains where the WKB-approximations hold. On the plane with two cuts (drawn between the points λ_1, λ_2 and λ_3, λ_4 so that we could choose a unique branch of the function μ), we consider six domains Ω_k, $k = 1, \dots, 6$, that are pairwise disjoint and satisfy the condition $\bigcup_k \overline{\Omega}_k = \mathbb{C}$. They are shown in Figure 6, where the cuts are denoted by dashed lines.

In each of the domains Ω_k there exists a canonical solution Ψ_k with asymptotics at infinity given by formula (5.4). This solution can be extended to a wider domain containing the boundary common for Ω_k and Ω_{k-1}. The structure of the Stokes matrix (5.5) shows that $\Psi_1 \equiv \Psi_4$, $\Psi_2 \equiv \Psi_3$, and $\Psi_5 \equiv \Psi_6$, and hence it does not matter whether the solution Ψ_1 or Ψ_4 is chosen near the zero.

To find g, it suffices to determine only the Stokes matrix S_1 after Ψ_1 and Ψ_2 are calculated.

Let us outline further calculations. First, we define

$$\begin{aligned}\Psi_k^{\mathrm{WKB}} &= \frac{1}{\sqrt{\mu a_{12}}}\begin{pmatrix} a_{12} & a_{12} \\ -(\mu + a_{11}) & \mu - a_{11}\end{pmatrix} \\ &\times \begin{pmatrix} \exp\left(-\frac{1}{\varepsilon}\int_{\lambda_1}^{\lambda}\mu\,d\lambda - \int_{\lambda_1}^{\lambda} b\,d\lambda\right) & 0 \\ 0 & \exp\left(\frac{1}{\varepsilon}\int_{\lambda_1}^{\lambda}\mu\,d\lambda + \int_{\lambda_1}^{\lambda} b\,d\lambda\right)\end{pmatrix} \\ &\times \Big(I + O(\varepsilon) + \sum_{j=1}^{4} O(\varepsilon(\lambda - \lambda_j)^{-3/2})\Big),\end{aligned}$$

where $k = 1, 2$. Now we choose matrices C_k, which are constant with respect to λ, so that $\Psi_k^{\mathrm{WKB}} C_k = \Psi_k$. Here Ψ_k, $k = 1, 2$, are canonical solutions with asymptotics (5.4) as $\lambda \to \infty$ in the corresponding domain Ω_k.

On the other hand, $\Psi_1^{\mathrm{WKB}} = \Psi N_1$ and $\Psi_2^{\mathrm{WKB}} = \Psi N_2$, where Ψ is the solution of system (5.2) near the turning point λ_1 and N_1, N_2 are matrices constant with respect to λ. Then we have $S_1 = \Psi_1^{-1}\Psi_2 = C_1^{-1}N_1^{-1}N_2C_2$. So the problem of calculating g is reduced to finding the matrices C_1, C_2, N_1, N_2.

Detailed calculations according to this plan were performed in [**14**]. Here we do not present these calculations completely. We only construct the most important asymptotics of Ψ near the turning point λ_1 at which the WKB-approximation cannot be used. To this end, we reduce the first order system (5.2) to second order differential equations. Differentiating the equations in (5.2) with respect to λ, we obtain four equations, which can readily be written as

$$\varepsilon^2 \frac{d^2\Psi_{nm}}{d\lambda^2} = (\mu^2 \pm 8i\varepsilon\lambda)\Psi_{nm} \pm 4\varepsilon y \Psi_{kl}. \tag{5.17}$$

Here Ψ_{nm}, Ψ_{kl} are entries of the matrix Ψ.

Now we introduce a new independent variable $\xi = \theta^{1/3}(\lambda - \lambda_1)/\varepsilon^{2/3}$, where $\theta \overset{\mathrm{def}}{=} -64\lambda_1(\lambda_1^2 + x/4)$. We look for the asymptotics of equation (5.17) as

$$\Psi_{nm} = \varepsilon^{-1/6}\Psi_{nm}^0(\xi) + O(\varepsilon^{1/6}). \tag{5.18}$$

Then $\Psi_{nm}^0(\xi)$ will satisfy the Airy equation

$$\frac{d^2\Psi_{nm}^0}{d\xi^2} - \xi\Psi_{nm}^0 = 0, \qquad \xi \in \mathbb{C}. \tag{5.19}$$

The solution of (5.19) is a linear combinations of the Airy functions: $\Psi_{nm}^0 = v_{nm}\,\mathrm{Ai}(\xi) + w_{nm}\,\mathrm{Bi}(\xi)$. Here v_{nm}, w_{nm} are constants.

For $1 \ll |\xi| \ll \varepsilon^{-2/3}$ both the expansion (5.18) and the WKB-approximation hold. Hence for such ξ we can match the asymptotics (5.18) near the turning point λ_1 and the WKB-approximation in the domain Ω_2 [**32**]. Here we use the well-known asymptotic expansions of the Airy functions $\mathrm{Ai}(\xi)$, $\mathrm{Bi}(\xi)$ in sectors of the complex plane [**33**], [**14**]. These asymptotics are different in different sectors (the Stokes phenomenon takes place). Therefore, after the matrices C_1, C_2, N_1, N_2 are calculated, we see that the canonical solutions Ψ_1, Ψ_2 are different and the Stokes matrix relating them has the form $S_1 = \begin{pmatrix} 1 & 0 \\ g & 1 \end{pmatrix}$, where as $\varepsilon \to +0$ the function $g = g(x, y, \varepsilon y')$ can be represented as

$$\begin{aligned} g = \frac{2\lambda_1^0}{y}(1 + O(\varepsilon^{1/4}))\exp\bigg\{ &-\frac{2i}{\varepsilon}\Big(\frac{4}{3}(\lambda_1^0)^3 + \lambda_1^0 x\Big) \\ &+ \frac{2}{\varepsilon}\int_{\lambda_1^0}^{+\infty}\Big[\mu - 4i\Big(\lambda^2 + \frac{x}{4}\Big)\Big]\,d\lambda + \int_{\lambda_1^0}^{+\infty}\Big(\frac{1}{\lambda} + \frac{(2\lambda + i\varepsilon y^{-1}y')b}{\lambda^2 + i\varepsilon y^{-1}y'/2}\Big)\,d\lambda\bigg\}. \end{aligned} \tag{5.20}$$

Here $\lambda_1^0 = (\sqrt{-x + p^2} + ip)/2$ and the functions μ and b are determined by formulas (5.8) and (5.15).

Next, instead of y, we substitute the Kuzmak–Whitham expansion for the Painlevé equation (5.1) into formula (5.20) and calculate the elliptic integrals thus obtained. There very cumbersome calculations were carried out in [**14**]. As a result, the expression for g is significantly simplified.

THEOREM 5.2. *Let $y(x, \varepsilon)$ be a solution of the Painlevé equation* (5.1) *with the Kuzmak–Whitham asymptotics* (2.2). *Then the value of g parametrizing the*

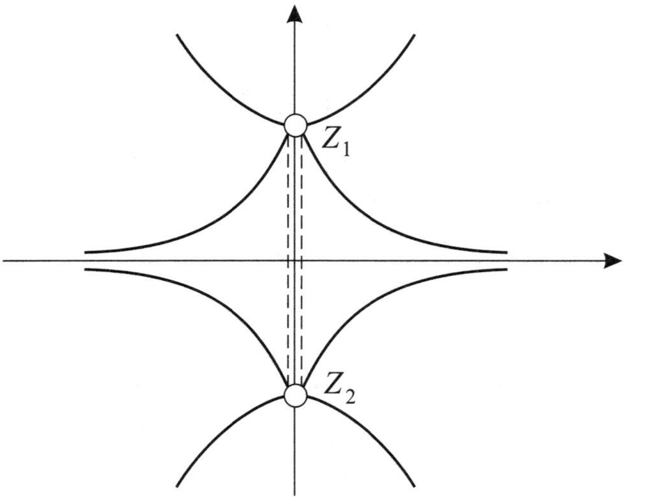

FIGURE 7

monodromy data for system (5.2) *has the form*

$$g = \exp\left\{2\left(\frac{c}{\varepsilon} - c_1\right) + i\left(\frac{S(x_*)}{\varepsilon} + \eta(x_*) + \frac{\pi}{2}\right) + O(\varepsilon^{1/4})\right\} \tag{5.21}$$

in the domain where this asymptotics holds.

5.3. Calculation of the first integral for the exponentially decreasing solution. Now we consider the domain to the right of the turning point $x_* = (3c)^{2/3}$, where the solution of the Painlevé equation (5.1) decreases exponentially. In this case, the following expansion for the solution of equation (5.1) was found in [**14**]:

$$y = \varepsilon^{1/2}\tilde{y}_0 + \varepsilon^{3/2}\tilde{y}_1 + O(\varepsilon^{5/2}\tilde{y}_0), \qquad x > x_*. \tag{5.22}$$

Here the functions $\tilde{y}_0, \tilde{y}_1$ have the form

$$\tilde{y}_0 = \frac{\varkappa_0}{x^{1/4}} \exp\left(\frac{2c}{\varepsilon} - \frac{2}{3}\frac{x^{3/2}}{\varepsilon}\right),$$

$$\tilde{y}_1 = \frac{1}{x^{1/4}}\left(\varkappa_1 + \frac{5\varkappa_0}{144c} - \frac{5\varkappa_0}{48x^{3/2}}\right)\exp\left(\frac{2c}{\varepsilon} - \frac{2}{3}\frac{x^{3/2}}{\varepsilon}\right) - \frac{\varkappa_0^3}{4x^{7/4}}\exp\left(\frac{6c}{\varepsilon} - \frac{2x^{3/2}}{\varepsilon}\right),$$

where $c, \varkappa_0, \varkappa_1$ are constants. In what follows, we consider only x such that $x - x_* > \varepsilon^{1-h}$, where $h > 0$ is arbitrary. In this domain as $\varepsilon \to +0$ the estimate $y = O(\varepsilon^\infty)$ holds.

We again apply the isomonodromy deformation method outlined above with the only difference that the exponentially decreasing asymptotics (5.22) rather than the Kuzmak–Whitham expansion holds for the function y in system (5.2). Since system (5.2) preserves its form, all expressions obtained in Subsection 5.2 that concern the WKB-approximation and do not take into account the special features of the function y, remain valid. In particular, formula (5.16) remains valid for Ψ.

Now we determine the domain where this formula can be used. We begin by analyzing expression (5.8) for μ. Since $y = O(\varepsilon^\infty)$, we have $\varepsilon^2(y')^2 - xy^2 + y^4 =$

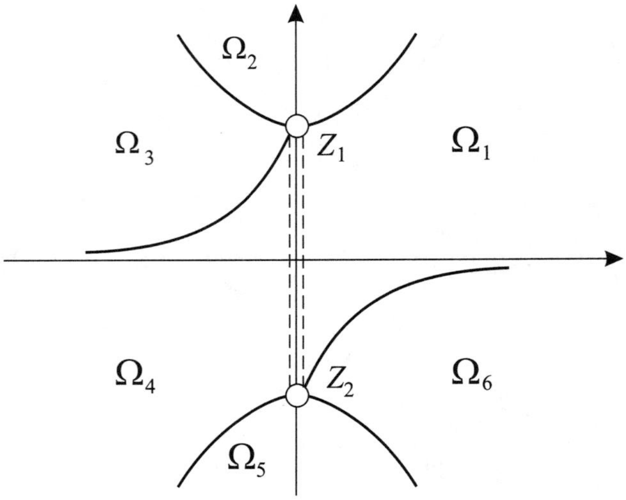

FIGURE 8

$O(\varepsilon^\infty)$ and the four turning points determined by the relation $\mu = 0$ as $\varepsilon \to +0$ with accuracy $O(\varepsilon^\infty)$ merge into the two points $z_1 = i\sqrt{x}/2$ and $z_2 = -i\sqrt{x}/2$. Hence the structure of the Stokes lines and the cuts behave as it is shown in Figure 7.

The position of the domains Ω_k, $k = 1, \dots, 6$, is shown in Figure 8. In each of these domains there exists a canonical solution Ψ_k whose behavior at infinity is determined by formula (5.4).

We note that the position of the domains Ω_k remains the same (only the boundaries were deformed after the confluence of the turning points). The structure of the Stokes matrices (5.5) also remains the same. The WKB-approximation (5.16) cannot be used near the turning points z_1, z_2.

In [**14**] the Stokes matrix S_1 was calculated in the case where y decreases exponentially. Here we construct the asymptotic solution of (5.2) only near the turning point z_1. This construction plays the key role in finding the matrix S_1.

We introduce the new independent variable

$$\eta = 2^{3/2} x^{1/4} (\lambda - i\sqrt{x}/2)/\sqrt{\varepsilon}$$

and consider, for instance, Ψ_{22}, which is an entry of the matrix Ψ for $n = 2$ and $m = 2$. The entry Ψ_{22} satisfies equation (5.17) (with the + sign in parenthesis). Expanding the coefficients in (5.17) by the Taylor formula, we pass to the variable η. By setting $\Psi_{kl} = O(y\Psi_{22})$, we obtain the equation

$$\frac{d^2\Psi_{22}}{d\eta^2} + \left\{\frac{1}{2} - \frac{\eta^2}{4} + \frac{i\sqrt{\varepsilon}}{2^{3/2}x^{3/4}}\left(\frac{\eta^3}{2} - \eta\right) + O(\varepsilon\eta^4)\right\}\Psi_{22} = 0,$$

whose solution will be sought in the form $\Psi_{22} = \Psi_{22}^0 + O(\sqrt{\varepsilon}\Psi_{22}^1)$.

For the leading term Ψ_{22}^0 we have

$$\frac{d^2\Psi_{22}^0}{d\eta^2} + \left(\frac{1}{2} - \frac{\eta^2}{4}\right)\Psi_{22}^0 = 0, \qquad \eta \in \mathbb{C}.$$

Solutions of this equation are parabolic cylinder functions, which have been studied in detail. In particular, the asymptotics of parabolic cylinder functions are known in different sectors of the complex plane for large values of the argument [**34**]. Moreover, the Stokes phenomenon also holds.

Using the expansions of parabolic cylinder functions, it is possible to match the WKB-approximation and the asymptotics of the solution of (5.2) near the turning point z_1 and to find the matrices C_1, C_2, N_1, N_2. The form of these matrices implies

$$S_1 = C_1^{-1} N_1^{-1} N_2 C_2 = \begin{pmatrix} 1 & 0 \\ g & 1 \end{pmatrix},$$

where as $\varepsilon \to +0$ the function $g = g(x, y, \varepsilon y')$ can be represented as

$$g = -\frac{2\sqrt{\pi}\sqrt[4]{x}y}{\sqrt{\varepsilon}} \exp\left(\frac{2x^{3/2}}{3\varepsilon}\right)\left(1 + O(\varepsilon^{1/5})\right). \tag{5.23}$$

Finally, substituting, instead of y, the exponentially decreasing asymptotics (5.22) into formula (5.23), we obtain the following assertion.

THEOREM 5.3. *Let $y(x, \varepsilon)$ be the solution of the Painlevé equation* (5.1) *with the asymptotics* (5.22)*. Then in the domain where this asymptotics holds, the value of g parametrizing the monodromy data for system* (5.2) *has the form*

$$g = -2\sqrt{\pi}\varkappa_0 \exp\left(2c/\varepsilon + O(\varepsilon^{1/5})\right). \tag{5.24}$$

5.4. Connection formulas for the second Painlevé transcendent. Since g is the first integral of the Painlevé equation (5.1), then, comparing formulas (5.21) and (5.24) for the asymptotic expansions in different domains, it is possible to connect the constants in these expansions. We have the following final assertion.

THEOREM 5.4. *Let $y(x, \varepsilon)$ be a smooth solution of the Painlevé equation* (5.1) *exponentially decreasing as $x > x_* + h$, where $h > 0$, and having the Kuzmak–Whitham asymptotics for $x < x_* - h$. If the expansion*

$$y = -\frac{\nu\sqrt{\varepsilon}}{2\sqrt{\pi}\sqrt[4]{x}}\left(1 + O(\varepsilon^{1/5})\right) \exp\left(\frac{2c}{\varepsilon} - 2c_1 - \frac{2x^{3/2}}{3\varepsilon}\right), \qquad \varepsilon \to +0, \tag{5.25}$$

where $\nu = \pm 1$, holds for $x > x_* + h$, $h > 0$, then for $x < x_* - h$ the Kuzmak–Whitham asymptotics (2.2) *determined by formulas* (2.5), (2.6), (2.10), *and* (2.15)–(2.18), *where $V(x) = -x$ and $c > 0$, c_1 are the constants contained in* (5.25), *has the phase shift*

$$\delta(x_*) = -\frac{S(x_*)}{\varepsilon} - \eta(x_*) = \frac{\nu\pi}{2} + O(\varepsilon^{1/4}) \qquad (\text{modulo } 2\pi). \tag{5.26}$$

REMARK 5.1. The Miles solution is a separatrix solution. Hence, as shown in [**1**], it is also necessary to study the stability of this solution, so that the results of this section imply sufficient conditions under which the solution having the Kuzmak–Whitham asymptotics for $x < x_* - h$ will exponentially decrease for $x > x_* + h$.

Next, in Sections 6 and 10 the connection formulas (5.25) and (5.26) will be used to construct global asymptotic solutions of nonlinear differential equations with cubic nonlinearity.

6. Matching the asymptotic solutions for equations with cubic nonlinearity. Quantization rule

6.1. Construction of the asymptotics. We consider the equation with cubic nonlinearity

$$\varepsilon^2 \frac{d^2 y}{dx^2} + V(x)y + 2y^3 = 0, \qquad x \in \mathbb{R}^1. \tag{6.1}$$

We normalize the desired function y by the condition

$$\int_{-\infty}^{\infty} y^2\, dx = I, \tag{6.2}$$

where I is a constant. It is required to determine the discrete values $\varepsilon = \varepsilon_n$ for which there exist real solutions of problem (6.1), (6.2). We obtain the quantization rule of Bohr–Sommerfeld type, which allows us to calculate the asymptotics of the numbers ε_n for large n (as $\varepsilon_n \to +0$).

For equation (6.1), in Section 2 we constructed the Kuzmak–Whitham expansion that holds between the turning points x_- and x_+. In what follows, we find the asymptotic solutions admissible near the turning points and outside the interval $[x_-, x_+]$. To construct the global asymptotics, it remains to match the local expansions.

For brevity, if the result is independent of the choice of the turning point x_- or x_+, we shall write x_* instead of $x_\pm$. Near the turning point, we replace equation (6.1) with the model equation

$$\varepsilon^2 \frac{d^2 y_p}{dx^2} + (V(x_*) + V'(x_*)(x - x_*))\, y_p + 2y_p^3 = 0, \tag{6.3}$$

which after the change

$$y_p = |V'(x_*)|^{1/3} u, \qquad -\xi = |V'(x_*)|^{-2/3} (V(x_*) + V'(x_*)(x - x_*))$$

is reduced to the Painlevé equation studied in detail in Section 5,

$$\varepsilon^2 \frac{d^2 u}{d\xi^2} - \xi u + 2u^3 = 0. \tag{6.4}$$

In particular, the phase shift of the Miles solution of equation (6.4) is equal to $\pi/2$ (modulo π) and the turning point has the form $\xi_* = -V(x_*)|V'(x_*)|^{-2/3} = (3c_p)^{2/3}$. Here c_p is the constant contained in the Kuzmak–Whitham expansion for $u(\xi)$. (Introducing the subscript p, we point out that we deal with solutions of equation (6.3) rather than with solutions of (6.1).) To simplify the notation, we assume without loss of generality (which follows from Remark 1.3) that in what follows, the constant c_1 in the Kuzmak–Whitham formulas is zero.

The global asymptotics of equation (6.1) consists of the Kuzmak–Whitham expansion between the turning points x_- and x_+ and the exponentially decreasing solutions of the Painlevé equation (6.3) near the turning points and outside the interval $x \in [x_-, x_+]$.

Near the turning points, we shall study the asymptotics of the functions contained in the Kuzmak–Whitham expansion (2.2). We begin with equation (2.5) for the amplitude $p(x)$. In Section 2 it was shown that the turning points for equation (2.1) are determined by relation (2.9) and the value of the function p at the turning point is $p(x_*) = (3c)^{1/3}$. We expand the left-hand side of equation (2.5) as $x \to x_*$. This allows us to find the next terms in the asymptotics of the function $p(x)$. We have the following assertion.

LEMMA 6.1 ([**14**]). *We have the following asymptotic expansions as $x \to x_*$:*

$$p(x) = (3c)^{1/3} + \frac{(x-x_*)V'(x_*)}{2(3c)^{1/3}}\left(-1-\frac{2}{\ln|x-x_*|}+O\left(\frac{\ln|\ln|x-x_*||}{(\ln|x-x_*|)^2}\right)\right), \tag{6.5}$$

$$V(x)+p^2(x) = -\frac{2V'(x_*)(x-x_*)}{\ln|x-x_*|}+O\left(\frac{|x-x_*|\ln|\ln|x-x_*||}{(\ln|x-x_*|)^2}\right). \tag{6.6}$$

Furthermore, we calculate the asymptotics of S'. It follows from (2.6) that as $x \to x_*$

$$\frac{\pi}{2S'} = \frac{1}{p}\int_0^1 \frac{du}{\sqrt{\gamma+u^2}} + O(1) = \frac{1}{p}\ln\frac{1}{\sqrt{\gamma}} + O(1),$$

where, by (6.5) and (6.6),

$$\gamma(x) \stackrel{\text{def}}{=} \frac{p^2(x)+V(x)}{p^2(x)} = -\frac{2V'(x_*)(x-x_*)}{(3c)^{2/3}\ln|x-x_*|}\left(1+O\left(\frac{\ln|\ln|x-x_*||}{\ln|x-x_*|}\right)\right).$$

Hence we have the relation

$$\ln\gamma = \ln|x-x_*| - \ln|\ln|x-x_*|| + \ln\left(\frac{2|V'(x_*)|}{(3c)^{2/3}}\right) + O\left(\frac{\ln|\ln|x-x_*||}{\ln|x-x_*|}\right).$$

So we have proved the following lemma.

LEMMA 6.2. *We have the following asymptotic expansion for S' as $x \to x_*$:*

$$S'(x) = -\frac{\pi(3c)^{1/3}}{\ln|x-x_*|} - \frac{\pi(3c)^{1/3}\ln|\ln|x-x_*||}{(\ln|x-x_*|)^2} + O\left((\ln|x-x_*|)^{-2}\right). \tag{6.7}$$

6.2. Matching the asymptotics. Quantization rule. Let us match [**32**] the asymptotic Kuzmak–Whitham expansion and the expansion for the Painlevé transcendent near the point x_1 lying at a distance of the order of $\varepsilon^{4/7}$ from the turning point x_*. In both expansions we deal with the Kuzmak–Whitham asymptotics, but we deal with the function $V(x)$ in one case, and we have $V_p(x) \stackrel{\text{def}}{=} V(x_*) + V'(x_*)(x-x_*)$ in the other. The difference between them is $\Delta V \stackrel{\text{def}}{=} V(x) - V_p(x) = O((x-x_*)^2)$, $x \to x_*$.

We represent the functions p and S as $p = p_p + \Delta p$ and $S = S_p + \Delta S$, where Δp is the perturbation of $p_p(x)$ and ΔS is the perturbation of $S_p(x)$ caused by adding ΔV to the function $V_p(x)$.

We have the following assertion.

LEMMA 6.3. *Near the turning point x_* we have the estimates*

$$\Delta p = O\left((x-x_*)^2\right), \tag{6.8}$$

$$\frac{\Delta S}{\varepsilon} = O\left(\frac{(x-x_*)^2}{\varepsilon\ln|x-x_*|}\right). \tag{6.9}$$

PROOF. Differentiating (2.5) with respect to x, we obtain

$$\int_0^p \frac{(p^2(V+p^2))' - V'z^2}{\sqrt{(p^2-z^2)(V+p^2+z^2)}}\,dz = 0 \tag{6.10}$$

and thus find

$$\frac{dp^2}{dV}=\frac{1}{(V+2p^2)}\Big(-p^2+\frac{2S'}{\pi}\int_0^p\frac{z^2\,dz}{\sqrt{(p^2-z^2)(V+p^2+z^2)}}\Big).$$

Hence $dp/dV=O(1)$ as $x\to x_*$. Since $\Delta V=O((x-x_*)^2)$, the estimate (6.8) holds. The estimate (6.9) follows from the relation $dS/dV=S'/V'$ and expansion (6.7) for S'. The proof of the lemma is complete. □

Formulas (6.8) and (6.9) show that the Kuzmak–Whitham expansions to be compared differ by a value of order

$$O\big((x-x_*)^2\big)+O\Big(\frac{(x-x_*)^2}{\varepsilon\ln|x-x_*|}\Big).$$

Near the point x_1 such that $|x_1-x_*|\sim\varepsilon^{4/7}$ we obtain the estimate $O(\varepsilon^{1/7}/\ln(1/\varepsilon))$.

We shall use the connection formulas obtained in Section 5. For matching the asymptotic Kuzmak–Whitham expansion and the Painlevé transcendent near the turning point, the phase shift δ at the turning point must be equal to

$$\delta=\frac{\pi}{2}+O(\varepsilon^{1/7-h})\qquad(\text{modulo }\pi),\tag{6.11}$$

where $h>0$ is arbitrary.

We assume that if $V(\overline{x})>0$, then $c>0$, but if $V(\overline{x})\le 0$, then $c>|V(\overline{x})|^{3/2}/3+h$, where the number $h>0$ is arbitrary. Then the conditions imposed on the function $V(x)$ in Section 2 guarantee that equation (6.1) has two turning points $x_-<x_+$. If we know the phase shifts at the turning points, then we can write the quantization rule, which allows us to find the asymptotics of the numbers $\varepsilon=\varepsilon_n$ for which the solution of problem (6.1), (6.2) exists.

Let the phase shifts at the points x_- and x_+ be equal to $\delta(x_-)$ and $\delta(x_+)$. Without loss of generality, we restrict ourselves to the case $\eta'(x)=0$ (i.e., $c_1=0$).

LEMMA 6.4. *For the global existence on the interval (x_-,x_+) of the leading term of the asymptotic Kuzmak–Whitham expansion* (2.2) *of the form $y_0=y_0^-$, where*

$$y_0^-=Y\Big(\frac{1}{\varepsilon}\int_{x_-}^{x}S'(\xi)\,d\xi-\delta(x_-),E(x),x\Big)$$

near x_- and $Y=Y(\tau,E(x),x)$ is determined by (2.15), *and of the form $y_0=\pm y_0^+$, where*

$$y_0^+=Y\Big(\frac{1}{\varepsilon}\int_{x}^{x_+}S'(\xi)\,d\xi-\delta(x_+),E(x),x\Big)$$

near x_+, it is necessary and sufficient that the following quantization rule be satisfied:

$$\frac{1}{\varepsilon}\int_{x_-}^{x_+}S'(\xi)\,d\xi-\delta(x_-)-\delta(x_+)=\pi n,\qquad\textit{where } n \textit{ is an integer.}\tag{6.12}$$

PROOF. Note (see (2.15)) that the function $Y(\tau,E,x)$ has the following properties: $Y(\tau+2\pi,E,x)=Y(\tau,E,x)$, $Y(-\tau,E,x)=Y(\tau,E,x)$, and $Y(2\pi-\tau,E,x)=Y(\tau,E,x)$. Hence the Fourier series of $Y(\tau,E,x)$ has the form

$$Y(\tau,E(x),x)=\sum_{k=0}^{\infty}b_{2k+1}(x)\cos((2k+1)\tau),\tag{6.13}$$

where b_{2k+1} are the Fourier coefficients.

Together with the function y_0^-, we can take the function y_0^+ to be the leading term of the asymptotics (2.2). For the function y_0^+ to coincide identically with y_0^- on (x_-, x_+), the following condition is necessary and sufficient:

$$\frac{1}{\varepsilon}\int_{x_-}^{x_+} S'(\xi)\,d\xi - \delta(x_-) - \delta(x_+) = 2\pi n, \tag{6.14}$$

where n is an integer. Indeed, using (6.13), we can rewrite the relation $y_0^- = y_0^+$ as

$$\sum_{k=0}^{\infty} b_{2k+1}(x)\sin\left[\frac{(2k+1)}{2}\left(\frac{1}{\varepsilon}\int_{x_-}^{x_+} S'(\xi)\,d\xi - \delta(x_-) - \delta(x_+)\right)\right]$$
$$\times \sin\left[\frac{(2k+1)}{2}\left(\frac{1}{\varepsilon}\int_{x}^{x_+} S'(\xi)\,d\xi - \frac{1}{\varepsilon}\int_{x_-}^{x} S'(\xi)\,d\xi + \delta(x_-) - \delta(x_+)\right)\right] \equiv 0,$$

which implies condition (6.14).

Similarly, together with the function y_0^-, we consider the function $-y_0^+$, which also can be the leading term of the asymptotics. For the identity $y_0^+ \equiv -y_0^-$ to hold, it is necessary and sufficient that

$$\frac{1}{\varepsilon}\int_{x_-}^{x_+} S'(\xi)\,d\xi - \delta(x_-) - \delta(x_+) = \pi + 2\pi n, \tag{6.15}$$

where n is an integer. Combining conditions (6.14) and (6.15), we obtain the quantization rule (6.12). The proof of the lemma is complete. □

Substituting expression (6.11) for the phase shift at the turning points into the quantization rule (6.12), we obtain

$$\varepsilon_n^0 = \frac{1}{\pi n}\int_{x_-}^{x_+} S'(\xi)\,d\xi, \qquad \text{where } n \text{ is an integer.} \tag{6.16}$$

Here S' is defined by (2.6), the function $p = p(x)$ is a solution of equation (2.5), and the turning points $x_\pm$ are given by relation (2.9). Relation (6.16) can be used to find the asymptotic eigenvalues ε_n^0.

6.3. Normalization condition. Statement of the final theorem. Let us now calculate the norm of the global asymptotics of (6.1). We calculate the norm of the Kuzmak–Whitham expansion on the interval $[x_{1,-}, x_{1,+}]$, where $x_{1,-} = x_- + \Delta x_-$, $x_{1,+} = x_+ - \Delta x_+$, and $\Delta x_\pm \sim \varepsilon^{4/7}$, and in the remaining region the contribution of the Painlevé transcendents is taken into account.

Relation (6.10) implies that

$$\frac{1}{2\pi}\int_0^{2\pi} y_0^2\,d\tau = \frac{4S'}{2\pi}\int_0^p \frac{z^2\,dz}{\sqrt{(p^2 - z^2)(V + p^2 + z^2)}} = \frac{(p^2(V + p^2))'}{V'}. \tag{6.17}$$

We choose the endpoints of the interval $[x_{1,-}, x_{1,+}]$ on which the Kuzmak–Whitham expansion $y^{\mathrm{KW}} = y_0 + \varepsilon y_1$ is considered so that $\partial y_0/\partial\tau = 0$ at these endpoints. The following relation holds:

$$\int_{x_{1,-}}^{x_{1,+}} (y^{KW})^2\,dx = \int_{x_{1,-}}^{x_{1,+}} (y_0^2 + 2\varepsilon y_0 y_1 + \varepsilon^2 y_1^2)\,dx. \tag{6.18}$$

Next, since y_0 is an even function and y_1 is an odd function with respect to τ about the point $-\eta$, we obtain

$$\int_0^{2\pi} y_0 y_1 \, d\tau = 0. \tag{6.19}$$

The oscillating terms in the Fourier series for y_0^2 and $y_0 y_1$ in (6.18) can be twice integrated by parts. As a result, they are multiplied by a small factor. Since y_0, y_1 are smooth functions of τ, their Fourier coefficients decrease fast with increasing number, and it is possible to estimate the behavior of the Fourier series. So we see that the function $(y^{\mathrm{KW}})^2$ in (6.18) can be replaced by the average with respect to the fast variable τ:

$$\int_{x_{1,-}}^{x_{1,+}} (y^{KW})^2 \, dx = \int_{x_{1,-}}^{x_{1,+}} (p^2(V+p^2))'(V')^{-1} \, dx + O(\varepsilon^{8/7-h}). \tag{6.20}$$

Here $h > 0$ is arbitrary.

Let us study the contribution of the Painlevé transcendents to the norm. It follows from (6.3) that

$$\big(\varepsilon^2 (y_p')^2 + y_p^4\big)' = -\big(V(x_*) + V'(x_*)(x - x_*)\big)(y_p^2)'. \tag{6.21}$$

Integrating (6.21), we obtain

$$\int_{-\infty}^{x_{1,-}} y_p^2 \, dx = (V'(x_-))^{-1} (\varepsilon^2 (y_p')^2 + V_p y_p^2 + y_p^4)\Big|_{-\infty}^{x_{1,-}}. \tag{6.22}$$

Since we have matched expansions near the point $x_{1,-}$, calculating the right-hand side in (6.22), we obtain

$$\begin{aligned} \int_{-\infty}^{x_{1,-}} y_p^2 \, dx &= p^2(V+p^2)(V')^{-1}\Big|_{x=x_{1,-}} + O(\Delta x_-^2) \\ &= \int_{x_-}^{x_{1,-}} (p^2(V+p^2))'(V')^{-1} \, dx + O(\Delta x_-^2). \end{aligned} \tag{6.23}$$

For $\Delta x_- \sim \varepsilon^{4/7}$, the estimate $O(\varepsilon^{8/7})$ holds for the remaining terms in (6.23).

Similarly, we can calculate the contribution of the asymptotics near the point x_+ to the norm. Combining relations (6.20) and (6.23), we obtain the following assertion.

LEMMA 6.5. *For $\varepsilon = \varepsilon_n^0$, the global asymptotic solution of (6.1), $y = Y_n(x, \varepsilon_n^0)$, consisting of the Kuzmak–Whitham expansion and the Painlevé transcendents, satisfies the relation*

$$\int_{-\infty}^{\infty} Y_n^2 \, dx = \int_{x_-}^{x_+} (p^2(V+p^2))'(V')^{-1} \, dx + O((\varepsilon_n^0)^{8/7-h}). \tag{6.24}$$

Here the function $p(x)$ can be found from (2.5), the points x_-, x_+ are determined by (2.9), and the number $h > 0$ is arbitrary.

Relations (6.2) and (6.24) imply the equation

$$\int_{x_-(c)}^{x_+(c)} \Big(p^2(x,c)\big(V(x) + p^2(x,c)\big)\Big)' (V'(x))^{-1} \, dx = I,$$

which allows us to find the value of $c = c(I)$ from the normalization constant I given in (6.2). We shall consider only the values of I for which (6.1) has two turning points, which follows from the conditions on c given above.

If the value of c is known, then the quantization rule (6.16) determines the numbers

$$\varepsilon_n^0 = \frac{1}{\pi n}\int_{x_-(c)}^{x_+(c)} S'(x, p(x,c))\, dx, \qquad \text{where } n \text{ is an integer.} \tag{6.25}$$

It remains to estimate the discrepancy arising after the substitution of $y = Y_n(x, \varepsilon_n^0)$ in (6.1). For this discrepancy in the $L_2(\mathbb{R}^1)$-norm the estimate $O((\varepsilon_n^0)^{8/7-h})$, where $h > 0$ is arbitrary, was obtained in [**14**].

So we have the final statement.

THEOREM 6.1. *The number ε_n^0 given by (6.25) and the function $Y_n(x, \varepsilon_n^0)$ consisting of the Kuzmak–Whitham expansion and the Painlevé transcendents are the asymptotic solution of the eigenvalue problem (6.1), (6.2). Namely, for $\varepsilon = \varepsilon_n^0$ the function $y = Y_n(x, \varepsilon_n^0)$ satisfies (6.1) with accuracy $O(n^{-8/7+h})$ in the $L_2(\mathbb{R}^1)$-norm, satisfies the normalization condition (6.2) with accuracy $O(n^{-8/7+h})$, and is concentrated modulo $O(n^{-\infty})$ on the interval $[x_-, x_+]$. Here $n \to +\infty$ and the number $h > 0$ is arbitrary.*

7. Cauchy problem for the second order nonlinear equation. Asymptotics near the turning point

In this section we construct the asymptotic solution of an equation with nonlinearity more general than that in equation (6.1). In addition to the cubic nonlinearity, this equation also contains other odd powers of y. In this case, the model equation will not have the Painlevé property any more.

We consider the Cauchy problem [**35**]

$$\varepsilon^2 \frac{d^2 y}{dx^2} + f(x, y) = 0, \qquad -\beta \le x \le \beta, \tag{7.1}$$

$$y(\beta, \varepsilon) = y^0, \qquad \varepsilon \frac{dy}{dx}(\beta, \varepsilon) = y^1. \tag{7.2}$$

Here $\beta > 0$ is a constant, $\varepsilon > 0$ is a small parameter, $y \in \mathbb{R}^1$, and the function $f(x, y)$ has the form

$$f(x, y) = a_1(x)y + a_2(x)y^3 + \cdots + a_l(x)y^{2l-1}. \tag{7.3}$$

The functions a_j, $j = 1, \ldots, l$, contained in (7.3) have the form $a_j(x) = x^{2n_j}\gamma_j(x)$, where n_j are positive integers and $\gamma_j(x) \ge 0$ are real smooth functions such that $\gamma_j(0) > 0$ and $\gamma_s(x) > 0$ for some $s \ge 2$. Then we can study problem (7.1), (7.2) in the framework of the Kuzmak–Whitham method outside a small neighborhood of the turning point $x = 0$. The form of the function $f(x, y)$ ensures Condition 1 of the Kuzmak–Whitham method. Moreover, we assume that Conditions 2 and 3 of the Kuzmak–Whitham method are satisfied for problem (7.1), (7.2). (In what follows, Condition 2 is specified for this problem.)

In the Kuzmak–Whitham method the asymptotic solution is sought in the form

$$y(x, \varepsilon) = \sum_{i=0}^{n} \varepsilon^i y_i(\tau, x) + O(\varepsilon^{n+1}), \qquad \tau = \frac{S(x)}{\varepsilon}, \tag{7.4}$$

where y_i, $i = 0, 1, \dots, n$, are 2π-periodic in τ. We denote $p(x) = z^+(x, E(x))$. In what follows, we are not looking for the function $E(x)$. Instead, we look for the amplitude of oscillations $p(x)$ in terms of which the function E is expressed by the formula $E(x) = H(x, p(x))$. Here

$$H(x,y) = 2\int_0^y f(x,z)\,dz = \sum_{j=1}^{l} a_j(x) y^{2j}/j.$$

Since $z^- = -p(x)$, we have the following equation for $p(x)$:

$$\int_0^{p(x)} \sqrt{H(x,p(x)) - H(x,z)}\,dz = c, \tag{7.5}$$

where c is a constant. Next, we find $S'(x)$ and $y_0(\tau, x)$ from the relations

$$\begin{gathered} \frac{\pi}{2S'(x)} = \int_0^{p(x)} \frac{dz}{\sqrt{H(x,p(x)) - H(x,z)}}, \\ (S'(x))^2\left(\frac{\partial y_0}{\partial \tau}\right)^2 = H(x,p(x)) - H(x,y_0). \end{gathered} \tag{7.6}$$

We set

$$\mu = \min_{1\le j\le l} \frac{n_j}{j+1}, \tag{7.7}$$

and by I we denote the set of values of j for which the expression in (7.7) attains its minimum. Since the estimates $y_i(\tau, x) = O(x^{-\mu-\lambda i})$, where $\lambda = 1 + 2\mu$ and $i = 0, 1, \dots, n$, hold as $x \to 0$, the asymptotic expansion (7.4) can be used only for $|x| > \varepsilon^{1/\lambda - h}$. Here h is an arbitrary number such that $0 < h < 1/\lambda$.

For small x we seek the asymptotic solution in the form

$$y(x,\varepsilon) = x^{-\mu}\left[\rho_0(\xi) + \sum_{i=1}^{n} \varepsilon^{\sigma_i}\rho_i(\xi) + O(\varepsilon^{\sigma_{n+1}})\right], \tag{7.8}$$

where $\xi = x^\lambda/\varepsilon$ and the constants σ_i satisfy the conditions: $\sigma_i > 0$, $\sigma_{i+1} > \sigma_i$, $\sigma_i \to +\infty$ as $i \to +\infty$. We obtain the following equation for the function $\rho_0(\xi)$:

$$\lambda^2\frac{d^2\rho_0}{d\xi^2} + \frac{\mu(\mu+1)}{\xi^2}\rho_0 + \sum_{j\in I}\gamma_j(0)\rho_0^{2j-1} = 0. \tag{7.9}$$

The equations for higher approximations have the form

$$\lambda^2\frac{d^2\rho_i}{d\xi^2} + \frac{\mu(\mu+1)}{\xi^2}\rho_i + \left(\sum_{j\in I}\gamma_j(0)(2j-1)\rho_0^{2j-2}\right)\rho_i = f_i, \qquad i \ge 1, \tag{7.10}$$

where we have $f_i = f_i(\xi, \rho_0, \dots, \rho_{i-1})$ in the right-hand side.

REMARK 7.1. If the set I consists of a single element, then equation (7.9) can be reduced to the well-known Emden–Fowler equation. The behavior of solutions of this equation was studied both numerically and analytically in numerous papers (e.g., see [**36**]). But if $I = \{1\}$, then equation (7.9) becomes linear and its solutions can be expressed in terms of the Bessel functions.

The solutions of equation (7.9) are bounded as $\xi \to \infty$, and the solutions of equation (7.10) have at most polynomial growth as $\xi \to \infty$. If we also study these solutions as $\xi \to 0$, then we can see that expansion (7.8) is an asymptotic solution of equation (7.1) for $|x| \le \varepsilon^{1/\lambda - h}$, where the number $h > 0$ is sufficiently small. Moreover, we can show that expansions (7.4) and (7.8) can be matched on the domain where they hold simultaneously.

Now we refine Condition 2 of the Kuzmak–Whitham method according to which not all the solutions of equation (1.13) are 2π-periodic in τ. Let $\omega_1 = \partial y_0/\partial\tau$. Then for the second independent solution

$$\omega_2(\tau, x) = \omega_1(\tau, x) \int \frac{d\tau}{\omega_1^2(\tau, x)}$$

of (1.13) to be not 2π-periodic in τ, the function

$$\int \frac{d\tau}{\omega_1^2(\tau, x)} \tag{7.11}$$

should be nonperiodic. Next, expressing $(\partial y_0/\partial\tau)^2$ from (7.6), we pass in the integral (7.11) from the variable τ to y_0. Because the solution $p(x)$ of equation (7.5) has the limit

$$\lim_{x\to 0} p(x)x^{\mu} = p_0,$$

as $x \to 0$, we have

$$H(x, p(x)) - H\left(x, s\frac{p_0}{x^{\mu}}\right) \sim Q(s)x^{2\mu}.$$

Here the function $Q(s)$ has the form

$$Q(s) \stackrel{\text{def}}{=} \sum_{j\in I} \gamma_j(0)(1 - s^{2j})p_0^{2j}/j.$$

Hence, in the limit as $x \to 0$, after the regularization, Condition 2 takes the form

$$\oint_L \frac{ds}{(Q(s))^{3/2}} \ne 0, \tag{7.12}$$

where L is a closed contour in the complex plane, that goes only once around the points $s = \pm 1$ in the positive direction and does not enclose any other roots of $Q(s)$.

To state the final theorem, we introduce some notation. Suppose that $Y(x, \varepsilon)$ is an exact solution of problem (7.1), (7.2), λ and h are positive constants, and the intervals Ω_+, Ω_0, and Ω_- are given by the formulas $\Omega_+ = \{x : \varepsilon^{1/\lambda - h} \le x \le \beta\}$, $\Omega_0 = \{x : |x| \le \varepsilon^{1/\lambda - h}\}$, $\Omega_- = \{x : -\beta \le x \le -\varepsilon^{1/\lambda - h}\}$.

We have the following statement.

THEOREM 7.1 ([**35**]). *Suppose that condition* (7.12) *is satisfied. Then for sufficiently small* $\beta > 0$ *and* $h > 0$, *there exist sets of functions* $S^{\pm}(x)$, $y_i^{\pm}(x)$, *and* $\rho_i(\xi)$, $i \ge 0$, *such that for any number* $N > 0$ *there exists a large number* $N_1 > 0$ *such that the following estimates hold for this number*:

$$\left| Y(x, \varepsilon) - \sum_{i=0}^{N_1} \varepsilon^i y_i^{(\pm)}\left(\frac{S^{(\pm)}(x)}{\varepsilon}, x\right)\right| \le C\varepsilon^N,$$

$$\left|\frac{d}{dx}\left\{ Y(x, \varepsilon) - \sum_{i=0}^{N_1} \varepsilon^i y_i^{(\pm)}\left(\frac{S^{(\pm)}(x)}{\varepsilon}, x\right)\right\}\right| \le C\varepsilon^N$$

for $x \in \Omega_\pm$ and

$$\left|Y(x,\varepsilon) - x^{-\mu}\left[\rho_0\left(\frac{x^\lambda}{\varepsilon}\right) + \sum_{i=1}^{N_1} \varepsilon^{\sigma_i}\rho_i\left(\frac{x^\lambda}{\varepsilon}\right)\right]\right| \le C\varepsilon^N,$$

$$\left|\frac{d}{dx}\left\{Y(x,\varepsilon) - x^{-\mu}\left[\rho_0\left(\frac{x^\lambda}{\varepsilon}\right) + \sum_{i=1}^{N_1} \varepsilon^{\sigma_i}\rho_i\left(\frac{x^\lambda}{\varepsilon}\right)\right]\right\}\right| \le C\varepsilon^N$$

for $x \in \Omega_0$. Here C is a constant.

Part 3. Asymptotic Solutions of Self-consistent Field Equations

8. Radially symmetric asymptotic solutions of the Hartree equation

8.1. Statement of the problem. In $L_2(\mathbb{R}^3)$ we consider the Hartree equation with Coulomb interaction

$$\left(\frac{\hbar^2}{2m}\Delta + \frac{\beta}{|X|} + \alpha \int_{\mathbb{R}^3} \frac{|\varphi(\widetilde{X})|^2}{|X-\widetilde{X}|}\, d\widetilde{X}\right)\varphi(X) + \lambda\varphi(X) = 0. \tag{8.1}$$

We impose the normalization condition

$$\|\varphi\|_{L_2} = 1 \tag{8.2}$$

on the desired function φ. Here $X \in \mathbb{R}^3$, Δ is the Laplace operator, $\lambda < 0$ is an eigenvalue, and α, β are real parameters such that $\alpha + \beta > 0$. Equation (8.1) appears in quantum mechanics in the description of a self-consistent field in the case where a particle is under the action of both the external field and the field induced by the particle itself. Thus an integral nonlinearity in the form of the convolution of the squared modulus of the solution $|\varphi|^2$ and the function $1/|X|$ appears in (8.1).

The eigenvalue problem (8.1), (8.2) is homogeneous. The change of variables

$$\varphi(X) = \left(\frac{|\lambda|}{\alpha+\beta}\right)^{3/2}\psi\left(\frac{|\lambda| X}{\alpha+\beta}\right)$$

and parameters

$$\varepsilon \stackrel{\text{def}}{=} \frac{\hbar}{\alpha+\beta}\sqrt{\frac{|\lambda|}{2m}}, \qquad \mu \stackrel{\text{def}}{=} \frac{\beta}{\alpha+\beta}$$

leads to the following problem for the function $\psi = \psi(x)$ and the number $\varepsilon > 0$:

$$\left(\varepsilon^2\Delta + \frac{\mu}{|x|} + (1-\mu)\int_{\mathbb{R}^3} \frac{|\psi(\tilde{x})|^2}{|x-\tilde{x}|}\, d\tilde{x} - 1\right)\psi(x) = 0, \qquad x \in \mathbb{R}^3, \tag{8.3}$$

$$\int_{\mathbb{R}^3} |\psi(x)|^2\, dx = 1. \tag{8.4}$$

We shall construct a series of asymptotic solutions of this problem as $\varepsilon \to +0$. So asymptotic solutions of the original problem (8.1), (8.2) as $\hbar \to 0$ or $|\lambda| \to 0$ or $\alpha + \beta \to \infty$ will be found.

REMARK 8.1. Since the function $-1/(4\pi|x|)$ is a fundamental solution of the Laplace operator in $\mathbb{R}^3$, the Hartree equation (8.3) can be written as the system of differential equations

$$\varepsilon^2\Delta\psi(x) + U(x)\psi(x) = 0, \qquad x \in \mathbb{R}^3,$$
$$\Delta U(x) = -(1-\mu)4\pi|\psi(x)|^2.$$

Here Δ is the Laplace operator.

Since the operator in the left-hand side of equation (8.3) has an invariant subspace of functions depending only on $r = |x|$, we can consider radially symmetric solutions of equation (8.3). We pass to spherical coordinates. Then the Laplace operator and the integral in (8.3) take the form

$$\Delta = \frac{1}{r^2}\frac{\partial}{\partial r}\left(r^2\frac{\partial}{\partial r}\right) + \frac{1}{r^2\sin\theta}\frac{\partial}{\partial\theta}\left(\sin\theta\frac{\partial}{\partial\theta}\right) + \frac{1}{r^2\sin^2\theta}\frac{\partial^2}{\partial\varphi^2},$$

$$\int_{\mathbb{R}^3}\frac{|\psi(|\tilde{x}|)|^2}{|x-\tilde{x}|}\,d\tilde{x} = \int_0^\infty G(r,\tilde{r})|\psi(\tilde{r})|^2 4\pi\tilde{r}^2\,d\tilde{r}$$

with the function

$$G(r,\tilde{r}) = \frac{1}{4\pi}\int_0^{2\pi}\int_0^{\pi}\sin\tilde{\theta}\Big[r^2+\tilde{r}^2 - 2r\tilde{r}\big(\sin\theta\sin\tilde{\theta}\cos(\varphi-\tilde{\varphi}) + \cos\theta\cos\tilde{\theta}\big)\Big]^{-1/2}\,d\tilde{\theta}\,d\tilde{\varphi}. \tag{8.5}$$

Moreover, the integral in the right-hand side of (8.5) does not depend on θ and on φ. Hence it suffices to calculate this integral only for $\theta = 0$. We have

$$G(r,\tilde{r}) = \frac{1}{2}\int_0^\pi \frac{\sin\tilde{\theta}\,d\tilde{\theta}}{\sqrt{r^2+\tilde{r}^2-2r\tilde{r}\cos\tilde{\theta}}} = \begin{cases}1/r & \text{for } r \ge \tilde{r},\\ 1/\tilde{r} & \text{for } r < \tilde{r}.\end{cases} \tag{8.6}$$

So we obtain the following problem for $\psi = \psi(r)$ on the halfline:

$$\begin{gathered}\left(\frac{\varepsilon^2}{r^2}\frac{d}{dr}\left(r^2\frac{d}{dr}\right) + \frac{\mu}{r} + (1-\mu)\int_0^\infty G(r,\tilde{r})|\psi(\tilde{r})|^2 4\pi\tilde{r}^2\,d\tilde{r} - 1\right)\psi(r) = 0, \quad r > 0,\\ \int_0^\infty |\psi(r)|^2 4\pi r^2\,dr = 1.\end{gathered} \tag{8.7}$$

The function G in (8.7) is determined by (8.6).

The properties of solutions of this problem depend on the sign of μ. First, we consider the case $\mu < 0$. In this case the asymptotic solution is concentrated modulo $O(\varepsilon^\infty)$ on some finite interval $[\alpha, \beta]$ with the left endpoint $\alpha > 0$. Therefore, the fact that for $\mu < 0$ the coefficients in (8.7) have a singularity at zero does not play any significant role.

After the change of the variable $\psi(r) = y(r)/(2\sqrt{\pi}r)$, the problem for $y = y(r,\varepsilon)$ takes the form

$$\varepsilon^2\frac{d^2y}{dr^2} + V(r,\varepsilon)y = 0, \qquad r > 0, \tag{8.8}$$

$$\int_0^\infty |y|^2\,dr = 1. \tag{8.9}$$

Here the function V is determined by the equation

$$V(r,\varepsilon) = \frac{\mu}{r} + (1-\mu)\int_0^\infty G(r,\tilde{r})|y(\tilde{r})|^2\,d\tilde{r} - 1. \tag{8.10}$$

8.2. System of equations for the self-consistent potential. We assume that there exists $V_0(r) = \lim_{\varepsilon \to +0} V(r, \varepsilon)$. Our goal is to obtain the problem for $V_0(r)$, since, knowing $V_0(r)$, we can approximately replace equation (8.8) by the linear equation

$$\varepsilon^2 \frac{d^2 y}{dr^2} + V_0(r) y = 0, \qquad r > 0, \tag{8.11}$$

whose properties have been studied in detail (e.g., see [**31**]). In particular, for equation (8.11) to have solutions localized modulo $O(\varepsilon^\infty)$ on the interval $[\alpha, \beta]$, α and β must be turning points and $V_0(r) > 0$ for $r \in (\alpha, \beta)$.

The asymptotic solution $y = Y_n(r, \varepsilon_n^0)$ of equation (8.11) can be written in terms of the Airy function $\mathrm{Ai}(z)$ [**33**]. However, if we exclude small neighborhoods of the turning points from our consideration, then we have $Y_n(r, \varepsilon_n^0) = O((\varepsilon_n^0)^\infty)$ outside the interval $[\alpha, \beta]$ and between the turning points $Y_n(r, \varepsilon_n^0)$ has the form

$$Y_n(r, \varepsilon_n^0) = \frac{c}{\sqrt{\pi}\sqrt[4]{V_0(r)}} \cos\left(\frac{1}{\varepsilon_n^0} \int_r^\beta \sqrt{V_0(r)}\, dr - \frac{\pi}{4}\right) + O(\varepsilon_n^0). \tag{8.12}$$

Here c is a constant. Hence in (8.9), (8.10) we can bound the domain of integration by the interval $[\alpha, \beta]$ and replace $|y(r)|^2$ in (8.9), (8.10) by the function $c^2/(2\pi\sqrt{V_0(r)}\,)$. We obtain this function by squaring the function $Y_n(r, \varepsilon_n^0)$ determined by formula (8.12) and truncating the fast oscillating part. This is possible, because the integrals of the fast oscillating part of Y_n^2 are small (which is verified by integrating by parts). Finally, by (8.9), we have

$$\frac{c^2}{2\pi} \int_\alpha^\beta \frac{d\tilde{r}}{\sqrt{V_0(\tilde{r})}} = 1, \tag{8.13}$$

and (8.10) implies

$$V_0(r) = \frac{\mu}{r} - k \int_\alpha^\beta \frac{d\tilde{r}}{\tilde{r}\sqrt{V_0(\tilde{r})}} - 1 \qquad \text{for } 0 \le r \le \alpha, \tag{8.14}$$

$$V_0(r) = \frac{\mu}{r} - k \left[\frac{1}{r} \int_\alpha^r \frac{d\tilde{r}}{\sqrt{V_0(\tilde{r})}} + \int_r^\beta \frac{d\tilde{r}}{\tilde{r}\sqrt{V_0(\tilde{r})}}\right] - 1 \qquad \text{for } \alpha < r < \beta, \tag{8.15}$$

$$V_0(r) = \frac{\mu}{r} - k \int_\alpha^\beta \frac{d\tilde{r}}{\sqrt{V_0(\tilde{r})}} = \frac{1}{r} - 1 \qquad \text{for } r \ge \beta. \tag{8.16}$$

Here the number k has the form $k \overset{\text{def}}{=} c^2(\mu - 1)/(2\pi)$.

So we have obtained a system of equations for the function $V_0(r)$ and the constant c. Since $-V(r, \varepsilon)$ is a self-consistent potential, system (8.13)–(8.16) is called [**37**] the system of equations for the *self-consistent potential.*

We denote $\omega(r) \overset{\text{def}}{=} rV_0(r)$.

LEMMA 8.1 ([**37**]). *For $\mu < 0$ the constant β is equal to* 1 *and in the class of continuously differentiable functions problem* (8.13)–(8.16) *is equivalent to the following boundary value problem for the function $\omega(r)$ and the numbers $k < 0$, $\alpha \in (0, 1)$:*

$$\omega''(r) = \frac{k}{\sqrt{r\omega(r)}}, \qquad r \in (\alpha, 1), \tag{8.17}$$

$$\begin{aligned} &\omega(\alpha) = 0, && \omega(1) = 0, \\ &\alpha\omega'(\alpha) + \mu = 0, && \omega'(1) = -1. \end{aligned} \tag{8.18}$$

PROOF. By (8.16) we have $V_0(1) = 0$ and $V_0(r) < 0$ for $r > 1$. Hence the point $r = 1$ coincides with the right endpoint of the interval (α, β) such that $\omega(r) > 0$ for $\alpha < r < \beta$, $\omega(\alpha) = 0$, and $\omega(\beta) = 0$. In this case for the left endpoint we have $\alpha > 0$, because, by (8.14),

$$\omega(0) = \lim_{r \to 0} \left(\mu - kr \int_\alpha^1 \frac{d\tilde{r}}{\sqrt{\tilde{r}\omega(\tilde{r})}} - r \right) = \mu < 0.$$

Furthermore, differentiating (8.15), for $\alpha < r < 1$ we have

$$\omega'(r) = -1 - k \int_r^1 \frac{d\tilde{r}}{\sqrt{\tilde{r}\omega(\tilde{r})}}, \tag{8.19}$$

whence, by continuity, we obtain $\omega'(1) = -1$ and

$$\omega'(\alpha) = -1 - k \int_\alpha^1 \frac{d\tilde{r}}{\sqrt{\tilde{r}\omega(\tilde{r})}}. \tag{8.20}$$

By (8.15), we can rewrite the relation $\omega(\alpha) = 0$ as

$$\mu - k\alpha \int_\alpha^1 \frac{d\tilde{r}}{\sqrt{\tilde{r}\omega(\tilde{r})}} - \alpha = 0. \tag{8.21}$$

Then it follows from (8.20), (8.21) that $\alpha\omega'(\alpha) + \mu = 0$. So we have found all the boundary conditions given in (8.18). Equation (8.17) is derived by differentiating (8.19).

Conversely, on the interval $(\alpha, 1)$ the solution of problem (8.17), (8.18) determines the number k and the function $V_0(r)$ satisfying equations (8.13), (8.15). Then the function $V_0(r)$ is uniquely continued outside the interval $(\alpha, 1)$ by formulas (8.14), (8.16). The proof of the lemma is complete. □

REMARK 8.2. Equation (8.17) is the well-known Langmuir–Blodgett equation [**38**].

The boundary value problem (8.17), (8.18) was studied in detail in [**37**], where, in particular, it was proved that the solution of the problem exists and is a smooth function everywhere on the interval $(\alpha, 1)$. At the endpoints of the domain of definition the following expansions hold:

$$\omega(r) = 1 - r + \frac{4k}{3}(1-r)^{3/2} + O((1-r)^2) \qquad \text{as } r \to 1 - 0, \tag{8.22}$$

$$\omega(r) = \gamma(r - \alpha) + \frac{4k}{3\sqrt{\alpha\gamma}}(r-\alpha)^{3/2} + O((r-\alpha)^2) \qquad \text{as } r \to \alpha + 0. \tag{8.23}$$

Here γ is a constant.

8.3. Statement of the main theorem. The solution of problem (8.17), (8.18) in the phase space

$$\{(r, p_r) : r > 0, \ -\infty < p_r < +\infty\}$$

determines the closed curve

$$M = \{(r, p_r) : p_r^2 = V_0(r)\}$$

and the measure on it

$$d\sigma = \frac{k\,dr}{2(\mu-1)\sqrt{V_0(r)}}.$$

The pair $(M, d\sigma)$ allows us to find the asymptotic eigenvalues

$$\varepsilon_n^0 = \frac{1}{\pi(n+1/2)} \int_\alpha^1 \sqrt{V_0(r)}\,dr, \qquad n = 1, 2, \dots \tag{8.24}$$

and the asymptotic eigenfunctions

$$\Psi_n(r, \varepsilon_n^0) = \frac{1}{2\sqrt{\pi}r} \left(\mathscr{K}_{M,\sigma}^{1/\varepsilon_n^0} 1\right)(r). \tag{8.25}$$

Here $\mathscr{K}_{M,\sigma}^{h^{-1}}$ is the Maslov canonical operator [**39, 40**] corresponding to M, σ, and the parameter $h \to 0$.

REMARK 8.3. In what follows, in the case $\mu = 0$ we write the asymptotic eigenfunctions explicitly constructed using the Fourier transformation (see (8.36)). For $\mu < 0$ the formula for the asymptotic eigenfunctions (8.25) is constructed similarly.

Next, it is necessary to substitute $\varepsilon = \varepsilon_n^0$ and $\psi = \Psi_n(r, \varepsilon_n^0)$ in (8.7) and to estimate the remainders thus obtained. It follows from general theorems that Ψ_n must satisfy the Hartree equation with accuracy $O(n^{-2})$ as $n \to \infty$. However, formulas (8.22), (8.23) imply that at the turning points $r = \alpha$ and $r = 1$ the function $V_0(r)$ in (8.11) belongs only to the class C^1 but not to the class C^2. Therefore, the estimates used in the standard smooth case [**40**] cannot be applied here. It turns out that the estimate of the remaining terms becomes less sharp and is $O(n^{-7/6})$. We have the following statement [**37**].

THEOREM 8.1. *For $\mu < 0$ the number ε_n^0 and the function $\Psi_n(|x|, \varepsilon_n^0)$ given by formulas* (8.24), (8.25) *are the asymptotic solution of the eigenvalue problem* (8.3), (8.4). *Namely, for $\varepsilon = \varepsilon_n^0$ the function $\psi = \Psi_n(|x|, \varepsilon_n^0)$ satisfies equation* (8.3) *with accuracy $O(n^{-7/6})$ in the $L_2(\mathbb{R}^3)$-norm and the normalization condition* (8.4) *with accuracy $O(n^{-7/6})$ and, moreover, is concentrated on the set $\{x \in \mathbb{R}^3 : \alpha \le |x| \le 1\}$ modulo $O(n^{-\infty})$. Here $n \to \infty$.*

REMARK 8.4. There is another approach to the construction of the asymptotics in the nonsmooth case [**41**]. In small (of order $O(\varepsilon^{2/3})$) neighborhoods of the turning points α and 1 we replace the function $V_0(r)$ by fifth order polynomials in r with coefficients singularly depending on ε. As a result, we obtain a function $V_\varepsilon(r)$ with the same turning points, which coincides with $V_0(r)$ outside small neighborhoods of the points α and 1 but already belongs to the class C^2 for $r > 0$.

Furthermore, if in (8.11) we replace the function V_0 by V_ε, then the well-known formulas obtained by Olver [**42, 31**] can be used to construct the asymptotic solution. These formulas contain Airy functions and remain valid near the turning points.

8.4. Logarithmic correction to the quantization rule. The polaron spectrum. In the polaron model [**43, 44**], as well as in some models of nonlinear optics [**45**], the following nonlinear equation appears:

$$\Delta\varphi(X) + \int_{\mathbb{R}^3} \frac{|\varphi(\widetilde{X})|^2}{|X - \widetilde{X}|}\,d\widetilde{X}\,\varphi(X) + \lambda\varphi(X) = 0, \qquad X \in \mathbb{R}^3, \tag{8.26}$$

where Δ is the Laplace operator and $\lambda = \text{const}$. Equation (8.26) admits $L_2(\mathbb{R}^3)$-normalized solutions corresponding to the bounded below sequence of negative eigenvalues λ. Later, following [**46**], we construct an asymptotics of radially symmetric eigenfunctions, as well as an asymptotics of the corresponding series of eigenvalues $\lambda = \lambda_n$ for large values of the number n, i.e., as $\lambda_n \to -0$.

We perform the change of variables

$$\varphi(X) = |\lambda|^{3/2}\psi(|\lambda|X), \qquad \varepsilon = \sqrt{|\lambda|}. \tag{8.27}$$

Then equation (8.26) takes the form (8.3), where $\mu = 0$. Moreover, the function ψ satisfies the normalization condition (8.4). Then we once more perform the change of variables $\psi(r) = y(r)/(2\sqrt{\pi}r)$, $r = |x|$, and obtain the following problem for $y = y(r, \varepsilon)$:

$$\varepsilon^2 \frac{d^2y}{dr^2} + Vy = 0, \qquad r > 0, \tag{8.28}$$

$$y(0) = 0, \tag{8.29}$$

$$\int_0^\infty |y|^2\, dr = 1. \tag{8.30}$$

Here the function V determines the equation

$$V = \frac{1}{r}\int_0^r |y(z)|^2\, dz + \int_r^\infty \frac{|y(z)|^2}{z}\, dz - 1. \tag{8.31}$$

The boundary condition (8.29) ensures that the function ψ is bounded at zero.

In contrast to the case $\mu < 0$ in which the solutions are modulo $O(\varepsilon^\infty)$ supported by the intervals $[\alpha, 1]$, $0 < \alpha < 1$, for $\mu = 0$ the solutions of problem (8.28)–(8.31) are modulo $O(\varepsilon^\infty)$ concentrated on the interval $[0, 1]$. Therefore, the behavior of the self-consistent potential near zero becomes significant. In the limit as $\varepsilon \to +0$ the potential becomes singular. On radially symmetric solutions this potential has a logarithmic singularity at zero (more precisely,

$$V \sim A\Big(\ln\frac{1}{r}\Big)^{2/3} \tag{8.32}$$

as $r \to +0$) and the turning point $r = 1$,

$$V \sim 1 - r - \frac{2}{3}c^2(1-r)^{3/2} \qquad \text{as } r \to 1-0.$$

Here A and c are some positive constants.

We construct the asymptotic solution of problem (8.28)–(8.31) using the method of matching asymptotic expansions [**32**]. First, we consider the neighborhood of zero $0 < r < kr_\varepsilon$, where

$$r_\varepsilon \stackrel{\text{def}}{=} \varepsilon\Big(\ln\frac{1}{\varepsilon}\Big)^{-1/12} \tag{8.33}$$

and $k > 0$ is an arbitrary constant. Here the function V is given by the approximate formula (8.32) and has a logarithmic singularity at zero. The local asymptotic solution on this interval has the form

$$y \sim y_0 + \frac{y_1}{x_*}, \tag{8.34}$$

where

$$y_0 = \frac{d_0}{x_*^{1/6}} \sin\left(\frac{\sqrt{A} x_*^{1/3} r}{\varepsilon}\right),$$

$$x_* = \ln\frac{1}{\varepsilon} + \frac{1}{3}\ln\ln\frac{1}{\varepsilon} + \ln(6^{1/3}c^{2/3}) + O\left(\frac{\ln\ln(1/\varepsilon)}{\ln(1/\varepsilon)}\right), \tag{8.35}$$

and d_0 is a constant. Formulas for the correction y_1 are given in [**46**].

The next interval $r > r_\varepsilon$ contains the turning point $r = 1$. Here we construct the semiclassical eigenfunctions using the Fourier transformation

$$\begin{aligned} y \sim{}& \frac{c\eta(r)}{\sqrt[4]{V_{\mathrm{as}}(r)}} \cos\left(\frac{1}{\varepsilon}\int_0^r \sqrt{V_{\mathrm{as}}(z)}\,dz - \delta\right) \\ &+ \frac{c(-1)^n}{2\sqrt{\pi\varepsilon}} \int_{-\infty}^{\infty} \frac{(1-\eta(r(p)))}{\sqrt{|V'_{\mathrm{as}}(r(p))|}} \exp\left\{\frac{i}{\varepsilon}\left(-xp + \int_0^p r(z)\,dz\right)\right\} dp. \end{aligned} \tag{8.36}$$

In this case the quantization rule

$$\frac{1}{\varepsilon}\int_0^1 \sqrt{V_{\mathrm{as}}(r)}\,dr - \delta - \frac{\pi}{4} = \pi n, \qquad \text{where } n \text{ is an integer}, \tag{8.37}$$

must be satisfied. Here δ is a constant (the phase shift at zero), and $\eta \in C^\infty(r_\varepsilon, 1)$ is an arbitrary function identically equal to 1 near $r = r_\varepsilon$ and equal to zero near $r = 1$. By $r = r(p)$ we denote the solution of the equation

$$p^2 - V_{\mathrm{as}}(r) = 0. \tag{8.38}$$

In (8.36)–(8.38) the function V_{as} as well as the constant c are the solution of the nonlinear boundary value problem

$$\left(r^2 V'_{\mathrm{as}}(r)\right)' = -\frac{c^2}{2\sqrt{V_{\mathrm{as}}(r)}}, \qquad r_\varepsilon < r < 1, \tag{8.39}$$

$$V_{\mathrm{as}}(1) = 0, \qquad V'_{\mathrm{as}}(1) = -1, \tag{8.40}$$

$$\begin{aligned} &\frac{V_{\mathrm{as}}(r_\varepsilon)}{3x_*^{4/3}}\left(\ln\frac{1}{r_\varepsilon}\right)^{1/3} - r_\varepsilon V'_{\mathrm{as}}(r_\varepsilon) \\ &\quad = \frac{c^{4/3}}{6^{4/3}x_*^{4/3}}\left(8x_* + \ln\frac{1}{x_*} - 2\right) + \frac{c^{2/3}\pi\varepsilon}{48^{2/3} r_\varepsilon x_*^{5/3}}, \end{aligned} \tag{8.41}$$

where the numbers r_ε, x_* are determined by formulas (8.33), (8.35). The function V_{as} is continuously extended to the interval $(0, r_\varepsilon)$.

LEMMA 8.2 ([**46**]). *Problem* (8.39)–(8.41) *is solvable. The asymptotics of the solution as* $\varepsilon \to +0$ *has the form*

$$V_{\mathrm{as}} = \frac{\omega_0(r)}{r} + \frac{\varepsilon\omega_1(r)}{x_*^2 r} + O\left(\frac{\varepsilon((\ln(1/r))^{1/3}+1)}{r(\ln(1/\varepsilon))^{3-h}}\right),$$

$$c = c_0 + \frac{\varepsilon}{x_*^2}c_1 + O\left(\frac{\varepsilon}{(\ln(1/\varepsilon))^{3-h}}\right),$$

where $h > 0$ is arbitrary. The leading term of the asymptotics is given by the solution of the following boundary value problem for the Langmuir–Blodgett equation:

$$\omega_0''(r) = -\frac{c_0^2}{2\sqrt{r\omega_0(r)}}, \qquad 0 < r < 1, \tag{8.42}$$
$$\omega_0(1) = 0, \qquad \omega_0'(1) = -1, \qquad \omega_0(0) = 0.$$

The corrections ω_1 and c_1 are determined by the variational equation

$$\omega_1''(r) - \frac{c_0^2}{4}\frac{\omega_1(r)}{\sqrt{r}(\omega_0(r))^{3/2}} = -\frac{c_0 c_1}{\sqrt{r\omega_0(r)}}, \qquad r_\varepsilon < r < 1,$$

with the boundary conditions

$$\omega_1(1) = 0, \qquad \omega_1'(1) = 0, \qquad -r_\varepsilon^2\Big(\frac{\omega_1}{r}\Big)'(r_\varepsilon) = \frac{c_0^{2/3}\pi x_*^{1/3}}{48^{2/3}}.$$

Far from the turning point $r = 1$, the expression (8.36) is the standard WKB-approximation exponentially decreasing on the right of the turning point and fast oscillating on the left of the turning point. The procedure of matching asymptotics (8.34) and (8.36) allows us to find that the phase shift at zero is

$$\delta = \frac{\pi}{2} + O\Big(\frac{1}{(\ln(1/\varepsilon))^{1+1/4-h}}\Big) \qquad (\text{modulo } \pi),$$

where the number $h > 0$ is arbitrary. The integral in (8.37) can be expressed in terms of the solution of the universal problem (8.42). As a result, the quantization rule (8.37) takes the form

$$\frac{1}{\varepsilon}\int_0^1 \sqrt{\frac{\omega_0(r)}{r}}\,dr - \frac{3\pi}{4} - \frac{\pi}{24\ln(1/\varepsilon)} + O\Big(\frac{1}{(\ln(1/\varepsilon))^{1+1/4-h}}\Big) = \pi n, \tag{8.43}$$

where n is an integer. The fact that the self-consistent potential has a logarithmic singularity yields a logarithmic correction in (8.43). The following statement holds [**46**].

THEOREM 8.2. *The numbers $\varepsilon = \varepsilon_n$ given by the quantization rule (8.43) are with accuracy $O(n^{-2}(\ln n)^{-1-1/4+h})$ asymptotic eigenvalues of problem (8.28)–(8.31) as $n \to \infty$. Here $h > 0$ is arbitrary.*

The rule (8.43) serves as an equation for approximate values of the parameter $\varepsilon = \varepsilon_n$ and if we recall the change (8.27), then this is also an equation for λ_n. As a result, the rule (8.43) implies the following approximate formula for the eigenvalues (8.26):

$$\lambda_n \sim -\sigma^2\Big(n + \frac{3}{4} + \frac{1}{24\ln(n+2)}\Big)^{-2}, \qquad n \to \infty, \tag{8.44}$$

where the constant σ is given by

$$\sigma = \frac{1}{\pi}\int_0^1 \sqrt{\frac{\omega_0(r)}{r}}\,dr \approx 0.218.$$

The first nine eigenvalues λ_n were found in [**44, 47**] as a direct numerical solution of problem (8.28)–(8.31). The comparison between the numerical values and the asymptotic formulas carried out in [**46**] shows that formula (8.44) containing the logarithmic correction approximates the spectrum well enough already for quantum numbers $n \geq 3$.

9. Asymptotic solution of the Hartree equation concentrated near plane disks

In this section we briefly consider the methods for constructing not radially symmetric asymptotic solutions of the Hartree equation

$$\varepsilon^2\Delta\psi(x)+\Big(-V(x)+\int_{\mathbb{R}^3}\frac{|\psi(\tilde{x})|^2}{|x-\tilde{x}|}\,d\tilde{x}+\lambda\Big)\psi(x)=0,\qquad x\in\mathbb{R}^3, \tag{9.1}$$

$$\|\psi\|_{L_2}=1. \tag{9.2}$$

Here Δ is the Laplace operator, $V(x)$ is a smooth potential well (which ensures that the spectrum is discrete), and λ, ψ are the desired eigenvalue and the desired eigenfunction. We follow [**48**] where all details can be found.

The theory of the semiclassical approximation as $\varepsilon\to 0$ for nonlinear Hartree type equations has been systematically developed since [**49, 50**]. First, the case of a singular kernel that appears in actual physical models was not considered. In this smooth case the solutions concentrated near a point for the Hartree equation were first constructed in [**51, 52**]. Then an attempt to study the case of a singular interaction kernel by the same methods was made in [**37**]. The radially symmetric asymptotics were constructed [**37, 46**]. We presented these results in Section 8.

To construct nonradially symmetric asymptotic solutions for Hartree type equations is a serious problem. For example, as became clear, the main difficulty in the construction of solutions concentrated near plane disks is the fact that the kernel of the two-dimensional Coulomb interaction has a singularity of logarithmic character.

The existence of nonsymmetric solutions was proved in several cases in [**53, 54**]. Moreover, such solutions were studied by numerical methods (e.g., see [**55**]). Finally, a series of asymptotic eigenvalues and eigenfunctions concentrated near plane disks was constructed in [**56, 48**] for problem (9.1), (9.2) in the case $V(x)=a|x|^\alpha$. Here a and α are positive constants.

To solve this problem, a tedious study of model equations with singular kernels was required. In this case there is no close analogy with the picture typical of the usual WKB-approximation [**51**].

The first distinction is that the asymptotic solution along the normal to the disk is not a simple Gaussian packet (i.e., is not determined by the oscillator equation as in the linear case [**51**]), but it is determined by the solution of the following nonlinear problem (the *one-dimensional Bogolyubov polaron*):

$$\begin{gathered}\frac{d^2t(\tau)}{d\tau^2}-\Big(\pi\int_{-\infty}^{\infty}|\tau-\tau'|t^2(\tau')\,d\tau'+\rho\Big)t(\tau)=0,\\ \int_{-\infty}^{\infty}t^2(\tau)\,d\tau=1,\qquad \int_{-\infty}^{\infty}\tau t^2(\tau)\,d\tau=0.\end{gathered} \tag{9.3}$$

Here ρ is a constant.

The second distinction is that on the disk itself, near its edge (the caustic), the model equation for (9.1) with $V(x)=a|x|^\alpha$ is not the usual Airy equation, but its nonlinear modification with additional logarithmic interaction kernel, i.e., *the Airy version of the two-dimensional polaron equation*

$$\frac{\partial^2 G}{\partial \xi^2} + \frac{\partial^2 G}{\partial \eta^2} + \left(\xi - \iint_{-\infty}^{\infty} \ln\left[\frac{(\xi-\xi')^2 + (\eta-\eta')^2}{(\xi')^2}\right] G^2(\xi',\eta')\, d\xi' d\eta'\right) G = 0. \tag{9.4}$$

We are interested in solutions $G = G(\xi,\eta)$ of equation (9.4) which exponentially decrease as $|\eta| \to \infty$ and $\xi \to -\infty$. Such solutions can be considered as new special functions (requiring a special study) somewhat similar to the Airy function. Their asymptotic expansions as $\xi \to \pm\infty$ were found in [**56**].

So the nonlinear model equations (9.3) and (9.4) respectively determine the character of the global asymptotic solution decreasing with the distance either from the disk in the normal direction or from its edge.

On the disk itself, the expansion is mainly given by the usual WKB-formulas, but the corrections to the leading part are determined by the nonlinear interaction with the polaron along the normal and with the Airy polaron near the turning points. As a result, the quantization rule determining the asymptotics of the eigenvalues λ has the form that is relatively different from that of the standard Bohr–Sommerfeld rule and, for example, contains expansions in fractional powers of the parameter ε (for more detail, see [**48**]).

REMARK 9.1. The methods presented in Section 9 can also be used to construct asymptotic solutions of a two-dimensional Hartree type equation localized near intervals [**57**].

10. One-dimensional self-consistent field equation with additional cubic nonlinearity

10.1. Statement of the problem. Differential equations containing both local and integral nonlinearities are typical in many areas of physics. In this section we study the model arising in the Thomas–Fermi–von Weizäcker theory [**58**] of atoms and molecules and in nonlinear optics [**45**]. This model is a combination of the cubic nonlinearity and a Hartree type nonlinearity. We consider the one-dimensional case.

In the one-dimensional case the system has the form

$$\psi'' + U\psi + 2\psi^3 = 0, \qquad x \in \mathbb{R}^1, \tag{10.1}$$

$$U'' = -\psi^2, \tag{10.2}$$

where the prime denotes differentiation with respect to x. The system consists of equation (10.1) with cubic nonlinearity in which the function U satisfies the Poisson equation (10.2). It follows from Remark 8.1 that it is the Poisson equation (10.2) that determines the Hartree nonlinearity in the one-dimensional case.

We seek real solutions of (10.1), (10.2) satisfying the conditions

$$\psi, \psi' \in L_2(\mathbb{R}^1). \tag{10.3}$$

We choose a point $\overline{x}$ which is the "center of the solution," i.e.,

$$\int_{-\infty}^{\overline{x}} \psi^2\, dx = \int_{\overline{x}}^{\infty} \psi^2\, dx. \tag{10.4}$$

Since the problem is translationally invariant, without loss of generality we set

$$\overline{x} = 0 \tag{10.5}$$

and impose the zero condition for the self-consistent potential at the point $\overline{x}$:

$$U(0) = 0. \tag{10.6}$$

We construct the series of asymptotic solutions $\psi = \Psi_n$ of problem (10.1)–(10.6) whose L_2-norm $\|\Psi_n\|$ tends to ∞ as $n \to \infty$. In this case the functions $\Psi_n(x)$ are concentrated between two turning points. They oscillate fast between these points and Ψ_n has $n-1$ zeros.

We write an integro-differential equivalent of the system of differential equations (10.1)–(10.6). Namely, from (10.1) and (10.3) we find

$$\int_{-\infty}^{\infty} U'\psi^2\,dx = 0.$$

Here and in (10.4) we replace the function ψ^2 by $-U''$ and obtain

$$U'(0) = 0. \tag{10.7}$$

Then it follows from (10.2), (10.6), and (10.7) that

$$U = \frac{1}{2}\int_{-\infty}^{\infty} (|\xi| - |x-\xi|)\psi^2(\xi)\,d\xi \tag{10.8}$$

and equation (10.1) can be written as

$$\psi''(x) + \frac{1}{2}\Big(\int_{-\infty}^{\infty} (|\xi| - |x-\xi|)\psi^2(\xi)\,d\xi\Big)\psi(x) + 2\psi^3(x) = 0. \tag{10.9}$$

We construct asymptotic solutions of this integro-differential equation with the additional conditions (10.3). The construction of the asymptotics can be divided into several steps.

First step. We calculate the limit self-consistent potential

$$-V_0 = \lim_{\|\psi\|\to\infty} (-U/\|\psi\|^2).$$

For this potential we obtain an equation generalizing equation (8.17) for the potential of the Hartree equation. In our case the equation for V_0 is linked with the equation for the amplitude in the Kuzmak–Whitham method. Nevertheless, the function V_0 can be calculated in quadratures.

Second step. We replace the function U in equation (10.1) by the function $U_0 = \|\psi\|^2 V_0$. We obtain a new (approximate) equation for $\psi(x)$, which contains only a local cubic nonlinearity. After the renormalization $y = \psi/\|\psi\|$, this equation takes the form

$$\varepsilon^2 y'' + V_0 y + 2y^3 = 0, \tag{10.10}$$

where $\varepsilon = \|\psi\|^{-1}$ is a small parameter. In Section 6 its asymptotic solution was globally constructed taking into account two turning points $x_\pm$.

The asymptotic of equation (10.10) on the interval between the turning points has the Kuzmak–Whitham form $y^{\mathrm{KW}} = y_0\big(S(x)/\varepsilon, x\big) + \varepsilon y_1\big(S(x)/\varepsilon, x\big)$. Outside the interval $[x_-, x_+]$, the asymptotics decreases exponentially. Finally, near the turning points, the asymptotics is determined by the second Painlevé transcendent, which satisfies equation (6.4). In this case the phase shift at the turning points for the Kuzmak–Whitham expansion is equal to $\pi/2$ (modulo π). This fact allows us to write an analog of the Bohr–Sommerfeld quantization rule in this problem. This rule separates a discrete series of solutions $\psi = \Psi_n(x)$ as $n \to \infty$.

Third step. We substitute the asymptotic solutions Ψ_n constructed above into the right-hand side of (10.8) and prove that the contribution of the fast oscillating part of the solution is small as $n \to \infty$. As a result, we obtain the equation for the limit potential $-V_0$ from which we started at the beginning. So it will be proved that Ψ_n are asymptotic solutions of the self-consistent problem (10.9), (10.3).

10.2. Problem for the limit self-consistent potential. After the renormalization $y = \psi/\|\psi\|$, $V = U/\|\psi\|^2$, equation (10.8) takes the form

$$V(x) = \frac{1}{2}\int_{-\infty}^{\infty} (|\xi| - |x-\xi|)y^2(\xi)\,d\xi. \tag{10.11}$$

In (10.11) we replace y by the asymptotic solution of (10.10) constructed in Section 6. Because outside the interval $[x_-, x_+]$ the asymptotic solution decreases exponentially, we replace it by zero. Moreover, in the right-hand side of (10.11), in the domain between the turning points, we omit the fast oscillating part of $(y^{\mathrm{KW}})^2$ whose contribution is small. As a result, by formulas (6.17) and (6.19) for V_0, we obtain the equation

$$V_0(x) = \frac{1}{2}\int_{x_-}^{x_+} (|\xi| - |x-\xi|)\frac{d}{d\xi}\left(p^2(V_0+p^2)\right)\left(\frac{dV_0}{d\xi}\right)^{-1} d\xi. \tag{10.12}$$

Here the function $p = p(V_0, c)$ is determined by the equation

$$\int_0^p \sqrt{(p^2-z^2)(V_0+p^2+z^2)}\,dz = c, \tag{10.13}$$

and the constant c satisfies the condition

$$\int_{x_-(c)}^{x_+(c)} \{p^2(V_0,c)(V_0+p^2(V_0,c))\}'(V_0')^{-1}\,dx = 1. \tag{10.14}$$

Moreover, the turning points $x_- < 0$ and $x_+ > 0$ are determined by the relation

$$V_0(x_\pm) = -(3c)^{2/3}. \tag{10.15}$$

Note that (10.13) is the equation for the amplitude $p(x)$ of the Kuzmak–Whitham expansion and condition (10.14) follows from the normalization of the function $y = \psi/\|\psi\|$, i.e.,

$$\int_{-\infty}^{\infty} y^2\,dx = 1 \tag{10.16}$$

(see (6.24)).

In the class of twice continuously differentiable functions, equation (10.12) for $V_0(x)$ is equivalent to the differential equations

$$V_0'' + \left(p^2(V_0+p^2)\right)'/V_0' = 0, \qquad x \in [x_-, x_+], \tag{10.17}$$

$$V_0'' = 0, \qquad x \notin [x_-, x_+], \tag{10.18}$$

with the conditions

$$V_0(0) = 0, \tag{10.19}$$

$$V_0'(0) = 0. \tag{10.20}$$

So we have obtained the *problem for the limit self-consistent potential* V_0, which consists of relations (10.13)–(10.15) and (10.17)–(10.20).

We emphasize that this problem is absolutely universal. Its solution (i.e., the function $V_0(x)$, the points $x_\pm$, and the constant c) is a model function and consists of universal constants in terms of which the asymptotic solutions of (10.9), (10.3) can be expressed.

We show that problem (10.13)–(10.15), (10.17)–(10.20) can be integrated in quadratures. For this, we write (10.17) as

$$(V_0')^2 + 2p^2(V_0 + p^2) = \theta. \tag{10.21}$$

Here θ is a constant. Relations (10.14), (10.17) imply that $V_0'(x_-) - V_0'(x_+) = 1$, and relations (10.15), (10.21) imply that $V_0'(x_+) = -V_0'(x_-)$. Hence we have

$$V_0'(x_-) = \frac{1}{2}, \qquad V_0'(x_+) = -\frac{1}{2}, \tag{10.22}$$

and the constant of integration is $\theta = 1/4$, because $V_0 + p^2 = 0$ at the turning points (see (2.7)). Integrating equation (10.21) once more and taking into account condition (10.19), we obtain the following assertion.

LEMMA 10.1. *On the interval* $[x_-, x_+]$ *the following formula holds for the function* V_0:

$$|x| = \int_{V_0}^{0} \frac{dz}{\sqrt{1/4 - 2p^2(z)(z + p^2(z))}}, \tag{10.23}$$

where p *is determined by equation* (10.13).

It follows from (10.18) that outside $[x_-, x_+]$, the function V_0 is a linear function. Taking (10.15), (10.22) into account, we obtain

$$V_0 = (x - x_-)/2 - (3c)^{2/3} \qquad \text{for } x < x_-, \tag{10.24}$$

$$V_0 = -(x - x_+)/2 - (3c)^{2/3} \qquad \text{for } x > x_+. \tag{10.25}$$

Here, according to (10.15) and (10.23), the turning points x_-, x_+ are given by the relation

$$|x_\pm| = \int_{-(3c)^{2/3}}^{0} \frac{dz}{\sqrt{1/4 - 2p^2(z)(z + p^2(z))}}. \tag{10.26}$$

It remains to find the value of the constant c contained in formulas (10.23)–(10.26). To this end, we use relation (10.21) with $x = 0$. It follows from conditions (10.19), (10.20) that $p(0) = 8^{-1/4}$. Finally, substituting the value of $p(0)$ thus obtained into (10.13), we see that

$$c = 8^{-3/4} \int_0^1 \sqrt{1 - z^4}\, dz. \tag{10.27}$$

The integral in (10.27) can be expressed in terms of the gamma function. Namely,

$$c = \{\Gamma(1/4)\}^2 / (3\sqrt{\pi} 2^{15/4}). \tag{10.28}$$

Approximately, we have $c \approx 0.18$. So we have proved the following lemma.

LEMMA 10.2. *Let us find the constant* c *according to* (10.28) *and the function* $p = p(V_0)$ *from* (10.13). *Then the limit self-consistent potential* $-V_0$ *is determined by formula* (10.23) *on the interval* $[x_-, x_+]$, *where* $x_\pm$ *has the form* (10.26), *and by formulas* (10.24), (10.25) *outside* $[x_-, x_+]$.

Note that according to (10.23)–(10.25), the function $V_0(x)$ is twice continuously differentiable and even with respect to x. Its maximum value is attained for $x = 0$ and is equal to zero.

10.3. Statement of the theorem. After the limit self-consistent potential $-V_0$ is calculated, equation (10.9) is approximately reduced to the nonlinear problem studied above, which consists of equation (10.10) and the normalization condition (10.16). In Section 6 we determined the discrete series of asymptotic eigenvalues $\varepsilon = \varepsilon_n^0$ for which there exist real functions $y = Y_n(x, \varepsilon_n^0)$ satisfying this problem as $n \to \infty$ with accuracy $O(n^{-8/7+h})$ for any $h > 0$ in the $L_2(\mathbb{R}^1)$-norm. The numbers ε_n^0 are given by the quantization rule (6.25), and the asymptotic eigenfunctions $Y_n(x, \varepsilon_n^0)$ consist of the Kuzmak–Whitham expansion between the turning points and the Painlevé transcendents near the turning points and outside the interval $[x_-, x_+]$. Since $\psi = y/\varepsilon$, the asymptotic solution of problem (10.9), (10.3) has the form

$$\Psi_n = Y_n/\varepsilon_n^0. \tag{10.29}$$

The following theorem holds [**59**].

THEOREM 10.1. *The functions $\psi = \Psi_n$ given by* (10.29) *satisfy equation* (10.9) *as $n \to \infty$ with accuracy $O(n^{2-1/7+h})$ for any $h > 0$ in the $L_2(\mathbb{R}^1)$-norm and satisfy conditions* (10.3)*. Moreover, the functions Ψ_n are concentrated modulo $O(n^{-\infty})$ on the interval $[x_-, x_+]$ and the L_2-norms of the solutions constructed have the form $\|\Psi_n\| = (\varepsilon_n^0)^{-1} + O(n^{-1/7+h})$.*

REMARK 10.1. It follows from the estimate $\Psi_n = O(n)$ that the leading terms in (10.9) increase as $O(n^3)$ as $n \to \infty$. Therefore, although the discrepancy increases as $O(n^{2-1/7+h})$, the functions $\psi = \Psi_n$ are asymptotic solutions of (10.9).

To prove Theorem 10.1, it remains to prove that it is admissible to replace the function U in (10.9) as $n \to \infty$ by the function $U_0 = \|\Psi_n\|^2 V_0$, where the limit self-consistent potential $-V_0$ is determined by equations (10.13)–(10.15), (10.17)–(10.20). To this end, it is necessary to substitute the asymptotics $\psi = \Psi_n$ constructed above into the right-hand side of (10.8) and then to study the integrals obtained. In [**59**] it is shown that such a change is really possible.

REMARK 10.2. In [**59**] the self-consistent problem is studied not only in the case of zero energy level (as in (10.1)–(10.6)) but also in the case of an arbitrary energy level λ where condition (10.6) is absent. Then the problem takes the form

$$\varepsilon^2 y''(x) + \left\{\frac{1}{2}\int_{-\infty}^{\infty} (|z| - |x - z|) y^2(z)\, dz + \lambda\right\} y(x) + 2y^3(x) = 0,$$

$$\|y\|_{L_2} = 1.$$

Part 4. Global Asymptotic Solution of Equations with a Saturating Nonlinearity

11. Statement of the problem

In Section 6, using the isomonodromy deformation method, we constructed the global asymptotic solution of equation (6.1) with cubic nonlinearity. However, the equations containing strong nonlinearities in general form are absolutely inaccessible for the methods based on the Painlevé property of model equations.

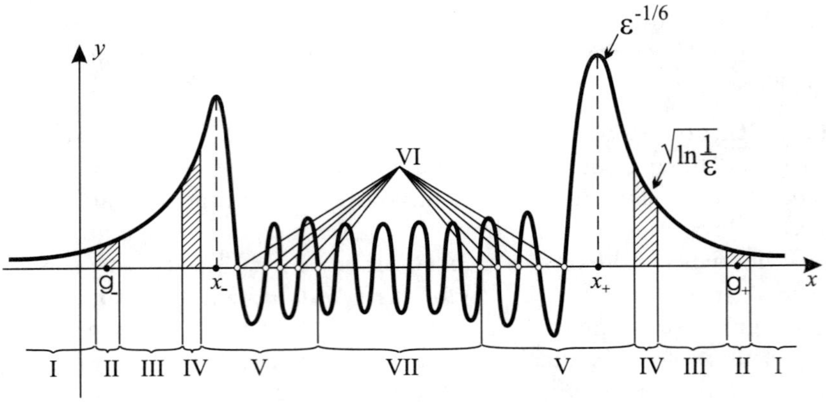

FIGURE 9

In this part we consider a case of nonlinearities in general form in which the asymptotics can be constructed. This is the case of a saturating nonlinearity tending to a constant with increasing amplitude of the solution. For example, the saturating (or decreasing) local nonlinearities of the form

$$ye^{-y^2}, \quad y\Big(\frac{\pi}{2} - \arctan y^4\Big), \quad \frac{y}{(1+y^2)^4}$$

etc., are of physical interest. In what follows, we illustrate this situation with the example of the equation [**12, 60**]

$$\varepsilon^2 \frac{d^2 y}{dx^2} + \big(U(x) + a(x) f(y^2)\big) y = 0, \qquad x \in \mathbb{R}^1, \tag{11.1}$$

where $\varepsilon > 0$ is a small parameter and $a(x), U(x), f(\sigma)$ are smooth real functions. Here $\overline{a}_0 \geq a(x) \geq a_0 > 0$, where a_0, $\overline{a}_0$ are constants and $f(\sigma)$ is a bounded monotone function (i.e., $df/d\sigma < 0$ for $\sigma \geq 0$) rapidly decreasing as $\sigma \to \infty$. We consider only the case of an exponentially decreasing function in which

$$f(\sigma) = e^{-\omega^2 \sigma}\big(1 + O(1/\sigma)\big) \qquad \text{as } \sigma \to \infty, \tag{11.2}$$

where $\omega > 0$ is a constant. The case in which f decreases as a sufficiently large power $1/\sigma^N$ can be considered similarly.

Let $-U(x)$ be a smooth potential well. This means that the function U has simple turning points $x_- < x_+$ such that $U(x_\pm) = 0$, $U'(x_\pm) \neq 0$, $U(x) > 0$ for $x \in (x_-, x_+)$, and $U(x) < 0$ outside $[x_-, x_+]$. Moreover, we assume that there exist points g_-, g_+ such that

$$U(x) + a(x) f(0) > 0 \tag{11.3}$$

for $g_- < x < x_-$, $x_+ < x < g_+$ and

$$U(g_\pm) + a(g_\pm) f(0) = 0; \tag{11.4}$$

$$U(x) + a(x) f(0) < 0 \tag{11.5}$$

for $-\infty < x < g_-$, $g_+ < x < +\infty$. It is required to construct real functions $y = y(x, \varepsilon)$ decreasing at infinity and satisfying equation (11.1) with accuracy $o(\varepsilon)$ in the $L_2(\mathbb{R}^1)$-norm.

The solution of this problem is not unique. There exists a series of functions $y = y_n$ numbered by the integers n (these numbers appear from a Bohr–Sommerfeld type quantization rule). These solutions differ from one another, for example, by the values $\|y_n\|$ of the L_2-norm.

If in the original statement of the problem the norm of the solution

$$\|y\|_{L_2} = I \tag{11.6}$$

is fixed, then the quantization conditions force the values of the parameter $\varepsilon = \varepsilon_n$ to be discrete.

The behavior of the solutions of equation (11.1) is significantly different inside and outside the interval $[g_-, g_+]$. (The asymptotic solutions are localized modulo $O(\varepsilon^\infty)$ on the interval $[g_-, g_+]$.)

The character of the solution is shown in Figure 9. The entire axis is divided into several zones (numbered by I, ..., VII) where the asymptotic solution is determined by its model equations. In what follows, we write these model equations and explain how the asymptotics merge passing from one zone to another.

12. Structure of the asymptotic solution

12.1. The Kuzmak–Whitham expansion. In zone VII (zone of "light"), we seek the asymptotic solution of equation (11.1) in the form

$$y = y_0\left(\frac{S(x)}{\varepsilon}, x\right) + \varepsilon y_1\left(\frac{S(x)}{\varepsilon}, x\right) + O(\varepsilon^2), \tag{12.1}$$

where $y_i(\tau, x)$, $i = 0, 1$, are smooth real functions 2π-periodic in τ. We assume that the conditions of the Kuzmak–Whitham method (see Section 1) are satisfied for equation (11.1). Then the leading term of the asymptotics has the form

$$y_0(\tau, x) = Y(\tau + \eta(x), x),$$

where the function $Y(\tau, x)$ is even with respect to the argument $\tau \in [-\pi, \pi]$ and is determined by the equation

$$\tau = \int_{-p(x)}^{Y} \frac{S'(x)\, dz}{\sqrt{D(x, z)}}. \tag{12.2}$$

Here $D(x, z) \overset{\text{def}}{=} U(x)(p^2 - z^2) + a(x)\big(H(p^2) - H(z^2)\big)$, and the function H is determined by the relations

$$\frac{dH}{d\sigma}(\sigma) = f(\sigma), \qquad H(\infty) = 0. \tag{12.3}$$

The phase $S(x)$ in (12.1), (12.2) is determined by the differential equation

$$S'(x) = \frac{\pi}{2}\left(\int_0^p \frac{dz}{\sqrt{D(x, z)}}\right)^{-1} \tag{12.4}$$

and the amplitude $p = p(x)$ by the equation

$$\int_0^p \sqrt{D(x, z)}\, dz = \frac{c}{4}, \tag{12.5}$$

where c is a constant. The correction to the phase $\eta = \eta(x)$ is determined by the relation

$$\eta'(x) = -\frac{2e_0(S')^3 a(x)}{\pi p^2(U + a(x)f(p^2))}\int_0^p \frac{(H(p^2) - H(z^2) + f(z^2)z^2 - f(p^2)p^2)\,dz}{\sqrt{D^3(x,z)}}$$

where e_0 is a constant.

We can also write the next term $y_1\big(S(x)/\varepsilon, x\big) = Y_1\big(S(x)/\varepsilon + \eta(x), x\big)$ of expansion (12.1) in quadratures. For this, we need to study the equation

$$(S'(x))\Big(\frac{\partial Y}{\partial \tau}\Big)^2 + U(x)Y^2 + a(x)H(Y^2) = E,$$

where the value of E is independent of τ. Then the function Y_1 has the form

$$Y_1(\tau, x) = 2\frac{\partial Y}{\partial \tau}\int_0^\tau \Pi\frac{\partial Y}{\partial E}\,d\tau + 2\frac{\partial Y}{\partial E}\Big(e_0 S' - \int_0^\tau \Pi\frac{\partial Y}{\partial \tau}\,d\tau\Big) + e_1(x)\frac{\partial Y}{\partial \tau},$$

where

$$\Pi \stackrel{\text{def}}{=} 2S'\frac{\partial^2 Y}{\partial \tau\,\partial x} + S''\frac{\partial Y}{\partial \tau},$$

and $e_1(x)$ is a smooth function.

We emphasize that expansion (12.1) cannot be used in small neighborhoods of the turning points.

In what follow, it is convenient to seek the phase shift at the turning point

$$\delta(x_*) = -\frac{S(x_*)}{\varepsilon} - \eta(x_*)$$

as the difference $\delta(x_*) = \delta^+(x_*) - \delta^-(x_*)$, where $\delta^+(x_*)$ and $\delta^-(x_*)$ are the left and right shifts determined by the contributions of the domains where $U(x) > 0$ and $U(x) < 0$, respectively.

REMARK 12.1. Since the problem under consideration has not one but two turning points, in what follows, in any definition or statement that is independent of which turning point is chosen, this point will be denoted by x_*. If it is important to distinguish these points, then we denote them by $x_\pm$ (the same notation is used for the points $g_\pm$).

We introduce the number

$$c_{0,\max} = \max\{c_{0,+}, c_{0,-}\}, \tag{12.6}$$

where

$$c_{0,\pm} = 4\int_0^\infty \sqrt{-a(x_\pm)H(z^2)}\,dz \tag{12.7}$$

and the function H is given by relations (12.3). In what follows, we assume that

$$c - c_{0,\max} \gg \varepsilon.$$

Here c is a constant in equation (12.5). It turns out that under this assumption the situation near the turning points x_-, x_+ is rather similar to that in the linear case.

Next, to describe the asymptotics in different zones, we use the Roman numerals as subscripts to indicate the zone in Figure 9 to which the solution belongs.

12.2. Zone of "shadow". For definiteness, in what follows, we consider solutions y that are positive for $x > x_*$, where x_* is the turning point and $U(x) < 0$ for $x > x_*$. In the zone of "shade" (i.e., in zone I) for $g_* < x < +\infty$ we are looking for the asymptotics of equation (11.1) in the form

$$y(x,\varepsilon) = \varepsilon^{1/2} y_I\Big(\frac{S_I(x)}{\varepsilon}, x\Big) + O(\varepsilon^{3/2}),$$

where $y_I = y_I(\tau, x)$ is a function exponentially decreasing as $\tau \to +\infty$. Then for the leading term of the asymptotics we obtain the equation

$$(S_I')^2 \frac{\partial^2 y_I}{\partial \tau^2} + \big(U(x) + a(x)f(0)\big)y_I = 0,$$

where $(S_I')^2 + U(x) + a(x)f(0) = 0$. Analyzing the next approximation, we obtain the expression

$$y_I = \frac{c_I}{\sqrt[4]{-\Gamma_I(x)}} \exp\Big\{-\frac{1}{\varepsilon}\int_{g_*}^{x} \sqrt{-\Gamma_I(z)}\,dz\Big\}. \tag{12.8}$$

Here $\Gamma_I(x) = U(x) + a(x)f(0)$ is a function and $c_I = |\Gamma_I'(g_*)|^{1/2}(2\pi a(g_*)|f'(0)|)^{-1/2}$ is a constant. The choice of this constant in (12.8) is determined by matching asymptotics.

12.3. The Painlevé equation. In zone II, near the point g_*, we seek the asymptotics of equation (11.1) in the form

$$y(x,\varepsilon) = \varepsilon^{1/3} y_{\mathrm{II}}(\xi) + O(\varepsilon),$$

where $\xi = \varepsilon^{-2/3}|\Gamma_I'(g_*)|^{1/3}(x - g_*)$. Then the function $y_{\mathrm{II}}(\xi)$ satisfies the equation

$$\frac{d^2 y_{\mathrm{II}}}{d\xi^2} + \Big(-\xi - a(g_*)|f'(0)||\Gamma_I'(g_*)|^{-2/3} y_{\mathrm{II}}^2\Big) y_{\mathrm{II}} = 0. \tag{12.9}$$

The change $y_{\mathrm{II}}(\xi) = \beta_{\mathrm{II}} u(\xi)$, where $\beta_{\mathrm{II}} = \sqrt{2}|\Gamma_I'(g_*)|^{1/3}\big(a(g_*)|f'(0)|\big)^{-1/2}$, transforms equation (12.9) into the model second Painlevé equation

$$\frac{d^2 u}{d\xi^2} = \xi u + 2u^3$$

whose solutions were studied in detail (see [**17**]). We choose the smooth real solution of equation (12.9) with the asymptotics

$$\begin{aligned} y_{\mathrm{II}}(\xi) &= \beta_{\mathrm{II}}\sqrt{-\xi/2}\big(1 + O((-\xi)^{-3})\big) \quad \text{as } \xi \to -\infty, \\ y_{\mathrm{II}}(\xi) &= \frac{\beta_{\mathrm{II}}}{2\sqrt{\pi}\sqrt[4]{\xi}} \exp\Big(-\frac{2}{3}\xi^{3/2}\Big)\big(1 + O(\xi^{-3/2})\big) \\ &\quad + O\big(\xi^{-7/4}\exp(-2\xi^{3/2})\big) \quad \text{as } \xi \to +\infty. \end{aligned}$$

12.4. Transient zone. The interval $(x_* + \overline{v}, g_*)$, where $\overline{v} = \varepsilon^{2/3} T |U'(x_*)|^{-1/3}$, is considered. Here $T = T(\varepsilon)$ is chosen so that

$$T = \Big(\frac{1}{4}\ln\frac{1}{\varepsilon}\Big)^{2/3} + O\Big(\frac{\ln\ln(1/\varepsilon)}{(\ln(1/\varepsilon))^{1/3}}\Big) \qquad \text{as } \varepsilon \to +0.$$

In the transient zone III (on this interval) we seek the asymptotic solution of equation (11.1) in the form

$$y(x,\varepsilon) = y_{0,\mathrm{III}}(x) + l(\varepsilon) y_{1,\mathrm{III}}\Big(\frac{S_{\mathrm{III}}(x)}{\varepsilon}, x\Big) + O(l^2),$$

where $y_{1,\mathrm{III}} = y_{1,\mathrm{III}}(\tau, x)$ is a function exponentially decreasing as $\tau \to +\infty$ and

$$l(\varepsilon) = \varepsilon^{1/6} T^{1/4} \Big(\ln \frac{a(x_*)}{|U'(x_*)|^{2/3}\varepsilon^{2/3} T}\Big)^{-1/4}.$$

The functions $y_{0,\mathrm{III}}$, $y_{1,\mathrm{III}}$ are determined by the equations

$$U(x) + a(x) f(y_{0,\mathrm{III}}^2) = 0,$$

$$(S'_{\mathrm{III}})^2 \frac{\partial^2 y_{1,\mathrm{III}}}{\partial \tau^2} + 2a(x) y_{0,\mathrm{III}}^2 f'(y_{0,\mathrm{III}}^2) y_{1,\mathrm{III}} = 0,$$

where $(S'_{\mathrm{III}})^2 + 2a(x) y_{0,\mathrm{III}}^2 f'(y_{0,\mathrm{III}}^2) = 0$. Solving these equations, we find

$$y_{0,\mathrm{III}} = \sqrt{f^{-1}\Big(-\frac{U(x)}{a(x)}\Big)}, \tag{12.10}$$

$$\begin{aligned} y_{1,\mathrm{III}} = {} & \frac{c_{\mathrm{III}}}{\sqrt[4]{-2a(x)\big(y_{0,\mathrm{III}}(x)\big)^2 f'\big((y_{0,\mathrm{III}}(x))^2\big)}} \\ & \times \exp\Big\{-\frac{1}{\varepsilon}\int_{x_*+\overline{v}}^{x} \sqrt{-2a(x)(y_{0,\mathrm{III}}(z))^2 f'\big((y_{0,\mathrm{III}}(z))^2\big)}\,dz\Big\}. \end{aligned} \tag{12.11}$$

Here c_{III} is a constant. We note that the leading term of the asymptotics (12.10) does not contain the fast variable $S_{\mathrm{III}}/\varepsilon$.

12.5. Intermediate boundary layer. The key role in matching asymptotics is played by the expansion in the intermediate boundary layer (i.e., in zone IV). We seek the asymptotics of equation (11.1) in the form $y = y_{\mathrm{IV}}(x,\varepsilon) + O(\varepsilon^{2/3-h})$ for any $h > 0$. Expanding the functions U and a according to the Taylor formula near x_* and omitting small terms, we obtain the equation

$$\varepsilon^2 \frac{d^2 y_{\mathrm{IV}}}{dx^2} + \Big(-|U'(x_*)|(x - x_*) + a(x_*) f\big(y_{\mathrm{IV}}^2\big)\Big) y_{\mathrm{IV}} = 0. \tag{12.12}$$

After the change of the variable $|U'(x_*)|^{1/3}(x - x_*) = \varepsilon^{2/3} T(1 + rT^{-3/2}K^{-1/2})$, where

$$K = \ln \frac{a(x_*)}{|U'(x_*)|^{2/3}\varepsilon^{2/3} T},$$

equation (12.12) becomes

$$K \frac{d^2 y_{\mathrm{IV}}}{dr^2} + \big(-(1 + rT^{-3/2}K^{-1/2}) + e^K f(y_{\mathrm{IV}}^2)\big) y_{\mathrm{IV}} = 0. \tag{12.13}$$

Equation (12.13) cannot be integrated in quadratures. However, this equation contains a large parameter K ($K \sim (2/3)\ln(1/\varepsilon)$ as $\varepsilon \to +0$) and, in turn, we can expand with respect to this large parameter. We seek the asymptotics of the solution in the form

$$y_{\mathrm{IV}} = \frac{1}{\omega}\Big(\sqrt{K} + \frac{\varkappa_{\mathrm{IV}}(r)}{\sqrt{K}} + O(K^{-3/2})\Big). \tag{12.14}$$

Substituting this expansion into (12.13) and taking into account the asymptotic expansion (11.2) for f, for $\varkappa_{\rm IV}$ we obtain the equation

$$\frac{d^2\varkappa_{\rm IV}}{dr^2} - 1 + e^{-2\varkappa_{\rm IV}} = 0.$$

Solving this equation, we see that the function $\varkappa_{\rm IV} = \varkappa_{\rm IV}(r)$ is determined by the implicit formula

$$r = \int_{\varkappa_{\rm IV}}^{\varkappa_{\rm IV}(0)} \frac{dz}{\sqrt{-1+2z+e^{-2z}}}, \tag{12.15}$$

where $\varkappa_{\rm IV}(0)$ is the constant of integration, $-\infty < r < +\infty$, and $0 < \varkappa_{\rm IV} < +\infty$.

12.6. Connection formulas. On the half-interval $[x_*, x_* + \overline{v})$, in zone V, the fact that the solution increases at the turning point x_* and f decreases fast at infinity allows us to neglect the contribution of the nonlinearity in (11.1). Then the leading term of the asymptotics of $y_{\rm V}$ satisfies the model Airy equation [**33**]

$$\frac{d^2 y_{\rm V}}{d\xi^2} - \xi y_{\rm V} = 0,$$

whose solutions have the form

$$y_{\rm V} = c_1^1 \,\mathrm{Ai}(\xi) + c_2^1 \,\mathrm{Bi}(\xi). \tag{12.16}$$

Here c_1^1, c_2^1 are constants, $\mathrm{Ai}(\xi), \mathrm{Bi}(\xi)$ are the Airy functions, and

$$\xi = \varepsilon^{-2/3} |U'(x_*)|^{1/3} (x - x_*).$$

We assume that c_1^1 is related to the parameter c in (12.5) by the formula

$$\varepsilon^{1/6} c_1^1 = \sqrt{c - c_{0,\pm}} + O(\varepsilon^{1/6-h}),$$

where $h > 0$ is arbitrary. Then for $x < x_*$ the function (12.16) is matched with the Kuzmak–Whitham expansion (12.1).

Studying the asymptotics of the function $\varkappa_{\rm IV} = \varkappa_{\rm IV}(r)$ as $r \to \pm\infty$, we can establish the relation between the constants contained in the expansions in zones III, IV, and V. Namely, the constants in (12.11) and (12.15) are related as follows:

$$\frac{\varkappa_{\rm IV}(0)}{\omega} e^{L(\varkappa_{\rm IV}(0))} = \frac{c_{\rm III}}{\sqrt[4]{2}} |U'(x_*)|^{-1/6} + O(K^{-1/2+h}),$$

where $h > 0$ is arbitrary, the function $L(\varkappa)$ has the form

$$L(\varkappa) = \int_0^{\varkappa} \left(\frac{\sqrt{2}}{\sqrt{-1+2z+e^{-2z}}} - \frac{1}{z} \right) dz,$$

and the numbers in formulas (12.15) and (12.16) satisfy the conditions

$$\begin{aligned} c_1^1 &= \sqrt{\pi}\sqrt[4]{T} e^{2T^{3/2}/3} \left(\frac{1}{\omega} \sqrt{\ln \frac{1}{\varepsilon^{2/3} T}} + \overline{r}(\varkappa_{\rm IV}(0)) + O\left(\frac{1}{\sqrt{\ln(1/\varepsilon)}} \right) \right), \\ c_2^1 &= \frac{\sqrt{\pi}}{2} \sqrt[4]{T} e^{-2T^{3/2}/3} \left(\frac{1}{\omega} \sqrt{\ln \frac{1}{\varepsilon^{2/3} T}} - \overline{r}(\varkappa_{\rm IV}(0)) + O\left(\frac{1}{\sqrt{\ln(1/\varepsilon)}} \right) \right). \end{aligned} \tag{12.17}$$

Here

$$\overline{r}(\varkappa) = \begin{cases} \sqrt{2\varkappa - 1} + L_1(\varkappa) & \text{for } \varkappa \geq \dfrac{1}{2}, \\ \displaystyle\int_{1/2}^{\varkappa} \frac{dz}{\sqrt{-1 + 2z + e^{-2z}}} + L_1\Big(\frac{1}{2}\Big) & \text{for } 0 < \varkappa < \dfrac{1}{2}, \end{cases}$$

$$L_1(\varkappa) = \int_{\infty}^{\varkappa} \left(\frac{1}{\sqrt{-1 + 2z + e^{-2z}}} - \frac{1}{\sqrt{-1 + 2z}} \right) dz \qquad \text{for } \varkappa \geq \frac{1}{2}.$$

So all the above local asymptotics are matched with one another.

REMARK 12.2. Relations (12.17) allow us to calculate the right phase shift $\delta^-(x_*)$ at the turning point

$$\delta^-(x_*) = \frac{\pi(\ln(1/\varepsilon))^{4/3}}{(c_1^1)^2 \omega^2 2^{2/3} 3} \left(1 - \frac{7}{3} \frac{\ln\ln(1/\varepsilon)}{\ln(1/\varepsilon)} + O\Big(\frac{1}{\ln(1/\varepsilon)}\Big)\right),$$

which implies the estimate $\delta^-(x_*) = O\big(\varepsilon^{1/3}(\ln(1/\varepsilon))^{4/3}\big)$ as $\varepsilon \to +0$.

12.7. Asymptotics near the turning point. In the construction of the asymptotic solution of equation (11.1) near the turning point x_*, where $U(x_*) = 0$, we can approximately replace the potential $U(x)$ by a linear function. After the transformation

$$\xi = \varepsilon^{-2/3} \big(U'(x_*)\big)^{1/3} (x - x_*)$$

(we assume that the turning point x_* is simple), we obtain the model equation

$$\frac{d^2 y}{d\xi^2} + \big(\xi + \varepsilon^{-2/3} a_1(\varepsilon^{2/3}\xi) f(y^2)\big) y = 0, \qquad \xi \in \mathbb{R}^1, \tag{12.18}$$

where $a_1(\varepsilon^{2/3}\xi) = a(x)|U'(x_*)|^{-2/3}$. If $f \equiv 0$, then this is the usual Airy equation (after ξ is replaced by $-\xi$) whose solution is the linear combination of the Airy functions

$$y = c_1^i \operatorname{Ai}(-\xi) + c_2^i \operatorname{Bi}(-\xi), \tag{12.19}$$

where c_1^i, c_2^i are some constants. In the general case, if $y \sim \varepsilon^{-h}$ $(h > 0)$ in equation (12.18), then we have $\varepsilon^{-2/3} a_1 f(y^2) y = O(\varepsilon^\infty)$, because the function f is exponentially decreasing. Since the amplitude of the solution near the turning point is large, the solution of (12.18) also has the form (12.19) outside small neighborhoods of the zeros γ_i of the function y $(i = 1, 2, \dots)$.

Now we construct asymptotic solutions of equation (12.18) near the zeros γ_i (in zone VI). The leading term of the asymptotics of y_i near the ith zero satisfies the equation with constant coefficients

$$\frac{d^2 y_i}{d\xi^2} + \xi_i y_i + \varepsilon^{-2/3} a_1(\varepsilon^{2/3}\xi_i) f(y_i^2) y_i = 0,$$

where ξ_i is a point near γ_i. This equation can be integrated in quadratures. As a result, we see that the function y_i is determined by the relation

$$\xi - \xi_i = \int_{y_i}^{y_i(\xi_i)} \frac{dz}{\sqrt{P(z)}}, \tag{12.20}$$

where

$$P(z) \overset{\text{def}}{=} \left(\frac{dy_i}{d\xi}(\xi_i)\right)^2 + \xi_i\big((y_i(\xi_i))^2 - z^2\big) + \varepsilon^{-2/3} a_1(\varepsilon^{2/3}\xi_i)\big(H((y_i(\xi_i))^2) - H(z^2)\big).$$

Here we fix some initial value $y_i(\xi_i) \sim \varepsilon^{-h}$, where h is an arbitrarily small positive number.

13. Scattering problem. Calculation of the phase shift. Final theorems

13.1. Calculation of the phase shift. We consider the following problem of the scattering problem type. Let $c_1^i(\varepsilon)$ and $c_2^i(\varepsilon)$ be the coefficients of the Airy functions forming the solution of (12.19) on the interval between γ_{i-1} and γ_i (here we denote $\gamma_0 = 0$). It is required to calculate the coefficients $c_1^{i+1}(\varepsilon)$ and $c_2^{i+1}(\varepsilon)$ on the next interval between γ_i and γ_{i+1}. Formula (12.20) for the function y_i allows us to solve this "scattering problem" explicitly.

Obviously, for $i = 1$ we need to choose c_1^1 and c_2^1 so that the conditions imposed in Subsection 12.6 are satisfied. (Recall that in this case we have $c_1^1 \sim \overline{c}\varepsilon^{-1/6}$ and $c_2^1 \sim \overline{\overline{c}}\varepsilon^{1/6}(\ln(1/\varepsilon))^{4/3}$ with some constants $\overline{c}$ and $\overline{\overline{c}}$.) So, starting from c_1^1 and c_2^1, we can successively find c_1^i and c_2^i for any $i > 1$. We are interested in the behavior of c_1^i and c_2^i for large i, when we are in the domain where the Kuzmak–Whitham expansion (12.1) is also valid.

As $i \to \infty$, in formulas for c_1^i and c_2^i there are sums that can be grouped as Riemann integral sums. We pass from sums to integrals, using formulas (see [**12**]) similar to the well-known Euler–Maclaurin summation formulas and the asymptotic expansions for the Airy functions. As a result, we obtain an expression for the phase of the solution. Note that this is the phase itself rather than its derivative, which allows us to calculate the phase shift at the turning point.

REMARK 13.1. In the study of the "scattering problem" it is convenient to consider the numbers η^i, Δ^i given by the relations

$$c_1^i = \eta^i\varepsilon^{-1/6}\cos\Delta^i, \qquad c_2^i = \eta^i\varepsilon^{-1/6}\sin\Delta^i$$

rather than the coefficients c_1^i, c_2^i present in (12.19). Similarly, we use the phase $\theta(\xi)$ and the amplitude $M(\xi)$ for the Airy functions [**33**] given by the formulas

$$\mathrm{Ai}(-\xi) = M(\xi)\cos\theta(\xi), \qquad \mathrm{Bi}(-\xi) = M(\xi)\sin\theta(\xi).$$

Our calculations give the following formulas for the phase shift. In the leading term we obtain the linear shift [**61**]

$$\delta(x_\pm) = \frac{\pi}{4} + o(1), \qquad \varepsilon \to 0, \tag{13.1}$$

and then a small significantly nonlinear correction arises. The phase shift is no longer a constant and becomes a composite function of the parameter ε. We have the following statement.

LEMMA 13.1 ([**12**]). *The more precise formula for the phase shift has the form*

$$\begin{aligned}\delta(x_\pm) &= \frac{\pi}{4}\\ &- \sum_{j=1}^{m-1} 2\theta'(k_j)\varepsilon^{1/3}F\Big(-\frac{\theta'(k_j)(c-c_{0,\pm})\varepsilon^{1/3}}{\pi}, |U'(x_\pm)|^{-2/3}a(x_\pm), \varepsilon^{-h}\Big)\\ &- \int_0^{k_m-\pi/(2\sqrt{k_m})} \frac{2\xi\varepsilon^{1/3}}{\pi}F\Big(\frac{\sqrt{\xi}(c-c_{0,\pm})\varepsilon^{1/3}}{\pi}, |U'(x_\pm)|^{-2/3}a(x_\pm), \varepsilon^{-h}\Big)\,d\xi\\ &+ O(\varepsilon^{1/5-h}), \qquad \varepsilon \to 0.\end{aligned} \tag{13.2}$$

Here the function F is determined by the formula

$$F(b_1, b_2, b_3) = \int_0^{b_3} \left(\frac{1}{\sqrt{b_1 - b_2 H(z^2)}} - \frac{1}{\sqrt{b_1}} \right) dz,$$

$h > 0$ is arbitrary, k_j is the jth zero of the Airy function Ai*, $k_m \sim \varepsilon^{-2/15}$, θ is the phase of the function* Ai*, and the numbers $c_{0,\pm}$ are determined by relation* (12.7).

13.2. Statement of the final theorems on asymptotic solutions. We begin with a weaker result when the phase shifts $\delta(x_+)$ and $\delta(x_-)$ are given by formula (13.1). We assume that

$$c - c_{0,\max} \gg \varepsilon, \tag{13.3}$$

where the number $c_{0,\max}$ is determined by relation (12.6).

We write the quantization rule

$$\frac{1}{2\varepsilon} \oint_\Gamma p\,dx - \delta(x_-) - \delta(x_+) = \pi n, \qquad \text{where } n \text{ is an integer}, \tag{13.4}$$

on a closed curve $\Gamma = \{|p| = S'(x)\}$ in the phase (x, p)-space, where S' is determined by formulas (12.4), (12.5).

If the quantization rule (13.4) is satisfied, then from (12.5) we can find discrete values of the parameter $c = c_n(\varepsilon)$ for which equation (11.1) has asymptotic solutions $Y = Y_n$ exponentially decreasing at infinity. Their structure is described in detail in Section 12. We have the following statement [**12, 60**].

THEOREM 13.1. *Suppose that the values $c = c_n(\varepsilon)$ are determined by the quantization rule* (13.4) *with accuracy $o(\varepsilon)$ and condition* (13.3) *is satisfied. Then there exist asymptotic solutions $Y = Y_n$ satisfying equation* (11.1) *with accuracy $o(\varepsilon)$ in the $L_2(\mathbb{R}^1)$-norm. These solutions are modulo $O(\varepsilon^\infty)$ concentrated on the interval $[g_-, g_+]$, where the boundaries g_-, g_+ of "light" are determined by formulas* (11.3)–(11.5).

This theorem can be generalized as follows. We write the quantization condition

$$\frac{1}{2\varepsilon} \oint_\Gamma p\,dx - \delta(x_-) - \delta(x_+) = \pi n, \qquad \text{where } n \text{ is an integer}, \tag{13.5}$$

where the curve Γ is determined similarly to (13.4) and the phase shifts $\delta(x_-)$ and $\delta(x_+)$ are given by the more precise formula (13.2). In addition, we assume that in the intermediate boundary layer (in zone IV) the asymptotic solution of equation (12.12) exists with accuracy $O(\varepsilon^{1+1/5-h})$ for any $h > 0$ in the $L_2(\mathbb{R}^1)$-norm and the leading terms of this solution are described by expansion (12.14). Then we have the following assertion [**12, 60**].

THEOREM 13.2. *Suppose that the values $c = c_n(\varepsilon)$ are determined by the quantization rule* (13.5) *with accuracy $O(\varepsilon^{1+1/5-h})$ for any $h > 0$. Suppose also that condition* (13.3) *is satisfied. Then there exist asymptotic solutions $Y = Y_n$ satisfying equation* (11.1) *with accuracy $O(\varepsilon^{1+1/5-h})$ for any $h > 0$ in the $L_2(\mathbb{R}^1)$-norm. These solutions are modulo $O(\varepsilon^\infty)$ concentrated on $[g_-, g_+]$, where the boundaries g_-, g_+ of "light" are determined by formulas* (11.3)–(11.5).

So, after a saturation nonlinearity is added to the linear operator in (11.1), the form of the curve Γ in the quantization rule changes and small nonlinear corrections are added to the linear shifts at the turning points. The position of the turning points x_-, x_+ remains the same for $c - c_{0,\max} \gg \varepsilon$, but the support of the solution modulo $O(\varepsilon^\infty)$ increases from $[x_-, x_+]$ in the linear case to $[g_-, g_+]$.

Finally, we discuss the situation in which the norm of the solution of the original problem (11.1), (11.6) is fixed. Then the normalization condition (11.6) allows us to find the relation between the parameters I and c. Suppose that the quantization rule

(13.6) $$\frac{1}{2\pi\varepsilon}\oint_\Gamma p\,dx = n + \frac{1}{2}, \qquad \text{where } n \text{ is an integer,}$$

is satisfied on the closed curve $\Gamma = \{|p| = S'(x)\}$ in the phase (x,p)-space, where S' is determined by equations (12.4), (12.5). The explicit formulas for the asymptotics $Y = Y_n$ (see Section 12) allow us to calculate

$$\int_{-\infty}^{\infty} Y_n^2(x,\varepsilon)\,dx = I_\Sigma^2(c,\varepsilon) + o(\varepsilon).$$

Since each of zones II–VIII shown in Figure 9 makes a contribution to $I_\Sigma(c,\varepsilon)$, the expressions for I_Σ are very cumbersome (see [**12**]).

The normalization condition (11.6) leads to the equation

(13.7) $$I_\Sigma(c,\varepsilon) = I,$$

which shows that there is a one-to-one dependence between I and c. In particular, the number $I_{0,\max}$ corresponding to the number $c_{0,\max}$ can be determined. Then condition (13.3) is equivalent to the condition

$$I - I_{0,\max} \gg \varepsilon,$$

which we assume to be satisfied.

Now we express the parameter $c = c(I)$ from equation (13.7). Then from the quantization rule (13.6) we find the number $\varepsilon = \varepsilon_n^0$ and define the function $Y = Y_n$ as in Theorem 13.1. The following assertion holds [**12, 60**].

THEOREM 13.3. *The number $\varepsilon = \varepsilon_n^0$ and the function $Y = Y_n$ are an asymptotic solution of the eigenvalue problem* (11.1), (11.6) *as $n \to \infty$. More precisely, for $\varepsilon = \varepsilon_n^0$ as $n \to \infty$ the function $Y = Y_n$ satisfies equation* (11.1) *with accuracy $o(n^{-1})$ in the $L_2(\mathbb{R}^1)$-norm and the normalization condition* (11.6) *with accuracy $o(n^{-1})$. This function is concentrated modulo $O(n^{-\infty})$ on the interval $[g_-, g_+]$, where the boundaries g_-, g_+ of "light" are determined by formulas* (11.3)–(11.5).

REMARK 13.2. The asymptotic solution of the eigenvalue problem for Thomas–Fermi–von Weizäcker type equations containing both integral and local saturating nonlinearities was found in [**62**].

References

[1] A. V. Kitaev, *Elliptic asymptotics of the first and second Painlevé transcendents*, Uspekhi Mat. Nauk **49** (1994), no. 1, 77–140; English transl. in Russian Math. Surveys. **49** (1994).

[2] G. E. Kuzmak, *Asymptotic solutions of nonlinear second order differential equations with variable coefficients*, Prikl. Mat. Mekh. **23** (1959), no. 3, 515–526; English transl. in J. Appl. Math. Mekh. **23** (1959).

[3] N. N. Krylov, *Study of the longitudinal stability of airplanes*, Gos. Aviats. i Avtotrakt. Izd-vo, Moscow–Leningrad, 1932. (Russian).

[4] G. B. Whitham, *Linear and nonlinear waves*, Wiley, New York, 1974.

[5] J. C. Luke, *A perturbation method for nonlinear dispersive wave problems*, Proc. Roy. Soc. **A292** (1966), no. 1430, 403–412.

[6] S. Yu. Dobrokhotov and V. P. Maslov, *Finite-gap almost periodic solutions in the WKB-approximations*, Itogi Nauki i Tekhniki, Sovremennye Problemy Matematiki. Fundamental'nye Naprav. Vol. 15, VINITI, Moscow, 1980, pp. 3–94; English transl., J. Soviet Math. **16** (1981), 1433–1487.

[7] M. V. Fedoryuk, *Ordinary differential equations*, Nauka, Moscow, 1985. (Russian).

[8] ———, *WKB-method for a nonlinear second order equation*, Zh. Vychisl. Mat. i Mat. Fiz. **26** (1986), no. 2, 198–210; English transl. in U.S.S.R. Comput. Math. and Math. Phys. **26** (1986).

[9] F. S. Rofe-Beketov, *Constants of Kneser type and effective masses for gap potentials*, Dokl. Akad. Nauk SSSR **276** (1984), no. 2, 356–359; English transl., Soviet Phys. Dokl. **29** (1984), no. 5, 391–393.

[10] S. Yu. Dobrokhotov, *Integration in quadratures of* $2n$*-dimensional linear Hamiltonian systems with* n *known skew-orthogonal solutions*, Uspekhi Mat. Nauk **53** (1998), no. 2, 143–144; English transl. in Russian Math. Surveys **53** (1998).

[11] A. V. Pereskokov, *Asymptotic solutions of ordinary differential equations*, Moscow Power Engineering Institute, Moscow, 1997. (Russian).

[12] M. V. Karasev and A. V. Pereskokov, *Turning points, phase shifts, and quantization rules in ordinary differential equations with local fast decreasing nonlinearity*, Trudy Moskov. Mat. Obshch. **56** (1995), 107–176; English transl., Trans. Moscow Math. Soc. **56** (1995), 81–135.

[13] H. Bateman and A. Erdélyi, *Higher transcendental functions*, vol. I, McGraw–Hill, New York–Toronto–London, 1953.

[14] M. V. Karasev and A. V. Pereskokov, *On connection formulas for the second Painlevé transcendent. Proof of the Miles conjecture, and a quantization rule*, Izv. Ross. Akad. Nauk Ser. Mat. **57** (1993), no. 3, 92–151; English transl. in Russian Acad. Sci. Izv. Math. **42** (1994), no. 3, 501–560.

[15] M. J. Ablowitz and H. Segur, *Solitons and the inverse scattering transform*, SIAM, Philadelphia, 1981.

[16] H. Flashka and A. C. Newell, *Monodromy and spectrum preserving deformation.* I, Commun. Math.Phys. **76** (1980), no. 1, 65–116.

[17] A. R. Its and V. Yu. Novokshenov, *The isomonodromy deformation method in the theory of Painlevé equations*, Lect. Notes in Math., vol. 1991, Springer-Verlag, Berlin, 1986.

[18] V. Yu. Novokshenov, *The Boutroux ansatz for the second Painlevé equation in the complex plane*, Izv. Akad. Nauk SSSR Ser. Mat. **54** (1990), no. 6, 1229–1251; English transl., Math. USSR-Izv. **37** (1991), no. 3, 587–609.

[19] S. P. Novikov, *Quantization of finite-gap potentials and nonlinear semiclassics arising in the nonperturbed theory of strings*, Funktsional. Anal. i Prilozhen. **24** (1990), no. 4, 43–53; English transl. in Functional Anal. Appl.

[20] V. L. Vereshchagin, *Quantization of one-phase potentials and Painlevé equations*, Differentsial'nye Uravneniya **35** (1999), no. 6, 837–839; English transl. in Differential Equations **35** (1999).

[21] ———, *Asymptotic behavior of singular-at-zero solutions of the sine-Gordon equation*, Mat. Zametki **67** (2000), no. 3, 329–342; English transl. in Math. Notes **67** (2000).

[22] L. A. Kalyakin, *Perturbation of solutions with moving singular points*, Teoret. Mat. Fiz. **127** (2001), no. 3, 401–410; English transl. Theoret. and Math. Phys. **127** (2001), no. 3, 751–759.

[23] V. Yu. Novokshenov and V. G. Shagalov, *Bound state type solutions of the elliptic sine-Gordon equation*, Teoret. Mat. Fiz. **111** (1997), no. 1, 15–31; English transl. in Theoret. and Math. Phys. **111** (1997).

[24] M. J. Ablowitz and H. Segur, *Asymptotic solutions of the Korteweg-de Vries equation*, Stud. Appl. Math. **57** (1977), no. 1, 13–44.

[25] J. W. Miles, *The asymptotic solution of the Korteweg-de Vries equation in the absence of solitons*, Stud. Appl. Math. **LX** (1979), no. 1, 59–72.

[26] J. A. Giannini, *The role of the second Painlevé transcendent in the nonlinear optics*, Phys. Lett. A **141** (1989), no. 8–9, 417–419.

[27] P. G. Daniels, *Effects of geometrical imperfections at the onset of convection in a shallow two-dimensional cavity*, Int. J. Heat Mass Transfer **25** (1982), no. 3, 337–343.

[28] J. W. Miles, *On the second Painlevé transcendent*, Proc. Roy. Soc. **361A** (1978), 277–291.

[29] ———, *The second Painlevé transcendent: a nonlinear Airy functions*, Mech. Today **5** (1980), no. 5, 297–313.

[30] A. R. Its and A. A. Kapaev, *The isomonodromy deformation method and connection formulas for asymptotics of the second Painlevé transcendent*, Izv. Akad. Nauk SSSR Ser. Mat. **51** (1987), no. 4, 878–892; English transl. in Math. USSR-Izv.

[31] M. V. Fedoryuk, *Asymptotic methods for linear ordinary differential equations*, Nauka, Moscow, 1983. (Russian).

[32] A. M. Il'in, *Matching of asymptotic expansions of solutions of boundary value problems*, Nauka, Moscow, 1989; English transl., Amer. Math. Soc., Providence, RI, 1992.

[33] M. Abramowitz and I. Stegan (eds.), *Handbook of mathematical functions with formulas, graphs and mathematical tables*, National Bureau of Standards, Washington, and Wiley, New York, 1972.

[34] H. Bateman and A. Erdélyi, *Higher transcendental functions*, vol. II, McGraw–Hill, New York–Toronto–London, 1953.

[35] L. Yu. Motylev, *Asymptotics near the turning points for a second order nonlinear equation*, Mat. Zametki **48** (1990), no. 6, 145–146; English transl. in Math. Notes **48** (1990).

[36] A. L. Dyshko, M. P. Karpent'er, N. B. Konyukhova, and P. M. Lima, *On singular problems for nonlinear second order ordinary differential equations of Emden–Fowler type*, Zh. Vychisl. Mat. i Mat. Fiz. **41** (2001), no. 4, 595–619; English transl. in Comput. Math. and Math. Phys.

[37] M. V. Karasev and V. P. Maslov, *Algebras with general commutation relations and their applications.* II, *Unitary-nonlinear operator equations*, Itogi Nauki i Tekhniki, Sovremennye Problemy Matematiki. Fundamental'nye Naprav. Vol. 13, VINITI, Moscow, 1979, pp. 145–267; English transl., J. Soviet Math. **15** (1981), 273–368.

[38] I. Langmuir and K. Blodgett, *Currents limited by space charge between concentric spheres*, Phys. Rev. **24** (1924), no. 1, 49–59.

[39] V. P. Maslov, *Perturbation theory and asymptotic methods*, Moscow State Univ., Moscow, 1965; French transl., Dunod, Paris, 1972.

[40] V. P. Maslov and M. V. Fedoryuk, *Semiclassical approximation in quantum mechanics*, Nauka, Moscow, 1976; English transl., Reidel, Dordrecht, 1981.

[41] S. I. Chernykh, *A quasiclassical particle in a one-dimensional self-consistent field*, Teoret. Mat. Fiz. **52** (1982), no. 3, 491–494; English transl., Theoret. and Math. Phys. **52** (1982), 939–941.

[42] F. Olver, *Asymptotics and special functions*, Academic Press, New York, 1974.

[43] S. I. Pekar, *Studies in electron theorem of crystals*, Gostekhizdat, Moscow, 1951. (Russian).

[44] N. K. Balabaev and V. D. Lakhno, *On the structure of a polaron with strong relation*, preprint, NTsBI AN SSSR, Pushchino, 1979. (Russian).

[45] S. A. Akhmanov and N. I. Zheludev, *Spectroscopy of nonlinear optical activity in crystals*, Singapore, World Sci., 1984.

[46] M. V. Karasev and A. V. Pereskokov, *Logarithmic corrections in the quantization rule. The polaron spectrum*, Teoret. Mat. Fiz. **97** (1993), no. 1, 78–93; English transl., Theoret. and Math. Phys. **97** (1993), 1160–1170.

[47] A. A. Bogolyubskaya and I. L. Bogolyubskii, *Study of barion-similar bound states of non-relativistic quarks in the self-consistent approximation*, Teoret. Mat. Fiz. **54** (1983), no. 2, 258–267; English transl. in Theoret. and Math. Phys. **54** (1983).

[48] M. V. Karasev and A. V. Pereskokov, *Asymptotic solutions of the Hartree equations concentrated near small-dimensional submanifolds.* II. *Localization on plane disks*, Izv. Ross. Akad. Nauk Ser. Mat. **65** (2001), no. 6, 57–98; English transl. in Russian Acad. Sci. Izv. Math. **65** (2001).

[49] V. P. Maslov, *Complex markov chains and the Feynman pass integral*, Nauka, Moscow, 1976. (Russian).

[50] ———, *Equations of self-consistent field*, Itogi Nauki i Tekhniki, Sovremennye Problemy Matematiki. Fundamental'nye Naprav. Vol. 11, VINITI, Moscow, 1978, pp. 153–234; English transl., J. Soviet Math. **11** (1979), 123–195.

[51] ———, *Complex WKB-Method in Nonlinear Equations*, Nauka, Moscow, 1977; English transl. of first part, *The Complex WKB-Method for Nonlinear Equations.* I, *Linear Theory*, Birkhäuser, Basel, 1994.

[52] I. V. Simenog, *On the asymptotics of solutions of the stationary nonlinear Hartree equation*, Teoret. Mat. Fiz. **30** (1977), no. 3, 408–411; English transl., Theoret. and Math. Phys. **30** (1977), no. 3, 263–268.

[53] M. V. Karasev and Yu. A. Osipov, *Eigenfunctions of the Hartree–Fock equation without spherical symmetry*, Teoret. Mat. Fiz. **52** (1982), no. 2, 263–269; English transl., Theoret. and Math. Phys. **52** (1982), 789–793.

[54] M. V. Karasev, *Quantum reduction to the orbits of algebras of symmetries and the Ehrenfest problem*, Preprint ITF-87-157P, ITF AN SSSR, Kiev, 1987. (Russian).

[55] R. R. Gabdullin, *Small-dimensional approximation of solutions of the Pekar polaron equation*, preprint, NTsBI AN SSSR, Pushchino, 1991. (Russian).

[56] M. V. Karasev and A. V. Pereskokov, *Asymptotic solutions of the Hartree equations concentrated near small-dimensional submanifolds.* I. *Model with a logarithmic singularity*, Izv. Ross. Akad. Nauk Ser. Mat. **65** (2001), no. 5, 33–72; English transl., Russian Acad. Sci. Izv. Math., **65** (2001), no. 5, 883–921.

[57] A. V. Pereskokov, *Asymptotic solutions of two-dimensional Hartree-type equations localized in the neighborhood of line segments*, Teoret. Mat. Fiz. **131** (2002), no. 3, 389–406; English transl., Theoret. and Math. Phys. **131** (2002), no. 3, 775–790.

[58] R. Bengurua, H. Brezis, and E. H. Lieb, *The Thomas–Fermi–von Weizäcker theory of atoms and molecules*, Comm. Math. Phys. **79** (1981), no. 2, 167–180.

[59] M. V. Karasev and A. V. Pereskokov, *One-dimensional equations of self-consistent field with cubic nonlinearity in quasiclassical approximation*, Mat. Zametki **52** (1992), no. 2, 66–82; English transl., Math. Notes. **52** (1992), 801–814.

[60] ———, *Turning point and phase shifts in ordinary differential equations with saturating nonlinearity*, Asymptotic Analysis **7** (1993), no. 1, 49–66.

[61] H. A. Kramers, *Wellenmechanik und halbzahlige Quantisierung*, Z. Physik **39** (1926), no. 1, 828–840.

[62] M. V. Karasev and A. V. Pereskokov, *Quantization rules for self-consistent field equations with local and nonlocal nonlinearities*, Excited Polaron States in Condensed Media, V. D. Lakno (ed.), Manchester Univ. Press, Manchester, 1991, pp. 157–170.

Applied Mathematics Department, Moscow Institute of Electronics and Mathematics, B. Trekhsvyatitel'skii per., 3/12, Moscow 109028, Russia

E-mail address: `karasev@miem.edu.ru`

Mathematics Modelling Department, Moscow Power Engineering Institute (technical university), Krasnokazarmennaya, 14, Moscow 111250, Russia

E-mail address: `pereskokov@mm.mpei.ac.ru`

Amer. Math. Soc. Transl.
(2) Vol. **208**, 2003

Asymptotics of Eigenfunctions in Shallow Potential Wells and Related Problems

P. Zhevandrov and A. Merzon

ABSTRACT. We propose a simple method for constructing asymptotics of eigenfunctions for the Schrödinger equation with a shallow potential well and its generalization to the problem of water waves trapped by an underwater ridge.

CONTENTS

1. Introduction
2. Heuristic considerations
3. Proof of part 1) of Theorem 1.1 (nondegenerate case)
4. Proof of part 2) of Theorem 1.1 (degenerate case)
5. Proof of part 1) of Theorem 1.3 for water waves (nondegenerate case). Reduction of the problem to an integral equation
6. The form of the solution and some preliminary expansions
7. Asymptotics of the residue part of the main equation as $\varepsilon \to 0$
8. Asymptotics of the integral $\int_\gamma G(p, x, \varepsilon)\, dp$ as $\varepsilon \to 0$
9. System of equations for the unknown terms of the expansions of the eigenvalue and eigenfunction
10. Solution of the system of equations for the unknown terms of the expansions of the eigenvalue and eigenfunction in the nondegenerate case
11. Completion of the proof of Theorem 1.3
12. Proof of part 2) of Theorem 1.3 (degenerate case)
References

1. Introduction

Consider the Schrödinger equation

$$(-\Delta + U)\Psi = E\Psi \tag{1.1}$$

2000 *Mathematics Subject Classification.* Primary 81Q15; Secondary 76B15.
Key words and phrases. Schrödinger equation, asymptotics, eigenfunctions.
The authors were supported in part by CONACYT and CIC-UMSNH (Mexico).

and assume that the function U describes a shallow potential well (i.e., $U = \varepsilon V(x)$, $V(x) \in C_0^\infty(\mathbb{R}^d)$, and $\varepsilon \to 0$; see Figure 1).

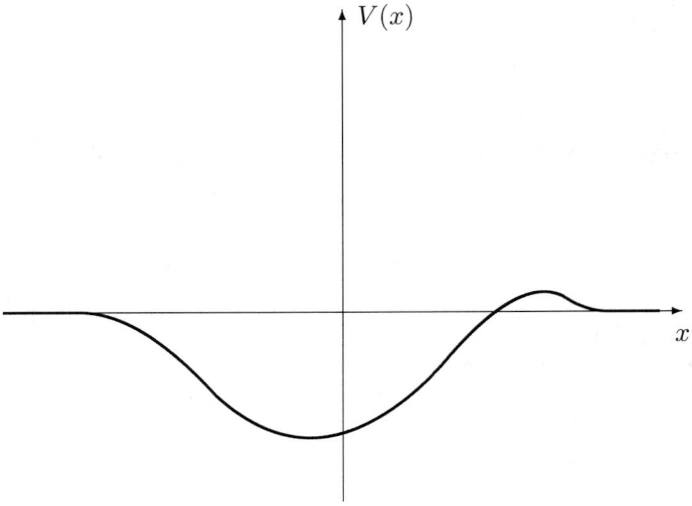

Figure 1

In the case where $\int_{\mathbb{R}^d} V(x)\,dx \le 0$ and the dimension d of the configuration space is 1 or 2, the well-known phenomenon occurs. Namely, for $\varepsilon < \varepsilon_0$ with some positive ε_0, (1.1) has exactly one eigenvalue $E_\varepsilon = -\beta^2$, $\beta \in \mathbb{R}$ and $\beta \to 0$ as $\varepsilon \to 0$, below the essential spectrum $[0, \infty)$ (see Figure 2).

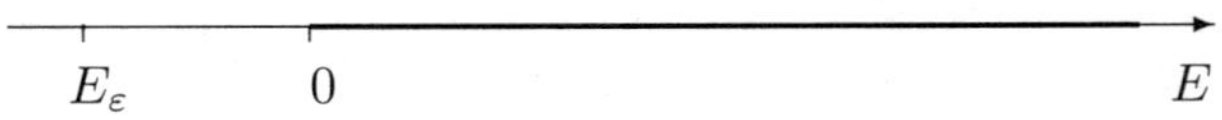

Figure 2

It is also known that E_ε is simple. Let us denote the corresponding normalized eigenfunction by Ψ_ε; so that

$$\|\Psi_\varepsilon\| = 1 \quad \text{in } L_2(\mathbb{R}^d), \quad d = 1, 2. \tag{1.2}$$

These facts were established for $d = 1$ and, in the radially symmetric case for $d = 2$, already in the famous textbook of Landau and Lifshitz [**6**] and later were demonstrated in the general case in dimension 2 by Simon [**9**]. The methods used by those authors are quite different and consist, in brief, in the following. Landau and Lifshitz construct the asymptotics of the eigenfunction in the domains where $V \equiv 0$ and $V \not\equiv 0$ separately and then glue them together; thus, the asymptotics of the eigenfunction is nonuniform and the method *per se* is applicable only in the radially symmetric case for $d = 2$. The asymptotics of the eigenvalues is obtained from the gluing conditions. On the other hand, Simon reduces the problem to an equation for the eigenvalues (the secular equation) which he solves by means of a Taylor expansion using the implicit function theorem; thus in his approach the asymptotics of the eigenfunction does not appear at all. Moreover, Simon's method is by no means trivial because it uses, for example, the theory of nuclear operators. Close results on the limiting behavior of the resolvent can be found in [**1**].

Our goal here is to construct a uniform asymptotics of the eigenfunction in this situation. Our approach is quite standard for the asymptotic theory: we choose an appropriate ansatz in order to construct a formal asymptotic solution. Afterwards, by means of standard reasoning, we prove that the formal asymptotics indeed approximates the eigenfunction. It turns out that this construction is completely elementary when one passes to the momentum representation.

Moreover, our method is equally efficient for the Schrödinger equation and the problem of water waves trapped by a submarine ridge, which, as is known in folklore, is analogous to the Schrödinger equation with a potential well. The corresponding problem after the passage to dimensionless variables reads as follows:

$$\begin{aligned} &\Delta\Phi - \Phi = 0, && -h(x) < y < 0, \\ &\partial\Phi/\partial n = 0, && y = -h(x), \\ &\Phi_y = \omega^2\Phi, && y = 0; \end{aligned} \tag{1.3}$$

here $\Phi \in H_1 = (-h < y < 0, x \in \mathbb{R})$ is the velocity potential, H_1 is the standard Sobolev space of functions with finite energy, $h(x)$ is the depth, x and y are the horizontal and vertical coordinates, respectively, and ω is the frequency and at the same time the spectral parameter. We assume that

$$h(x) = h_0 - \varepsilon V(x), \qquad V(x) \in C_0^\infty(\mathbb{R}) \tag{1.4}$$

(see Figure 3).

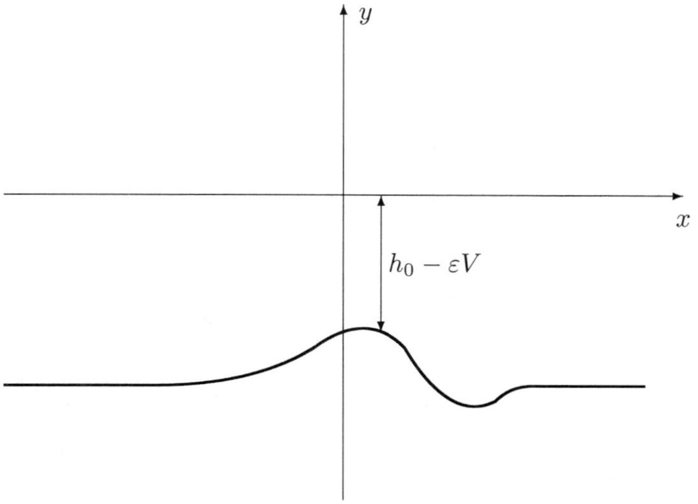

Figure 3

From the results of [**2**] it follows that for sufficiently small ε there exists exactly one eigenvalue ω^2 below the essential spectrum $[\tanh h_0, \infty)$ when $\int_{\mathbb{R}} V(x) \le 0$. The asymptotics of this eigenvalue was obtained in [**4**] for the strict inequality in the last formula and in a closely related but different asymptotic regime (long-wave approximation).

We prove the following theorems. Denote

$$\tilde{V}(p) = (2\pi)^{-1/2} \int_{\mathbb{R}} e^{-ipx} V(x)\,dx.$$

THEOREM 1.1 (the Schrödinger equation). *1) Let*

$$\int_{\mathbb{R}} V(x)\,dx < 0. \tag{1.5}$$

Then

$$\Psi_n(x) = c(\varepsilon)\mu_n \int_{\mathbb{R}} e^{ipx} \frac{a_0(p) + \varepsilon a_1(p) + \cdots + \varepsilon^n a_n(p)}{p^2 + \mu_n^2}\,dp, \tag{1.6}$$

$n = 0, 1, \ldots,$ is the asymptotics of the eigenfunction Ψ satisfying condition (1.2) and belonging to the eigenvalue

$$E = -\mu_n^2 + O(\varepsilon^{n+5/2}), \tag{1.7}$$

i.e.,

$$\|\Psi - \Psi_n\| = O(\varepsilon^{n+1/2}) \qquad \text{as } \varepsilon \to +0. \tag{1.8}$$

Moreover,

$$\mu_n = \varepsilon(\beta_0 + \varepsilon\beta_1 + \cdots + \varepsilon^n \beta_n), \quad \beta_0 = -\sqrt{\frac{\pi}{2}}\tilde{V}(0), \quad a_0(p) = \frac{\tilde{V}(p)}{\tilde{V}(0)}, \tag{1.9}$$

$$c(\varepsilon) = \varepsilon^{1/2}(d_0 + d_1\varepsilon + \cdots + d_n\varepsilon^n), \quad d_0 = \sqrt{\frac{2\beta_0}{\pi}}, \tag{1.10}$$

and the remaining values $\beta_1, \ldots, \beta_n$ and functions $a_1, \ldots, a_n$ are determined from system (3.17)–(3.22).

2) *Let $\int_{\mathbb{R}} V(x)\,dx = 0$. Then*

$$\Psi_n(x) = c(\varepsilon)\mu_n \int_{\mathbb{R}} e^{ipx} \frac{a_0(p) + \varepsilon a_1(p) + \cdots + \varepsilon^n a_n(p)}{p^2 + \mu_n^2}\,dp, \tag{1.11}$$

$n = 1, 2, \ldots,$ is the asymptotics of the eigenfunction Ψ satisfying condition (1.2) and belonging to the eigenvalue

$$E = -\mu_n^2 + O(\varepsilon^{n+2}) \tag{1.12}$$

in the sense that

$$\|\Psi - \Psi_n\| = O(\varepsilon^{n-2}). \tag{1.13}$$

Moreover,

$$\begin{aligned}
&\mu_n = \varepsilon^2(\beta_1 + \varepsilon\beta_2 + \cdots + \varepsilon^n\beta_n), \qquad \beta_1 = \frac{1}{2}\int_{\mathbb{R}} \frac{|V(t)|^2}{t^2}\,dt,\\
&\beta_2 = -\frac{1}{2\sqrt{2\pi}} \int_{\mathbb{R}+i}\int_{\mathbb{R}+i} \frac{\tilde{V}(t-s)\tilde{V}(s)\tilde{V}(-t)}{t^2 s^2}\,dtds,\\
&a_0(p) = -\frac{\sqrt{\pi/2}\tilde{V}(p)}{\beta_1},\\
&a_1(p) = \sqrt{\pi/2}\frac{\beta_2}{\beta_1} + \frac{i\pi}{2\beta_1}\tilde{V}'(0)\tilde{V}(p) + \frac{1}{2\beta_1}\int_{\mathbb{R}+i} \frac{\tilde{V}(p-t)\tilde{V}(t)}{t^2}\,dt,
\end{aligned} \tag{1.14}$$

$c(\varepsilon)$ is determined from (4.31) and the remaining values of $\beta_3, \ldots, \beta_n$ and the functions $a_2, \ldots, a_n$ are determined from system (3.17)–(3.22).

REMARK 1.2. Similar results are valid also in the case $d = 2$ (the detailed formulation of the corresponding theorem can be found in [**12**]). Nevertheless, due to the exponential smallness of β in this case, the proof requires additional arguments which are not connected with the main idea of our method and are rather cumbersome. Therefore, we do not include them here and we postpone their exposition until a forthcoming publication.

THEOREM 1.3 (water waves). 1) (*Nondegenerate case*). *Let*

$$\int_{\mathbb{R}} V(x)\,dx < 0. \tag{1.15}$$

Then

$$\Psi_n(x,y) = d(\varepsilon)\mu_n \int_{\mathbb{R}} e^{ipx}\Phi_n(p,\varepsilon)\Big(\frac{\omega^2}{\tau}\sinh(\tau y) + \cosh(\tau y)\Big)\,dp, \tag{1.16}$$

where

$$\begin{aligned} \Phi_n(p,\varepsilon) &= \frac{\tilde{a}_0(p) + \varepsilon\tilde{a}_1(p) + \cdots + \varepsilon^n\tilde{a}_n(p)}{\cosh h_0[L(p) + \mu_n^2]}, \\ L(p) &= \sqrt{1+p^2}\tanh(\sqrt{1+p^2}h_0) - \tanh h_0, \qquad \tau = \sqrt{1+p^2}, \end{aligned} \tag{1.17}$$

is the asymptotics of the trapped wave Ψ *satisfying condition* (1.2) *and corresponding to the frequency*

$$\omega_n^2 = \tanh h_0 - \mu_n^2 + O(\varepsilon^{n+5/2}) \tag{1.18}$$

in the sense of (1.8). *Moreover,*

$$\begin{aligned} d(\varepsilon) &= \varepsilon^{1/2}\Big[\frac{|\beta_0|^{1/2}(2l)^{1/4}\cosh h_0}{\sqrt{\pi}} + \varepsilon d_1 + \cdots + \varepsilon^n d_n + O(\varepsilon^{n+1})\Big], \\ \mu_n &= \varepsilon(\beta_0 + \varepsilon\beta_1 + \cdots + \varepsilon^n\beta_n), \\ \beta_0 &= \frac{1}{\sqrt{2l}\cosh^2 h_0}\int_{\mathbb{R}} V(x)dx, \qquad a_0(x) = \frac{\sqrt{2\pi}V(x)}{\int_{\mathbb{R}} V(x)dx}, \\ \beta_1 &= -\frac{\int_{\mathbb{R}} V^2(x)dx}{l\sqrt{2l}\cosh^4 h_0} + \frac{\int_{\mathbb{R}} V(x)dx\int_{\mathbb{R}} xV(x)dx}{l\sqrt{2l}\cosh^4 h_0} \\ &\quad + \frac{1}{2\sqrt{l\pi}\cosh^2 h_0}\int_{\mathbb{R}}\int_{\mathbb{R}}\frac{e^{itx}\mathcal{I}_1(t,x)\tilde{V}(t)dtdx}{L(t)\cosh(\tau h_0)}, \\ a_1(x) &= \frac{1}{\beta_0}\Big\{ -\beta_1 a_0(x) - \frac{\sqrt{2\pi}\beta_0 V(x)}{l\cosh^2 h_0} - \frac{\sqrt{2\pi}\beta_0 V'(x)}{l\cosh^2 h_0} \\ &\quad + \frac{i\sqrt{2\pi}a_{01}\beta_0 V(x)}{l\cosh^2 h_0} + \frac{\beta_0}{\sqrt{2\pi}}\Big[\int_{\mathbb{R}}\frac{e^{itx}\mathcal{I}_1(t,x)a_0(t)\,dt}{L(t)\cosh(\tau h_0)}\Big]\Big\}, \end{aligned} \tag{1.19}$$

and the remaining values $\beta_1, \ldots, \beta_n$ *and functions* $a_1, \ldots, a_n$ *are determined from the corresponding system* (9.3)–(9.6).

Here

$$
\begin{aligned}
l &= \tanh h_0 - h_0(\tanh h_0)^2 + h_0,\\
\mathcal{I}_1(p,x) &= V(x)\iota_{1,1}(p) + V'(x)\iota_{1,2},\\
\iota_{1,1}(p) &= \tau^2\cosh(\tau h_0) - t_0\tau\sinh(\tau h_0),\\
\iota_{1,2}(p) &= -ip\cosh(\tau h_0) + \frac{ipt_0\sinh(\tau h_0)}{\tau},\\
a_{01} &= -i\int_{\mathbb{R}} xV(x)\,dx.
\end{aligned}
\tag{1.20}
$$

2) (*Degenerate case*). *Let* $\int_{\mathbb{R}} V(x)\,dx = 0$. *Then*

$$
\Psi_n(x) = d(\varepsilon)\mu_n\int_{\mathbb{R}} e^{ipx}\frac{\tilde{a}_0(p)+\varepsilon\tilde{a}_1(p)+\cdots+\varepsilon^n\tilde{a}_n(p)}{\cosh h_0[L(p)+\mu_n^2]}\,dp
\tag{1.21}
$$

is the asymptotics of the trapped wave Ψ *satisfying condition* (1.2) *and corresponding to the frequency*

$$
\omega^2 = \tanh h_0 - \mu_n^2 + O(\varepsilon^{n+2})
\tag{1.22}
$$

in the sense of (1.13). *Moreover,*

$$
\begin{aligned}
&\mu_n = \varepsilon^2(\beta_1+\varepsilon\beta_2+\cdots+\varepsilon^n\beta_n),\\
&\beta_1 = \frac{1}{\sqrt{2l}\cosh^2 h_0}\int_{\mathbb{R}}|\tilde{V}(p)|^2 f(p)\,dp,\quad \tilde{a}_0(p) = \frac{\sqrt{\pi}\tilde{V}(p)}{\beta_1\sqrt{l}\cosh^2 h_0},\\
&a_1(x) = -\frac{\beta_2}{\beta_1^2}\frac{\sqrt{\pi}V(x)}{\sqrt{l}\cosh^2 h_0}\\
&\qquad + \frac{1}{\beta_1\sqrt{2l}\cosh^2 h_0}\left[\int_{\mathbb{R}}\frac{e^{itx}\tilde{V}(t)\mathcal{I}_1(t,x)dt}{L(t)\cosh(\tau h_0)}\right] - \frac{\pi\sqrt{2}\tilde{V}'(0)V(x)}{\beta_1 l\sqrt{l}\cosh^2 h_0},
\end{aligned}
\tag{1.23}
$$

where $f(p)$ *is a positive function,*

$$
f(p) = \frac{\tau - \tanh h_0\tanh(\tau h_0)}{\tau^2\tanh\tau h_0 - \tanh h_0\tau},\qquad \tau = \sqrt{1+p^2},
$$

l *is defined by* (1.20), *and the remaining values* $\beta_2,\ldots,\beta_n$ *and functions* $a_2,\ldots,a_n$ *are determined from the corresponding system,* (9.3)–(9.6); $\mathcal{I}_1(t,x)$ *is determined from* (8.20), $d(\varepsilon)$ *is determined from* (12.21).

2. Heuristic considerations

Before we go further, we would like to give some heuristic considerations which explain the specific form of the asymptotics of the eigenfunctions in Theorems 1.1 and 1.3.

We will present these arguments only in the simplest case of the Schrödinger equation in dimension 1; their generalizations for other cases are straightforward. Thus we would like to construct an approximate solution of

$$
-\Psi'' + \varepsilon V(x)\Psi = E\Psi.
\tag{2.1}
$$

We already know (although we will also obtain this fact later) that if $\int V(x)\,dx < 0$, then for the energy we have $E = O(\varepsilon^2)$, say $E = -\mu^2$. After performing the Fourier transform in (2.1), we obtain

$$\tilde{\Psi}(p)(p^2+\mu^2) = -\varepsilon \int \tilde{V}(p-p')\tilde{\Psi}(p')\,dp', \tag{2.2}$$

where the tilde denotes the Fourier transform. Obviously, for $x \notin \operatorname{supp} V(x)$, we have $\Psi \sim e^{-\mu|x|}$ with appropriate constants and the Fourier transform of this function is a δ-type sequence. Hence the integral in the right-hand side of (2.2) is approximately equal to

$$-\varepsilon C\tilde{V}(p)$$

with some normalization constant C. Therefore, by (2.2), $\tilde{\Psi}(p)$ is approximately equal to

$$\mathrm{Const}\,\frac{A(p)}{p^2+\mu^2}, \tag{2.3}$$

where $A(p) = \tilde{V}(p)$ is a function from $S(\mathbb{R})$. The singular dependence of the eigenfunction on ε is reflected in the denominator in (2.3). We note that the structure of (2.3) is identical to formulas of Theorems 1.1 and 1.3. Furthermore, expanding A and μ in regular series in ε, calculating the asymptotics of the integral in (2.2) and equating to zero the coefficients of like powers of ε, one obtains the complete asymptotic series for Ψ. The theorem on closeness of formal asymptotics to the exact solution [**7**] provides the final step of the proof.

In conclusion, a few words about the water wave problem. As shown in [**11**], problem (1.3) reduces to the following integral equation for the function $\phi(x) = \Phi|_{y=0}$:

$$\tilde{\phi}(p)(L(p)+\mu^2) = \varepsilon \int_\gamma M(\varepsilon,p,p')\tilde{\phi}(p')\,dp', \tag{2.4}$$

where $\mu^2 = \tanh h_0 - \omega^2$, $L(p)$ is defined in (1.16), γ is an appropriate contour in the complex plane, and the function $M(\varepsilon,p,p')$ is analytic in p, p' along γ and linear in ε. Since $L(p) \sim \mathrm{Const}\,p^2$ for small p, we see that (2.4) is similar to (2.2) and our arguments are still valid if we change the denominator in (2.3) to $L(p)+\mu^2$.

In fact, our method can be applied to more general equations of the type

$$K(-i\partial/\partial x, x, \varepsilon)\psi = \mu\psi$$

with a pseudodifferential operator K whose symbol satisfies

$$K(p,x,\varepsilon) = p^2 + \varepsilon V(x) + O(p^3);$$

for example, the discrete Schrödinger equation satisfies this condition. Here we prefer to explain the details of the method by means of two examples described above.

3. Proof of part 1) of Theorem 1.1 (nondegenerate case)

In this section we will give the proof of the first part of Theorem 1.1.

Passing to the Fourier transform in (1.1), we obtain

$$(p^2 - E)\tilde{\Psi}(p) = -\frac{\varepsilon}{\sqrt{2\pi}} \int_{\mathbb{R}} \tilde{V}(p-p')\tilde{\Psi}(p')\,dp'. \tag{3.1}$$

According to the scheme outlined in the previous section, we look for the approximate solution of this equation in the form

$$\tilde{\psi}_n(p) = \varepsilon B_n \frac{A_n(p)}{p^2 + \varepsilon^2 B_n^2}, \quad A_n(p) = a_0(p) + \varepsilon a_1(p) + \cdots + \varepsilon^n a_n(p). \tag{3.2}$$

We assume that $a_0(p) \not\equiv 0$ and

$$B_n = \beta_0 + \varepsilon\beta_1 + \cdots + \varepsilon^n \beta_n; \tag{3.3}$$

here a_k, β_k are unknown. The approximate energy level is

$$E_n = -\varepsilon^2 B_n^2. \tag{3.4}$$

We look for the solution satisfying the normalization conditions

$$a_0(0) = 1, \qquad a_k(0) = 0, \quad k = 2, \dots, n. \tag{3.5}$$

Afterwards we will multiply this solution by a normalization constant in order that the norm of the eigenfunction be of order $1 + O(\varepsilon^{n+1})$. This is necessary for the application of two basic lemmas from [**7**] which provide a rigorous foundation of our method.

Our goal is to construct $\beta_0, \beta_1 \dots, \beta_n$ and functions $a_0(p), \dots, a_n(p)$ such that $\tilde{\psi}_n(p)$ satisfy equation (3.1) up to $O(\varepsilon^{n+2})$, where $\|O(\varepsilon^{n+2})\| \le \mathrm{Const}\, \varepsilon^{n+2}$, and the norm here is understood in the sense of L_2.

Substituting (3.2) and (3.4) in (3.1), we obtain an equivalent equation

$$\varepsilon B_n A_n(p) = -\frac{\varepsilon^2 B_n}{\sqrt{2\pi}} \int_{\mathbb{R}} \frac{\tilde{V}(p-p')A_n(p')\,dp'}{p'^2 + \varepsilon^2 B_n^2}. \tag{3.6}$$

We will need an auxiliary lemma on the asymptotic expansions of integrals of the form

$$\int_{\mathbb{R}} \frac{\phi(p,t)}{t^2 + \mu^2}\, dt \tag{3.7}$$

as $\mu \to 0$, where $\phi(p,t)$ is an entire function in t and belongs to $S(\mathbb{R}_t)$ uniformly in $p \in \mathbb{R}$. There are several methods for calculating such asymptotics (see [**3**]); for our case the method based on the calculus of residues turns out to be more convenient. Introduce the contour in the complex plane $\mathbb{C}$ (see Figure 4):

$$\gamma := (-\infty, -1] \cup \{x + iy : x^2 + y^2 = 1, |x| \le 1, y > 0\} \cup [1, +\infty).$$

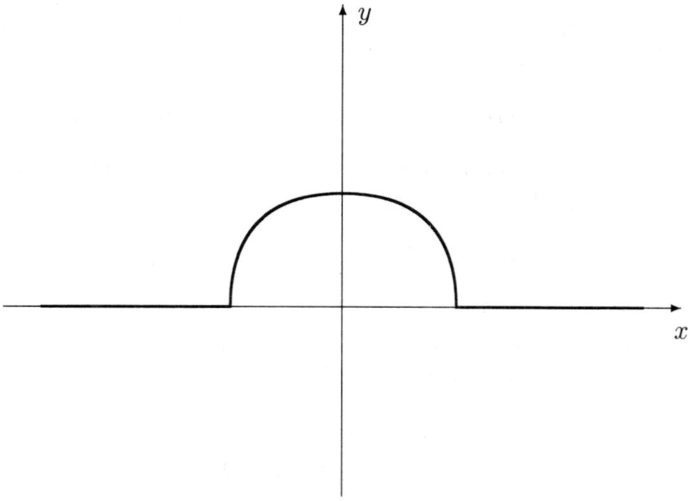

Figure 4

LEMMA 3.1. *Let $\phi(t)$ be an entire function and $\phi(t) \in S(\mathbb{R})$, $t \in \mathbb{R}$. Then as $\mu \to +0$*

$$\int_{\mathbb{R}} \frac{\phi(t)\,dt}{t^2+\mu^2} = \sum_{k=0}^{n} \alpha_k \mu^k \int_\gamma \frac{\phi(t)\,dt}{t^{k+2}} + \mu^{2([n/2]+1)} \int_\gamma \frac{\phi(t)\,dt}{t^{2([n/2]+1)}(t^2+\mu^2)} + \frac{\pi}{\mu}\left\{ \sum_{k=0}^{n} (i\mu)^k \frac{\phi^k(0)}{k!} + \alpha_{2([n/2]+1)} \frac{(i\mu)^{n+1}}{n!} \int_0^1 (1-t)^n \phi^{(n+1)}(ti\mu)\,dt \right\}, \tag{3.8}$$

where $\alpha_k = (-1)^{k/2}(1+(-1)^k)/2$.

PROOF. By the Cauchy residue theorem

$$\int_{\mathbb{R}} \frac{\phi(t)}{t^2+\mu^2} = \int_\gamma \frac{\phi(t)\,dt}{t^2+\mu^2} + \frac{\pi}{\mu}\phi(i\mu). \tag{3.9}$$

From the geometric series expansion

$$\frac{1}{t^2+\mu^2} = \sum_{k=0}^{n} \alpha_k \frac{\mu^k}{t^{k+2}} + \alpha_{2([n/2]+1)} \frac{\mu^{2([n/2]+1)}}{t^{2([n/2]+1)}(t^2+\mu^2)} \tag{3.10}$$

with the coefficients $\alpha_k = (-1)^{k/2}(1+(-1)^k)$, we obtain the following expansion for the first summand in the right-hand side of (3.9):

$$\int_\gamma \frac{\phi(t)\,dt}{t^2+\mu^2} = \sum_{k=0}^{n} \alpha_k \mu^k \int_\gamma \frac{\phi(t)\,dt}{t^{k+2}} + \alpha_{2([n/2]+1)} \mu^{2([n/2]+1)} \int_\gamma \frac{\phi(t)\,dt}{t^{2([n/2]+1)}(t^2+\mu^2)}. \tag{3.11}$$

Using the analyticity of the function ϕ in $\mathbb{C}$, expanding $\phi(i\mu)$ in the Taylor series at the point 0 and writing the remainder term in the form of Cauchy, we obtain

$$\frac{\pi}{\mu}\phi(i\mu) = \frac{\pi}{\mu}\left\{\sum_{k=0}^{n}\frac{\phi^{(k)}(0)}{k!}(i\mu)^k + \frac{(i\mu)^{n+1}}{n!}\int_0^1 (1-t)^n\phi^{(m)}(ti\mu)\,dt\right\}. \tag{3.12}$$

From this formula, (3.9) and (3.11) we get (3.8). The lemma is proved. □

Let us continue the proof of Theorem 1.1. Expanding the left-hand side of (3.6) in ε, using (3.2) and (3.3), we obtain

$$\varepsilon B_n A_n(p) = \sum_{k=1}^{n+1}\varepsilon^k\left(\sum_{l=0}^{k}\beta_l a_{k-l}(p)\right) + \varepsilon^{n+2}R_{n+2}(\varepsilon,\bar\beta,\bar a), \tag{3.13}$$

where $\bar\beta = (\beta_0,\beta_1,\dots,\beta_n)$, $\bar a = (a_0,a_1,\dots,a_n)$, and $R_{n+2}(\cdot,\cdot,\cdot)$ is a polynomial in its arguments. Substituting in Lemma 3.4

$$\mu \equiv \mu_n = \varepsilon B_n, \tag{3.14}$$

$\phi(t) = \varepsilon B_n\tilde V(p-t)A_n(t)$ and calculating the coefficients of $\varepsilon^0,\varepsilon^1,\varepsilon^2$ and also observing how $\beta_0,\dots,\beta_{n-1},a_{n-2},a_{n-1},a_n$ enter the coefficient of ε^n, we obtain the expansion of the integral in the right-hand side of (3.6):

$$\begin{aligned}
\varepsilon B_n\int_{\mathbb R}\frac{\tilde V(p-t)A_n(t)\,dt}{t^2+\varepsilon^2B_n^2} &= \pi a_0(0)\tilde V(p)\\
&- \varepsilon\pi\left\{\beta_0\left[ia_0(0)\tilde V'(p) - ia_0'(0)\tilde V(p) - \frac{1}{\pi}\int_\gamma\frac{\tilde V(p-t)}{t^2}a_0(t)\,dt\right] - a_1(0)\tilde V(p)\right\}\\
&- \varepsilon^2\pi\left\{\beta_0\left[ia_1(0)\tilde V'(p) - ia_1'(0)\tilde V(p) - \frac{1}{\pi}\int_\gamma\frac{\tilde V(p-t)}{t^2}a_1(t)\,dt\right]\right.\\
&\qquad + \beta_0^2\left[\frac{1}{2}a_0(0)\tilde V''(p) - a_0'(0)\tilde V'(p) + \frac{1}{2}a_0''(0)\tilde V(p)\right]\\
&\qquad \left. + \beta_1\left[ia_0(0)\tilde V'(p) - ia_0'(0)\tilde V(p) - \frac{1}{\pi}\int_\gamma\frac{\tilde V(p-t)}{t^2}a_0(t)\,dt\right] - a_2(0)\tilde V(p)\right\}\\
&- \sqrt{2\pi}\sum_{k=3}^{n-1}\varepsilon^k\left\{\int_\gamma\frac{\tilde V(p-t)S_k(\bar\beta_{0,k-1},\bar a_{0,k-1}(t),t)\,dt}{t^{l_k}}\right.\\
&\qquad \left. + P_k\left(\bar\beta_{0,k-1},\bar a^{(i)}_{0,k-1}(0),\tilde V(p),\tilde V'(p),\dots,\tilde V^{(k)}(p)\right)\right\}\\
&- \varepsilon^n\pi\left\{\beta_0\left[ia_{n-1}(0)\tilde V'(p) - ia_{n-1}'(0)\tilde V(p) - \frac{1}{\pi}\int_\gamma\frac{\tilde V(p-t)}{t^2}a_{n-1}(t)\,dt\right]\right.\\
&\qquad + \beta_0^2\left[\frac{1}{2}a_{n-2}(0)\tilde V''(p) - a_{n-2}'(0)\tilde V'(p) + \frac{1}{2}a_{n-2}''(0)\tilde V(p)\right]\\
&\qquad + \beta_1\left[ia_{n-2}(0)\tilde V'(p) - ia_{n-2}'(0)\tilde V(p) - \frac{1}{\pi}\int_\gamma\frac{\tilde V(p-t)}{t^2}a_{n-2}(t)\,dt\right]\\
&\qquad + \int_\gamma\frac{\tilde V(p-t)T_n(\bar\beta_{3,n-2},\bar a_{2,n-3}(t))\,dt}{t^{l_n}}\\
&\qquad + R_n\left(\bar\beta_{3,n-2},\bar a^{(i)}_{2,n-3}(0),\tilde V(p),\dots,\tilde V^{(n)}(p)\right)
\end{aligned} \tag{3.15}$$

$$+ \beta_{n-1}\left[ia_0(0)\tilde{V}'(p) - ia_0'(0)\tilde{V}(p) - \frac{1}{\pi}\int_\gamma \frac{\tilde{V}(p-t)}{t^2}a_0(t)\,dt\right] - a_n(0)\tilde{V}(p)\Big\}$$

$$+ \varepsilon^{n+1}\Bigg[\int_\gamma \frac{\tilde{V}(p-t)S_{n+1}(\bar{\beta},\varepsilon,\bar{a}(t),t)\,dt}{t^{l_{n+1}}(t^2+\varepsilon^2 B_n^2)}$$

$$+ P_{n+1}\left(\bar{\beta},\varepsilon,\bar{a}^{\bar{i}}(0),\tilde{V}(p),\tilde{V}'(p),\dots,\tilde{V}^{(n)}(p)\right)$$

$$+ \int_0^1 (1-t)^n Q_{n+1}\left(\bar{\beta},\varepsilon,\bar{a}^{(\bar{i})}(ti\varepsilon B_n),\tilde{V}(p-ti\varepsilon B_n),\dots,\tilde{V}^{(n)}(p-ti\varepsilon B_n)\right)dt\Bigg],$$

where

$$\begin{aligned}
&\bar{\beta}_{p,r} = (\beta_p,\dots,\beta_r), && \bar{a}_{p,r}(t) = \big(a_p(t),\dots,a_r(t)\big), \quad p<r,\\
&\bar{\beta} = (\beta_0,\dots,\beta_n), && \bar{a}(t) = \big(a_0(t),\dots,a_n(t)\big),\\
&\bar{a}^{(\bar{i})}_{p,r}(t) = (a_p^{(i_1)},\dots,a_r^{(i_{r-p+1})}), && \bar{a}^{(\bar{i})}(t) = (a_0^{(i_1)},\dots,a_n^{(i_n)}),\\
&\bar{i} = (i_1,i_2,\dots,i_n), \quad i_k \le n+1;
\end{aligned} \tag{3.16}$$

S_k, P_k for $k=3,\dots,n-1$, $T_n, R_n, S_{n+1}, P_{n+1}, Q_{n+1}$ are polynomials in their arguments which contain only the terms proportional to powers of $\tilde{V}(p)$ (or to powers of $\tilde{V}(p-ti\varepsilon B_n)$ for Q_{n+1}) and their derivatives.

Multiplying the obtained expression by $-\varepsilon/\sqrt{2\pi}$ and equating in (3.6) the coefficients of ε^k, $k=1,\dots,n+1$, to zero, we obtain the system for β_k, a_k, $k=0,1,\dots,n$:

$$\beta_0 a_0(p) = -\sqrt{\frac{\pi}{2}}a_0(0)\tilde{V}(p), \tag{3.17}$$

$$\begin{aligned}
&\beta_0 a_1(p) + \beta_1 a_0(p)\\
&\quad = \sqrt{\frac{\pi}{2}}\Big\{\beta_0\left[ia_0(0)\tilde{V}'(p) - ia_0'(0)\tilde{V}(p) - \frac{1}{\pi}\int_\gamma\frac{\tilde{V}(p-t)}{t^2}a_0(t)\,dt\right]\\
&\qquad - a_1(0)\tilde{V}(p)\Big\},
\end{aligned} \tag{3.18}$$

$$\begin{aligned}
&\beta_0 a_2(p) + \beta_1 a_1(p) + \beta_2 a_0(p)\\
&\quad = \sqrt{\frac{\pi}{2}}\Big\{\beta_0\left[ia_1(0)\tilde{V}'(p) - ia_1'(0)\tilde{V}(p) - \frac{1}{\pi}\int_\gamma\frac{\tilde{V}(p-t)}{t^2}a_1(t)\,dt\right]\\
&\quad + \beta_0^2\left[\frac{1}{2}a_0(0)\tilde{V}''(p) - a_0'(0)\tilde{V}'(p) + \frac{1}{2}a_0''(0)\tilde{V}(p)\right]\\
&\quad + \beta_1\left[ia_0(0)\tilde{V}'(p) - ia_0'(0)\tilde{V}(p) - \frac{1}{\pi}\int_\gamma\frac{\tilde{V}(p-t)}{t^2}a_0(t)\,dt\right]\\
&\qquad - a_2(0)\tilde{V}(p)\Big\},
\end{aligned} \tag{3.19}$$

$$\begin{aligned}
&\beta_0 a_3(p) + \beta_1 a_2(p) + \beta_2 a_1(p) + \beta_3 a_0(p)\\
&\quad = \sqrt{\frac{\pi}{2}}\Big\{\beta_0\left[ia_2(0)\tilde{V}'(p) - ia_2'(0)\tilde{V}(p) - \frac{1}{\pi}\int_\gamma\frac{\tilde{V}(p-t)}{t^2}a_2(t)\,dt\right]\\
&\quad + \beta_0^2\left[\frac{1}{2}a_2(0)\tilde{V}''(p) - a_1'(0)\tilde{V}'(p) + \frac{1}{2}a_1''(0)\tilde{V}(p)\right]
\end{aligned} \tag{3.20}$$

$$
\begin{aligned}
&+\beta_1\left[ia_1(0)\tilde V'(p)-ia_1'(0)\tilde V(p)-\frac{1}{\pi}\int_\gamma \frac{\tilde V(p-t)}{t^2}a_1(t)\,dt\right]\\
&+\cdots+\beta_2\left[ia_0(0)\tilde V'(p)-ia_0'(0)\tilde V(p)-\frac{1}{\pi}\int_\gamma \frac{\tilde V(p-t)}{t^2}a_0(t)\,dt\right]\\
&\qquad\qquad -a_3(0)\tilde V(p)\Bigg\}
\end{aligned}
$$

. . . .

$$
\begin{aligned}
&\beta_0 a_k(p)+\beta_1 a_{k-1}(p)+\cdots+\beta_k a_0(p)\\
&\quad=\int_\gamma \frac{\tilde V(p-t)S_k(\bar\beta_{0,k-1},\bar a_{0,k-1}(t),t)\,dt}{t^{l_k}}\\
&\qquad+P_k\left(\bar\beta_{0,k-1},\bar a^{(\bar i)}_{0,k-1}(0),\tilde V(p),\tilde V'(p),\dots,\tilde V^{(k)}(p)\right)
\end{aligned}
\tag{3.21}
$$

. . . .

$$
\begin{aligned}
&\beta_0 a_n(p)+\beta_1 a_{n-1}(p)+\beta_2 a_{n-2}(p)+\cdots+\beta_{n-1}a_1(p)+\beta_n a_0(p)\\
&\quad=\sqrt{\frac{\pi}{2}}\Bigg\{\beta_0\bigg[ia_{n-1}(0)\tilde V'(p)-ia_{n-1}'(0)\tilde V(p)\\
&\qquad\qquad-\frac{1}{\pi}\int_\gamma \frac{\tilde V(p-t)}{t^2}a_{n-1}(t)\,dt\bigg]\\
&\quad+\beta_0^2\left[\frac12 a_{n-2}(0)\tilde V''(p)-a_{n-2}'(0)\tilde V'(p)+\frac12 a_{n-2}''(0)\tilde V(p)\right]\\
&\quad+\beta_1\left[ia_{n-2}(0)\tilde V'(p)-ia_{n-2}'(0)\tilde V(p)-\frac{1}{\pi}\int_\gamma \frac{\tilde V(p-t)}{t^2}a_{n-2}(t)\,dt\right]\\
&\quad+\int_\gamma \frac{\tilde V(p-t)T_n(\bar\beta_{3,n-2},\bar a_{2,n-3}(t))\,dt}{t^{l_n}}\\
&\quad+R_n\left(\bar\beta_{3,n-2},\bar a^{(\bar i)}_{2,n-3}(0),\tilde V(p),\dots,\tilde V^{(n)}(p)\right)\\
&\quad+\beta_{n-1}\left[ia_0(0)\tilde V'(p)-ia_0'(0)\tilde V(p)-\frac{1}{\pi}\int_\gamma \frac{\tilde V(p-t)}{t^2}a_0(t)\,dt\right]\\
&\qquad\qquad-a_n(0)\tilde V(p)\Bigg\}.
\end{aligned}
\tag{3.22}
$$

Note that equation (3.22) gives the first four equtions (3.17)–(3.20) if we substitute 0, 1, 2, 3 in place of n, respectively, and set to zero the terms with two indices with the first being less than the second and the terms with negative indices.

LEMMA 3.2. *System* (3.17)–(3.22) *for* $\beta_0,\dots,\beta_n$, $a_0(p),\dots,a_n(p)$ *is uniquely solvable under conditions* (3.5), *and its solutions* $a_0,\dots,a_n$ *belong to* $S(\mathbb{R})$.

PROOF. Setting $p=0$ in (3.17) and taking into account that by (1.5) $\tilde V(0)\neq 0$, we obtain

$$
\beta_0=-\sqrt{\frac{\pi}{2}}\tilde V(0). \tag{3.23}
$$

By the condition $a_0(0) = 1$, we obtain from (3.18)

$$a_0(p) = \frac{\tilde{V}(p)}{\tilde{V}(0)}. \tag{3.24}$$

Set $p = 0$ in (3.18). By (3.24) and the condition $a_1(0) = 0$, we obtain

$$\beta_1 = \frac{1}{2}\int_\gamma \frac{\tilde{V}(t)\tilde{V}(-t)}{t^2}\,dt. \tag{3.25}$$

Now let us find $a_1(p)$ from (3.18). Substituting (3.23), (3.24) and (3.25) in (3.18) and taking into account the fact that $a_0(0) = 1$, we obtain

$$\begin{aligned} a_1(p) = {} & \frac{i}{\tilde{V}(0)}\sqrt{\frac{\pi}{2}}\left[\tilde{V}'(p)\tilde{V}(0) - \tilde{V}'(0)\tilde{V}(p)\right] \\ & - \frac{1}{\sqrt{2\pi}\tilde{V}(0)}\int_\gamma \frac{\tilde{V}(t)\tilde{V}(p-t)}{t^2}\,dt + \frac{\tilde{V}(p)}{\sqrt{2\pi}\tilde{V}(0)^2}\int_\gamma \frac{\tilde{V}(t)\tilde{V}(-t)}{t^2}\,dt. \end{aligned} \tag{3.26}$$

We see that indeed $a_1(0) = 0$. Proceeding similarly, we obtain β_n and a_n assuming that $\beta_0, \beta_1, \dots, \beta_{n-1}, a_0, a_1, \dots, a_{n-1}$ are known and $a_k(0) = 0, k = 2, \dots, n-1$. We look for $a_n(p)$ such that $a_n(0) = 0$. Setting $p = 0$ in (3.22) and taking into account the fact that $a_0(0) = 1$, we obtain

$$\begin{aligned} \beta_n = {} & \beta_0\left[-i\sqrt{\frac{\pi}{2}}a'_{n-1}(0)\tilde{V}(0) - \frac{1}{\sqrt{2\pi}}\int_\gamma \frac{\tilde{V}(-t)a_{n-1}(t)}{t^2}\,dt\right] + \dots \\ & + \beta_{n-1}\left[i\sqrt{\frac{\pi}{2}}\tilde{V}'(0) - i\sqrt{\frac{\pi}{2}}a'_0(0)\tilde{V}(0) - \frac{1}{\sqrt{2\pi}}\int_\gamma \frac{\tilde{V}(p-t)a_0(t)}{t^2}\,dt\right]; \end{aligned} \tag{3.27}$$

that is, β_n is uniquely determined. Substituting (3.27) and $\beta_0, \dots, \beta_{n-1}$, $a_0, \dots$, a_{n-1} in (3.22), we see that $a_n(p)$ is uniquely determined since $\beta_0 \neq 0$. It is not hard to see that indeed $a_n(0) = 0$.

Thus we are left with the proof of the fact that all the $a_k \in S(\mathbb{R})$. This is obvious for $k = 0$. For the function a_1 this follows from (3.26), the Peetre inequality [**10**]

$$(1 + |\theta|^2)^s \leq 2^{|s|}(1 + |\theta - \theta'|^2)^{|s|}(1 + |\theta'|^2)^s$$

valid for all $\theta, s \in \mathbb{R}$, and the condition $\tilde{V}(p - t) \in S(\mathbb{R})$. In fact, the first and the third summands in the right-hand side of (3.26) belong to $S(\mathbb{R})$ because $\tilde{V}$ and $\tilde{V}'$ belong to $S(\mathbb{R})$. The second summand belongs to $S(\mathbb{R})$ too by the Peetre inequality. Let us prove it in more detail. For arbitrary $N \in \mathbb{N}$ from the condition $\tilde{V} \in S$, we have

$$\begin{aligned} \left|\partial_p^k \int \gamma \frac{\tilde{V}(t)\tilde{V}(p-t)\,dt}{t^2}\right| &= \left|\int \gamma \frac{\tilde{V}(t)\partial_p^k\tilde{V}(p-t)\,dt}{t^2}\right| \\ &\leq \int \gamma \frac{|\tilde{V}(t)\partial_p^k\tilde{V}(p-t)\,dt|}{|t|^2} \leq C_{k,N}\int \gamma \frac{|dt|}{(1+|t|^2)^N(1+|p-t|^2)^N}. \end{aligned} \tag{3.28}$$

By the Peetre inequality for $s = N$, $\theta = p$, and $\theta' = t$, we obtain the desired statement from (3.28).

The corresponding assertions for a_k, $k > 1$, are proved by induction. Lemma 3.2 is proved. □

Let us complete the proof of Theorem 1.1. From Lemma 3.2 and (3.17)–(3.22) it follows that B_n and A_n expressed in terms of values $\beta_0, \dots, \beta_n$ and functions $a_0 \dots, a_n$ by means of (3.2), (3.3) solve equation (3.6) up to $O(\varepsilon^{n+2})$ in the norm of $L_2(\mathbb{R})$. To prove this, consider the coefficient of ε^{n+1} in (3.15). We have $P_{n+1} \in S(\mathbb{R})$ uniformly with respect to $\varepsilon \in (0,1]$ because $\tilde{V}(p) \in S(\mathbb{R})$. The first summand in the coefficient at ε^{n+1} belongs to $S(\mathbb{R})$ by the Peetre inequality and the same inclusion for $\tilde{V}$ and a_i.

Finally, the third summand in the same coefficient belongs to $S(\mathbb{R})$ by the following inequalities for $t, \varepsilon \in (0,1]$ and some N:

$$\left|Q_{n+1}\left(\bar{\beta}, \varepsilon, \bar{a}^{(i)}(ti\varepsilon B_n), \tilde{V}(p - ti\varepsilon B_n), \dots, \tilde{V}^{(n)}(p - ti\varepsilon B_n)\right)\right| \le \frac{1}{(1 + |p - \varepsilon t B_n|)^N} \le \frac{1}{|p|^N}. \tag{3.29}$$

This means that the function $\tilde{\psi}_n(p)$ in (3.2) solves equation (3.1) with E_n determined from (3.4), (3.3) up to $O(\varepsilon^{n+2})$. More precisely,

$$A_\varepsilon \psi_n = E_n \psi_n + O(\varepsilon^{n+2}), \tag{3.30}$$

where $O(\varepsilon^{n+2})$ is understood in the sense of the L_2-norm.

Now we are going to use an assertion from [**7**].

LEMMA 3.3. *Let A be a self-adjoint operator in a Hilbert space H. Let μ be some point on the real line, and let r be the distance from μ to the spectrum of the operator A. Then for each $g \in D(A)$ the following inequality is valid:*

$$r\|g\| \le \|(A - \mu)g\|. \tag{3.31}$$

We are going to apply this lemma to our situation. To this end, we consider the Schrödinger operator (1.1) $A_\varepsilon = -\Delta + \varepsilon V$ depending of the parameter ε as the operator A. The Hilbert space H in our case is the space $L_2(\mathbb{R})$. We use the following important fact proved in [**9**]: There exists $\varepsilon_0 > 0$ such that for all ε, $0 < \varepsilon < \varepsilon_0$, the operator A_ε has a unique eigenvalue E_ε, which is located below the essential spectrum $[0, \infty)$, and a unique (up to a constant factor) eigenfunction Ψ_ε from the space $L_2(\mathbb{R})$, i.e.,

$$A_\varepsilon \Psi_\varepsilon = E_\varepsilon \Psi_\varepsilon. \tag{3.32}$$

Our reasoning is as follows. Instead of finding an exact solution of equation (3.32), we solve it approximately, i.e., we find an approximate eigenvalue E_n (see (3.4)) and eigenfunction ψ_n (we omit here the index ε but keep in mind that these symbols correspond to (3.32)). Namely, we have proved equality (3.30).

We are going to substitute the distance from the approximate eigenvalue E_n to the spectrum instead of r in inequality (3.31) and the approximate eigenfunction ψ_n instead of g. To obtain the required estimate for r which is really the difference between the exact eigenvalue $E \equiv E_\varepsilon$ and the approximate eigenvalue E_n, it is necessary only to normalize the eigenfunction ψ_n according to (1.2). Namely, we construct an eigenfunction Ψ_n satisfying the following normalization condition:

$$\|\Psi_n\| - 1 = O(\varepsilon^{n+1}). \tag{3.33}$$

We do this by choosing a constant $c(\varepsilon)$ such that, after multiplying ψ_n by it, we obtain an eigenfunction Ψ_n satisfying estimate (3.33).

From (3.2) and the Parseval equality it follows that

$$\|\psi_n(x)\| = \varepsilon|B_n|\left(\int_{\mathbb{R}} \frac{|A_n(p)|^2 dp}{|p^2+\varepsilon^2 B_n^2|^2}\right)^{1/2}. \tag{3.34}$$

Note that $\beta_0 > 0$ by (1.9) and (1.5). Thus for sufficiently small ε we have $|B_n| = B_n$.

From (3.3) it follows that the expression $(B_n)^{-1/2}$ is analytic in ε for small ε. We also need the asymptotics of the integral in (3.34). Applying arguments similar to those of Lemma 3.1 and taking into account the analyticity of $B_n^{-1/2}$, we can write the asymptotic expansion of the norm (3.34):

$$\|\psi_n(x)\| = \varepsilon^{-1/2}\left[l_0(\bar\beta, \bar a(p)) + \cdots + \varepsilon^n l_n(\bar\beta, \bar a(p)) + O(\varepsilon^{n+1})\right], \quad \varepsilon \to 0. \tag{3.35}$$

To obtain the principal term $l_0(\bar\beta, \bar a(p))$ of this expansion, we use the following method. Changing the variable in (3.34), $p \equiv p(\varepsilon, t) = \mu_n t$, where μ_n is determined from (3.14), we obtain

$$\|\psi_n(x)\| = \varepsilon^{-1/2}(B_n)^{-1/2}\left(\int_{\mathbb{R}} \frac{|A_n(t\mu_n)|^2 dt}{|t^2+1|^2}\right)^{1/2}. \tag{3.36}$$

Taking into account (3.5) and the fact that

$$\int_{\mathbb{R}} \frac{dt}{(t^2+1)^2} = \pi/2, \tag{3.37}$$

we have

$$l_0(\bar\beta, \bar a(p)) = \sqrt{\frac{\pi}{2\beta_0}}. \tag{3.38}$$

Now we are going to normalize the eigenfunction ψ_n selecting the appropriate constant $c(\varepsilon)$. From (3.35) it follows that this constant has the form

$$c(\varepsilon) = \varepsilon^{1/2}(d_0 + d_1\varepsilon + \cdots + d_n\varepsilon^n) + O(\varepsilon^{n+3/2}). \tag{3.39}$$

Multiplying expression (3.35) by the expression for d, we obtain

$$c(\varepsilon)\|\psi_n\| = d_0 l_0 + (d_0 l_1 + d_1 l_0)\varepsilon + \cdots + (d_0 l_n + \cdots + d_n l_0)\varepsilon^n + O(\varepsilon^{n+1}). \tag{3.40}$$

Now it is clear that we can make this product equal to $1 + O(\varepsilon^{n+1})$. In fact, to this end, it is sufficient to solve the system of equations with unknowns d_i, $i = 0, \dots, n$:

$$d_0 l_0 = 1, \qquad d_0 l_k + \cdots + d_k l_0 = 0, \quad k = 1, \dots, n. \tag{3.41}$$

Obviously, this is possible since $l_0 \neq 0$ by (3.38). Moreover,

$$d_0 = \sqrt{\frac{2\beta_0}{\pi}}. \tag{3.42}$$

Thus we have found the constant $c(\varepsilon)$ such that the approximate eigenfunction

$$\Psi_n(x) \equiv c(\varepsilon)\psi_n(x) \tag{3.43}$$

satisfies condition (3.33).

It is evident that this normalized eigenfunction Ψ_n satisfies (3.30) with a new residual term $O(\varepsilon^{n+2}\varepsilon^{1/2})$, i.e.,

$$A_\varepsilon \Psi_n = E_n \Psi_n + O(\varepsilon^{n+5/2}), \quad n = 0, \dots. \tag{3.44}$$

Now substituting $g = \Psi_n$ in Lemma 3.3, we obtain from estimate (3.31) the estimate (1.7).

To prove that Ψ_n is the asymptotics of the eigenfunction Ψ, i.e., to prove the estimate (1.8), we use an assertion from the same book [**7**].

LEMMA 3.4. *Let λ_0 be an isolated point of the spectrum of a self-adjoint operator A, let M_{λ_0} be the remaining spectrum (i.e., $M_{\lambda_0} = \sigma_A \setminus \{\lambda_0\}$), and let d_{λ_0} be the distance from the point μ to the set M_{λ_0}.*

Let P_{λ_0} be the projection operator onto the subspace generated by the eigenfunctions corresponding to the point λ_0.

Then for $g \in D(A)$ the following inequality is valid:

$$d_{\lambda_0}\|(1 - P_{\lambda_0})g\| \le \|(A - \mu)g\|. \tag{3.45}$$

We apply this lemma to our situation in the following way. The point λ_0 is the energy E which is the isolated eigenvalue of the Schrödinger operator A_ε; M_{λ_0} is the semiaxis $[0, \infty)$, $d_{\lambda_0} = |E_n|$. From (1.7), (1.9), (1.5) it follows that

$$c_1\varepsilon^2 \le |E_n| \le c_2\varepsilon^2, \qquad c_{1,2} > 0. \tag{3.46}$$

Furthermore, we take again the approximate eigenfunction Ψ_n as g. Then from (3.46), estimate (3.45) and (3.44), it follows that

$$\|\Psi_n - \langle\Psi_n, \Psi_\varepsilon\rangle\Psi_\varepsilon\| \le C\varepsilon^{n+1/2}, \tag{3.47}$$

where $\langle\Psi_n, \Psi_\varepsilon\rangle$ means the scalar product in L_2.

Let $c_\varepsilon \equiv \langle\Psi_n, \Psi_\varepsilon\rangle$. Note that by the Schwarz inequality, (1.2), and (3.33) $|c_\varepsilon| \le 1 + O(\varepsilon^{n+1})$. From this inequality and from (3.47) we obtain

$$\begin{aligned}\|\Psi_n - \Psi_\varepsilon\| &= \|\Psi_n - c_\varepsilon\Psi_\varepsilon + c_\varepsilon\Psi_\varepsilon - \Psi_\varepsilon\| \\ &\le O(\varepsilon^{n+1/2}) + |c_\varepsilon - 1| \le O(\varepsilon^{n+1/2}).\end{aligned} \tag{3.48}$$

This completes the proof of the first part of Theorem 1.1. □

4. Proof of part 2) of Theorem 1.1 (degenerate case)

Here we consider the case $\int_{\mathbb{R}} V(x)dx = 0$ which is equivalent to the condition

$$\tilde{V}(0) = 0. \tag{4.1}$$

We call this case degenerate in order to distinguish it from the case $\tilde{V}(0) \ne 0$.

In this case it is impossible to solve system (3.17)–(3.22) as directly as in the nondegenerate case because, substituting $p = 0$ in (3.17), we obtain

$$\beta_0 a_0(0) = 0.$$

Let us show that in this case

$$\beta_0 = 0, \qquad a_0(0) = 0. \tag{4.2}$$

Let us assume that $\beta_0 \ne 0$. Then by (3.17) we have $a_0(0) = 0$. Thus (using the same equation) $\beta_0 a_0(p) = 0$ for all $p \in \mathbb{R}$ and so $a_0(p) \equiv 0$. But it contradicts (3.2). Hence $\beta_0 = 0$. Since $\tilde{V} \not\equiv 0$, it follows that $a_0(0) = 0$.

Thus in this case $\beta_0 = 0$ in all equations of the system (3.17)–(3.22) (in particular, (3.17) disappears); therefore it is necessary to modify the reasoning which we applied in the nondegenerate case.

We look for a solution of system (3.18)–(3.22) (with $\beta_0 = 0$) which satisfies the conditions

$$a_0(0) = a_2(0) = \cdots = a_n(0) = 0, \quad a_1(0) = 1. \tag{4.3}$$

LEMMA 4.1. *There exists a unique solution of the system (3.18)–(3.22) satisfying conditions (4.3). Moreover, $\beta_1, \beta_2, a_0, a_1$ are determined from (1.14) and the remaining values of $\beta_3, \dots, \beta_n$ and functions $a_2, \dots, a_n$ are determined from the system (3.18)–(3.22). All the functions $a_k(x) \in S(\mathbb{R})$, $k = 0, 1 \dots, n$.*

PROOF. Let us consider equation (3.19). Since $\beta_0 = 0$, $a_1(0) = 1$, this equation is equivalent to the equation

$$\beta_1 a_0(p) = -\sqrt{\frac{\pi}{2}}\tilde{V}(p). \tag{4.4}$$

Let us assume that $\beta_1 \neq 0$. Afterwards we will see that this is indeed true. Therefore,

$$a_0(p) = -\frac{\sqrt{\pi/2}}{\beta_1}\tilde{V}(p). \tag{4.5}$$

Now we substitute this expression in (3.19). Setting $p = 0$ in this equation and taking into account the fact that $a_1(0) = 1$ (we will verify that this is true later), $a_0(0) = 0$, we obtain

$$\beta_1 = \frac{1}{2}\int_{\gamma 1} \frac{\tilde{V}(t)\tilde{V}(-t)}{t^2}\,dt. \tag{4.6}$$

Now we see that $\beta_1 \neq 0$ indeed because

$$\int_{\gamma 1} \frac{\tilde{V}(t)\tilde{V}(-t)}{t^2}\,dt = \int_{\mathbb{R}} \frac{\tilde{V}(t)\tilde{V}(-t)}{t^2}\,dt = \int_{\mathbb{R}} \frac{|\tilde{V}(t)|^2}{t^2}\,dt,$$

and $V(t)$ is a real nonzero function. (Note that in this case we can change the contour γ to $\mathbb{R}$ because $\tilde{V}(0) = 0$.)

Thus we have found β_1 under the assumptions

$$a_0(0) = 0, \quad a_1(0) = 1. \tag{4.7}$$

Expressing $a_0(p)$ from (3.18), we get

$$a_0(p) = \frac{-\sqrt{\pi/2}\tilde{V}(p)}{\beta_1}, \tag{4.8}$$

from which it follows that $a_0(0) = 0$ indeed.

Now we are going to find an expression for $a_1(p)$ in terms of β_2 under the assumptions

$$a_0(0) = 0, \quad a_1(0) = 1, \quad a_2(0) = 0. \tag{4.9}$$

We will check later that the second condition indeed holds.

From equation (3.19) and conditions (4.3) we obtain

$$\beta_1 a_1(p) + \beta_2 a_0(p) = \beta_1\left[-i\sqrt{\frac{\pi}{2}}a_0'(0)\tilde{V}(p) - \frac{1}{\sqrt{2\pi}}\int_\gamma \frac{\tilde{V}(p-t)}{t^2}a_0(t)\,dt\right]. \tag{4.10}$$

From this equation $a_1(p)$ is expressed by means of β_2 in the following way:

$$a_1(p) = -\frac{\beta_2}{\beta_1}a_0(p) - i\sqrt{\frac{\pi}{2}}a_0'(0)\tilde{V}(p) - \frac{1}{\sqrt{2\pi}}\int_\gamma a_0(t)\frac{\tilde{V}(p-t)}{t^2}\,dt. \tag{4.11}$$

Substituting expression (4.8) in (4.11), we obtain

$$a_1(p) = \sqrt{\frac{\pi}{2}}\frac{\beta_2}{\beta_1^2}\tilde{V}(p) + \frac{i\pi}{2\beta_1}\tilde{V}'(0)\tilde{V}(p) + \frac{1}{2\beta_1}\int_{\mathbb{R}+i}\frac{\tilde{V}(p-t)\tilde{V}(t)}{t^2}\,dt. \tag{4.12}$$

From (4.1) and (4.6) it follows that $a_1(0) = 1$ indeed.

Now we are going to find β_2 and, consequently, $a_1(p)$. To this end, let us consider equation (3.20) for $n = 3$. Taking into account the fact that $\beta_0 = 0$ and $a_1(0) = 1$, we see that this equation is equivalent to the following one:

$$\begin{aligned} &\beta_1 a_2(p) + \beta_2 a_1(p) + \beta_3 a_0(p) \\ &\quad = \beta_1\left[i\sqrt{\frac{\pi}{2}}\tilde{V}'(p) - i\sqrt{\frac{\pi}{2}}a_1'(0)\tilde{V}(p) - \frac{1}{\sqrt{2\pi}}\int_\gamma \frac{\tilde{V}(p-t)a_1(t)}{t^2}\,dt\right] \\ &\quad + \beta_2\left[-i\sqrt{\frac{\pi}{2}}a_0'(0)\tilde{V}(p) - \frac{1}{\sqrt{2\pi}}\int_\gamma \frac{\tilde{V}(p-t)a_0(t)}{t^2}\,dt\right] - \sqrt{\frac{\pi}{2}}a_3(0)\tilde{V}(p). \end{aligned} \tag{4.13}$$

Substituting $p = 0$ in (4.13) and taking into account the fact that $a_2(0) = 0$, we obtain

$$\begin{aligned} \beta_2 = \beta_1&\left[i\sqrt{\frac{\pi}{2}}\tilde{V}'(0) - \frac{1}{\sqrt{2\pi}}\int_\gamma \frac{\tilde{V}(-t)a_1(t)}{t^2}\,dt\right] \\ &+ \beta_2\left[-\frac{1}{\sqrt{2\pi}}\int \gamma\frac{\tilde{V}(-t)a_0(t)}{t^2}\,dt\right]. \end{aligned} \tag{4.14}$$

Substituting expression (4.12) instead of a_1 and expression (4.8) instead of $a_0(t)$ in (4.14), we obtain an equation for β_2. Solving this equation, we obtain expression (1.14).

Thus we have found β_2 under assumptions (4.9).

Substituting the expression obtained for β_2 into expression (4.12) for $a_1(p)$, we obtain the explicit expression (1.14) for $a_1(p)$, which determines a_1 uniquely (and not up to β_2).

Thus we determine $\beta_1, \beta_2, a_0, a_1$ under assumptions (4.9). Moreover, two of these conditions are fulfilled.

This process is continued inductively, and now we are going to show how β_n and a_{n-1} are found, considering that $\beta_1, \ldots, \beta_{n-1}, a_0, \ldots, a_{n-2}$ were already found.

We assume that

$$a_0(0) = 0, \quad a_1(0) = 1, \quad a_2(0) = \cdots = a_{n-2}(0) = 0. \tag{4.15}$$

Assume that

$$a_{n-1}(0) = a_n(0) = 0; \tag{4.16}$$

after finding β_n and a_{n-1}, we will verify that this is indeed true.

We consider equation (3.22), where, as always in the degenerate case, $\beta_0 = 0$. From this equation we express a_{n-1} by means of β_n taking into account (4.15) and the equality $a_n(0) = 0$:

(4.17)

$$\begin{aligned} a_{n-1}(p) = \frac{1}{\beta_1}\Bigg\{&[-\beta_2 a_{n-2}(p) - \cdots - \beta_{n-1}a_1(p) - \beta_n a_0(p)] \\ &+ \left[-\sqrt{\frac{\pi}{2}}a_{n-2}'(0)\tilde{V}(p) - \frac{1}{\sqrt{2\pi}}\int_\gamma \frac{\tilde{V}(p-t)a_{n-2}(t)}{t^2}\,dt\right] \end{aligned}$$

$$+\cdots+\beta_{n-1}\Big[-i\sqrt{\frac{\pi}{2}}a_0'(0)\tilde V(p)-\frac{1}{\sqrt{2\pi}}\int_\gamma\frac{\tilde V(p-t)a_0(t)}{t^2}\,dt\Big]\Big\}.$$

Let us rewrite (4.17) in the form

$$a_{n-1}(p)=-\frac{\beta_n}{\beta_1}a_0(p)+M\left(p,\bar\beta_{1,n-1},\bar a^{(\bar i)}_{0,n-2}\right), \tag{4.18}$$

where M does not depend of β_n.

Writing equation (3.22) for $n+1$, taking into account that again $\beta_0=0$ and (4.15), we obtain

$$\begin{aligned}(4.19)\quad &\beta_1a_n(p)+\beta_2a_{n-1}(p)+\cdots+\beta_{n-1}a_2(p)+\beta_na_1(p)+\beta_{n+1}a_0(p)\\ &=\sqrt{\frac{\pi}{2}}\Big\{\beta_1\Big[-ia_{n-1}'(0)\tilde V(p)-\frac{1}{\pi}\int_\gamma\frac{\tilde V(p-t)}{t^2}a_{n-1}(t)\,dt\Big]\\ &\quad+\int_\gamma\frac{\tilde V(p-t)T_n(\bar\beta_{3,n-1},\bar a_{2,n-2}(t))\,dt}{t^{l_{n+1}}}\\ &\quad+R_{n+1}\left(\bar\beta_{3,n-1},\bar a^{(\bar i)}_{2,n-2}(0),\tilde V(p),\dots,\tilde V^{(n+1)}(p)\right)\\ &\quad+\beta_n\Big[-ia_0'(0)\tilde V(p)-\frac{1}{\pi}\int_\gamma\frac{\tilde V(p-t)}{t^2}a_0(t)\,dt\Big]-a_n(0)\tilde V(p)\Big\}.\end{aligned}$$

Let us substitute $p=0$ in (4.19). From conditions (4.15) and (4.16) we obtain

$$\begin{aligned}(4.20)\quad \beta_n=\sqrt{\frac{\pi}{2}}\Big\{&\beta_1\Big[-\frac{1}{\pi}\int_\gamma\frac{\tilde V(-t)}{t^2}a_{n-1}(t)\,dt\Big]\\ &+\int_\gamma\frac{\tilde V(-t)T_n(\bar\beta_{3,n-1},\bar a_{2,n-2}(t))\,dt}{t^{l_{n+1}}}\\ &+R_{n+1}\left(\bar\beta_{3,n-1},\bar a^{(\bar i)}_{2,n-2}(0),\tilde V(0),\dots,\tilde V^{(n+1)}(0)\right)\\ &+\beta_n\Big[-\frac{1}{\pi}\int_\gamma\frac{\tilde V(-t)}{t^2}a_0(t)\,dt\Big]\Big\}.\end{aligned}$$

Substituting the expression obtained for β_n in expression (4.18) for $a_{n-1}(p)$, we determine $a_{n-1}(p)$. We have only to prove that

$$a_{n-1}(0)=0. \tag{4.21}$$

Of course, this follows from the fact that β_n is found from condition (4.21). Let us explain this in more detail.

We note that it is possible to write the system of equations (4.18) and (4.19) in the form

$$\beta_1a_n(p)+\beta_2a_{n-1}(p)+\beta_na_1(p)+\beta_{n+1}a_0(p)=F(p,a_{n-1}(p)), \tag{4.22}$$

where the structure of F is determined by the same equations (4.18) and (4.19) and a_{n-1}, in fact, depends on β_n according to (4.18). Substituting in this equation $p=0$ and assuming that $a_{n-1}(0)=0$ for all β_n, we find β_n from the equation $\beta_n=F(0,a_{n-1}(0))$. Now it is obvious from (4.22) that $a_{n-1}(0)=0$ indeed.

To complete the proof of Lemma 4.1, it is necessary to prove that all the functions a_k, $k=0,\dots,n$, belong to the space $S(\mathbb{R})$. But the proof of this fact is completely analogous to the proof of the corresponding statement of part 1) of Theorem 1.1. Lemma 4.1 is proved. □

Let us continue the proof of part 2) of Theorem 1.1. We solved system (3.17)–(3.22) in the degenerate case with the same correction $O(\varepsilon^{n+2})$, i.e., equality (3.30) is valid in this case too.

Furthermore, we use here (as in the first part) the two basic Lemmas 3.3 and 3.4. The difference is only in the normalization constant $c(\varepsilon)$ because in the degenerate case $\beta_0 = 0$ and $\beta_1 \neq 0$ and the approximate eigenvalue E_n has the following form (see (3.4), (4.2)):

$$E_n = -\varepsilon^4(\beta_1 + \varepsilon\beta_2 + \cdots + \varepsilon^{n-1}\beta_n). \tag{4.23}$$

We select the normalization constant $c(\varepsilon)$ here starting from the same inequality (3.33) as in the nondegenerate case. Repeating the arguments used for finding $c(\varepsilon)$ in the nondegenerate case, we obtain formula (3.34), where by (3.3), (4.2), (4.6)

$$B_n = \varepsilon\beta_1 + \cdots + \varepsilon^n\beta_n, \quad \beta_1 \neq 0. \tag{4.24}$$

Also, $A_n(p)$ is determined from (3.2) but, in contrast to the nondegenerate case, we have by (4.2) and (4.11)

$$a_0(0) = 0, \quad a_1(0) = 1 \tag{4.25}$$

From (3.34) and (4.24) it follows that

$$\|\psi_n(x)\| = \varepsilon^2(\beta_1 + \varepsilon\beta_2 + \cdots + \varepsilon^{n-1}\beta_{n-1}) \times \left(\int_{\mathbb{R}} \frac{|a_0(p) + \varepsilon a_1(p) + \cdots + \varepsilon^n a_n(p)|^2 dp}{|p^2 + \varepsilon^4(\beta_1 + \varepsilon\beta_2 + \cdots + \varepsilon^{n-1}\beta_n)^2|^2}\right)^{1/2}. \tag{4.26}$$

As in (3.35) we can write the asymptotics of the integral in (4.26) using a formula similar to that from Lemma 3.1. But in contrast to the nondegenerate case, the principal term of this asymptotics coincides with the principal term of the asymptotics of the expression

$$\left(\int_{\mathbb{R}} \frac{\varepsilon^2 a_1^2(p)\, dp}{(p^2 + \varepsilon^4\beta_1^2)^2}\right)^{1/2} \tag{4.27}$$

because of (4.25). Using the equality $a_1(0) = 1$, we calculate the principal term of (4.27) by means of the change of the variable described in the nondegenerate case. It turns out that this term is equal to

$$\frac{1}{\varepsilon^2\beta_1^{3/2}}\left(\int \frac{dt}{(t^2+1)^2}\right)^{1/2}.$$

Therefore the asymptotics of the norm (4.26) takes the form

$$\|\psi_n(x)\| = l_0(\bar\beta, \bar a(p)) + \cdots + \varepsilon^n l_n(\bar\beta, \bar a(p)) + O(\varepsilon^{n+1}), \quad \varepsilon \to 0, \tag{4.28}$$

where, taking into account (3.37),

$$l_0(\bar\beta, \bar a(p)) = \sqrt{\frac{\pi}{2\beta_1}}. \tag{4.29}$$

Furthermore, reasoning as in the nondegenerate case, we find the constant $c(\varepsilon)$ such that the norm of the function

$$\Psi_n \equiv c(\varepsilon)\psi_n \tag{4.30}$$

satisfies condition (3.33). It is clear that

$$c(\varepsilon) = \sqrt{\frac{2\beta_1}{\pi}} + d_1\varepsilon + \cdots + d_n\varepsilon^n. \tag{4.31}$$

From (4.31), (4.30), (3.2), and (3.14) we obtain (1.11).

Since $c(\varepsilon)$ is $O(1)$ (see (4.31)) and also from (4.30) and (3.30), Ψ_n satisfies equation (3.32) with the same correction as ψ_n, i.e.,

$$A_\varepsilon \Psi_n = E_n \Psi_n + O(\varepsilon^{n+2}). \tag{4.32}$$

Thus, substituting $g = \Psi_n$ in Lemma 3.3, from (4.32) and (3.31) we obtain estimate (1.12).

Finally, to prove that Ψ_n is the asymptotics of the eigenfunction in the degenerate case, we note that

$$c_1\varepsilon^4 \le E \le c_2\varepsilon^4, \qquad c_{1,2} > 0. \tag{4.33}$$

Therefore, estimate (1.13) follows now from (4.33) and (3.45).

Part 2) of Theorem 1.1 is proved. □

We pass to the proof of Theorem 1.3 concerning water waves.

5. Proof of part 1) of Theorem 1.3 for water waves (nondegenerate case). Reduction of the problem to an integral equation

We divide the proof of this theorem into several steps. Each step corresponds to one section. First, we reduce problem (1.3) to some integral equation.

We look for the solution of (1.3) in the form

$$\Psi(x,y) = \int_{\mathbb{R}} \big(A(p)\sinh(\tau y) + B(p)\cosh(\tau y)\big) e^{ipx}\, dp, \tag{5.1}$$

where

$$\tau := \tau(p) := \sqrt{p^2+1}. \tag{5.2}$$

It is evident that Ψ satisfies the first equation of system (1.3). Differentiating $\Psi(x,y)$ in the form (5.1) in y, we obtain from the third equation of system (1.3) that

$$A(p) = \frac{\omega^2 B(p)}{\tau}. \tag{5.3}$$

Substituting this expression in (5.1) and denoting the function $B(p)$ by $f(p)$, we obtain for the unknown function Ψ the following expression:

$$\Psi(x,y) = \int_{\mathbb{R}} f(p)\Big(\frac{\omega^2}{\tau}\sinh(\tau y) + \cosh(\tau y)\Big) e^{ipx}\, dp. \tag{5.4}$$

Note that if $f(p) \in S(\mathbb{R}_p)$, then this function satisfies two equations of system (1.3): the first and the third. Therefore we are left with solving the second equation of (1.3).

In order to do this, we express the normal derivative on the curve $y = -h(x)$ in the second equation of (1.3). We have

$$\frac{\partial\Psi}{\partial n}\bigg|_{y=-h} = \big(\partial_y\Psi + h'(x)\partial_x\Psi\big)\bigg|_{y=-h}. \tag{5.5}$$

Differentiating expression (5.4) with respect to y and then with respect to x and taking its restriction to the curve $y = -h(x)$, we get

$$\partial_y \Psi\Big|_{y=-h(x)} = \int_{\mathbb{R}} f(p)\left(-\tau \sinh(\tau h(x)) + \omega^2 \cosh(\tau h(x))\right) e^{ipx}\, dp,$$

$$h'(x)\partial_x \Psi\Big|_{y=-h(x)} = ih'(x)\int_{\mathbb{R}} pf(p)\Big(-\frac{\omega^2}{\tau}\sinh(\tau h(x)) + \cosh(\tau h(x))\Big)e^{ipx}\, dp.$$

From here we see that the second equation in (1.3) is equivalent to the following equation for $f(p)$

$$\int_{\mathbb{R}} \Big[\omega^2 \cosh(\tau h(x)) - \tau \sinh(\tau h(x)) + ih'(x)p\Big(-\frac{\omega^2}{\tau}\sinh(\tau h(x)) + \cosh(\tau h(x))\Big)\Big] f(p)e^{ipx}\, dp = 0. \tag{5.6}$$

We look for the solution of this equation in the form

$$\omega^2 = t_0 - (\beta\varepsilon)^2, \quad t_0 \equiv \tanh h_0, \quad \tau(p) \equiv \tau = \sqrt{p^2+1}, \tag{5.7}$$

where β is some unknown number which will be specified later in Section 6, and

$$f(p) \equiv f(p,\varepsilon) \equiv \frac{\tilde{\phi}(p,\varepsilon)}{\cosh(\tau h_0)(L(p) + \varepsilon^2\beta^2)}, \tag{5.8}$$

where

$$L(p) = \sqrt{1+p^2}\tanh(h_0\sqrt{1+p^2}) - \tanh h_0. \tag{5.9}$$

In (5.8) $\tilde{\phi}(p,\varepsilon)$ is an unknown function, which is the Fourier transform of the function $\phi(x,\varepsilon)$:

$$F_{x\to p}[\phi(x)](p) := \tilde{\phi}(p) = \frac{1}{\sqrt{2\pi}}\int_{\mathbb{R}} e^{-ipx}\phi(x)\, dx. \tag{5.10}$$

Now we are going to transform equation (5.6). To this end, we introduce the following function

$$\begin{aligned} I(p,x,\varepsilon,) \equiv I(p,x) \equiv I &\equiv \tau(p)\left[\sinh(h_0\tau(p)) - \sinh(h(x)\tau(p))\right] \\ &\quad + \omega^2\left[\cosh(h(x)\tau(p)) - \cosh(h_0\tau(p))\right] \\ &\quad + iph'(x)\left[\cosh(h(x)\tau(p)) - \omega^2\tau^{-1}(p)\sinh(h(x)\tau(p))\right]. \end{aligned} \tag{5.11}$$

It is not difficult to check the following identity:

$$\begin{aligned} &\omega^2\cosh(h\tau) - \tau\sinh(h\tau) + ih'p\Big(-\frac{\omega^2}{\tau}\sinh(h\tau) + \cosh(h\tau)\Big) \\ &= I + \left(\omega^2\cosh(h_0\tau) - \tau\sinh(h_0\tau)\right). \end{aligned} \tag{5.12}$$

From (5.12), (5.8), and (5.7) it follows that equation (5.6) is equivalent to the following one:

$$\int_{\mathbb{R}} G(p,x,\varepsilon)dp = \int_{\mathbb{R}} \tilde{\phi}(p)e^{ipx}\, dp, \tag{5.13}$$

where

$$G(p,x,\varepsilon) := \frac{g(p,x,\varepsilon)}{L(p)+\varepsilon^2\beta^2}, \quad g(p,x,\varepsilon) := g(p,x) \equiv \frac{I(p,x,\varepsilon)\tilde{\phi}(p)e^{ipx}}{\cosh(\tau h_0)}. \tag{5.14}$$

It is not hard to see that this equation is equivalent to (2.4) and hence for our heuristic scheme is applicable.

We assume that $\tilde{\phi}(p)$ is analytic in the upper half-plane. Later we will prove that this is indeed the case. Then from (5.14), (5.9) it follows that the function $G(p,x,\varepsilon)$ is analytic in the same region. Therefore we consider equation (5.13) with the new contour γ which is the same that was used in the case of the Schrödinger equation (see Figure 4).

By definition and from (5.11), (5.7), and (1.4) it follows that the integrals in (5.13) converge for sufficiently small ε and for the analytic function $\tilde{\phi}$ which decreases sufficiently rapidly on γ. We will see that the function $\tilde{\phi}$ which we will find satisfies this condition.

Let us consider the left-hand side in equation (5.13). We proceed analogously to the corresponding situation for the Schrödinger equation. Namely, we represent the integral in the left-hand side of this equation in the form

$$\int_\gamma G(p,x,\varepsilon)\,dp = \int_\gamma G(p,x,\varepsilon)\,dp + \frac{2\pi i g(p(\beta,\varepsilon),x,\varepsilon)}{L'(p(\beta,\varepsilon))}, \tag{5.15}$$

where $\gamma = \gamma$ in $\{|p| \geq 1/2\}$ and $\gamma = \{e^{iv}, 0 \leq v \leq \pi\}$. Let $p(\beta,\varepsilon)$ be a solution of the equation

$$L(p) = -\varepsilon^2\beta^2. \tag{5.16}$$

It is evident that this solution exists for sufficiently small $\varepsilon\beta$.

Thus equation (5.13) is equivalent (after multiplying by $1/\sqrt{2\pi}$) to the equation

$$\frac{1}{\sqrt{2\pi}}\int_\gamma \tilde{\phi}(p)e^{ipx}dp = \frac{1}{\sqrt{2\pi}}\int_\gamma \frac{g(p,x,\varepsilon)dp}{L(p)+\varepsilon^2\beta^2} + \frac{\sqrt{2\pi}ig(p(\beta,\varepsilon),x,\varepsilon)}{L'(p(\beta,\varepsilon))}. \tag{5.17}$$

If the solution $\tilde{\phi}(p)$ is analytic in $\mathbb{C}$ and decreases sufficiently rapidly on γ, then equation (5.17) is equivalent to the equation

$$\phi(x,\varepsilon) = \frac{1}{\sqrt{2\pi}}\int_\gamma G(p,x,\varepsilon)\,dp + \frac{i\sqrt{2\pi}g(p(\beta,\varepsilon),x,\varepsilon)}{L'(p(\beta,\varepsilon))}, \tag{5.18}$$

where $G(p,x,\varepsilon)$ is defined in (5.14).

So we have proved the following lemma.

LEMMA 5.1. *Problem* (1.3) *is equivalent to the problem of finding the function* $\phi(x,\varepsilon)$ *and the number* β *which satisfy equation* (5.18). *Moreover, the solution* $(\Phi,\ \omega)$ *of problem* (1.3) *is connected with* ϕ *and* β *by formulas* (5.4), (5.8), *and* (5.7).

In the next section we specify the form of the approximate solution ϕ and β and obtain some preliminary expansions.

6. The form of the solution and some preliminary expansions

We are going to find some approximate solutions of equation (5.18). The unknown function ϕ and the unknown parameter ω^2 (frequency) are contained in this equation. Namely, these unknowns are contained in g (see (5.14), (5.11)). In turn, ω^2 is expressed by means of β^2 as in (5.7).

We look for the approximate solution ϕ of equation (5.18) (more correctly, its Fourier transform) in the form

$$\tilde{\phi} := \tilde{\phi}_n(p, \varepsilon) := \mu_n \tilde{A}_n(p) := \varepsilon B_n \tilde{A}_n(p), \tag{6.1}$$

where

$$\tilde{A}_n(p) = \tilde{a}_0(p) + \varepsilon \tilde{a}_1(p) + \cdots + \varepsilon^{n-2}\tilde{a}_{n-2}(p) + \varepsilon^{n-1}\tilde{a}_{n-1}(p) + \varepsilon^n \tilde{a}_n(p), \tag{6.2}$$

and now β in equation (5.18) is determined as

$$\beta := B_n = \beta_0 + \varepsilon\beta_1 + \cdots + \varepsilon^n \beta_n. \tag{6.3}$$

The functions $\tilde{a}_k(p)$, $k = 0, \dots, n$, and coefficients β_k of the expansion of $\beta := B_n$ are unknown.

Moreover, we look for the approximate value of the unknown parameter ω^2 in the form

$$\omega_n^2 = t_0 - B_n^2. \tag{6.4}$$

Thus the function g in (5.14) takes the form

$$g(p, x, \varepsilon) := g(p, x) \equiv \frac{I(p, x, \varepsilon)\tilde{\phi}_n(p)e^{ipx}}{\cosh(\tau h_0)}. \tag{6.5}$$

According to (6.3), $p(\beta, \varepsilon)$ now is the solution (with the positive imaginary part) of the equation

$$L(p) = -\varepsilon^2\beta^2 = -\varepsilon^2 B_n^2. \tag{6.6}$$

After finding this approximate solution of equation (5.18), we find an approximate solution of equation (5.6) in the form

$$f_n(p) \equiv f(p, \varepsilon) \equiv \frac{\tilde{\phi}_n(p, \varepsilon)}{\cosh(\tau h_0)(L(p) + \varepsilon^2 B_n^2)}. \tag{6.7}$$

After finding this approximate solution f_n, we will find at first the approximate solution ψ_n, substituting the function f_n obtained previously in (5.4). Then we will normalize the function $\psi_n(x, 0)$ in order to use Lemmas 3.3 and 3.4 for the corresponding operator T (see Section 12). Then, dividing ψ_n by some normalization constant we will obtain the approximate solution Ψ_n from (1.17).

Our plan coincides with the plan for the Schrödinger equation and consists in the following. We substitute these unknowns in equation (5.18) and obtain some system of equations with the unknowns a_k, $k = 0, 1 \dots, n$ and β_k, $k = 0, 1, \dots, n$. Then we prove that this system has a unique solution under some normalization conditions. Also we will prove that the obtained values of β_k and functions ϕ_k, $k = 0, \dots, n$, satisfy equation (5.18) up to ε^{n+2} in $L_2(\mathbb{R}_x)$ in the sense of (3.30). Then we again apply some general facts of the spectral theory as in the case of the Schrödinger equation (namely, Lemmas 3.3 and 3.4).

Since we will need expansions of the left-hand side and the right-hand side of equation (5.13), we expand representation (6.1) in ε. We obtain after some calculations

$$\begin{aligned}\tilde{\phi}_n(p, \varepsilon) &= \varepsilon(\beta_0\tilde{a}_0) + \varepsilon^2(\beta_0\tilde{a}_1 + \beta_1\tilde{a}_0) + \varepsilon^3(\beta_0\tilde{a}_2 + \beta_1\tilde{a}_1 + \beta_2\tilde{a}_0) \\ &+ \cdots + \varepsilon^{n+1}(\beta_0\tilde{a}_n + \beta_1\tilde{a}_{n-1} + \beta_2\tilde{a}_{n-2} + \cdots + \beta_{n-1}\tilde{a}_1 + \beta_n\tilde{a}_0) \\ &+ \varepsilon^{n+2}P(\bar{\beta}, \bar{\tilde{a}}, \varepsilon),\end{aligned} \tag{6.8}$$

where $P(\bar{\beta}, \bar{\bar{\phi}}, \varepsilon)$ is a linear combination of $\tilde{a}_k$, $k = 0, \dots, n$, with coefficients being polynomials of β_k, $k = 0, \dots, n$, and ε. (Here we use the same notation as in (3.16).)

This expression is equivalent to the following one in the x-representation:

$$\begin{aligned}(6.9)\quad \phi_n(x,\varepsilon) &= \varepsilon(\beta_0 a_0) + \varepsilon^2(\beta_0 a_1 + \beta_1 a_0) + \varepsilon^3(\beta_0 a_2 + \beta_1 a_1 + \beta_2 a_0)\\ &\quad + \cdots + \varepsilon^{n+1}(\beta_0 a_n + \beta_1 a_{n-1} + \beta_2 a_{n-2} + \cdots + \beta_{n-1} a_1 + \beta_n a_0)\\ &\quad + \varepsilon^{n+2} P_*(\bar{\beta}, \bar{a}, \varepsilon),\end{aligned}$$

where $P_*(\bar{\beta}, \bar{\bar{\phi}}, \varepsilon)$ is a linear combination of a_k, $k = 0, \dots, n$, with coefficients being polynomials of β_k, $k = 0, \dots, n$, and ε.

Also, we will need the expansion of the solution of equation (5.16) for sufficiently small ε. In order to obtain it, we have to write the expansion for $L(p)$ from (5.9) in powers of p as $p \to 0$. From (5.9) we have

$$(6.10)\qquad L(p) = l_1 p^2 + l_2 p^4 + \cdots + l_{n+1} p^{n+1} + l_{n+2}(p) p^{n+2}, \qquad p \to 0,$$

where

$$(6.11)\qquad l_1 = (1/2)(t_0 - h_0 t_0^2 + h_0) := \frac{l}{2},$$

where l is defined in (1.20), $l > 0$ (by (5.7)), and

$$(6.12)\qquad |l_{n+2}(p)| \le C, \qquad |p| \le \varepsilon_0$$

for sufficiently small ε_0. Now using (6.10), we can write the expansion for the solution of equation (6.6) in ε. To this end, we use the same notation as for the Schrödinger equation, namely that given after formulas (3.13) and (3.15).

Choosing the root of the solution with the positive imaginary part and denoting this solution by $p(\beta, \varepsilon)$, we obtain

$$(6.13)\qquad p(\beta, \varepsilon) = \varepsilon\beta\left[p_1 + p_3 \varepsilon^2 \beta^2 + \cdots + p_n \varepsilon^n \beta^n + \varepsilon^{n+1} P_{n+1}(\bar{\beta}, \varepsilon)\right],$$

where

$$(6.14)\qquad p_1 = \frac{i}{\sqrt{l_1}} = \frac{i\sqrt{2}}{\sqrt{l}}, \qquad |P_{n+1}(\bar{\beta}, \varepsilon)| \le C$$

for sufficiently small ε.

From (6.13) it also follows that

$$(6.15)\qquad p(\beta, \varepsilon) = \varepsilon\beta\left(p_1 + \varepsilon^2 p_3 \beta_0^2 + \cdots + \varepsilon^n P_n(\bar{\beta}_{0,n-2}) + \varepsilon^{n+1} P_{n+1}(\bar{\beta}, \varepsilon)\right),$$

where P_n is a polynomial of its arguments and

$$(6.16)\qquad |P_{n+1}(\bar{\beta}, \varepsilon)| \le C$$

for sufficiently small ε. Here and in what follows, we will follow notation (3.16).

From (6.15) it also follows that

$$\begin{aligned}(6.17)\qquad p(\beta, \varepsilon) &= \varepsilon\beta_0 p_1 + \varepsilon^2 \beta_1 p_1 + \varepsilon^3(\beta_0^3 p_3 + \beta_2 p_1) + \cdots + \varepsilon^m P_n(\bar{\beta}_{0,m-1})\\ &\quad + \cdots + \varepsilon^{n+1} P_{n+1}(\bar{\beta}) + \varepsilon^{n+2} P_{n+2}(\bar{\beta}, \varepsilon), \qquad \varepsilon \to 0,\end{aligned}$$

where p_1 is determined in (6.14); P_m, $m = 0, \dots, n+1$, are polynomials of its arguments,

$$(6.18)\qquad |P_{n+2}(\bar{\beta}, \varepsilon)| \le C \quad \text{for } \varepsilon \le \varepsilon_0.$$

7. Asymptotics of the residue part of the main equation as $\varepsilon \to 0$

In this section we obtain the asymptotics of the residue part of the main equation (5.18), namely,

$$R(x,\varepsilon) := \frac{i\sqrt{2\pi}g(p(\beta,\varepsilon),x,\varepsilon)}{L'(p(\beta,\varepsilon))}. \tag{7.1}$$

LEMMA 7.1. *The function $R(x,\varepsilon)$ has the following expansion:*

(7.2)

$$\begin{aligned} R(x,\varepsilon) = \frac{i\sqrt{2\pi}}{L_0'}\Bigg\{ &\varepsilon\left[\frac{V(x)a_{00}}{\cosh^2 h_0}\right] \\ &+\varepsilon^2\left[\frac{iV(x)a_{00}\beta_0 p_1}{\cosh^2 h_0} + \frac{i_2(x)a_{00}}{\cosh h_0} + \frac{V(x)(a_{01}\beta_0 p_1 + a_{10})}{\cosh^2 h_0}\right] \\ &+\varepsilon^3 R_3\left(\bar\beta_{0,1}, V(x), V'(x), \bar a_{k,i}, x\right) \\ &+\cdots+\varepsilon^m R_m\left(\bar\beta_{0,m-2}, V(x), V'(x), \bar a_{k,i}, x\right) \\ &+\cdots+\varepsilon^{n+1} R_{n+1}\left(\bar\beta_{0,n-1}, V(x), V'(x), \bar a_{k,i}, x\right) + \varepsilon^{n+2}R_{n+2}(x,\varepsilon)\Bigg\}, \end{aligned}$$

$$i \le n-m-1, \quad k = 0,\dots,m-1, \quad m = 0,1,\dots,n+1,$$

where R_3 is determined from (7.45); all the coefficients R_m of ε^m, $m \le n+1$, are polynomials of their arguments, which do not contain the term proportional to $V^0 V'^0$ and therefore belong to $C_0^\infty(\mathbb{R}_x)$, and $R_{n+2} \in S(\mathbb{R}_x)$ satisfies the uniform with respect to ε estimate for ε sufficiently small:

$$\|R_{n+2}\|_{L_2(\mathbb{R}_x)} \le C. \tag{7.3}$$

PROOF. From (6.10) it follows that

$$L'(p) = p\left(2l_1 + 4l_2 p^2 + \cdots + l_n' p^n + l_{n+1}'(p)p^{n+1}\right), \qquad p \to 0, \tag{7.4}$$

where

$$|l_{n+1}'(p)| \le C \tag{7.5}$$

for sufficiently small p.

From (7.4), (6.15), (6.11) it follows that $L'(p(\beta,\varepsilon))$ has the expansion

$$\begin{aligned} L'(p(\beta,\varepsilon)) = \varepsilon\beta\big[&L_0' + \varepsilon^2 L_2' + \varepsilon^3 L_3' + \cdots + \varepsilon^m L_m'(\beta_{0,m-2}) \\ &+\cdots+\varepsilon^n L_n'(\bar\beta_{0,n-2}) + \varepsilon^{n+1}L_{n+1}'(\bar\beta,\varepsilon)\big], \end{aligned} \tag{7.6}$$

where L_m', $m = 0,\dots,n$, are polynomials of its arguments,

$$L_0' = 2i\sqrt{l_1} = i\sqrt{2l}, \tag{7.7}$$

l is defined from (1.20) and

$$|L_{n+1}'(\bar\beta,\varepsilon)| \le C \tag{7.8}$$

for sufficiently small ε.

From (7.6) it follows that

$$\begin{aligned} \frac{1}{L_1'(p(\beta,\varepsilon))} = \frac{1}{\varepsilon\beta L_0'}\Bigg[&1 - \varepsilon^2\frac{L_2'}{L_0'} - \cdots - \varepsilon^m P_m(\bar\beta_{0,m-2}) \\ &- \cdots - \varepsilon^n P_n(\bar\beta_{0,n-2}) + \varepsilon^{n+1}P_{n+1}(\bar\beta,\varepsilon)\Bigg], \end{aligned} \tag{7.9}$$

$$|P_{n+2}(\bar{\beta},\varepsilon)| \le C \tag{7.10}$$

for sufficiently small ε.

Now, we pass to finding the asymptotics of $I(p(\beta,\varepsilon))$, where I is determined from (5.11).

Let

$$I \equiv I_1 + I_2 + I_3, \tag{7.11}$$

where

$$\begin{aligned} &I_1(p,x,\varepsilon) \equiv \tau(p)\left[\sinh(\tau h_0) - \sinh(\tau h(x))\right], \\ &I_2(p,x,\varepsilon) \equiv \omega^2\left[\cosh(\tau h(x)) - \cosh(\tau h_0)\right], \\ &I_3(p,x,\varepsilon) \equiv iph'(x)\left[\cosh(\tau h(x)) - \omega^2\tau^{-1}\sinh(\tau h(x))\right]. \end{aligned} \tag{7.12}$$

From (6.13) it follows that

$$\begin{aligned} e^{ip(\beta,\varepsilon)x} &= 1 + i\varepsilon x\beta_0 p_1 + \varepsilon^2\left(ix\beta_1 p_1 - \frac{x^2}{2}\beta_0^2 p_1^2\right) \\ &+ \varepsilon^3\left(ix(\beta_0^3 p_3 + \beta_2 p_1) - x^2\beta_0\beta_1 p_1^2 - \frac{ix^3}{6}p_1^2\right) \\ &+ \cdots + \varepsilon^m E_m(\bar{\beta}_{0,m-1},x) + \cdots + \varepsilon^n E_n(\bar{\beta}_{0,n-1},x) \\ &+ \varepsilon^{n+1}E_{n+1}(\bar{\beta},x,\varepsilon), \end{aligned} \tag{7.13}$$

where E_m, $m = 0,\dots,n$, are polynomials of their arguments, $E_{n+1}(x,\varepsilon)$ is a smooth function in x uniformly bounded on compact sets with respect to sufficiently small ε. Now we expand $\tilde{A}_n(p(\beta,\varepsilon))$ with respect to ε up to ε^{n+1}. From (6.17) it follows that

$$\begin{aligned} \tilde{a}_k(p(\beta,\varepsilon)) &= a_{k0} + \varepsilon a_{k1}\beta_0 p_1 + \varepsilon^2\left[a_{k1}\beta_1 p_1 + \frac{a_{k2}}{2}\beta_0^2 p_1^2\right] \\ &+ \cdots + \varepsilon^m P_{k,m}(\bar{\beta}_{0,m-1},\bar{a}_{k,0,m}) + \cdots + \varepsilon^n P_{k,n}(\bar{\beta}_{0,n-1},\bar{a}_{k,0,n}) \\ &+ \cdots + \varepsilon^{n+1}P_{k,n+1}(\bar{\beta},\bar{a}_{k,0,n},\varepsilon), \qquad k = 0,1,\dots,n, \end{aligned} \tag{7.14}$$

where all the coefficients $P_{k,m}$, $m = 0,\dots,n$, are polynomials of their arguments; $\bar{a}_{k,0,m}$ is a vector $a_{k,0},\dots,a_{k,m}$,

$$a_{kl} := \tilde{a}_k^l(0), \qquad k,l = 0,1,\dots, \tag{7.15}$$

and $P_{k,n+1}$ satisfies estimate (6.18). Substituting (7.14) in (6.2), we obtain

$$\begin{aligned} \tilde{A}_n(p(\beta,\varepsilon)) = \varepsilon\beta\Big[&a_{00} + \varepsilon(a_{01}\beta_0 p_1 + a_{10}) \\ &+ \varepsilon^2\left(a_{01}\beta_1 p_1 + \frac{a_{02}}{2}\beta_0^2 p_1^2 + a_{11}\beta_0 p_1 + a_{20}\right) \\ &+ \cdots + \varepsilon^m(P_m(\bar{\beta}_{0,m-1},\bar{a}_{k,i})) \\ &+ \cdots + \varepsilon^n P_n(\bar{\beta}_{0,n-1},\bar{a}_{k,i}) + \varepsilon^{n+1}P_{n+1}(\bar{\beta},\bar{a}_{k,i},\varepsilon)\Big], \\ &i \le m-k, \quad k = 0,\dots,m, \quad m = 0,1\dots,n, \end{aligned} \tag{7.16}$$

where all the coefficients P_m, $m = 0,\dots,n$, are polynomials of their arguments; P_{n+1} satisfies the estimate (6.18). From (5.7) and (6.13) it follows that

$$\tau(p(\beta,\varepsilon)) = 1 + \varepsilon^2\tau_2 + \varepsilon^3\tau_3 + \cdots + \varepsilon^m\tau_m(\bar\beta_{0,m-2}) + \cdots + \varepsilon^n\tau_n(\bar\beta_{0,n-2}) + \varepsilon^{n+1}\tau_{n+1}(\bar\beta,\varepsilon), \tag{7.17}$$

where

$$\tau_2 \equiv \frac{1}{2}\beta_0^2 p_1^2, \qquad \tau_3 \equiv \beta_0\beta_1 p_1^2; \tag{7.18}$$

τ_m, $m = 2, \ldots, n$, are polynomials of their arguments, and

$$|\tau_{n+1}| \le C \tag{7.19}$$

for ε sufficiently small. From (7.17) it follows that

$$\begin{aligned} \tau(p(\beta,\varepsilon))h(x) &:= \tau(p(\beta,\varepsilon))(h_0 - \varepsilon V(x)) \\ &= h_0 + \varepsilon(-V(x)) + \varepsilon^2 h_0\tau_2 + \varepsilon^3(-\tau_2 V(x) + h_0\tau_3) \\ &+ \cdots + \varepsilon^m(\tau_m h_0 - \tau_{m-1}V(x)) \\ &+ \cdots + \varepsilon^n(\tau_n h_0 - \tau_{n-1}V(x)) + \varepsilon^{n+1}(\tau_{n+1}h_0 - \tau_n V(x)). \end{aligned} \tag{7.20}$$

Expanding $\sinh(\tau(p(\beta,\varepsilon))h(x))$ in the Taylor series at the point $\tau(p(\beta,\varepsilon))$ in powers of $-\varepsilon\tau V(x)$, we obtain from (1.4)

$$\begin{aligned} &\sinh(\tau(p(\beta,\varepsilon))h_0) - \sinh(\tau(p(\beta,\varepsilon))h(x)) \\ &= \cosh(\tau(p(\beta,\varepsilon))h_0)\tau(p(\beta,\varepsilon))\varepsilon V(x) \\ &- \frac{\sinh(\tau(p(\beta,\varepsilon))h_0)(\tau(p(\beta,\varepsilon))\varepsilon V(x))^2}{2} \\ &+ \frac{\cosh(\tau(p(\beta,\varepsilon))h_0)}{6}(\tau(p(\beta,\varepsilon))\varepsilon V(x))^3 \\ &+ \cdots + \frac{\mathrm{trh}_{n+1}(\tau(p(\beta,\varepsilon))h_0)(\tau(p(\beta,\varepsilon))\varepsilon V(x))^{n+1}}{(n+1)!} \\ &+ \frac{(\tau\varepsilon V(x))^{n+2}}{(n+1)!}\int_0^1 (1-t)^{n+1}\,\mathrm{trh}_{n+2}(t\tau\varepsilon V(x))\,dt, \end{aligned} \tag{7.21}$$

where $\mathrm{trh}_n(\cdot) = \cosh(\cdot)$ if n is odd and $\mathrm{trh}_n(\cdot) = -\sinh(\cdot)$ if n is even. Using (7.18), we expand $\cosh(\tau(p(\beta,\varepsilon))h_0)$ and $\sinh(\tau(p(\beta,\varepsilon))h_0)$ in powers of ε. Then, using the obtained expression for $\cosh(\tau(p(\beta,\varepsilon))h_0)$ and (7.20), we obtain the expansion for $\cosh(\tau(p(\beta,\varepsilon))h_0)\tau(p(\beta,\varepsilon)\varepsilon V(x))$. Analogously, we obtain the expressions for

$$\begin{gathered} \frac{\sinh(\tau(p(\beta,\varepsilon))h_0)(\tau(p(\beta,\varepsilon))\varepsilon V(x))^2}{2}, \qquad \frac{\cosh(\tau h_0)}{6}(\tau\varepsilon V)^3, \\ \frac{\mathrm{trh}_{n+1}(\tau(p(\beta,\varepsilon))h_0)(\tau(p(\beta,\varepsilon))\varepsilon V(x))^{n+1}}{(n+1)!} \quad \text{and} \\ \frac{(\tau\varepsilon V(x))^{n+2}}{(n+1)!}\int_0^1 (1-t)^{n+1}\,\mathrm{trh}_{n+2}(t\tau\varepsilon V(x))\,dt. \end{gathered} \tag{7.22}$$

Substituting the obtained expressions in (7.21), we obtain

$$\begin{aligned} &\sinh(\tau(p(\beta,\varepsilon))h_0) - \sinh(\tau(p(\beta,\varepsilon))h(x)) \\ &= \varepsilon V(x)\cosh h_0 - \varepsilon^2\frac{V^2(x)\sinh h_0}{2} \\ &+ \varepsilon^3\left[V(x)(\tau_2\cosh h_0 + h_0\tau_2\sinh h_0) + \frac{V^3(x)}{6}\cosh h_0\right] \\ &+ \cdots + \varepsilon^{n+1}V(x)P_{n+1}(\bar\beta_{0,n-2}, V(x)) + \varepsilon^{n+2}P_{n+2}(\bar\beta, V(x), \varepsilon), \end{aligned} \tag{7.23}$$

where P_{n+1} is a polynomial of its arguments and $P_{n+2} \in S(\mathbb{R}_x)$ satisfies the estimate of (7.3). From (7.23), (7.12), and (7.17) it follows that

$$(7.24)\qquad \begin{aligned} I_1(p(\beta,\varepsilon),x,\varepsilon) &= \varepsilon\cosh h_0 V(x) + \varepsilon^2\Big(-\frac{1}{2}V^2(x)\sinh h_0\Big) \\ &+ \cdots + \varepsilon^3\Big[2V(x)\tau_2\cosh h_0 + V(x)h_0\tau_2\sinh h_0 + \frac{V^3(x)}{6}\cosh h_0\Big] \\ &+ \cdots + \varepsilon^m P_m(\bar\beta_{0,m-2},V(x)) \\ &+ \cdots + \varepsilon^{n+1}V(x)P_{n+1}(\bar\beta_{0,n-2},V(x)) + \varepsilon^{n+2}I_{1,n+2}(\bar\beta,x,\varepsilon), \end{aligned}$$

where $P_m(\bar\beta_{0,m-2},V(x))$, $m=0,\dots,n+1$, are polynomials of their arguments and do not contain terms proportional to V^0, $I_{1,n+2}(\bar\beta,x,\varepsilon)\in S(\mathbb{R}_x)$ with the estimate of the type (7.3). Analogously to (7.21), we obtain from (1.4)

$$(7.25)\qquad \begin{aligned} &\cosh(\tau(p(\beta,\varepsilon))h(x)) - \cosh(\tau((\beta,\varepsilon))h_0) \\ &= -\varepsilon\tau(p(\beta,\varepsilon))V(x)\sinh(\tau(p(\beta,\varepsilon))h_0) \\ &+ \frac{\varepsilon^2\tau^2(\beta,\varepsilon)V^2(x)}{2}\cosh(\tau(p(\beta,\varepsilon))h_0) - (\tau\varepsilon V)^3 6\sinh(\tau h_0) \\ &+ \cdots + \frac{\operatorname{trh}_n(\tau(p(\beta,\varepsilon))h_0)(\tau(p(\beta,\varepsilon))\varepsilon V(x))^{n+1}}{(n+1)!} \\ &+ \frac{(\tau\varepsilon V(x))^{n+1}}{(n+1)!}\int_0^1 (1-t)^{n+1}\operatorname{trh}_{n+2}(t\tau\varepsilon V(x))\,dt, \end{aligned}$$

where trh_n is defined as in (7.21). Using (7.17), we expand the following expressions in series in ε:

$$(7.26)\qquad \begin{gathered} \sinh(\tau(p(\beta,\varepsilon))h_0)\tau(p(\beta,\varepsilon))\varepsilon V(x), \quad \frac{\cosh(\tau(p(\beta,\varepsilon))h_0)(\tau(p(\beta,\varepsilon))\varepsilon V(x))^2}{2}, \\ \frac{\operatorname{trh}_n(\tau(p(\beta,\varepsilon))h_0)(\tau(p(\beta,\varepsilon))\varepsilon V(x))^{n+1}}{(n+1)!}, \quad \text{and} \\ \frac{(\tau\varepsilon V(x))^{n+2}}{(n+1)!}\int_0^1 (1-t)^{n+1}\operatorname{trh}_{n+1}(t\tau\varepsilon V(x))\,dt. \end{gathered}$$

Substituting the obtained expansions in (7.25), we have

$$(7.27)\qquad \begin{aligned} &\cosh(\tau(p(\beta,\varepsilon))h(x)) - \cosh(\tau(p(\beta,\varepsilon))h_0) \\ &= -\varepsilon V(x)\sinh h_0 + \varepsilon^2\frac{V^2(x)\cosh h_0}{2} \\ &- \varepsilon^3 V(x)\Big(\tau_2\sinh h_0 + h_0\tau_2\cosh h_0 + \frac{V^3(x)}{6}\sinh h_0\Big) \\ &+ \cdots + \varepsilon^m P_m(\bar\beta_{0,m-2},V(x)) + \cdots + \varepsilon^{n+1}V(x)P_{n+1}(\bar\beta_{0,n-2},V(x)) \\ &+ \varepsilon^{n+2}P_{n+2}(\bar\beta,V(x),\varepsilon), \end{aligned}$$

where P_m, $m=0,1,\dots,n+1$, are polynomials of their arguments and $P_{n+2}\in S(\mathbb{R}_x)$ satisfies the estimate of the type (7.3). From (5.7) it follows that

$$(7.28)\qquad \begin{aligned} \omega^2 = t_0 &- \varepsilon^2\beta_0^2 + \varepsilon^3(-2\beta_0\beta_1) + \cdots + \varepsilon^m P_m(\bar\beta_{0,m-2}) \\ &+ \cdots + \varepsilon^{n+1}P_{n+1}(\bar\beta_{0,n-1}) + \varepsilon^{n+2}P_{n+2}(\bar\beta,\varepsilon), \end{aligned}$$

where P_m, $m = 0,\dots,n+1$, are polynomials of their arguments and $P_{n+2}(\bar\beta,\varepsilon)$ satisfies the uniform estimate (6.18).

From (7.12), (7.28), (7.27) it follows that

$$
\begin{aligned}
(7.29)\qquad I_2(p(\beta,\varepsilon),x,\varepsilon) &= \varepsilon(-t_0V(x)\sinh h_0) + \varepsilon^2\left(\frac{1}{2}t_0V^2(x)\cosh h_0\right)\\
&+\varepsilon^3\left[-2t_0V(x)\sinh h_0 - \frac{t_0V(x)^3}{6}\sinh h_0 + V(x)\beta_0^2\sinh h_0\right]\\
&+\cdots+\varepsilon^m P_m(\bar\beta_{0,m-2},V(x)) + \cdots + \varepsilon^{n+1}P_{n+1}(\bar\beta_{0,n-2},V(x))\\
&+\varepsilon^{n+2}I_{2,n+2}(\bar\beta,x,\varepsilon),
\end{aligned}
$$

where $P_m(\bar\beta_{0,n-2},V(x))$, $m = 0,1\dots,n+1$, are polynomials of their arguments, $I_{2,n+2}(\bar\beta,x,\varepsilon) \in S(\mathbb{R}_x)$ with the estimate of the type (7.3). Finally, we find the expansion for I_3 from (7.12). From (6.17), (1.4) it follows that

$$
\begin{aligned}
h'(x) &= -\varepsilon V'(x);\\
ip(\beta,\varepsilon)h'(x) &= \varepsilon^2(-i\beta_0p_1V'(x)) + \varepsilon^3(-i\beta_1p_1V'(x))\\
(7.30)\qquad &+\cdots+\varepsilon^m(-iV'(x))P_m(\bar\beta_{0,n-2}) + \cdots + \varepsilon^n(-iV'(x))P_n(\bar\beta_{0,n-2})\\
&+\varepsilon^{n+1}V'(x)P_{n+1}(\bar\beta_{0,n-1}) + \varepsilon^{n+2}V'(x)P_{n+2}(\bar\beta,\varepsilon),
\end{aligned}
$$

where P_m, $m = 0,1,\dots,n+2$, coincide with polynomials from (6.17). From (7.30), (7.28) it follows that

$$
\begin{aligned}
(7.31)\qquad \frac{\omega^2}{\tau(p(\beta,\varepsilon))} &= t_0 + \varepsilon^2(-\beta_0^2 - t_0\tau_2) + \varepsilon^3(-2\beta_0\beta_1 - t_0\tau_3)\\
&+\cdots+\varepsilon^m P_m(\bar\beta_{0,-2}) + \cdots + \varepsilon^n P_n(\bar\beta_{0,n-2})\\
&+\varepsilon^{n+1}P_{n+1}(\bar\beta,\varepsilon),
\end{aligned}
$$

where P_m, $m = 0,1,\dots,n$, are polynomials of their arguments and P_{n+1} satisfies the estimate (6.18). Using (1.4), (7.17) and (7.31), we expand the following expressions in series in ε

$$
(7.32)\qquad \sinh(\tau(p(\beta,\varepsilon))h(x)),\qquad \cosh(\tau(p(\beta,\varepsilon))h(x)),\quad\text{and}\quad \frac{\omega^2}{\tau(p(\beta,\varepsilon))}\sinh(\tau(p(\beta,\varepsilon))h(x)).
$$

Using the obtained expansions of expressions (7.32), we have

$$
\begin{aligned}
(7.33)\quad &\cosh(\tau(p(\beta,\varepsilon))h(x)) - \frac{\omega^2}{\tau(p(\beta,\varepsilon))}\sinh(\tau(p(\beta,\varepsilon))h(x))\\
&= \frac{1}{\cosh h_0} + \varepsilon^2\left[\frac{V^2\cosh h_0}{2} + \tau_2h_0\sinh h_0 - \beta_0^2\sinh h_0\right.\\
&\qquad\left. - t_0\tau_2\sinh h_0 + \frac{t_0V^2\sinh h_0}{2} + t_0\tau_2h_0\cosh h_0\right]\\
&+\cdots+\varepsilon^n P_m(\bar\beta_{0,m-2},V(x)) + \cdots + \varepsilon^n P_n(\bar\beta_{0,n-2},V(x))\\
&+\varepsilon^{n+1}\left[P_{n+1}(\bar\beta,\varepsilon) + Q_{n+1}(\bar\beta,\varepsilon)\right],
\end{aligned}
$$

where P_m, $m = 0,\dots,n$, are polynomials of their arguments, $P_{n+1} \in S_{\mathbb{R}_x}$ with the estimate (7.3), and Q_{n+1} satisfies the estimate of the type (7.19). From (7.12),

(7.33), and (7.30) it follows that

$$(7.34)\qquad I_3(p(\beta,\varepsilon),x,\varepsilon) = \varepsilon^2\left(\frac{-i\beta_0 p_1 V'(x)}{\cosh h_0}\right) + \varepsilon^3\left(\frac{-i\beta_1 p_1 V'(x)}{\cosh h_0}\right) + \cdots + \varepsilon^{m+1} V'(x) P_{m+1}(\bar\beta_{0,m-1}, V(x)) + \cdots + \varepsilon^{n+1} V'(x) P_{n+1}(\bar\beta_{0,n-1}, V(x)) + \varepsilon^{n+2} V'(x)[P_{n+2}(\bar\beta, V(x),\varepsilon) + Q_{n+2}(\bar\beta,\varepsilon)],$$

where all the coefficients P_m of the powers ε^m, $m = 2,\dots,n+2$, are polynomials of their arguments and do not contain the term proportional to $V'(0)$ and therefore belong to $C_0^\infty(\mathbb{R}_x)$; $P_{n+2} \in S(\mathbb{R}_x)$ with the estimate (7.3) and Q_{n+2} satisfies the estimate of the type (7.19). Summing the three summands I_1, I_2, and I_3, we obtain from (7.11), (7.24), (7.29), and (7.34) the following expansion for $I(p(\beta,\varepsilon),x,\varepsilon)$:

$$(7.35)\qquad I(p(\beta,\varepsilon),x,\varepsilon) = \varepsilon\left(\frac{V(x)}{\cosh h_0}\right) + \varepsilon^2 i_2(x) + \varepsilon^3\Bigg(2V(x)\tau_2\cosh h_0 + V(x)h_0\tau_2\sinh h_0 + \frac{V^3(x)}{6\cosh h_0} - 2t_0 V(x)\sinh h_0 + V(x)\beta_0^2\sinh h_0 - \frac{i\beta_1 p_1 V'(x)}{\cosh h_0}\Bigg) + \cdots + \varepsilon^m P_m(\bar\beta_{0,m-1}, V(x), V'(x)) + \cdots + \varepsilon^{n+1} P_{n+1}(\bar\beta_{0,n-1}, V(x), V'(x)) + \varepsilon^{n+2} I_{n+2}(\bar\beta, x,\varepsilon),$$

where, taking into account (6.14),

$$(7.36)\qquad i_2(x) = \frac{\beta_0\sqrt{2}V'(x)}{\sqrt{l}\cosh h_0},$$

P_m, $m = 0,1\dots,n+1$, are polynomials of their arguments which do not contain the term proportional to $V^0 V'^0$ and therefore belong to $C_0^\infty(\mathbb{R}_x)$; $I_{n+2} \in S(\mathbb{R}_x)$ with the estimate (7.3).

So, we have found the asymptotics of $I(p(\beta,\varepsilon),x,\varepsilon)$. Now we can find the asymptotics of $g(p(\beta,\varepsilon),x,\varepsilon)$ from (5.14).

From (7.35), (7.16) it follows that

$$(7.37)\qquad I(p(\beta,\varepsilon),x,\varepsilon)\tilde A_n(p(\beta,\varepsilon)) = \varepsilon\beta\Bigg[\varepsilon\frac{V(x)a_{00}}{\cosh h_0} + \varepsilon^2\left(i_2(x)a_{00} + \frac{V(x)(a_{01}\beta_0 p_1 + a_{10})}{\cosh h_0}\right) + \varepsilon^3 P_3(\bar\beta_{0,1}, V(x), V'(x), \bar a_{k,i}) + \cdots + \varepsilon^m P_m(\bar\beta_{0,m-2}, V(x), V'(x), \bar a_{k,i}) + \cdots + \varepsilon^{n+1} P_{n+1}(\bar\beta_{0,n-1}, V(x), V'(x), \bar a_{k,i}) + \varepsilon^{n+2} P_{n+2}(x,\varepsilon)\Bigg],$$

$$0 \le i \le m-1-k,\quad k = 0,\dots,m-1,\quad m = 1,\dots,n+1,$$

where

$$(7.38)\qquad P_3(\bar\beta_{0,1}, V(x), V'(x), \bar a_{k,i}) = \frac{V(x)a_{01}\beta_1 p_1}{\cosh h_0} + \frac{V(x)a_{02}\beta_0^2 p_1^2}{2\cosh h_0} + \frac{V(x)a_{11}\beta_0 p_1}{\cosh h_0} + \frac{V(x)a_{20}}{\cosh h_0}$$

$$+ i_2(x)a_{01}\beta_0 p_1 + i_2(x)a_{10} + 2V(x)\tau_2 a_{00}\cosh h_0$$
$$+ V(x)h_0\tau_2 a_{00}\sinh h_0 + \frac{V^3(x)a_{00}}{6\cosh h_0} - 2t_0 V(x)\sinh h_0 a_{00}$$
$$+ V(x)\beta_0^2 \sinh h_0 a_{00} - \frac{i\beta_1 p_1 V'(x)a_{00}}{\cosh h_0},$$

P_m, $m = 0, \dots, n+1$, are polynomials of their arguments which do not contain the term proportional to $V^0 V'^0$; $\bar{a}_{k,i}$ is defined as in (7.16); $P_{n+2} \in S$ with the estimate (7.3).

From (7.13), (7.37) it follows that

$$\begin{aligned}(7.39)\quad & I(p(\beta,\varepsilon),x,\varepsilon)\tilde{A}_n(p(\beta,\varepsilon))e^{ip(\beta,\varepsilon)x} \\ &= \varepsilon\beta\Bigg[\varepsilon\frac{V(x)a_{00}}{\cosh h_0} \\ &+ \varepsilon^2\left(\frac{ixV(x)a_{00}\beta_0 p_1}{\cosh h_0} + i_2(x)a_{00} + \frac{V(x)(a_{01}\beta_0 p_1 + a_{10})}{\cosh h_0}\right) \\ &+ \varepsilon^3 Q_3(\bar{\beta}_{0,1}, V(x), V'(x), \bar{a}_{k,i}, x) \\ &+ \dots + \varepsilon^m Q_m(\bar{\beta}_{0,m-2}, V(x), V'(x), \bar{a}_{k,i}, x) \\ &+ \dots + \varepsilon^n Q_{n+1}(\bar{\beta}_{0,n-1}, V(x), V'(x), \bar{a}_{k,i}, x) + \varepsilon^{n+2}Q_{n+2}(x,\varepsilon)\Bigg], \\ & i \le m-k-1, \quad k = 0,\dots,m-1, \quad m = 0,1,\dots,n+1,\end{aligned}$$

where

$$\begin{aligned}(7.40)\quad & Q_3(\bar{\beta}_{0,1}, V(x), V'(x), \bar{a}_{k,i}, x) \\ &= \frac{V(x)a_{00}ix\beta_1 p_1}{\cosh h_0} - \frac{V(x)a_{00}x^2\beta_0^2 p_1^2}{2\cosh h_0} + i_2(x)a_{00}ix\beta_0 p_1 \\ &+ \frac{V(x)(a_{01}\beta_0 p_1 + a_{10})ix\beta_0 p_1}{\cosh h_0} + P_3(\bar{\beta}_{0,1}, V(x), V'(x), \bar{a}_{k,i}),\end{aligned}$$

P_3 is determined from (7.38), Q_m, $m = 0, 1, \dots, n+1$, are polynomials of their arguments which do not contain the term proportional to $V^0 V'^0$; $Q_{n+2} \in S(\mathbb{R}_x)$ with the estimate (7.3). Expanding again $\cosh(\tau(p(\beta,\varepsilon))h_0)$ in the series in powers of ε, we write

$$\begin{aligned}(7.41)\quad \frac{1}{\cosh(\tau(p(\beta,\varepsilon)h_0))} &= \frac{1}{\cosh h_0} - \varepsilon^2\left(\frac{h_0\tau_2 t_0}{\cosh h_0}\right) + \dots + \varepsilon^n P_m(\bar{\beta}_{0,m-2}) \\ &+ \dots + \varepsilon^n P_n(\bar{\beta}_{0,n-2}) + \varepsilon^{n+1}P_{n+1}(\bar{\beta}_{0,n-1},\varepsilon),\end{aligned}$$

where P_m, $m = 0, \dots, n$, are polynomials of their arguments, $P_{n+1} \in S(\mathbb{R}_x)$ with the estimate (7.3), and τ_2 is defined from (7.18). Now we can write the expansion of $g(p(\beta,\varepsilon),x,\varepsilon)$ from (5.14). From (5.14), (7.39), (7.41) it follows that

$$\begin{aligned}(7.42)\quad & g(p(\beta,\varepsilon),x,\varepsilon) \\ &= \varepsilon\beta\Bigg[\varepsilon\frac{V(x)a_{00}}{\cosh^2 h_0} \\ &+ \varepsilon^2\left(\frac{iV(x)a_{00}\beta_0 p_1}{\cosh^2 h_0} + \frac{i_2(x)a_{00}}{\cosh h_0} + \frac{V(x)(a_{01}\beta_0 p_1 + a_{10})}{\cosh^2 h_0}\right) \\ &+ \varepsilon^3 g_3(\bar{\beta}_{0,1}, V(x), V'(x), \bar{a}_{k,i}, x)\end{aligned}$$

$$+\cdots+\varepsilon^m g_m(\bar\beta_{0,m-2},V(x),V'(x),\bar a_{k,i},x)$$
$$\left.+\cdots+\varepsilon^{n+1}g_{n+1}(\bar\beta_{0,n-1},V(x),V'(x),\bar a_{k,i},x)+\varepsilon^{n+2}g_{n+2}(x,\varepsilon)\right],$$
$$i\le m-k-1,\quad k=0,\dots,m-1,\quad m=0,\dots,n+1,$$

where

$$g_3(\bar\beta_{0,1},V(x),V'(x),\bar a_{k,i},x) = \frac{Q_3(\bar\beta_{0,1},V(x),V'(x),\bar a_{k,i},x)}{\cosh h_0}-\frac{h_0\tau_2 t_0 V(x)a_{00}}{\cosh^2 h_0}, \tag{7.43}$$

Q_3 is determined from (7.40), g_m, $m=0,\dots,n+1$, are polynomials of their arguments, which do not contain the term proportional to $V^0{V'}^0$, and $g_{n+2}\in S(\mathbb{R}_x)$ with the estimate (7.3).

Finally, we obtain the asymptotics of the residue part R from (7.1). From (7.1), (7.42), (7.9) it follows that

$$R(x,\varepsilon)=\frac{i\sqrt{2\pi}}{L_0'}\left\{\varepsilon\frac{V(x)a_{00}}{\cosh^2 h_0}\right. \tag{7.44}$$
$$+\varepsilon^2\left[\frac{iV(x)a_{00}\beta_0 p_1}{\cosh^2 h_0}+\frac{i_2(x)a_{00}}{\cosh h_0}+\frac{V(x)(a_{01}\beta_0 p_1+a_{10})}{\cosh^2 h_0}\right]$$
$$+\varepsilon^3 R_3(\bar\beta_{0,1},V(x),V'(x),\bar a_{k,i},x)$$
$$+\cdots+\varepsilon^m R_m(\bar\beta_{0,m-2},V(x),V'(x),\bar a_{k,i},x)$$
$$\left.+\cdots+\varepsilon^{n+1}R_{n+1}(\bar\beta_{0,n-1},V(x),V'(x),\bar a_{k,i},x)+\varepsilon^{n+2}R_{n+2}(x,\varepsilon)\right\},$$
$$i\le n-m-1,\quad k=0,\dots,m-1,\quad m=0,1,\dots,n+1,$$

where

$$R_3(\bar\beta_{0,1},V(x),V'(x),\bar a_{k,i},x)=g_3(\bar\beta_{0,1},V(x),V'(x),\bar a_{k,i},x)-\frac{L_2'V(x)a_{00}}{L_0'\cosh h_0}, \tag{7.45}$$

g_3 is determined from (7.43), R_m, $m=0,1,\dots,n+1$, are polynomials of their arguments, which do not contain the term proportional to $V^0{V'}^0$, and $R_{n+2}\in S(\mathbb{R}_x)$ with the estimate (7.3). Lemma 7.1 is proved. □

So we have found the asymptotics of the "residue" part with respect to $\varepsilon\to 0$. Now we find the asymptotics of the first summand in the right-hand side of (5.15) with respect to ε on γ.

8. Asymptotics of the integral $\int_\gamma G(p,x,\varepsilon)\,dp$ as $\varepsilon\to 0$

In this section we prove the following lemma.

LEMMA 8.1. *If a_k, $k=0,\dots,n$, belong to $C_0^\infty(\mathbb{R}_x)$, then $\int_\gamma G(p,x,\varepsilon)\,dp$ admits the expansion*

$$\int_\gamma G(p,x,\varepsilon)\,dp \tag{8.1}$$
$$=\varepsilon\left\{\varepsilon\beta_0 V(x)\int_\gamma\frac{e^{ipx}\mathcal{I}_1(p,x)a_0(p)\,dp}{L(p)\cosh(\tau h_0)}\right.$$

$$
\begin{aligned}
&+\varepsilon^2\bigg[\int_\gamma \frac{e^{ipx}}{L(p)\cosh(\tau h_0)}\Big(\beta_0 a_0(p)\mathcal{I}_2(p,x)\\
&+\big(\beta_0 a_1(p)+\beta_1\tilde{a}_0(p)\big)\mathcal{I}_1(p,x)\Big)\,dp\bigg]\\
&+\cdots+\varepsilon^m\,\mathrm{IG}_m(\bar\beta_{0,m-1},\bar a_{0,m-1},x)\\
&+\cdots+\varepsilon^{n+1}\,\mathrm{IG}_{n+1}(\bar\beta_{0,n-1},\bar a_{0,n-1},x)\\
&\qquad\qquad+\varepsilon^{n+2}\,\mathrm{IG}_{n+2}(x,\varepsilon)\bigg\},
\end{aligned}
$$

where all the coefficients IG_m *of* ε^m, $m=1,\dots,n+1$, *are functions from* $C_0^\infty(\mathbb{R}_x)$ *and the function* $\mathrm{IG}_{n+2}(x,\varepsilon)\in S(\mathbb{R}_x)$ *satisfies the estimate* (7.3).

PROOF. Expanding the expression $\frac{1}{L(p)+\varepsilon^2\beta^2}$ in geometric series with respect to $\varepsilon^2\beta^2$, also taking the rest as the sum of the geometric series and using (6.3), we obtain

$$
\begin{aligned}
(8.2)\quad &\frac{e^{ipx}}{\cosh(\tau h_0)(L(p)+\varepsilon^2\beta^2)}\\
&=\frac{e^{ipx}}{L(p)\cosh(\tau h_0)}\bigg[1-\varepsilon^2\frac{\beta_0^2}{L(p)}-\varepsilon^3\frac{2\beta_0\beta_1}{L(p)}+\cdots+\varepsilon^m P_m(\bar\beta_{0,m-2},L^{-1})\\
&\qquad+\varepsilon^n P_n(\bar\beta_{0,n-2},L^{-1})+\varepsilon^{n+1}\frac{P_{n+1}(\bar\beta,L^{-1},\varepsilon)}{L(p)+\varepsilon^2\beta^2}\bigg],\\
&\qquad p\in\gamma,\quad \varepsilon\to 0,
\end{aligned}
$$

where P_m, $m=0,\dots,n+1$, are polynomials of their arguments.

From (6.1) it follows that

$$
\begin{aligned}
(8.3)\quad \tilde\phi_n(p,\varepsilon)&=\varepsilon\beta_0\tilde a_0(p)+\varepsilon^2\big(\beta_0\tilde a_1(p)+\beta_1\tilde a_0(p)\big)\\
&+\varepsilon^3\big(\beta_0\tilde a_2(p)+\beta_1\tilde a_1(p)+\beta_2\tilde a_0(p)\big)\\
&+\cdots+\varepsilon^{n-1}\big(\beta_0\tilde a_{n-2}(p)+\cdots+\beta_{n-2}\tilde a_0(p)\big)\\
&+\varepsilon^n\big(\beta_0\tilde a_{n-1}(p)+\cdots+\beta_{n-1}\tilde a_0(p)\big)\\
&+\varepsilon^{n+1}\big(\beta_0\tilde a_n(p)+\cdots+\beta_n\tilde a_0(p)\big)+\varepsilon^{n+2}P_{n+2}(\bar{\tilde a}(p),\bar\beta),
\end{aligned}
$$

where

$$
(8.4)\qquad \bar{\tilde a}(p):=\big(\tilde a_0(p),\dots,\tilde a_n(p)\big),\qquad \bar\beta:=(\beta_0,\dots\beta_n),
$$

and all the coefficients of ε^m, $m=0,1,\dots,n+2$, are linear combinations of the functions $\tilde a_0,\dots,\tilde a_m$ with the coefficients from the set $\{\beta_0,\dots,\beta_m\}$.

From (8.3), (8.2) we obtain

$$
\begin{aligned}
(8.5)\quad &\frac{e^{ipx}\tilde\phi_n(p,\varepsilon)}{\cosh(\tau h_0)(L(p)+\varepsilon^2\beta^2)}\\
&=\frac{e^{ipx}}{L(p)\cosh(\tau h_0)}\Big\{\varepsilon\beta_0\tilde a_0(p)+\varepsilon^2\big(\beta_0\tilde a_1(p)+\beta_1\tilde a_0(p)\big)\\
&+\varepsilon^3\Big(-\frac{\beta_0^3\tilde a_0(p)}{L(p)}+\beta_0\tilde a_2(p)+\beta_1\tilde a_1(p)+\beta_2\tilde a_0(p)\Big)\\
&+\cdots+\varepsilon^{n-1}P_{n-1}\big(\bar\beta_{0,n-2},\bar{\tilde a}_{0,n-2},L^{-1}\big)+\varepsilon^nP_n(\bar\beta_{0,n-1},\bar{\tilde a}_{0,n-1},L^{-1}(p))
\end{aligned}
$$

$$+ \varepsilon^{n+1} \frac{P_{n+1}(\bar{\beta}, \bar{\tilde{a}}, L^{-1}, \varepsilon)}{L(p) + \varepsilon^2 \beta^2} \Big\},$$

$$p \in \gamma, \quad \varepsilon \to 0,$$

where all the coefficients of ε^k, $k = 0, \dots, n$, are linear combinations of $\tilde{a}_0, \dots, \tilde{a}_n$ with the coefficients which are polynomials of $\bar{\beta}$, L^{-1} and P_{n+1} is a linear combination of $\tilde{a}_0, \dots, \tilde{a}_n$ with the coefficients which are polynomial of $\bar{\beta}$, L^{-1}, and ε. From this and from the estimate

$$|L^{-1}(p)| \geq C, \qquad p \in \gamma, \tag{8.6}$$

it follows that all the coefficients in expression (8.5) of ε^m, $0 \leq m \leq n$, belong to $S(\mathbb{R}_p)$ as functions of $p_1 := \Re p$, $p \in \gamma$ under the assumption that all $a_k \in S(\mathbb{R}_p)$ and the coefficient

$$\frac{P_{n+1}(\bar{\beta}, \bar{\tilde{a}}, L^{-1}, \varepsilon)}{L(p) + \varepsilon^2 \beta^2} \in S(\mathbb{R}_p)$$

uniformly with respect to ε for sufficiently small ε in the sense of (7.3).

Now we find the expansion of $I(p, x, \varepsilon)$ from (5.11) on γ with respect to ε.

We begin with $I_1(p, x, \varepsilon)$ from (7.12). From (1.4) it follows that

$$\begin{aligned} \sin(\tau(p)h(x)) &\equiv \sinh\big(\tau(p)(h_0 - \varepsilon V(x))\big) \\ &= \sinh(\tau h_0) - \varepsilon \tau V(x) \cosh(\tau h_0) + \varepsilon^2 \left(\frac{\tau^2 V^2(x) \sinh(\tau h_0)}{2} \right) \\ &\quad - \varepsilon^3 \left(\frac{\tau^3 V^3(x) \cosh(\tau h_0)}{6} \right) + \dots + \varepsilon^n \left(\frac{\tau^n V^n(x)}{n!} \operatorname{trh}_n(\tau h_0) \right) \\ &\quad + \varepsilon^{n+1} \mathcal{S}_{n+1}(p, x, \varepsilon), \qquad p \in \gamma. \end{aligned} \tag{8.7}$$

We use the Taylor formula with the remainder term in the form of Cauchy (cf. (3.12)):

$$\begin{aligned} f(x) &= f(x_0) + f'(x_0)(x - x_0) + \dots + \frac{f^{(n)}(x_0)}{n!} \\ &\quad + \frac{(x - x_0)^{n+1}}{n!} \int_0^1 (1-t)^n f^{n+1}(tx + (1-t)x_0)\, dt. \end{aligned} \tag{8.8}$$

Applying this formula to the expansion (8.7), we obtain that

$$S_{n+1}(p, x, \varepsilon) = \frac{(-\tau V(x))^{n+1}}{n!} \int_0^1 (1-t)^n \operatorname{trh}_{n+1}\big(\tau(h_0 - t\varepsilon V(x))\big)\, dt, \tag{8.9}$$

where $\operatorname{trh}_n(\cdot) = \sinh(\cdot)$ when n is even and $\operatorname{trh}_n(\cdot) = -\cosh(\cdot)$ when n is odd.

Also from (8.7) and (8.8) it follows that all the coefficients of ε^k, $1 \leq k \leq n+1$, in expansion (8.7) belong to the space $C_0^\infty(\mathbb{R}_x, C^\infty(\mathbb{R}_p))$ and satisfy the estimate which we write only for $S_{n+1}(p, x, \varepsilon)$ and which holds for all the coefficients of the expansion

$$|D_x^\alpha \mathcal{S}_{n+1}(p, x, \varepsilon)| \leq C_{n+1,\alpha,N} (1 + |x|)^{-N} (1 + |p|)^{n+1} e^{h_0 |p|}, \qquad p \in \gamma \tag{8.10}$$

for ε sufficiently small. From (7.12) and (8.7) it follows that

$$\begin{aligned} I_1(p, x, \varepsilon) &= \varepsilon \tau^2 V(x) \cosh(\tau h_0) + \varepsilon^2 \left(- \frac{\tau^3 V^2(x) \sinh(\tau h_0)}{2} \right) \\ &\quad + \dots + \varepsilon^n \left(\frac{\tau^{n+1} V^n(x)}{n!} \operatorname{trh}_n(\tau h_0) \right) + \tau \mathcal{S}_{n+1}(p, x, \varepsilon). \end{aligned} \tag{8.11}$$

From (8.11) and (8.10) it follows that all the coefficients $I_{1,k}$ of ε^k, $k \le n+1$, belong to $C_0^\infty(\mathbb{R}_x, C^\infty(\gamma))$ and satisfy the estimate

$$(8.12) \qquad |D^\alpha I_{1,k}(p,x,\varepsilon)| \le C_{n+1,\alpha,N}(1+|x|)^{-N}(1+|p|)^{n+2}e^{h_0|p|}, \qquad p \in \gamma$$

for ε sufficiently small.

Now let us get a similar expansion for $I_2(p,x,\varepsilon)$ from (7.12). Expanding analogously to the Taylor series of $\cosh(\tau h(x))$ with respect of ε and using (1.4), we obtain

$$(8.13) \quad \begin{aligned} \cosh(\tau(h(x))) &= \cosh(\tau h_0) - \varepsilon\tau V(x)\sinh(\tau h_0) + \varepsilon^2\frac{\tau^2 V^2(x)\cosh(\tau h_0)}{2} \\ &\quad - \varepsilon^3\frac{\tau^3 V^3(x)\sinh(\tau h_0)}{6} + \cdots + \varepsilon^n\frac{\tau^n V^n(x)\,\mathrm{trh}_{n+1}(\tau h_0)}{n!} \\ &\quad + \varepsilon^{n+1}\mathcal{C}_{n+1}(p,x,\varepsilon), \end{aligned}$$

where

$$(8.14) \qquad \mathcal{C}_{n+1}(p,x,\varepsilon) = \frac{(-1)^{n+1}\varepsilon^{n+1}\tau^{n+1}V^{n+1}(x)}{n!}\int_0^1 (1-t)\,\mathrm{trh}_n(t\tau\varepsilon V(x))\,dt,$$

and where all the coefficients of ε^k, $1 \le k \le n+1$, in the expansion (8.13) possess the same properties as the coefficients (8.7). From (8.13), (7.28), and (7.12) it follows that

$$(8.15) \quad \begin{aligned} I_2(p,x,\varepsilon) &= -\varepsilon t_0\tau V(x)\sinh(\tau h_0) + \varepsilon^2\left(\frac{t_0\tau^2 V(x)\sinh(\tau h_0)}{6}\right) \\ &\quad + \varepsilon^3\left(\beta_0^2\tau V(x)\sinh(\tau h_0) - \frac{t_0\tau^3 V(x)\sinh(\tau h_0)}{6}\right) \\ &\quad + \cdots + \varepsilon^k I_{2,k}\left(\bar\beta_{0,k-3},\tau,V,\sinh(\tau h_0),\cosh(\tau h_0)\right) \\ &\quad + \cdots + \varepsilon^n I_{2,n}\left(\bar\beta_{0,n-3},\tau,V,\sinh(\tau h_0),\cosh(\tau h_0)\right) \\ &\quad + \varepsilon^{n+1} I_{2,n+1}(p,x,\varepsilon), \end{aligned}$$

where $I_{2,k}$, $k \le n+1$, possess the same properties as the coefficients of (8.11).

Now we get the expansion for $I_3(p,x,\varepsilon)$ from (7.12). From (8.13), (7.28), and (8.7) it follows that

$$(8.16) \quad \begin{aligned} &\cosh(\tau h(x)) - \frac{t_0 - (\beta\varepsilon)^2}{\tau}\sinh(\tau h(x)) \\ &= \cosh(\tau h_0) - \frac{t_0}{\tau}\sinh(\tau h_0) \\ &\quad + \varepsilon\big(t_0 V(x)\cosh(\tau h_0) - \tau V(x)\sinh(\tau h_0)\big) \\ &\quad + \varepsilon^2\left(\frac{\cosh(\tau h_0)V^2(x)\tau^2}{2} - \frac{\beta_0^2\sinh(\tau h_0)}{\tau} + \frac{t_0\tau V^2(x)\sinh(\tau h_0)}{2}\right) \\ &\quad + \cdots + \varepsilon^n P_n\left(\bar\beta_{0,n-2},\tau,V,\sinh(\tau h_0),\cosh(\tau h_0)\right) \\ &\quad + \varepsilon^{n+1}\mathcal{B}_{n+1}(p,x,\varepsilon), \end{aligned}$$

where all the coefficients of ε^k, $1 \le k \le n+1$, depend on $\bar\beta_{0,k-2}$, belong to $C^\infty(\mathbb{R}_x, C^\infty(\gamma))$, and satisfy the following estimate (we write this estimate only for $\mathcal{B}_{n+1}(p,x,\varepsilon)$):

$$(8.17) \qquad |D_x^\alpha \mathcal{B}_{n+1}(p,x,\varepsilon)| \le C_{n+1,\alpha}e^{h_0|p|}, \qquad p \in \gamma.$$

REMARK 8.2. We see that this estimate differs from the estimate (8.12). This is because of the structure of the expression. We need the estimate (8.12). This estimate is provided by the multiplication of the expression $iph'(x)$ as in (7.12).

From (8.16) and (7.12) it follows that

$$
\begin{aligned}
(8.18)\qquad I_3(p,x,\varepsilon) &= \varepsilon\Big[-ipV'(x)\cosh(\tau h_0) + \frac{ipV'(x)t_0\sinh(\tau h_0)}{\tau}\Big] \\
&+ \varepsilon^2\left[-it_0pV'(x)V(x)\cosh(\tau h_0) + ip\tau V'(x)V(x)\sinh(\tau h_0)\right] \\
&+ \varepsilon^3\Big[\frac{-ip\tau^2V^2(x)V'(x)\cosh(\tau h_0)}{2} + \frac{i\beta_0^2V'(x)\sinh(\tau h_0)}{\tau} \\
&\qquad - \frac{it_0p\tau V'(x)V(x)\sinh(\tau h_0)}{2}\Big] \\
&+ \cdots + \varepsilon^n\big[-pV'(x)P_n(\bar\beta_{0,n-2},\tau,V,\sinh(\tau h_0),\cosh(\tau h_0))\big] \\
&+ \varepsilon^{n+1}I_{3,n+1}(p,x,\varepsilon), \qquad p\in\gamma,\ \varepsilon\to 0,
\end{aligned}
$$

where now all the coefficients $I_{3,k}$ of ε^k, $0\le k\le n+1$, possess the same properties as the coefficients in (8.11).

Finally, we can write the expansion of $I(p,x,\varepsilon)$. From (7.11), (8.11), (8.15), and (8.18) it follows that

$$
(8.19)\qquad I(p,x,\varepsilon) = \varepsilon\mathcal{I}_1(p,x) + \varepsilon^2\mathcal{I}_2(p,x) + \cdots + \varepsilon^n\mathcal{I}_n(p,x) + \varepsilon^{n+1}\mathcal{I}_{n+1}(p,x,\varepsilon),
$$

where

$$
(8.20)\qquad
\begin{aligned}
\mathcal{I}_1(p,x) &= V(x)\iota_{1,1}(p) + V'(x)\iota_{1,2}, \\
\iota_{1,1}(p) &= \tau^2\cosh(\tau h_0) - t_0\tau\sinh(\tau h_0), \\
\iota_{1,2}(p) &= -ip\cosh(\tau h_0) + \frac{ipt_0\sinh(\tau h_0)}{\tau}, \\
\mathcal{I}_2(p,x) &= V^2(x)\iota_{2,1}(p) + V(x)V'(x)\iota_{2,2}(p), \\
\iota_{2,1}(p) &= -\frac{\tau^3\sinh(\tau h_0)}{2} + \frac{t_0\tau^2\cosh(\tau h_0)}{2}, \\
\iota_{2,2}(p) &= -ipt_0\cosh(\tau h_0) + ip\tau\sinh(\tau h_0), \\
&\cdots\cdots\cdots\cdots\cdots\cdots \\
\mathcal{I}_n(p,x) &= P_n(\bar\beta_{0,n-2},p,\tau,V,V',\sinh(\tau h_0),\cosh(\tau h_0)),
\end{aligned}
$$

and $\mathcal{I}_i(p,x)$ are the $C_0^\infty(\mathbb{R}_x, C^\infty(\gamma))$ functions with respect to the variables p, x which satisfy the estimate (8.12) for $i=0,\dots,n+1$.

From (5.14), (8.5), (8.19), and (8.20) it follows that

(8.21)

$$
\begin{aligned}
G(p,x,\varepsilon) &= \frac{e^{ipx}}{L(p)\cosh(\tau h_0)}\Big\{\varepsilon^2\beta_0\tilde a_0(p)\mathcal{I}_1(p,x) \\
&+ \varepsilon^3\left[\beta_0\tilde a_0(p)\mathcal{I}_2(p,x) + \mathcal{I}_1(p,x)(\beta_0\tilde a_1(p) + \beta_1\tilde a_0(p))\right] \\
&+ \cdots + \varepsilon^k G_k\left(\bar\beta_{0,k-2},\bar{\tilde a}_{0,k-2},L^{-1},p,\tau,V,V',\sinh(\tau h_0),\cosh(\tau h_0)\right) \\
&+ \cdots + \varepsilon^{n+1}G_{n+1}\left(\bar\beta_{0,n-1},\bar{\tilde a}_{0,n-1},L^{-1},p,\tau,V,V',\sinh(\tau h_0),\cosh(\tau h_0)\right) \\
&\qquad + \varepsilon^{n+2}G_{n+2}(p,x,\varepsilon)\Big\},
\end{aligned}
$$

where

$$G_{n+2}(p,x,\varepsilon)=\frac{P_{n+2}(\bar\beta,\bar{\tilde a},L^{-1},\overline{\mathcal{I}_{1,n+1}})}{L(p)+\varepsilon^2\beta^2},$$

and all the coefficients G_k of ε^k, $k\le n+2$, and also P_{n+2} are linear combinations of a_k, $k=0,\dots,n$, with the coefficients which are polynomials of $\bar\beta$, L^{-1}, $\bar{\mathcal{I}}$.

Thus all the coefficients G_k of ε^k, $k=0,\dots,n+2$, in the expansion belong to $C_0^\infty(\mathbb{R}_x,C^\infty(\gamma))$ (under the assumption that $a_k\in C_0^\infty(\mathbb{R}_x)$), and from the estimate (8.12) it follows that the G_k satisfy the estimate

$$|D^\alpha G_k(p,x,\varepsilon)|\le C_{n+1,\alpha,N,M}(1+|x|)^{-N}(1+|p|)^{-M}e^{h_0|p|},\quad p\in\gamma,\quad N,M\in\mathbb{N},\tag{8.22}$$

for ε sufficiently small.

Integrating $G(p,x)$ along the contour γ, we obtain from (8.21)

(8.23)

$$\begin{aligned}\int_\gamma G(p,x,\varepsilon)\,dp &= \varepsilon^2\beta_0V(x)\int_\gamma\frac{e^{ipx}\mathcal{I}_1(p,x)\tilde a_0(p)\,dp}{L(p)\cosh(\tau h_0)}\\ &+\varepsilon^3\Big[\int_\gamma\frac{e^{ipx}}{L(p)\cosh(\tau h_0)}\Big(\beta_0\tilde a_0(p)\mathcal{I}_2(p,x)\\ &\qquad+\big(\beta_0\tilde a_1(p)+\beta_1\tilde a_0(p)\big)\mathcal{I}_1(p,x)\Big)\,dp\Big]\\ &+\dots+\varepsilon^k\Big[\int_\gamma\frac{e^{ipx}}{L(p)\cosh(\tau h_0)}\\ &\qquad\times G_k\big(\bar\beta_{0,k-2},\bar{\tilde a}_{0,k-2},L^{-1},p,\tau,V,V',\sinh(\tau h_0),\cosh(\tau h_0)\big)\,dp\Big]\\ &+\dots+\varepsilon^{n+1}\Big[\int_\gamma\frac{e^{ipx}}{L(p)\cosh(\tau h_0)}\\ &\qquad\times G_{n+1}\big(\bar\beta_{0,n-1},\bar{\tilde a}_{0,n-1},L^{-1},p,\tau,V,V',\sinh(\tau h_0),\cosh(\tau h_0)\big)\,dp\Big]\\ &+\varepsilon^{n+2}\Big[\int_\gamma\frac{e^{ipx}}{L(p)\cosh(\tau h_0)}G_{n+2}(p,x,\varepsilon)\,dp\Big].\end{aligned}$$

We note that from the fact that G_k, $k\le n+2$, belongs to $C_0^\infty(\mathbb{R}_x,C^{\infty(\gamma)})$ and from the estimate (8.12), it follows that the integrals in (8.23) of ε^k, $k\le n+2$, converge and determine the functions from $C_0^\infty(\mathbb{R}_x)$. Moreover, the integral of ε^{n+2} satisfies the estimate (7.3).

Thus we have found the asymptotics of the integral indicated in the title of Section 8. Lemma 8.1 is proved. □

9. System of equations for the unknown terms of the expansions of the eigenvalue and eigenfunction

Rename the coefficient R_3 of ε^3 in (7.44) and (7.45) by R_2, denoting the coefficient of ε^{n+1} in (8.23) which is

(9.1)

$$\int_\gamma\frac{e^{ipx}}{L(p)\cosh(\tau h_0)}P_{n+1}\left(\bar\beta_{0,n-1},\bar{\tilde a}_{0,n-1},L^{-1},p,\tau,V,V',\sinh(\tau h_0),\cosh(\tau h_0)\right)\,dp$$

by $\mathrm{IG}_n(\bar{\beta}_{0,n-1}, \bar{a}_{0,n-1}, x)$ and the coefficient of ε^{n+2} which is

$$\int_\gamma \frac{e^{ipx}}{L(p)\cosh(\tau h_0)} G_{n+2}(p, x, \varepsilon)\, dp$$

by $\mathrm{IG}_{n+1}(x, \varepsilon)$, we obtain from equations (5.18), (6.9), (8.23), and (7.2) after dividing by ε the following equation:

(9.2)

$$\begin{aligned}
&\beta_0 a_0(x) + \varepsilon\big[\beta_0 a_1(x) + \beta_1 a_0(x)\big] + \varepsilon^2\big[\beta_0 a_2(x) + \beta_1 a_1(x) + \beta_2 a_0(x)\big] \\
&\quad + \cdots + \varepsilon^n\big[\beta_0 a_n(x) + \beta_1 a_{n-1}(x) + \beta_2 a_{n-2}(x) + \cdots + \beta_{n-1} a_1(x) + \beta_n a_0(x)\big] \\
&\quad + \varepsilon^{n+1} P_*(\bar{\beta}, \bar{a}, \varepsilon) \\
&\quad = \frac{1}{\sqrt{2\pi}}\Big\{\varepsilon\beta_0 \int_\gamma \frac{e^{ipx}}{L(p)\cosh(\tau h_0)} \mathcal{I}_1(p, x)\tilde{a}_0(p)\, dp \\
&\qquad + \varepsilon^2 \int_\gamma \frac{e^{ipx}}{L(p)\cosh(\tau h_0)} \Big(\beta_0 \tilde{a}_0(p)\mathcal{I}_2(p, x) \\
&\qquad\qquad + \big(\beta_0\tilde{a}_1(p) + \beta_1\tilde{a}_0(p)\big)\mathcal{I}_1(p, x)\Big)\, dp \\
&\qquad + \cdots + \varepsilon^m \,\mathrm{IG}_n(\bar{\beta}_{0,m-1}, \bar{\tilde{a}}_{0,m-1}, x) \\
&\qquad\quad + \cdots + \varepsilon^n \,\mathrm{IG}_n(\bar{\beta}_{0,n-1}, \bar{\tilde{a}}_{0,n-1}, x) + \varepsilon^{n+1}\,\mathrm{IG}_{n+1}(x, \varepsilon)\Big\} \\
&\quad + \frac{\sqrt{2\pi} i}{L_0'}\Big\{\frac{V(x) a_{00}}{\cosh^2 h_0} \\
&\qquad + \varepsilon\Big[\frac{iV(x) a_{00}\beta_0 p_1}{\cosh^2 h_0} + \frac{i_2(x) a_{00}}{\cosh h_0} + \frac{V(x)(a_{01}\beta_0 p_1 + a_{10})}{\cosh^2 h_0}\Big] \\
&\qquad + \varepsilon^2 R_2(\bar{\beta}_{0,1}, V(x), V'(x), \bar{a}_{k,i}, x) \\
&\qquad + \cdots + \varepsilon^m R_m\big(\bar{\beta}_{0,m-1}, V(x), V'(x), \bar{a}_{k,i}, x\big) \\
&\quad + \cdots + \varepsilon^n R_n\big(\bar{\beta}_{0,n-1}, V(x), V'(x), \bar{a}_{k,i}, x\big) + \varepsilon^{n+1} R_{n+1}(x, \varepsilon)\Big\}, \\
&\qquad i \le m - k, \quad k = 0, \dots, m, \quad m = 0, \dots, n.
\end{aligned}$$

Here R_2 is determined from (7.45), (7.43), and (7.40); $\mathrm{IG}_n(\bar{\beta}_{0,n-1}, \bar{a}_{0,n-1}, x) \in C_0^\infty(\mathbb{R}_x)$, $\mathrm{IG}_{n+1}(x, \varepsilon) \in S(\mathbb{R}_x)$ uniformly in ε in the sense (7.3); R_m are polynomials of their arguments, which do not contain the term proportional to $V^0 V'^0$, i.e., belong to C_0^∞. Equating the coefficients of ε^k, $k = 0, 1, \dots, n$, in this equation and taking into account (7.7), we obtain the following system of equations for unknown coefficients $\beta_0, \dots, \beta_n$ and unknown functions $a_0, \dots, a_n$:

$$\beta_0 a_0(x) = \frac{\sqrt{\pi} a_{00} V(x)}{\sqrt{l}\cosh^2 h_0}, \tag{9.3}$$

$$\begin{aligned}
&\beta_0 a_1(x) + \beta_1 a_0(x) \\
&\quad = \frac{\sqrt{\pi}}{l\cosh^2 h_0}\Big\{-\sqrt{2} a_{00}\beta_0 V(x) + \sqrt{2}\beta_0 a_{00} V'(x) \\
&\qquad\qquad + V(x)\Big(i\sqrt{2} a_{01}\beta_0 + a_{10}\sqrt{l}\Big)\Big\}
\end{aligned} \tag{9.4}$$

$$+\frac{\beta_0}{\sqrt{2\pi}}\int_\gamma \frac{e^{ipx}}{L(p)\cosh(\tau h_0)}\mathcal{I}_1(p,x)\tilde{a}_0(p)\,dp,$$

$$\beta_0 a_2(x)+\beta_1 a_1(x)+\beta_2 a_0(x) \tag{9.5}$$
$$=\frac{1}{\sqrt{2\pi}}\int_\gamma \frac{e^{ipx}}{L(p)\cosh(\tau h_0)}\Big(\beta_0\tilde{a}_0(p)\mathcal{I}_2(p,x)$$
$$+\left(\beta_0\tilde{a}_1(p)+\beta_1\tilde{a}_0(p)\right)\mathcal{I}_1(p,x)\Big)\,dp$$
$$+i\sqrt{\frac{\pi}{l}}R_2\left(\bar{\beta}_{0,1},V(x),V'(x),\bar{a}_{k,i},x\right),$$

. . . .

$$\beta_0 a_n(x)+\beta_1 a_{n-1}(x)+\beta_2 a_{n-2}(x)+\dots+\beta_{n-1}a_1(x)+\beta_n a_0(x) \tag{9.6}$$
$$=\frac{1}{\sqrt{2\pi}}\mathrm{IG}_n\left(\bar{\beta}_{0,n-1},\bar{a}_{0,n-1},x\right)$$
$$+i\sqrt{\frac{\pi}{l}}R_n\left(\bar{\beta}_{0,n-1},V(x),V'(x),\bar{a}_{k,i},x\right),$$
$$i\le n-k,\quad k=0,1,\dots,n,$$

where $\mathrm{IG}_n(\bar{\beta}_{0,n-1},\bar{a}_{0,n-1},x)$, Q_n are the same as in (9.2), and $\mathcal{I}_1$ is determined from (8.20).

So we have obtained the system for the unknown coefficients of the expansions β of the eigenvalue ω^2 (see (5.7)) and the eigenfunction (see (6.1), (5.8)). In the next section we will show that this system has a unique solution in the space $C_0^\infty(\mathbb{R}_x)$ under appropriate conditions.

10. Solution of the system of equations for the unknown terms of the expansions of the eigenvalue and eigenfunction in the nondegenerate case

First we consider the nondegenerate case, i.e., (1.15) is fulfilled. In this case we find the solution of system (9.3)–(9.6) satisfying the following conditions which are analogous to conditions (3.5) in the case of the Schrödinger equation:

$$a_k(0)=0,\qquad k=1,2,\dots,n,\quad a_0(0)=1. \tag{10.1}$$

The following lemma is valid.

LEMMA 10.1. *There exists a solution of system* (9.3)–(9.6) *satisfying condition* (10.1); *each function* a_k, $k=0,\dots,n$, *belongs to* C_0^∞. *Moreover* β_0, β_1, a_0, a_1 *are determined from* (1.19).

PROOF. Passing to the Fourier transform in the first equation (9.3), substituting instead of L_0' its value (7.7), and using (7.15), we obtain the equivalent equation

$$\beta_0\tilde{a}_0(p)=\frac{\sqrt{\pi}a_{00}\tilde{V}(p)}{\sqrt{l}\cosh^2 h_0}. \tag{10.2}$$

Substituting $p=0$ in (10.2) and using (1.15) and (5.10), we obtain the expression for β_0 from (1.19). Substituting this expression in equation (10.2), we find $\tilde{a}_0(p)$ from (1.19). We see that $a_0(p)$ indeed satisfies (10.1). It is also clear that the function $a_0(x)\in C_0^\infty$. Now we consider the second equation (9.4). Passing to the

Fourier transform in this equation and taking into account that by (10.1) $a_{00} = 1$, $a_{10} = 0$, the expression for $\tilde{a}_0(t)$ from (10.2) and the expression for β_0 from (1.19), we obtain

$$(10.3)\quad \beta_0\tilde{a}_1(p) + \beta_1\tilde{a}_0(p) = \frac{\sqrt{\pi}}{l\cosh^2 h_0}\Big\{ -\sqrt{2}a_{00}\beta_0\tilde{V}(p) + \sqrt{2}\beta_0 a_{00} ip\tilde{V}(p) + \tilde{V}(p)\left(i\sqrt{2}a_{01}\beta_0 + a_{10}\sqrt{l}\right)\Big\} + \frac{1}{\sqrt{2l}\cosh^2 h_0}F_{x\to p}\left[\int_\gamma \frac{e^{itx}\mathcal{I}_1(t,x)\,dt}{L(t)\cosh(\tau h_0)}\right],$$

where the Fourier transform $F_{x\to p}$ is determined from (5.10). Substituting $p = 0$ in (10.3) and taking into account (10.1), (7.15), and the expression for a_0 from (1.19), we obtain the expression for β_1 from (1.19). Substituting this value β_1 in (10.3), taking into account that $a_{00} = 1$ and $a_{10} = 0$, we obtain

$$(10.4)\quad \tilde{a}_1(p) = -\frac{\beta_1}{\beta_0}\tilde{a}_0(p) + \frac{\sqrt{\pi}}{l\cosh^2 h_0}\left\{-\sqrt{2}\tilde{V}(p) + i\sqrt{2}p\tilde{V}(p) + i\sqrt{2}a_{01}V(p)\right\} + \frac{1}{\sqrt{2l}\beta_0\cosh^2 h_0}F_{x\to p}\left[\int_\gamma \frac{e^{itx}\mathcal{I}_1(t,x)\,dt}{L(t)\cosh(\tau h_0)}\right].$$

From this expression we get the x-representation for $a_1(x)$ from (1.19).

It is easy to see that (10.1) holds for $\tilde{a}_1$. Moreover, $a_1(x) \in C_0^\infty$. In fact, $V(x) \in C_0^\infty$ by (1.4), $i_2 \in C_0^\infty$ by (7.36), and the integral in (10.4) belongs to C_0^∞ by (8.20).

Let us consider the nth equation (9.6) of the system.

We assume that the coefficients β_0, $\beta_1,\dots,\beta_{n-1}$ and the functions $a_0, a_1,\dots, a_{n-1}$ are found already. Moreover, all the functions $a_k \in C_0^\infty(\mathbb{R}_x)$, $k = 0,\dots,n-1$, and conditions (10.1) hold. We will show that under these conditions it is possible to find β_n and a_n uniquely with the conditions

$$(10.5)\qquad a_n \in C_0^\infty(\mathbb{R}_x)$$

and

$$(10.6)\qquad a_n(0) = 0.$$

Passing to the Fourier transform in (9.6), we obtain

$$(10.7)\quad \begin{aligned} &\beta_0\tilde{a}_n(p) + \beta_1\tilde{a}_{n-1}(p) + \cdots + \beta_{n-1}\tilde{a}_1(p) + \beta_n\tilde{a}_0(p) \\ &= F_{x\to p}\left[\mathrm{IG}_n(\bar{\beta}_{0,n-1}, \bar{a}_{0,n-1}, x)\right] \\ &\quad + F_{x\to p}\left[R_n(\bar{\beta}_{0,n-1}, V(x), V'(x), \bar{a}_{k,i}, x)\right], \\ &i \le n-k, \quad k = 0,1,\dots,n, \end{aligned}$$

where functions L, τ coincide with (5.9), (7.18) with the change of p to t; IG_n and R_n are the same as in (9.6).

Let us substitute $p = 0$ in this equation. Then the function

$$(10.8)\quad R_n\left(\bar{\beta}_{0,n-1}, V(x), V'(x), \bar{a}_{k,i}, x\right), \qquad i \le n-k, \quad k = 0,1,\dots,n,$$

becomes well defined because this function depends only on $a_n(0)$, which is assumed to be equal to 0. After obtaining this function, we will verify that this condition holds.

Taking into account the condition (10.1), we obtain

$$\beta_n = \int_{\mathbb{R}} \Big[\mathrm{IG}_n(\bar\beta_{0,n-1}, \bar a_{0,n-1}, x) + R_n\left(\bar\beta_{0,n-1}, V(x), V'(x), \bar a_{k,i}, x\right) \Big]\, dx, \tag{10.9}$$
$$i \le n-k, \quad k = 0, 1, \dots, n,$$

where IG_n is determined from (9.1).

Now we are going to find $\tilde a_n$. Since $\beta_0 \neq 0$ by (1.19), we obtain from (10.7) the expression for $\tilde a_n$:

$$\begin{aligned} \tilde a_n(p) = \frac{1}{\beta_0}\Big\{ &- \beta_1 \tilde a_{n-1}(p) - \dots - \beta_{n-1}\tilde a_1(p) - \beta_n \tilde a_0(p) \\ &+ F_{x\to p}\left[\mathrm{IG}_n(\bar\beta_{0,n-1}, \bar a_{0,n-1}, x)\right] \\ &+ F_{x\to p}\left[R_n(\bar\beta_{0,n-1}, V(x), V'(x), \bar a_{k,i}, x)\right] \Big\}, \end{aligned} \tag{10.10}$$
$$i \le n-k, \quad k = 0, 1, \dots, n.$$

So we have found $\tilde a_n(p)$ and therefore $a_n(x)$. Now let us prove that conditions (10.5), (10.6) are fulfilled. Condition (10.6) is fulfilled by (10.1) and (10.8). From Lemmas 7.1 and 8.1 it follows that $a_n \in C_0^\infty(\mathbb{R}_x)$ under the conditions $a_k(x) \in C_0^\infty(\mathbb{R}_x)$, $k = 0, 1, \dots, n-1$. Lemma 10.1 is proved. □

11. Completion of the proof of Theorem 1.3

Thus we have found the approximate value of the parameter B_n which is found from (6.3) with β_0, β_1 defined from (1.19) and $\beta_2, \dots, \beta_n$ defined uniquely from system (9.3)–(9.6). Also we have found the approximate function $\phi_n(x)$ defined from (6.1) with the functions a_0, a_1 determined from (1.19) and other functions $a_3, \dots, a_n$ determined from system (9.3)–(9.6). From Lemmas 8.1 and 10.1 it follows that B_n, A_n solve the system up to ε^{n+1} in the sense of $L_2(\mathbb{R}_x)$. This means that the parameter ω_n^2 determined from (6.4) and the function ϕ_n solve equation (5.18) up to ε^{n+2} in $L_2(\mathbb{R}_x)$. As was already mentioned, the function (6.7) and the parameter ω_n^2 from (6.4) satisfy the following approximate equation:

$$\begin{aligned} &\int_\gamma \Big[\omega_n^2 \cosh(\tau h(x)) - \tau \sinh(\tau h(x)) \\ &\qquad + iph'(x)\Big(-\frac{\omega_n^2}{\tau}\sinh(\tau h(x)) + \cosh(\tau h(x))\Big)\Big] f_n(p) e^{ipx}\, dp \\ &= \varepsilon^{n+2}\Xi_{n+2}(x), \end{aligned} \tag{11.1}$$

where

$$\Xi_{n+2}(x) \in L_2(\mathbb{R}). \tag{11.2}$$

Moreover, $f_n \in S(\mathbb{R}_p)$ by Lemma 10.1, (6.1), (6.2), and (6.7). Let us consider the function

$$\psi_n(x,y) = \int_\gamma f_n(p)\Big(\frac{\omega_n^2}{\tau}\sinh(\tau y) + \cosh(\tau y)\Big) e^{ipx}\, dp. \tag{11.3}$$

It is easy to check that this function ψ_n is a solution of the system

$$\begin{aligned} &\Delta\psi_n - \psi_n = 0, && -h(x) < y < 0, \\ &\partial\psi_n/\partial n = \varepsilon^{n+2}\Xi_{n+2}, && y = -h(x), \\ &\psi_{ny}(x,0) = \omega_n^2 \psi_n(x,0), && y = 0. \end{aligned} \tag{11.4}$$

Now we repeat the arguments from [**8**]. Namely, let Γ_F be the free surface $\{(x,y) : y = 0\}$. We consider the operator T which associates functions f from $L_2(\Gamma_F)$ to the restriction of the normal derivative of the solution u of the problem

$$\begin{aligned} \Delta u - u &= 0, \qquad -h(x) < y < 0, \\ \partial u/\partial n &= 0, \qquad y = -h(x), \\ u\big|_{\Gamma_F} &= f(x). \end{aligned} \tag{11.5}$$

This problem has a unique generalized solution for any $f \in L_2(\Gamma_f)$ and satisfies

$$\|u\| \le \text{const}\, \|f\|\,\big|_{L_2}. \tag{11.6}$$

Also $\partial u/\partial y\big|_{\Gamma_F} \in L_2(\Gamma_f)$. So we obtain the linear operator

$$T : f \mapsto Tf := \left.\frac{\partial u}{\partial y}\right|_{\Gamma_F}. \tag{11.7}$$

It is self-adjoint; see [**5**]. The third equation of (1.3) now reads

$$Tf = \omega^2 f \tag{11.8}$$

and, obviously, (11.8) is equivalent to (1.3). It is well known (see, e.g., [**2**]) that the continuous spectrum of T coincides with $[\tanh h_0, \infty)$. Let us show that approximate solutions of (1.3) are approximate solutions of (11.8). More precisely, let $\Psi \in H_1$, and let ω_n^2 satisfy (1.3) approximately. Let us show that in this case $f_n := \Psi_n|\Gamma_F$ satisfies (11.8) up to $\varepsilon^{n+2}\Pi_{n+2}$, $\Pi_{n+2} \in L_2(\Gamma_f)$.

To do this, we will prove that

$$T\psi_n(x,0) = \omega_n^2\psi_n(x,0) + \varepsilon^{n+2}\Pi_{n+2}, \qquad \Pi_{n+2} \in L_2(\Gamma_F). \tag{11.9}$$

To prove this, let us calculate $T\psi_n(x,0)$. Let $\Phi(x,y) \in H_1$ be a solution of the problem

$$\begin{aligned} \Delta\Phi - \Phi &= 0, \qquad -h(x) < y < 0, \\ \partial\Phi/\partial n &= 0, \qquad y = -h(x), \\ \Phi(x,0) &= \psi_n(x,0), \qquad x \in \mathbb{R}. \end{aligned} \tag{11.10}$$

Then

$$T\psi_n(x,0) = \Phi_y(x,0), \qquad x \in \mathbb{R}. \tag{11.11}$$

Now let us consider $\Gamma := \psi_n - \Phi$. From (11.4), (11.10) it follows that Γ satisfies the system

$$\begin{aligned} \Delta\Gamma - \Gamma &= 0, \qquad -h(x) < y < 0, \\ \partial\Gamma/\partial n &= \varepsilon^{n+2}\Xi_{n+2}, \qquad y = -h(x), \\ \Gamma(x,0) &= 0, \qquad x \in \mathbb{R}. \end{aligned} \tag{11.12}$$

Since any solution of this problem admits the estimate analogous to (11.6), we obtain

$$\|\Gamma\|_{H_1} \le \varepsilon^{n+2} \tag{11.13}$$

and therefore

$$\left\|\frac{\partial\Gamma}{\partial y}(x,0)\right\|_{L_2} \le \varepsilon^{n+2}. \tag{11.14}$$

Now, from (11.4) and the definition of Γ we have

$$\Phi_y(x,0) = \frac{\partial\psi_n(x,0)}{\partial y}(x,0) - \frac{\partial\Gamma}{\partial y}(x,0) = \omega_n^2\psi_n(x,0) - \frac{\partial\Gamma}{\partial y}(x,0). \tag{11.15}$$

Denoting

$$\Pi := \frac{\partial\Gamma}{\partial y}(x,0)$$

and using (11.13), (11.11), and (11.4), we obtain (11.9). Now, we are going to use the arguments similar to those in the case of the Schrödinger equation. Namely, we use Lemmas 3.3 and 3.4 to prove that ω_n^2 and Ψ_n (after some normalization) are asymptotics of the frequency and the trapped wave, respectively. In order to apply these lemmas, we need only to normalize in the sense of (3.33) the function $\psi_n(x,0)$, which by (11.9) is an approximate solution of the equation of the type (11.8). From (11.3), (6.7), (6.1), and the Parseval equality it follows that

$$\begin{aligned} c(\varepsilon) :&= \|\psi_n(x,0)\|_{L_2} = \sqrt{2\pi}\,\|f_n(p)\| \\ &= \varepsilon\sqrt{2\pi}|B_n|\left\|\frac{A_n(p)}{\cosh(\tau h_0)(L(p)+\varepsilon^2B_n^2)}\right\|_{L_2}, \end{aligned} \tag{11.16}$$

where A_n is determined from (6.2); a_0, a_1 are determined from (1.19) and all the remaining functions are determined from system (9.3-9.6). Let us find at first the asymptotics of

$$\left\|\frac{A_n(p)}{\cosh(\tau h_0)(L(p)+\varepsilon^2B_n^2)}\right\|_{L_2} \tag{11.17}$$

as $\varepsilon \to 0$. From (6.10) and (6.11) it follows that there exists $\varepsilon_0 > 0$ and a function $\alpha(t) \in C^\infty$, $\alpha'(0) \neq 0$, $|t| < \varepsilon_0$, such that

$$L(\alpha(t)) = t^2, \quad |t| < \varepsilon_0, \quad \text{and} \quad \alpha(t) = \frac{t}{\sqrt{l_1}} + O(t^2) \quad \text{as } t \to 0.$$

From (6.10), (6.11) it follows that

$$\alpha'(0) = \frac{1}{\sqrt{l_1}} = \sqrt{\frac{2}{l}}. \tag{11.18}$$

Let us make the change of the variable

$$p = \alpha(t). \tag{11.19}$$

We calculate the norm in (11.17) partitioning the region of integration $\mathbb{R}$ in two parts: $|p| \le \varepsilon_0$ and $|p| \ge \varepsilon_0$. We get

$$\left\|\frac{A_n(p)}{\cosh(\tau h_0)(L(p)+\varepsilon^2B_n^2)}\right\|_{L_2} = (c_1(\varepsilon)+c_2(\varepsilon))^{1/2}, \tag{11.20}$$

where

$$\begin{aligned} c_1(\varepsilon) &= \int_{|p|\le\alpha(\varepsilon_0)} \left|\frac{A_n(p)}{\cosh(\tau h_0)(L(p)+\varepsilon^2B_n^2)}\right|^2 dp; \\ c_2(\varepsilon) &= \int_{|p|\ge\alpha(\varepsilon_0)} \left|\frac{A_n(p)}{\cosh(\tau h_0)(L(p)+\varepsilon^2B_n^2)}\right|^2 dp. \end{aligned} \tag{11.21}$$

We can expand the second integral in the regular series in ε and obtain

$$c_2(\varepsilon) = \int_{|p|\geq\alpha(\varepsilon_0)} \frac{|\tilde{a}_0(p)|^2\,dp}{\cosh^2(\tau h_0)L^2(p)} + \varepsilon c_{2,1} + \cdots + \varepsilon^n c_{2,n} + O(\varepsilon^{n+1}). \tag{11.22}$$

Let us calculate $c_1(\varepsilon)$. Making the change of variables (11.19) in (11.21), for c_1 we obtain

$$c_1(\varepsilon) = \int_{|t|\leq\varepsilon_0} \frac{|A_n(\alpha(t))|^2|\alpha'(t)|\,dt}{|\cosh(\tau(\alpha(t))h_0)(t^2+\varepsilon^2B_n^2)|^2}. \tag{11.23}$$

From the condition

$$\frac{|A_n(\alpha(0))|^2|\alpha'(0)|}{|\cosh(\tau(\alpha(0))h_0)|} \neq 0$$

it follows that $c_1(\varepsilon)$ has the asymptotics

$$c_1(\varepsilon) = \frac{1}{\varepsilon^3|\beta_0|^3}\left[c_{1,0} + \varepsilon c_{1,1} + \cdots + c_{n,1}\varepsilon^n + O(\varepsilon^{n+1})\right]. \tag{11.24}$$

Let us calculate the principal term of this asymptotics $c_{1,0}$. Let $\alpha(t)$, $|t|\geq t_0$, be some C^∞-extension of 0 to the whole real line. Then it is evident that the principal term of the asymptotics (11.23) coincides with the principal term of the asymptotics of the integral

$$\int_{\mathbb{R}} \frac{|A_n(\alpha(t))|^2|\alpha'(t)|\,dt}{|\cosh(\tau(\alpha(t))h_0)(t^2+\varepsilon^2B_n^2)|^2}. \tag{11.25}$$

Reasoning as in the case of the Schrödinger equation, we make the change of variable

$$t := t(u,\varepsilon) := \varepsilon B_n u. \tag{11.26}$$

Denoting

$$\alpha(t(u,\varepsilon)) := \alpha(u,\varepsilon), \qquad \tau(\alpha(t(u,\varepsilon))) := \tau(u,\varepsilon),$$

we obtain from (11.23) that $c_{1,0}$ coincides with the principal term of the asymptotics of the integral

$$\int_{\mathbb{R}} \frac{|A_n(\alpha(u,\varepsilon))|^2|\alpha'(t(u,\varepsilon))|\,du}{|\cosh(\tau(u,\varepsilon)h_0)(u^2+1)|^2}. \tag{11.27}$$

From (6.2) it follows that $c_{1,0}$ coincides with the main term of the asymptotics of the integral

$$\int_{\mathbb{R}} \frac{|\tilde{a}_0(\alpha(u,\varepsilon))|^2|\alpha'(t(u,\varepsilon))|\,du}{|\cosh(\tau(u,\varepsilon)h_0)(u^2+1)|^2}. \tag{11.28}$$

From the fact that $\tilde{a}_0(p)$ is analytic by Lemma 10.1 and from (11.18), it follows that the principal term of the asymptotics of (11.28) coincides with the principal term of the asymptotics of

$$c_{1,0}(\varepsilon) = \frac{|\tilde{a}_0(0)|^2}{\sqrt{l_1}\cosh^2 h_0}\int_{\mathbb{R}} \frac{du}{(u^2+1)^2}. \tag{11.29}$$

Using (10.1), (11.18), and (3.37), we obtain

$$c_{1,0} = \frac{\pi}{\sqrt{2l}\cosh^2 h_0}. \tag{11.30}$$

Substituting (11.25) in (11.24), we get

$$c_1(\varepsilon) = \frac{1}{\varepsilon^3|\beta_0|^3}\left[\frac{\pi}{\sqrt{2l}\cosh^2 h_0} + \varepsilon c_{1,1} + \cdots + c_{n,1}\varepsilon^n + O(\varepsilon^{n+1})\right]. \tag{11.31}$$

From (11.20), (11.22), and (11.31) it follows that

$$(11.32)\qquad \left\| \frac{A_n(p)}{\cosh(\tau h_0)(L(p)+\varepsilon^2 B_n^2)} \right\|_{L_2} = \frac{1}{\varepsilon^{3/2}|\beta_0|^{3/2}} \left[\frac{\sqrt{\pi}}{(2l)^{1/4}\cosh h_0} + \varepsilon c_1 + \cdots + \varepsilon^n c_n + O(\varepsilon^{n+1}) \right].$$

From (11.32) and (11.16), we have

$$(11.33)\qquad c(\varepsilon) := \|\psi_n(x,0)\|_{L_2} = \frac{1}{\varepsilon^{1/2}} \left[\left(\frac{2}{l}\right)^{1/4} \frac{\pi}{|\beta_0|^{1/2}\cosh h_0} + \varepsilon c_1 + \cdots + \varepsilon^n c_n + O(\varepsilon^{n+1}) \right].$$

From (11.33) it follows that

$$(11.34)\qquad d(\varepsilon) := \frac{1}{c(\varepsilon)} = \varepsilon^{1/2} \left[\left(\frac{l}{2}\right)^{1/4} \frac{|\beta_0|^{1/2}\cosh h_0}{\pi} + \varepsilon d_1 + \cdots + \varepsilon^n d_n + O(\varepsilon^{n+1}) \right].$$

Now, let us define the function $\Psi_n(x,y)$ as

$$(11.35)\qquad \Psi_n(x,y) := d(\varepsilon)\psi_n(x,y).$$

Substituting (11.3) in (11.35), we obtain

$$(11.36)\qquad \Psi_n(x,y) := d(\varepsilon) \int_\gamma f_n(p) \left(\frac{\omega_n^2}{\tau} \sinh(\tau y) + \cosh(\tau y) \right) e^{ipx}\, dp.$$

Substituting in (11.36) instead of f_n its expression (6.7), we get

$$(11.37)\qquad \Psi_n(x,y) := d(\varepsilon) \int_\gamma \frac{\tilde{\phi}_n(p,\varepsilon)}{\cosh(\tau h_0)\big(L(p)+\varepsilon^2 B_n^2\big)} \times \left(\frac{\omega_n^2}{\tau} \sinh(\tau y) + \cosh(\tau y) \right) e^{ipx}\, dp.$$

Finally, substituting in (11.37) instead of ϕ_n its expression (6.1), we obtain

$$(11.38)\qquad \Psi_n(x,y) := d(\varepsilon)\varepsilon B_n \int_\gamma \frac{\tilde{A}_n(p)}{\cosh(\tau h_0)\big(L(p)+\varepsilon^2 B_n^2\big)} \times \left(\frac{\omega_n^2}{\tau} \sinh(\tau y) + \cosh(\tau y) \right) e^{ipx}\, dp,$$

where

$$(11.39)\qquad \mu_n := \varepsilon B_n.$$

We see that the last expression coincides with the expression for Ψ_n from (1.16). Part 1) of Theorem 1.3 is proved. □

12. Proof of part 2) of Theorem 1.3 (degenerate case)

As in the case of the Schrödinger equation we are looking for the solution of system (9.3)–(9.6) which satisfies the conditions

$$(12.1)\qquad \begin{aligned} a_0(0) &= a_{00} = 0, \\ a_1(0) &= a_{10} = 1, \\ a_2(0) &= a_{20} = \cdots = a_n(0) = a_{n0} = 0. \end{aligned}$$

In this case equation (10.2) does not exist because $\beta_0 = a_{00} = 0$. (See the arguments in the case of the Schrödinger equation.) Let us recall that in this case

$$B_n = \varepsilon\beta_1 + \cdots - \varepsilon^n\beta_n. \tag{12.2}$$

Let us consider equation (9.4). Taking into account that $\beta_0 = 0$ and $a_{00} = 0$ and passing to the Fourier transform in this equation, we have

$$\beta_1\tilde{a}_0(p) = \frac{\sqrt{\pi}\tilde{V}(p)}{\sqrt{l}\cosh^2 h_0}. \tag{12.3}$$

From this equation we express $\tilde{a}_0(p)$ via the unknown parameter β_1. Then we substitute the obtained expression in the next equation. So, we have from (12.3):

$$\tilde{a}_0(p) = \frac{\sqrt{\pi}\tilde{V}(p)}{\beta_1\sqrt{l}\cosh^2 h_0}. \tag{12.4}$$

We will need the expression for $R_2(\bar{\beta}_{0,1}, V(x), V'(x))$ from (9.5) under conditions (12.1). Using (7.45), (7.43), (7.40), and the facts that $a_{00} = 0$, $a_{10} = 1$, $a_{20} = 0$, and $\beta_0 = 0$, we obtain

$$R_2(\bar{\beta}_{0,1}, V(x), V'(x)) = \frac{i\sqrt{2}\beta_1 a_{01} V(x)}{\sqrt{l}}. \tag{12.5}$$

From the third equation (9.4) and the conditions $a_{00} = 0$, $a_{10} = 1$, and $\beta_0 = 0$, we obtain the equation

$$\beta_1 a_1(x) + \beta_2 a_0(x) = \frac{\beta_1}{\sqrt{2\pi}}\int_\gamma \frac{e^{ipx}\tilde{a}_0(p)\mathcal{I}_1(p,x)\,dp}{L(p)\cosh(\tau h_0)} - \frac{\beta_1\sqrt{2\pi}a_{01}V(x)}{l}. \tag{12.6}$$

Passing to the Fourier transform in (12.6), we obtain

$$\beta_1\tilde{a}_1(p) + \beta_2\tilde{a}_0(p) = \frac{\beta_1}{\sqrt{2\pi}}F_{x\to p}\left[\int_\gamma \frac{e^{itx}\tilde{a}_0(t)\mathcal{I}_1(t,x)\,dt}{L(t)\cosh(\tau h_0)}\right] - \frac{\beta_1\sqrt{2\pi}a_{01}\tilde{V}(p)}{l}. \tag{12.7}$$

Substituting $p = 0$ in (12.7) and using the condition $a_{10}(0) = 1$, from (12.1) we obtain

$$1 = \frac{1}{2\pi}\int_{\mathbb{R}}\int_\gamma \frac{e^{itx}\tilde{a}_0(t)\mathcal{I}_1(t,x)\,dtdx}{L(t)\cosh(\tau h_0)}. \tag{12.8}$$

Substituting in (12.8) $\tilde{a}_0(t)$ in the form of (12.4), from (12.8) we get

$$\beta_1 = \frac{1}{2\sqrt{\pi l}\cosh^2 h_0}\int_{\mathbb{R}}\int_\gamma \frac{e^{itx}\tilde{V}(t)\mathcal{I}_1(t,x)\,dtdx}{L(t)\cosh(\tau h_0)}. \tag{12.9}$$

Now we are going to obtain β_1 in an explicit form. From (8.20) it follows that

$$\beta_1 = \frac{1}{2\sqrt{\pi l}\cosh^2 h_0}\int_{\mathbb{R}}\int_\gamma \frac{e^{itx}\tilde{V}(t)\big[V(x)\iota_{1,1}(p) + V'(x)\iota_{1,2}\big]\,dtdx}{L(t)\cosh(\tau h_0)}, \tag{12.10}$$

where ι_{11} and ι_{12} are defined from (8.20).

Substituting in (12.10) the values of ι_{11} and ι_{12} from (8.20) and integrating the obtained integrals over x, we get

$$\begin{aligned}\beta_1 = {} & \frac{1}{\sqrt{2l}\cosh^2 h_0}\int_\gamma \frac{|\tilde{V}(t)|^2\big(\tau^2 - t_0\tau\tanh(\tau h_0)\big)\,dt}{L(t)} \\ & + \frac{1}{\sqrt{2l}\cosh^2 h_0}\int_\gamma \frac{t^2|\tilde{V}(t)|^2(t_0\tanh(\tau h_0) - \tau)\,dt}{L(t)\tau}.\end{aligned} \tag{12.11}$$

From this equation, (5.9), and (5.7), we obtain

$$\beta_1 = \frac{1}{\sqrt{2l}\cosh^2 h_0}\int_\gamma \frac{|\tilde{V}(t)|^2(\tau - t_0\tanh(\tau h_0))\,dt}{\tau^2\tanh(\tau h_0) - t_0\tau}. \tag{12.12}$$

Let us express the function a_1 in terms of β_2. From (12.7) it follows that

$$\begin{aligned}\tilde{a}_1(p) = &-\frac{\beta_2}{\beta_1^2}\frac{\sqrt{\pi}\tilde{V}(p)}{\sqrt{l}\cosh^2 h_0} + \frac{1}{\beta_1\sqrt{2l}\cosh^2 h_0}F_{x\to p}\left[\int_\gamma \frac{e^{itx}\tilde{V}(t)\mathcal{I}_1(t,x)\,dt}{L(t)\cosh(\tau h_0)}\right]\\ &-\frac{\pi\sqrt{2}\tilde{V}'(0)\tilde{V}(p)}{\beta_1 l\sqrt{l}\cosh^2 h_0}.\end{aligned} \tag{12.13}$$

Now, it is clear that $a_{10} = 1$. So, we have found β_0, β_1, a_0 and a_1 from system (9.3)–(9.6). As above, in the degenerate case it is possible to prove that $a_0(x)$ and $a_1(x)$ belong to $C_0^\infty(\mathbb{R})$. Also, it is possible to prove that all the other values β_k and the functions a_k are determined uniquely under conditions (12.1) and $a_k \in C_0^\infty$, $k = 2, \dots, n$. Reasoning further as in the nondegenerate case, we see that to complete the proof of part 2) of Theorem 1.3, we need only to normalize the function $\psi_n(x, 0)$ from (11.3).

Let us calculate the constant $c(\varepsilon)$ from (11.16) in the degenerate case. To this end, we need to calculate the asymptotics of the constant

$$\left\|\frac{A_n(p)}{\cosh(\tau h_0)(L(p) + \varepsilon^2 B_n^2)}\right\|_{L_2} \tag{12.14}$$

as $\varepsilon \to 0$. Reasoning as in the nondegenerate case, we see that it is necessary to calculate the asymptotics of the constant $c_1(\varepsilon)$ from (11.23). This calculation is reduced, in turn, to the calculation of the asymptotics of the integral (11.25). Making the change of variables (11.26), we see that the principal term of this asymptotics coincides with the principal term of the asymptotics of the expression

$$\frac{1}{\varepsilon^3 B_n^3}\int_{\mathbb{R}} \frac{|A_n(\alpha(t(u)))|^2|\alpha'(t(u))|du}{|\cosh(\tau(\alpha(t(u)))h_0)(u^2+1)|^2}. \tag{12.15}$$

From the expansion

$$A_n(t)^2 = a_0(t)^2 + 2\varepsilon a_0(t)a_1(t) + \varepsilon^2 a_1(t)^2 + O(\varepsilon^3)$$

and the conditions $a_0(0) = 0$ and $a_1(0) = 1$ (see (12.1)), it follows that the principal term of this asymptotics (12.15) coincides with the principal term of the asymptotics of the expression

$$\frac{1}{\varepsilon B_n^3}\int_{\mathbb{R}} \frac{|\tilde{a}_1(\alpha(u,\varepsilon))|^2|\alpha'(t(u,\varepsilon))|\,du}{|\cosh(\tau(u,\varepsilon)h_0)(u^2+1)|^2}. \tag{12.16}$$

From the fact that $\tilde{a}_1(p)$ is analytic by (12.13) and is equal to 1 for $p = 0$ and from (11.18) and (3.37), it follows that the principal term of the asymptotics of this integral is equal to

$$\frac{\pi}{\sqrt{2l}\varepsilon^4|\beta_1|^3\cosh^2 h_0}. \tag{12.17}$$

Thus, $c_1(\varepsilon)$ from (11.21) in the degenerate case has the form

$$c_1(\varepsilon) = \frac{1}{\varepsilon^4}\left[\frac{\pi}{|\beta_1^3\sqrt{2l}\cosh^2 h_0|} + \varepsilon c_{1,1} + \cdots + c_{n,1}\varepsilon^n + O(\varepsilon^{n+1})\right]. \tag{12.18}$$

Analogously from (11.32), we get

$$(12.19)\qquad \left\| \frac{A_n(p)}{\cosh(\tau h_0)(L(p)+\varepsilon^2 B_n^2)} \right\|_{L_2} = \frac{1}{\varepsilon^2|\beta_1|^{3/2}} \left[\frac{\sqrt{\pi}}{(2l)^{1/4}\cosh h_0} + \varepsilon c_1 + \cdots + \varepsilon^n c_n + O(\varepsilon^{n+1}) \right].$$

Analogously from (11.33), from (12.2) we obtain

$$(12.20)\quad c(\varepsilon) = \|\psi_n(x,0)\|_{L_2} = \frac{2^{1/4}\pi}{l^{1/4}|\beta_1|^{1/2}\cosh h_0} + \varepsilon c_1 + \cdots + \varepsilon^n c_n + O(\varepsilon^{n+1}).$$

Analogously from (11.34), we have here

$$(12.21)\qquad d(\varepsilon) := \frac{1}{c(\varepsilon)} = \frac{l^{1/4}|\beta_1|^{1/2}\cosh h_0}{2^{1/4}\pi} + \varepsilon d_1 + \cdots + \varepsilon^n d_n + O(\varepsilon^{n+1}).$$

Reasoning further as in the nondegenerate case (see the reasoning after (11.34)), we obtain the solution Ψ_n in the degenerate case in the form (1.21) and the frequency ω_n^2 in the form (1.22). Part 2) of Theorem 1.3 is proved. □

Acknowledgments. The authors express their deep gratitude to CONACYT and Coordinacion de Investigacion Cientifica (UMSNH) for partial financial support.

References

[1] S. Albeverio, F. Gesztesy, R. Høegh-Krohn, and H. Holden, *Solvable models in quantum mechanics*, Springer-Verlag, 1988.

[2] A.-S. Bonnet-Ben Dhia and P. Joly, *Mathematical analysis of guided water waves*, SIAM J. Appl. Math., **53** (1993), 1507–1550.

[3] M. V. Fedoryuk, *Asymptotics integrals and series*, Nauka, Moscow, 1987. (Russian)

[4] R. Grimshaw, *Edge waves: a long-wave theory for oceans of finite depth*, J. Fluid Mech., **62** (1974), 775–791.

[5] R. M. Garipov, *On the linear theory of gravity waves: the theorems of existence and uniqueness*, Arch. Rational Mech. Anal., **4** (1967), 352–362.

[6] L. D. Landau and E. M. Lifshitz, *Quantum mechanics*, Pergamon, London, 1958.

[7] T. Kato, *Perturbation theory for linear operators*, Springer-Verlag, Berlin–Heidelberg–New York, 1976.

[8] A. E. Merzon and P. N. Zhevandrov, *High-frequency asymptotics of edge waves on a beach of nonconstant slope*, SIAM J. Appl. Math, **59**, no. 2 (1998), 529–546.

[9] B. Simon, *The bound state of weakly coupled Schrödinger operators in one and two-dimensions*, Ann. Phys. (NY), 97 (1976), pp. 279–288.

[10] F. Treves, *Introduction to pseudodifferential and Fourier integral operators*, vol. I, Plenum Press, New York–London, 1982.

[11] P. Zhevandrov, *Edge waves on a gently sloping beach: uniform asymptotics*, J. Fluid Mech., **233** (1991), 483–493.

[12] P. Zhevandrov and A. Merzon, *Asymptotics of eigenfunctions for the Schrödinger equation with a shallow potential well*, International Conference Dedicated to the Centenary Anniversary of I. G. Petrovskii, Book of Abstracts, Moscow, Moscow University Press, 2001, pp. 267–268.

Institute of Mathematics, UNAM (campus Morelia), Morelia, Michoacan, Mexico
Current address: On sabbatical leave from Institute of Physics and Mathematics, University of Michoacan, 58060 Morelia, Michoacan, Mexico
E-mail address: pzhevand@zeus.umich.mx

Institute of Physics and Mathematics, University of Michoacan, 58060 Morelia, Michoacan, Mexico
E-mail address: anatoli@ginette.ifm.umich.mx

Titles in This Subseries

53 **M. V. Karasev, Editor,** Asymptotic methods for wave and quantum problems (TRANS2/208)

52 **Yu. M. Suhov, Editor,** Analytic methods in applied probability (TRANS2/207)

51 **M. A. Shubin and M. S. Agranovich, Editors,** Partial differential equations (Mark Vishik's Seminar) (TRANS2/206)

50 **V. Turaev and A. Vershik, Editors,** Topology, ergodic theory, real algebraic geometry. Rokhlin's memorial (TRANS2/202)

49 **Michael Semenov-Tian-Shansky, Editor,** L. D. Faddeev's seminar on mathematical physics (TRANS2/201)

48 **L. Lerman, G. Polotovskiĭ, and L. Shilnikov, Editors,** Methods of qualitative theory of differential equations and related topics (TRANS2/200)

47 **R. A. Minlos, Senya Shlosman, and Yu. M. Suhov, Editors,** On Dobrushin's way. From probability theory to statistical physics (TRANS2/198)

46 **Vladimir Arnold, Maxim Kontsevich, and Anton Zorich, Editors,** Pseudoperiodic topology (TRANS2/197)

45 **Ya. Eliashberg, D. Fuchs, T. Ratiu, and A. Weinstein, Editors,** Northern California symplectic geometry seminar (TRANS2/196)

44 **Alexander Astashkevich and Serge Tabachnikov, Editors,** Differential Topology, Infinite-Dimensional Lie Algebras, and Applications (D. B. Fuchs' 60th Anniversary Collection) (TRANS2/194)

43 **A. Yu. Morozov and M. A. Olshanetsky, Editors,** Moscow Seminar in Mathematical Physics (TRANS2/191)

42 **S. Tabachnikov, Editor,** Differential and Symplectic Topology of Knots and Curves (TRANS2/190)

41 **V. Buslaev, M. Solomyak, and D. Yafaev, Editors,** Differential Operators and Spectral Theory (M. Sh. Birman's 70th anniversary collection) (TRANS2/189)

40 **M. V. Karasev, Editor,** Coherent Transform, Quantization, and Poisson Geometry (TRANS2/187)

39 **A. Khovanskiĭ, A. Varchenko, and V. Vassiliev, Editors,** Geometry of Differential Equations (TRANS2/186)

38 **B. Feigin and V. Vassiliev, Editors,** Topics in Quantum Groups and Finite-Type Invariants (Mathematics at the Independent University of Moscow) (TRANS2/185)

37 **Peter Kuchment and Vladimir Lin, Editors,** Voronezh Winter Mathematical Schools (Dedicated to Selim Krein) (TRANS2/184)

36 **V. E. Zakharov, Editor,** Nonlinear Waves and Weak Turbulence (TRANS2/182)

35 **G. I. Olshanski, Editor,** Kirillov's Seminar on Representation Theory (TRANS2/181)

34 **A. Khovanskiĭ, A. Varchenko, and V. Vassiliev, Editors,** Topics in Singularity Theory (TRANS2/180)

33 **V. M. Buchstaber and S. P. Novikov, Editors,** Solitons, Geometry, and Topology: On the Crossroad (TRANS2/179)

32 **R. L. Dobrushin, R. A. Minlos, M. A. Shubin, and A. M. Vershik, Editors,** Topics in Statistical and Theoretical Physics (F. A. Berezin Memorial Volume) (TRANS2/177)

31 **R. L. Dobrushin, R. A. Minlos, M. A. Shubin, and A. M. Vershik, Editors,** Contemporary Mathematical Physics (F. A. Berezin Memorial Volume) (TRANS2/175)

30 **A. A. Bolibruch, A. S. Merkur′ev, and N. Yu. Netsvetaev, Editors,** Mathematics in St. Petersburg (TRANS2/174)

29 **V. Kharlamov, A. Korchagin, G. Polotovskiĭ, and O. Viro, Editors,** Topology of Real Algebraic Varieties and Related Topics (TRANS2/173)

28 **L. A. Bunimovich, B. M. Gurevich, and Ya. B. Pesin, Editors,** Sinai's Moscow Seminar on Dynamical Systems (TRANS2/171)

27 **S. P. Novikov, Editor,** Topics in Topology and Mathematical Physics (TRANS2/170)

26 **S. G. Gindikin and E. B. Vinberg, Editors,** Lie Groups and Lie Algebras: E. B. Dynkin's Seminar (TRANS2/169)

25 **V. V. Kozlov, Editor,** Dynamical Systems in Classical Mechanics (TRANS2/168)

24 **V. V. Lychagin, Editor,** The Interplay between Differential Geometry and Differential Equations (TRANS2/167)

23 **Yu. Ilyashenko and S. Yakovenko, Editors,** Concerning the Hilbert 16th Problem (TRANS2/165)

22 **N. N. Uraltseva, Editor,** Nonlinear Evolution Equations (TRANS2/164)

Published Earlier as Advances in Soviet Mathematics

21 **V. I. Arnold, Editor,** Singularities and bifurcations, 1994

20 **R. L. Dobrushin, Editor,** Probability contributions to statistical mechanics, 1994

19 **V. A. Marchenko, Editor,** Spectral operator theory and related topics, 1994

18 **Oleg Viro, Editor,** Topology of manifolds and varieties, 1994

17 **Dmitry Fuchs, Editor,** Unconventional Lie algebras, 1993

16 **Sergei Gelfand and Simon Gindikin, Editors,** I. M. Gelfand seminar, Parts 1 and 2, 1993

15 **A. T. Fomenko, Editor,** Minimal surfaces, 1993

14 **Yu. S. Il′yashenko, Editor,** Nonlinear Stokes phenomena, 1992

13 **V. P. Maslov and S. N. Samborskiĭ, Editors,** Idempotent analysis, 1992

12 **R. Z. Khasminskiĭ, Editor,** Topics in nonparametric estimation, 1992

11 **B. Ya. Levin, Editor,** Entire and subharmonic functions, 1992

10 **A. V. Babin and M. I. Vishik, Editors,** Properties of global attractors of partial differential equations, 1992

9 **A. M. Vershik, Editor,** Representation theory and dynamical systems, 1992

8 **E. B. Vinberg, Editor,** Lie groups, their discrete subgroups, and invariant theory, 1992

7 **M. Sh. Birman, Editor,** Estimates and asymptotics for discrete spectra of integral and differential equations, 1991

6 **A. T. Fomenko, Editor,** Topological classification of integrable systems, 1991

5 **R. A. Minlos, Editor,** Many-particle Hamiltonians: spectra and scattering, 1991

4 **A. A. Suslin, Editor,** Algebraic K-theory, 1991

3 **Ya. G. Sinaĭ, Editor,** Dynamical systems and statistical mechanics, 1991

2 **A. A. Kirillov, Editor,** Topics in representation theory, 1991

1 **V. I. Arnold, Editor,** Theory of singularities and its applications, 1990